D1104110

ADVANCES IN CHEMICAL PHYSICS

VOLUME XL

ADVANCES IN CHEMICAL PHYSICS

EDITED BY

I. PRIGOGINE

University of Brussels
Brussels, Belgium
and
University of Texas
Austin, Texas

AND

STUART A. RICE

Department of Chemistry
and
The James Franck Institute
The University of Chicago
Chicago, Illinois

VOLUME XL

AN INTERSCIENCE® PUBLICATION

JOHN WILEY & SONS

New York • Chichester • Brisbane • Toronto

AN INTERSCIENCE® PUBLICATION

Library of Congress Catalog Card Number: 58-9935
ISBN 0-471 03884-9

Printed in the United States of America

10 9 8 7 6 5 4 3 2 1

CONTRIBUTORS TO VOLUME XL

DONALD M. BURLAND, IBM Research Laboratory, San Jose, California

MARC D. DONOHUE, Department of Chemical Engineering, Clarkson College of Technology, Potsdam, New York

GEORGE H. GILMER, Bell Laboratories, Murray Hill, New Jersey

M. R. HOARE, Department of Physics, Bedford College, University of London, London, England

JOSEPH L. KATZ, Department of Chemical Engineering, Clarkson College of Technology, Potsdam, New York

JOHN J. KOZAK, Department of Chemistry and Radiation Laboratory, University of Notre Dame, Notre Dame, Indiana

DAVID W. OXTOBY, Department of Chemistry and James Franck Institute, University of Chicago, Chicago, Illinois

JOHN D. WEEKS, Bell Laboratories, Murray Hill, New Jersey

AHMED H. ZEWAIL, A. A. Noyes Laboratory of Chemical Physics, California Institute of Technology, Pasadena, California

INTRODUCTION

Few of us can any longer keep up with the flood of scientific literature, even in specialized subfields. Any attempt to do more, and be broadly educated with respect to a large domain of science, has the appearance of tilting at windmills. Yet the synthesis of ideas drawn from different subjects into new, powerful, general concepts is as valuable as ever, and the desire to remain educated persists in all scientists. This series, *Advances in Chemical Physics*, is devoted to helping the reader obtain general information about a wide variety of topics in chemical physics, which field we interpret very broadly. Our intent is to have experts present comprehensive analyses of subjects of interest and to encourage the expression of individual points of view. We hope that this approach to the presentation of an overview of a subject will both stimulate new research and serve as a personalized learning text for beginners in a field.

ILYA PRIGOGINE

STUART A. RICE

CONTENTS

ADVANCES IN CHEMICAL PHYSICS

VOLUME XL

DEPHASING OF MOLECULAR VIBRATIONS IN LIQUIDS

DAVID W. OXTOBY*

Department of Chemistry
and
James Franck Institute
University of Chicago
Chicago, Illinois

CONTENTS

* Alfred P. Sloan Foundation Fellow

I. INTRODUCTION

Vibrational phase relaxation in liquids is a problem that has attracted considerable attention during the last 5 to 10 years; advances in both experiment and theory have led to a much deeper understanding of the microscopic processes that cause dephasing and have demonstrated the usefulness of vibrational phase relaxation studies as a probe of liquid state structure and dynamics. On the experimental side, the most dramatic advance has been the development of coherent time-resolved picosecond techniques, but increasingly accurate isotropic Raman lineshape studies have allowed systematic investigation of the temperature, density, and concentration dependence of dephasing in liquids. On the theoretical side, a wide variety of increasingly sophisticated theories and models have been developed and compared with experimental results. In spite of all this recent activity, there are no really comprehensive surveys of the problem of phase relaxation in liquids. Diestler[1] has reviewed vibrational relaxation (both phase and population) with primary emphasis on the application of projection operator techniques to the study of a harmonic oscillor in a heat bath; references are given to much of the earlier experimental work. Three surveys of picosecond spectroscopy[2,3] have included brief sections on vibrational dephasing. A number of surveys of rotational relaxation in liquids[4] have included depolarized Raman lineshape measurements, but discussed only briefly the effect of vibrational dephasing on these experiments.

Dephasing can be described most simply for dilute gases. The vibrational wave function of a molecule is determined by its quantum state (i.e., the occupation numbers of the various normal modes) and by an overall phase. Collisions with other molecules can be either elastic or inelastic; inelastic collisions give rise to population relaxation, whereas elastic collisions give only phase shifts. Because the gas is taken to be dilute, collisions are separate binary events; during a series of such collisions the phase will change by different amounts depending on the relative momenta, orientations, and impact parameters of the interacting molecules. After a sufficient number of such elastic collisions, the phase of the wave function will have lost correlation with its initial value and, for an ensemble of molecules, will have a random distribution of values between 0 and 2π; phase relaxation has taken place.

The gas phase description is straightforward (although of course the actual collision dynamics may be complicated) because collisions are separate and the effect of each is given entirely by the resulting phase shift. In a liquid, molecular vibrations are well described by localized wave functions, but a given molecule interacts continuously with many neighbors, thus the description of the dephasing is less simple. An understanding of dephasing in liquids is important for two reasons: first, probes of the vibrational motion provide information about the still poorly understood local structure and dynamics of molecular liquids; second, infrared or depolarized Raman studies of rotational relaxation also involve vibrational dephasing effects, which must be understood to separate out their contribution.

In Section II, the frequency- and time-domain experiments sensitive to dephasing are described and the correlation functions they determine are presented. In Section III a general theory for vibrational dephasing is given, which includes most formal theories developed to date as special cases; Section IV is devoted to the presentation and comparison of a number of theoretical models that have been developed. Section V summarizes recent experimental findings, emphasizing the five or six molecules that have been studied most extensively.

A number of related subjects are not discussed or are only mentioned briefly. Vibrational population relaxation can in some cases contribute to dephasing; studies of population relaxation in solid matrices have been reviewed by Legay[5]; Refs. 2 and 3 discuss picosecond investigations of liquids. Vibrational lineshapes of molecules in solids and matrices[6] are not discussed, although the problem is closely related to the liquid state dephasing problem, and comparative studies should prove interesting. Lineshapes in hydrogen-bonded liquids[7,8] are not covered. Coherence effects in nuclear magnetic resonance[9] are quite closely related to vibrational dephasing on a formal level, but fall outside the scope of this review, as do the recent coherent optical studies[10–12] of solids and gases.

II. EXPERIMENTAL PROBES OF VIBRATIONAL DEPHASING

Vibrational dephasing is accessible to experiment primarily through two types of measurement: spectral line shape determinations and coherent picosecond excitation studies. The former method is older and easier to carry out experimentally, but the interpretation is sometimes ambiguous; the latter technique is difficult experimentally, but in some cases gives cleaner results. In this section we outline the theory underlying these two types of experiment and determine the microscopic correlation functions that can be obtained from them.

A. Spectral Lineshape Studies

The scattering of light can be described either classically[13] or quantum mechanically.[14] The starting point here is the quantum mechanical expression derived by Placzek[15,16] for the differential scattering cross-section for Raman scattering into solid angle $d\Omega$ and frequency range $d\omega$ about ω:

$$\frac{d^2\sigma}{d\Omega\, d\omega} = \left(\frac{2\pi}{\lambda}\right)^4 \sum_{i,f} |\langle i|\, \hat{\varepsilon}_I \cdot \underset{\sim}{\alpha} \cdot \hat{\varepsilon}_s \,|f\rangle|^2 \rho_i\, \delta(\omega - \omega_{fi}) \tag{2.1}$$

Here λ is the wavelength of the scattered light, $\hat{e}_I$ and $\hat{e}_S$ are the polarization directions of incident and scattered light, and $\underset{\sim}{\alpha}$ is the polarizability tensor of the scattering medium. The initial and final states of the system with energies E_i and E_f are i and f; $\hbar\omega_{fi} = E_f - E_i$, and ρ_i is the probability that the system is initially in state i. The scattering intensity, which is proportional to this cross-section, can then be expressed in terms of time correlation functions using the procedure outlined by Gordon[16]: first, the delta function in (2.1) is Fourier transformed to give

$$\begin{aligned} I(\omega) &\propto \sum_{i,f} \int_{-\infty}^{\infty} dt e^{i(\omega-\omega_{fi})t} |\langle i|\, \hat{\varepsilon}_I \cdot \underset{\sim}{\alpha} \cdot \hat{\varepsilon}_s \,|f\rangle|^2 \rho_i \\ &= \int_{-\infty}^{\infty} dt e^{i\omega t} \sum_{i,f} \langle i|\, e^{iE_it/\hbar} \hat{\varepsilon}_I \cdot \underset{\sim}{\alpha} \cdot \hat{\varepsilon}_s e^{-iE_ft/\hbar} \,|f\rangle \\ &\quad \otimes \langle f|\, \hat{\varepsilon}_I \cdot \underset{\sim}{\alpha} \cdot \hat{\varepsilon}_s \,|i\rangle \rho_i \end{aligned} \tag{2.2}$$

The Heisenberg representation of an operator $\underset{\sim}{\alpha}$ is

$$\underset{\sim}{\alpha}(t) = e^{i\mathcal{H}t/\hbar} \underset{\sim}{\alpha} e^{-i\mathcal{H}/\hbar} \tag{2.3}$$

Equation 2.2 thus becomes

$$I(\omega) \propto \int_{-\infty}^{\infty} dt e^{i\omega t} \sum_{i,f} \langle i|\, \hat{\varepsilon}_I \cdot \underset{\sim}{\alpha}(t) \cdot \hat{\varepsilon}_s \,|f\rangle\langle f|\, \hat{\varepsilon}_I \cdot \underset{\sim}{\alpha} \cdot \hat{\varepsilon}_s \,|i\rangle \rho_i \tag{2.4}$$

The sum over i with weighting ρ_i is simply the equilibrium ensemble average $\langle\ \ \rangle_{eq}$ of the operator inside the brackets; thus

$$I(\omega) \propto \int_{-\infty}^{\infty} dt e^{i\omega t} \langle \hat{\varepsilon}_I \cdot \underset{\sim}{\alpha}(t) \cdot \hat{\varepsilon}_s \hat{\varepsilon}_I \cdot \underset{\sim}{\alpha} \cdot \hat{\varepsilon}_s \rangle_{eq} \tag{2.5}$$

The subscript "eq" is dropped except where confusion may result.

Two different experiments can be carried out, one in which $\hat{e}_I$ and $\hat{e}_S$ are parallel (giving $I_{\parallel}(\omega)$), and one in which they are perpendicular ($I_{\perp}(\omega)$). The isotropic and anisotropic lineshapes are defined as

$$\begin{aligned} I_{iso}(\omega) &= I_{\parallel}(\omega) - \tfrac{4}{3} I_{\perp}(\omega) \\ I_{aniso}(\omega) &= I_{\perp}(\omega) \end{aligned} \tag{2.6}$$

They are related to the mean polarizability

$$\alpha \equiv \tfrac{1}{3}\,\mathrm{Tr}\,\underline{\alpha} \tag{2.7}$$

and the polarizability anisotropy

$$\underline{\beta} = \underline{\alpha} - \alpha\underline{\mathcal{I}} \tag{2.8}$$

through[13]

$$\begin{aligned} I_{\mathrm{iso}}(\omega) &\propto \int_{-\infty}^{\infty} dt e^{i\omega t}\langle \alpha(t)\alpha(0)\rangle \\ I_{\mathrm{aniso}}(\omega) &\propto \int_{-\infty}^{\infty} dt e^{i\omega t}\langle \mathrm{Tr}\,\underline{\beta}(t)\cdot\underline{\beta}(0)\rangle \end{aligned} \tag{2.9}$$

The system polarizabilities α and $\underline{\beta}$ may be approximated as the sum of the mean polarizabilities or polarizability anisotropies of the molecules in the system. This corresponds to the neglect of pair and higher-order collision-induced polarizabilities,[14] which affect primarily the high-frequency wings of the spectrum.

The isotropic spectrum includes both low-frequency quasielastic scattering contributions (which are centered on the incident frequency and correspond to Rayleigh-Brillouin scattering) and higher-frequency vibrational Raman contributions (which arise from the dependence of α on vibrational coordinates and are therefore centered a distance $\pm\omega_0$ from the incident laser frequency, where ω_0 is the vibrational frequency). This latter part of the spectrum may be expressed as

$$I_{\mathrm{iso}}(\omega) \propto \left|\frac{\partial\alpha}{\partial Q}\right|^2 \int_{-\infty}^{\infty} dt e^{i\omega t} \sum_{j=1}^{N} \langle Q_i(t)Q_j(0)\rangle \tag{2.10}$$

where j is summed over the N molecules in the system and Q_i is the vibrational coordinate of interest on molecule i. It is clear from (2.10) that the isotropic Raman spectrum provides a direct probe of the vibrational dephasing process as it is reflected in the correlation function $\langle\sum_j Q_i(t)Q_j(0)\rangle$. It is frequently assumed that the phases of the vibrations on neighboring molecules are uncorrelated; it is shown in Section III.A that this is not generally true. If, however, this correlation is neglected as a first approximation, the lineshape becomes

$$I(\omega) \propto \int_{-\infty}^{\infty} dt e^{i\omega t}\langle Q_i(t)Q_i(0)\rangle \tag{2.11}$$

and depends only on the separate dephasing of each molecule. In the case where the correlation function of $Q_i(t)$ is exponential

$$\langle Q_i(t)Q_i(0)\rangle \approx \langle Q_i^2\rangle \exp(-t/\tau_v) \tag{2.12}$$

the isotropic spectrum is a Lorentzian with a full width at half height (FWHH) of $2/\tau_v$ $\sec^{-1}$ or $(\pi c \tau_v)^{-1}$ cm^{-1}.

The anisotropic spectrum includes low-frequency (rotational Raman) and higher-frequency (vibrational Raman) contributions. The latter can be written

$$I_{\text{aniso}}(\omega) \propto \int_{-\infty}^{\infty} dt e^{i\omega t} \left\langle \operatorname{Tr} \sum_{j=1}^{N} \frac{\partial \underline{\beta}_i(t)}{\partial Q_i} \cdot \frac{\partial \underline{\beta}_j(0)}{\partial Q_j} Q_i(t) Q_j(0) \right\rangle. \tag{2.13}$$

For a molecule with a threefold or higher axis in the direction $\mathbf{u}_i$, this expression can be simplified[13] to give

$$\begin{aligned} I_{\text{aniso}}(\omega) &\propto \int_{-\infty}^{\infty} dt e^{i\omega t} \sum_{j=1}^{N} \langle \tfrac{1}{2}[3(\mathbf{u}_i(t) \cdot \mathbf{u}_j(0))^2 - 1] Q_i(t) Q_j(0) \rangle \\ &= \int_{-\infty}^{\infty} dt e^{i\omega t} \sum_{j=1}^{N} \langle P_2(\mathbf{u}_i(t) \cdot \mathbf{u}_j(0)) Q_i(t) Q_j(0) \rangle \\ &\approx \int_{-\infty}^{\infty} dt e^{i\omega t} \langle P_2(\mathbf{u}_i(t) \cdot \mathbf{u}_i(0)) Q_i(t) Q_i(0) \rangle \end{aligned} \tag{2.14}$$

where P_2 is the second Legendre polynomial. In the last step of (2.14) we neglected the phase correlation between neighboring molecules (as discussed for the isotropic Raman line). The anisotropic Raman spectrum depends on rotational as well as vibrational relaxation; thus it is a less useful probe of the pure dephasing processes. On the other hand, the vibrational contribution must be understood and separated out if accurate information on rotational motion is to be extracted from measured depolarized Raman lineshapes.

The infrared absorption line shape $I_{\text{IR}}(\omega)$ is given[16] by an expression analogous to (2.1):

$$I(\omega) = 3 \sum_{i,f} |\langle f | \hat{\varepsilon} \cdot \boldsymbol{\mu} | i \rangle|^2 \rho_i \delta(\omega_{fi} - \omega) \tag{2.15}$$

where $\hat{\varepsilon}$ is a unit vector along the polarization direction of the incident radiation (this direction can be averaged over for ordinary unpolarized incident radiation), and $\boldsymbol{\mu}$ is the dipole moment operator for the system. An analogous transformation to that outlined above for the Raman lineshape gives[16]

$$I(\omega) = \frac{1}{2\pi} \int_{-\infty}^{\infty} dt e^{i\omega t} \langle \boldsymbol{\mu}(t) \cdot \boldsymbol{\mu}(0) \rangle \tag{2.16}$$

If the dipole moment of the system is approximated as the sum of dipole moments on the individual molecules, the absorption intensity near the vibrational frequency ω_0 is given by

$$\begin{aligned} I(\omega) &\propto \int_{-\infty}^{\infty} dt e^{i\omega t} \sum_{j=1}^{N} \left\langle \frac{\partial \boldsymbol{\mu}_i(t)}{\partial Q_i} \cdot \frac{\partial \boldsymbol{\mu}_j(0)}{\partial Q_j} Q_i(t) Q_j(0) \right\rangle \\ &\propto \int_{-\infty}^{\infty} dt e^{i\omega t} \sum_{j=1}^{N} \langle \mathbf{u}_i(t) \cdot \mathbf{u}_j(0) Q_i(t) Q_j(0) \rangle \\ &\approx \int_{-\infty}^{\infty} dt e^{i\omega t} \langle \mathbf{u}_i(t) \cdot \mathbf{u}_i(0) Q_i(t) Q_i(0) \rangle \end{aligned} \tag{2.17}$$

Here again, as in the anisotropic Raman scattering case, the lineshape is affected by both rotational and vibrational relaxation.

B. Coherent Picosecond Excitation Experiments

Although the study of isotropic Raman linewidths provides a direct and useful probe of vibrational dephasing, the interpretation of these experiments is sometimes clouded by the presence of hot bands, isotope splittings, and other sources of inhomogeneous broadening. In such cases, the recently developed picosecond pulse techniques[2,3] are very useful; a carefully designed experiment allows the determination of the contribution of purely homogeneous broadening to an inhomogeneous line. The principle of these experiments is as follows: a short (picosecond) laser pulse creates a coherent superposition of vibrationally excited molecules; a second probe pulse then measures the amplitude still present after a time t_D, through coherent anti-Stokes Raman scattering. In this section we outline the theory of pulsed picosecond dephasing experiments, with the emphasis less on experimental details than on the fundamental question of what microscopic properties are determined by a given experiment. A fuller discussion is given in Ref. 17.

The electric field of the incident laser beam is characterized by wave vector k_L and frequency ω_L; it has the form

$$\tfrac{1}{2}E_L \exp\left[i(k_L x - \omega_L t)\right] + \text{c.c.} \tag{2.18}$$

Suppose there are a number of distinct types of molecules (e.g., isotopes) in the system with frequencies ω_j, $j = 1, 2, \ldots, n$. Through stimulated Raman scattering, the laser field generates a Stokes field

$$\tfrac{1}{2}E_S \exp\left[i(k_S x - \omega_S t)\right] + \text{c.c.} \tag{2.19}$$

and a coherent superposition of vibrations q_j for each species j:

$$q_j = \tfrac{1}{2}Q_j \exp\left[i(k_j x - \omega_j t)\right] + \text{c.c.} \tag{2.20}$$

The stimulated excitation mechanism ensures that $k_j = k_L - k_S$ for all species present. E_S and Q_j are generated through the interaction of the laser field with the nonlinear polarization of the medium. They satisfy the equations[17]

$$\left(\frac{\partial}{\partial x}+\frac{1}{v_s}\frac{\partial}{\partial t}\right)E_s \propto iE_L \sum_j N_j Q_j^* \exp(-i\Delta\omega_j t)$$

$$\left(\frac{\partial}{\partial t}+\frac{1}{\tau_v^j}\right)Q_j \propto iE_L E_s^* \exp(-i\Delta\omega_j t) \tag{2.21}$$

where $\Delta\omega_j \equiv \omega_L - \omega_S - \omega_j$, N_j is the number of molecules of type j, and τ_v^j the dephasing time for molccules of type j; in writing (2.21) we assumed the dephasing to be described by a simple exponential decay. While the laser pulse is acting on the molecules, the time dependence resulting from (2.21) can be quite complicated; if, however, we consider the limiting case of an excitation pulse of very short duration, then the exponentials in (2.21) may be set to 1 and all the amplitudes Q_j build up with equal gain. Immediately after the pulse has passed, all the molecules have the same phase; they subsequently evolve in time according to

$$q_j(t) \propto \exp(-i\omega_j t) \exp(-t/\tau_v^j) \tag{2.22}$$

A second probe pulse is then sent in after a time delay t_D; the field of this pulse has the form

$$\tfrac{1}{2}E_{L2} \exp[i(k_{L2}x - \omega_{L2}t) + \text{c.c.} \tag{2.23}$$

Scattering of this probe field by the excited molecules of each species j gives rise to an anti-Stokes scattered field

$$\tfrac{1}{2}E_{ASj} \exp[i(k_{ASj}x - \omega_{ASj}t)] + \text{c.c.} \tag{2.24}$$

where, by energy conservation,

$$\omega_{ASj} = \omega_{L2} + \omega_j \tag{2.25}$$

E_{ASj} satisfies the equation[17]

$$\left(\frac{\partial}{\partial x}+\frac{1}{v_{AS}}\frac{\partial}{\partial t}\right)E_{ASj} \propto -iN_j E_{L2}(t-t_D)Q_j \exp(i\Delta k_j x) \tag{2.26}$$

where

$$\begin{aligned}\Delta k_j &= k_{L2} + k_j - k_{ASj}\\ &= k_{L2} + k_L - k_S - k_{ASj}\end{aligned} \tag{2.27}$$

is the wave vector mismatch, which differs for the individual species j. The detector is placed at the end of the sample, $x = l$. The anti-Stokes field at that point is obtained from eq. (2.26):

$$E_{ASj}(\ell, t) \propto N_j \int_0^{\ell} dx Q_j(x, t) \exp(i\Delta k_j x) \tag{2.28}$$

The amplitude $Q_j(x, t)$ grows exponentially with distance x due to its generation from a stimulated scattering process:

$$Q_j(x, t) \propto Q_j(\ell, t) \exp(x/2\Delta\ell) \tag{2.29}$$

where $\Delta\ell$ is the effective interaction length near the end of the gain medium. The probe pulse is taken to be very narrow in time compared to the dephasing processes of interest. The anti-Stokes probe signal measured at the detector is then

$$S(t_D) \propto \left| \sum_j E_{ASj}(\ell, t_D) \exp[i(k_{AS}\ell - \omega_{ASj}t_D)] \right|^2 \tag{2.30}$$

Combining (2.28), (2.29), and (2.30) and carrying out the integration over x then gives

$$S(t_D) \propto (\Delta\ell)^2 \left| \sum_j N_j Q_j(\ell, t_D)[1+(2\Delta k_j \Delta l)^2]^{-\frac{1}{2}} \exp[i(\Delta\omega_j t_D + \delta_j)] \right|^2 \tag{2.31}$$

where $\delta_j \equiv -\arctan(2\Delta k_j \Delta\ell)$.

We assumed in deriving (2.31) that the excitation and probe pulses have very short durations compared to the dephasing times. The more general calculation of Ref. 17 differs from our result in two ways: first, there is an additional contribution to the phase shift δ_j due to the dephasing of the different species during the excitation pulse; second, there is a convolution over the intensity envelope of the probe pulse. Using the more accurate expressions, values for τ_v can be measured even when they are several times shorter than the pulse duration.

There are two different limiting cases of interest for (2.31). Suppose first that $\Delta k_j \Delta\ell \ll 1$ for all j. Then (2.31) reduces to

$$\begin{aligned} S(t_D) &\propto (\Delta\ell)^2 \left| \sum_j N_j Q_j(\ell, t_D) \exp(i\Delta\omega_j t_D) \right|^2 \\ &\propto (\Delta\ell)^2 \left| \sum_j N_j \exp(-t_D/\tau_v^j) \exp(i\Delta\omega_j t_D) \right|^2 \end{aligned} \tag{2.32}$$

The measured signal of (2.32) is simply the square of the Fourier transform of the isotropic Raman lineshape, and features such as isotope splittings show up as beats in the decay. These have been observed and are discussed in Ref. 17. In the opposite limit, suppose $\Delta k_j \Delta\ell \ll 1$ for

$j = m$, whereas $\Delta k_j \Delta l \gg 1$ for all other species. In that case the only important contribution to (2.31) comes from $j = m$, giving

$$S(t_D) \propto (\Delta \ell)^2 \exp(-2t_D/\tau_v^m) \tag{2.33}$$

The result is a single exponential decay with time constant $\tau_v/2$. A comparison with (2.11) shows that the time constant from the picosecond experiments is equal to the inverse of the *full* width at half height (FWHH) of the corresponding homogeneously broadened Raman linewidth. There has been some controversy[18,19] on this point in the literature; according to the above analysis, Ref. 18 is in error in connecting the picosecond decay constant with the *half* width at half height. The essential point is that the picosecond experiments are *intensity* correlation experiments and measure not $\langle Q(t)Q(0)\rangle$ but $\langle Q(t)Q(0)\rangle^2$.

The analysis leading to (2.31) is predicated on the assumption of a finite set of distinct species j with frequencies ω_j. This is useful for isotope broadening, for example, but if the inhomogeneous broadening is due to slow relaxation of molecular environments, there can be a continuous distribution of "species" j, and selective phase matching for one particular component $j = m$ loses its meaning; in addition, different environments can have the same frequency but different dephasing times. In these cases the interpretation of the picosecond experiments is much less straightforward.

III. THEORY OF VIBRATIONAL DEPHASING

In this section a general theory is outlined which relates the isotropic Raman lineshape to the microscopic Hamiltonian of the system. A completely general Hamiltonian may be written in the form

$$\mathcal{H} = \mathcal{H}_0 + \mathcal{H}_B + D + R + V. \tag{3.1}$$

Here $\mathcal{H}_0$ is the Hamiltonian for the vibrational degrees of freedon of the isolated molecules; its precise form is specified later. Effects of the anharmonicity of the molecular vibrations are present in $\mathcal{H}_0$. $\mathcal{H}_B$ is the "bath" Hamiltonian which includes the rotational and translational degrees of freedom; they are eventually treated classically. D, R and V couple the vibrations to the bath and are defined in terms of the eigenfunctions of $\mathcal{H}_0$. If there are N molecules in the system, each eigenfunction of $\mathcal{H}_0$ is characterized by a set of N ordered labels $(\beta, \gamma, \ldots)$ where β is the vibrational state of the first molecule, γ of the second, and so forth (the index β describes the population of all the vibrational modes of a given molecule). D is then defined as that part of the coupling Hamiltonian diagonal in this basis, R couples only states whose labels are permutations of each other, and V couples states that differ in at least one index. As shown later, D gives rise to the environmental fluctuations in the vibrational frequency, R to resonant V—V transfer, and V to vibrational population changes. We assume that the Raman line of interest is well separated from its neighbors and therefore exclude from our consideration some of the interesting lineshape effects that arise due to the overlap of vibrational lines.[20]

A. General Theory of Dephasing

As was shown in Section II.A (see (2.9)), the isotropic Raman spectrum is given by

$$I(\omega) \propto \int_0^\infty dt e^{i\omega t} \langle \alpha(t)\alpha(0)\rangle \tag{3.2}$$

where $\alpha(t)$ is the sum of the mean polarizabilities of the N molecules in the system:

$$\alpha(t) = \sum_{i=1}^{N} \alpha^i(t) \tag{3.3}$$

We now focus our attention on one particular transition $\beta \rightarrow \gamma$, which may be a fundamental, overtone, hot band, or combination band. We project out the part of α that corresponds to this transition. Thus we replace the full operator α^i by one whose only nonzero matrix elements are $\alpha^i_{\beta\gamma}$ and $\alpha^i_{\gamma\beta}$. For simplicity we continue to refer to these projected operators as α^i and α.

In the Heisenberg representation,

$$\begin{aligned} \alpha(t) &= e^{i\hbar^{-1}\mathcal{H}t}\alpha e^{-i\hbar^{-1}\mathcal{H}t} \\ &\equiv e^{i\hbar^{-1}\mathcal{H}^\times t}\alpha \end{aligned} \tag{3.4}$$

where

$$\mathcal{H}^\times A \equiv \mathcal{H}A - A\mathcal{H} = [\mathcal{H}, A]_- \tag{3.5}$$

Generalizing the notation of Ref. 21, we introduce several different interaction representations through the use of barred variables:

$$\begin{aligned} e^{i\hbar^{-1}(\mathcal{H}_0+\mathcal{H}_B+D)^\times t}A &\equiv \bar{\bar{A}}(t) \\ e^{i\hbar^{-1}(\mathcal{H}_0+\mathcal{H}_B+D+R)^\times t}A &\equiv \bar{A}(t) \end{aligned} \tag{3.6}$$

Then

$$\bar{A}(t) = \exp_0\left[i\hbar^{-1}\int_0^t dt'\bar{\bar{R}}^\times(t')\right]\bar{\bar{A}}(t) \tag{3.7}$$

$$A(t) = \exp_0\left[i\hbar^{-1}\int_0^t dt'\bar{V}^\times(t')\right]\bar{A}(t)$$

where $\exp_0$ denotes a time-ordered exponential.[21–23] The time correlation function of interest can then be rewritten as follows:

$$\begin{aligned}\Phi(t) &\equiv \langle \alpha(t)\alpha \rangle \\ &= \langle \bar{\alpha}(t)\alpha\rangle \frac{\left\langle \exp_0\left[i\hbar^{-1}\int_0^t dt' \bar{V}^{\times}(t')\right]\bar{\alpha}(t)\alpha\right\rangle}{\langle\bar{\alpha}(t)\alpha\rangle} \\ &= \langle \bar{\bar{\alpha}}(t)\alpha\rangle \frac{\left\langle \exp_0\left[i\hbar^{-1}\int_0^t dt' \bar{\bar{R}}^{\times}(t')\right]\bar{\bar{\alpha}}(t)\alpha\right\rangle}{\langle\bar{\bar{\alpha}}(t)\alpha\rangle} \\ &\otimes \frac{\left\langle \exp_0\left[i\hbar^{-1}\int_0^t dt' \bar{V}^{\times}(t')\right]\bar{\alpha}(t)\alpha\right\rangle}{\langle\bar{\alpha}(t)\alpha\rangle} \\ &\equiv \Phi_1(t)\Phi_2(t)\Phi_3(t) \end{aligned} \tag{3.8}$$

Up to this point we have made no approximations. We now consider separately the functions Φ_1, Φ_2, and Φ_3.

1. *Environmental Fluctuations*

Consider first $\Phi_1(t)$. Introducing still another interaction representation gives

$$\bar{\bar{A}}_{mn}(t) = \exp_0\left[i\hbar^{-1}\int_0^t dt' \bar{\bar{\bar{D}}}^{\times}(t')\right]\bar{\bar{\bar{A}}}_{mn}(t) \tag{3.9}$$

where

$$\begin{aligned}\bar{\bar{\bar{A}}}_{mn}(t) &= [e^{i\hbar^{-1}(\mathcal{H}_0+\mathcal{H}_B)^{\times}t}A]_{mn} \\ &= e^{-i\omega_{nm}t}A_{mn}\end{aligned} \tag{3.10}$$

and

$$\begin{aligned}\hbar\omega_{nm} &= E_n - E_m \\ \mathcal{H}_0|n\rangle &= E_n|n\rangle\end{aligned} \tag{3.11}$$

Using the diagonal nature of D then gives

$$\bar{\bar{A}}_{mn}(t) = \exp\left[-i\hbar^{-1}\int_0^t dt'(D_{nn}(t') - D_{mm}(t'))\right]e^{-i\omega_{nm}t}A_{mn} \tag{3.12}$$

where the time dependence of a matrix element of D arises entirely from the bath degrees of freedom.

$$\Phi_1(t) = \sum_{i=1}^{N} \left\langle \exp\left[-i\hbar^{-1}\int_0^t dt'(D^i_{\gamma\gamma}(t') - D^i_{\beta\beta}(t'))\right] \otimes e^{-i\omega_{\gamma\beta}t} \alpha^i_{\beta\gamma}\alpha^i_{\gamma\beta} \right\rangle$$

$$= N\,|\alpha^i_{\beta\gamma}|^2\, e^{-i\omega_{\gamma\beta}t} \left\langle \exp\left[-i\hbar^{-1}\int_0^t dt'(D^i_{\gamma\gamma}(t') - D^i_{\beta\beta}(t'))\right]\right\rangle \qquad (3.13)$$

The angular brackets now refer to a (classical) ensemble average over the bath.

The physical meaning of (3.13) is straightforward: $D^i_{\gamma\gamma}(t)$ is the time-dependent level shift of level γ in molecule i, so that $\hbar^{-1}(D^i_{\gamma\gamma}(t) - D^i_{\beta\beta}(t))$ is the fluctuating frequency of the $\beta \to \gamma$ transition induced by the changing environment of molecule i. These fluctuations give a contribution to dephasing and line broadening. Carrying out a cumulant expansion[21,22] of (3.13) gives

$$\Phi_1(t) = N\,|\alpha^i_{\beta\gamma}|^2\, e^{-i\omega_{\gamma\beta}t} \exp\left[-i\hbar^{-1}\langle D^i_{\gamma\gamma} - D^i_{\beta\beta}\rangle t \right.$$

$$\left. -\hbar^{-2}\int_0^t dt' \int_0^{t'} dt'' \langle (D^i_{\gamma\gamma}(t') - D^i_{\beta\beta}(t'))(D^i_{\gamma\gamma}(t'') - D^i_{\beta\beta}(t''))\rangle_c + \ldots\right] \qquad (3.14)$$

where

$$\langle AB\rangle_c \equiv \langle AB\rangle - \langle A\rangle\langle B\rangle \qquad (3.15)$$

is a cumulant average. The first term in (3.14) gives only a line shift, which does not concern us here, whereas the second gives rise to line broadening. The expansion may be terminated after two terms for either of two reasons: (1) If molecule i is perturbed by a very large number $n \to \infty$ of neighbors, so that the central limit theorem[24] applies and $D^i_{\gamma\gamma} - D^i_{\beta\beta}$ will be a Gaussian random variable. In this case the higher-order cumulants vanish. (2) If D is small so that higher powers of D are negligible in comparison to those retained in (3.14).

2. *Resonant Transfer*

Consider next Φ_2; a cumulant expansion gives

$$\Phi_2(t) = \exp\left[i\hbar^{-1}\int_0^t dt' \frac{\langle [\bar{\bar{R}}(t'), \bar{\bar{\alpha}}(t)]_- \alpha\rangle}{\langle \bar{\bar{\alpha}}(t)\alpha\rangle} \right.$$

$$\left. -\hbar^{-2}\int_0^t dt' \int_0^{t'} dt'' \frac{\langle [\bar{\bar{R}}(t''), [\bar{\bar{R}}(t'), \bar{\bar{\alpha}}(t)]_-]_- \alpha\rangle_c}{\langle \bar{\bar{\alpha}}(t)\alpha\rangle} + \ldots\right] \qquad (3.16)$$

A truncation can be justified on the same grounds as those outlined at the end of the preceding section. As we have defined it, R includes resonant transfer of arbitrary numbers of quanta between molecules; in practice, only the lowest-order single-quantum exchanges are significant. Define 2^i as a state in which a normal mode of molecule i has two quanta. Then it should be remembered that a process such as

$$(2^i, 0^j) \rightarrow (1^i, 1^j)$$

is not a resonant transfer process because of anharmonicity; it is governed by V rather than R. A process such as

$$(2^i, 0^j) \rightarrow (0^i, 2^j)$$

involves a two-quantum exchange and is therefore very small. As a result of these two observations, we can take $\Phi_2(t) = 1$ for overtone and combination bonds (of course, in a higher-order calculation one could calculate $\Phi_2(t)$ for these transitions: see the discussion at the end of Section III.B). We take matrix elements of R to be zero, except those involving exchange of a ground state label 0 and a fundamental β.

For a $\beta \rightarrow \gamma$ hot band transition, where $0 \rightarrow \beta$ is a fundamental, there is a contribution from Φ_2. The first term in (3.16) vanishes, but the second gives

$$\Phi_2(t) = \exp\left[-\hbar^{-2}\int_0^t dt' \int_0^{t'} dt'' \sum_i \langle \bar{\bar{R}}_{i\beta j\beta}(t'')\bar{\bar{R}}_{j\beta i\beta}(t')\rangle_c\right] \tag{3.17}$$

For a fundamental $0 \rightarrow \gamma$ transition, both terms in (3.16) are present, giving

$$\Phi_2(t) = \exp\left[-i\hbar^{-1}\int_0^t dt' \frac{\sum_j \langle \bar{\bar{\alpha}}^j_{0\gamma}(t)\bar{\bar{R}}_{j\gamma i\gamma}(t')\alpha^i_{\gamma 0}\rangle}{\langle \bar{\bar{\alpha}}^i_{0\gamma}(t)\alpha\rangle}\right.$$
$$\left. -\hbar^{-2}\int_0^t dt' \int_0^{t'} dt'' \frac{\sum_{j,k} \langle \bar{\bar{\alpha}}^k_{0\gamma}(t)\bar{\bar{R}}_{k\gamma j\gamma}(t'')\bar{\bar{R}}_{j\gamma i\gamma}(t')\alpha^i_{\gamma 0}\rangle_c}{\langle \bar{\bar{\alpha}}^i_{0\gamma}(t)\alpha\rangle}\right] \tag{3.18}$$

$\Phi_2(t)$ is (approximately) the function obtained through isotopic dilution[21,25] When a molecule is diluted with a different isotope, the effect is to eliminate the resonant transfer Hamiltonian R. V is affected as well, and therefore $\Phi_3(t)$ is affected, but if these changes are small, as is usually assumed, then $\Phi_2(t)$ can be obtained by dividing the isotopically dilute correlation function into the pure fluid correlation function.

R is the only part of the coupling Hamiltonian that appears explicitly in (3.18), but it is important to remember that the double-barred quantities evolve under the influence of D. In fact, expanding (3.18) in D and combining it with $\Phi_1(t)$ gives

$$\Phi_1(t)\Phi_2(t) \approx N\,|\alpha^i_{\beta\gamma}|^2\, e^{-i\omega_{\gamma\beta}t} \exp\left[-i\langle\Delta\omega\rangle t - \int_0^t dt' \int_0^{t'} dt''\,\langle\Delta\omega(t')\Delta\omega(t'')\rangle_c\right] \quad (3.19)$$

where

$$\hbar\Delta\omega(t) \equiv D^i_{\gamma\gamma}(t) - D^i_{00}(t) + \sum_j R_{j\gamma i\gamma}(t) \quad (3.20)$$

The second term in (3.19) depends on cross correlations between D and R, whose effects are therefore not separable. This nonseparability has been discussed in Refs. 26 to 29.

3. *Population Relaxation*

Consider finally $\Phi_3(t)$. Trying first a cumulant expansion gives (the first-order cumulant vanishes)

$$\begin{aligned}
\Phi_3(t) &= \exp\left[-\hbar^2 \int_0^t dt' \int_0^{t'} dt''\, \frac{\langle[\bar{V}(t''), [\bar{V}(t'), \bar{\alpha}(t)]_-]_-\alpha\rangle}{\langle\bar{\alpha}(t)\alpha\rangle}\right] \\
&= \exp\left[-\hbar^{-2} \int_0^t dt' \int_0^{t'} dt''\, \frac{\sum_\delta \langle \bar{V}_{\beta\delta}(t'')\bar{V}_{\delta\beta}(t')\bar{\alpha}_{\beta\gamma}(t)\alpha_{\gamma\beta}\rangle}{\langle\bar{\alpha}_{\beta\gamma}(t)\alpha_{\gamma\beta}\rangle}\right. \\
&\qquad \left. -\hbar^{-2} \int_0^t dt' \int_0^{t'} dt''\, \frac{\sum_\delta \langle\bar{\alpha}_{\beta\gamma}(t)\bar{V}_{\gamma\delta}(t'')\bar{V}_{\delta\gamma}(t')\alpha_{\gamma\beta}\rangle}{\langle\bar{\alpha}_{\beta\gamma}(t)\alpha_{\gamma\beta}\rangle}\right]
\end{aligned} \quad (3.21)$$

Expanding in D and R gives

$$\begin{aligned}
\Phi_3(t) = \exp\Bigg[&-\hbar^{-2}\int_0^t dt' \int_0^{t'} dt'' \sum_\delta e^{-i\omega_{\delta\beta}(t''-t')} \langle V_{\beta\delta}(t''-t')V_{\delta\beta}(0)\rangle \\
&-\hbar^{-2}\int_0^t dt' \int_0^{t'} dt'' \sum_\delta e^{-i\omega_{\delta\gamma}(t''-t')} \langle V_{\gamma\delta}(t''-t')V_{\delta\gamma}(0)\rangle\Bigg]
\end{aligned} \quad (3.22)$$

There are two things to note about (3.22). First, there are no cross terms between V and D or R; thus to the extent to which a second-order cumulant expansion is valid, population relaxation is independent of "pure" dephasing. Second, the correlation functions of (3.22), unlike those of (3.19), contribute through their spectra at finite frequencies $\omega_{\delta\beta}$ and $\omega_{\delta\gamma}$. If β and γ are far from other energy levels, these Fourier transforms are small and $\Phi_3(t)$ close to unity. Physically, this corresponds to the fact that, in a population relaxation process induced by V, the excess energy $\hbar\omega_{\delta\beta}$ or $\hbar\omega_{\delta\gamma}$ must be taken up by the rotational and translational degrees of freedom, which slows the rate of such processes.

Let us now treat $\Phi_3(t)$ more generally, avoiding the cumulant expansion. It can be written as

$$\Phi_3(t) = \frac{\langle \bar{\mathcal{U}}(0, t)\bar{\alpha}(t)\bar{\mathcal{U}}(t, 0)\alpha\rangle}{\langle \bar{\alpha}(t)\alpha\rangle} \tag{3.23}$$

where the propagator $\bar{\mathcal{U}}$ is[23]

$$\bar{\mathcal{U}}(t, t') = 1 + \sum_{n=1}^{\infty} \frac{(-i)^n}{n!} \int_t^{t'} dt_n \int_t^{t_n} dt_{n-1} \ldots \int_t^{t_2} dt_1 \, \bar{V}(t_n) \otimes \bar{V}(t_{n-1}) \ldots \bar{V}(t_1) \tag{3.24}$$

We then have

$$\Phi_3(t) = \frac{\langle \bar{\mathcal{U}}_{\beta\beta}(0, t)\bar{\alpha}_{\beta\gamma}(t)\bar{\mathcal{U}}_{\gamma\gamma}(t, 0)\alpha_{\gamma\beta}\rangle}{\langle \bar{\alpha}_{\beta\gamma}(t)\alpha_{\gamma\beta}\rangle} \tag{3.25}$$

Now consider the population relaxation rate for level β, for example. It is

$$P_\beta(t) = \frac{\langle \bar{\mathcal{U}}_{\beta\beta}(0, t)\bar{n}_{\beta\beta}(t)\bar{\mathcal{U}}_{\beta\beta}(t, 0)n_{\beta\beta}\rangle}{\langle \bar{n}_{\beta\beta}(t)n_{\beta\beta}\rangle} \tag{3.26}$$

where $n_{\beta\beta}$ is the population fluctuation in level β. Since $n_{\beta\beta}$ is diagonal, it commutes with D and R, so that

$$P_\beta(t) = \langle \bar{\mathcal{U}}_{\beta\beta}(0, t)\bar{\mathcal{U}}_{\beta\beta}(t, 0)\rangle \tag{3.27}$$

Taking now the rapid modulation limit, where the bath correlation time is much less than t

$$\begin{aligned} P_\beta(t) &\to \langle \bar{\mathcal{U}}_{\beta\beta}(0, t)\rangle\langle \bar{\mathcal{U}}_{\beta\beta}(t, 0)\rangle = \langle \bar{\mathcal{U}}_{\beta\beta}(0, t)\rangle^2 \\ &= [\exp \langle \bar{\mathcal{U}}(0, t)\rangle_L]^2_{\beta\beta} = e^{-t/T_1^\beta} \end{aligned} \tag{3.28}$$

where $\langle \ \rangle_L$ is a linked cluster expansion[23] of $\bar{\mathcal{U}}$. T_1^β is the population relaxation rate of level β. In this same limit,

$$\Phi_3(t) \to \langle \bar{\mathcal{U}}_{\beta\beta}(0, t)\rangle\langle \bar{\mathcal{U}}_{\gamma\gamma}(t, 0)\rangle = \exp\left[-t\left(\frac{1}{2T_1^\beta} + \frac{1}{2T_1^\gamma}\right)\right] \tag{3.29}$$

so that there is a contribution to the dephasing rate equal to the *average* population relaxation rate of the two levels involved.

If the bath correlation time is *not* much shorter than the population relaxation times of the two levels, the connection between dephasing and population relaxation is no longer simple.

B. Nature of the Coupling Hamiltonian

A more precise specification of the Hamiltonian must now be given. The vibrational Hamiltonian for the isolated molecule i as taken as a sum over the vibrational modes β

$$\mathcal{H}_0^i = \sum_\beta \left(p_{i\beta}^2/2\mu_\beta + \tfrac{1}{2}\mu_\beta\omega_\beta^2 Q_{i\beta}^2 + \tfrac{1}{6}f_\beta Q_{i\beta}^3\right) \tag{3.30}$$

The cubic force constant f_β may be obtained for diatomics from the rotational constant B_e, the rotation-vibration interaction constant α_e, and the internuclear separation r_e[30]:

$$f = -\frac{3\omega_e^2}{2B_e r_e^3}\left[1 + \frac{\alpha_e\omega_e}{6B_e^2}\right] \tag{3.31}$$

where f is given in $\mathrm{cm}^{-1}/\mathrm{cm}^3$ and the spectroscopic constants in cm^{-1}. Note that it cannot be obtained from the usual anharmonicity $\omega_e x_e$[29,31] which depends as well on the quartic force constant. For polyatomic molecules, f_α can only be obtained through a careful multiparameter fit to high-resolution spectra with isotopic substitutions.[32] In general, there are also cubic terms coupling different vibrations, but they do not contribute in the present case. f is treated as a small perturbation, and only the terms of lowest order in f are retained.

Coupling between the vibrations and the bath arises from two sources: first, the interaction of a molecule with its neighbors, which depends on the internal vibrational coordinates, and second, the coupling between vibrations and rotations on a single molecule. The coupling Hamiltonian is then[33]

$$D + R + V = [V_S(\mathbf{Q}, t) - V_S(0, t)] + \sum_{i=1}^{N} \sum_{\alpha=x,y,z} \left(\frac{J_{i\alpha}^2(t)}{2I_{i\alpha}(\mathbf{Q})} - \frac{J_{i\alpha}^2(t)}{2I_{i\alpha}(0)}\right) \tag{3.32}$$

V_S is the intermolecular potential, which depends on the set of vibrational coordinates $\mathbf{Q}$; $J_{i\alpha}(t)$ is the α component of the angular momentum of molecule i, taken here to be a classical variable; $I_{i\alpha}(\mathbf{Q})$ is the moment of inertia of molecule i about the principle axis α, which depends on $\mathbf{Q}$. Expanding (3.32) in $\mathbf{Q}$ and retaining only the lowest-order terms (the series is generally quite rapidly convergent) gives

$$\begin{aligned} D + R + V = {} & \sum_{i,\beta}\left[\frac{\partial V_S}{\partial Q_{i\beta}}(0, t)Q_{i\beta} + \frac{1}{2}\frac{\partial^2 V_S}{\partial Q_{i\beta}^2}(0, t)Q_{i\beta}^2\right] \\ & + \frac{1}{2}\sum_{i,\beta}\sum_{\alpha=x,y,z} J_{i\alpha}^2(t)\left[\frac{\partial I_{i\alpha}^{-1}}{\partial Q_{i\beta}}(0)Q_{i\beta} + \frac{1}{2}\frac{\partial^2 I_{i\alpha}^{-1}}{\partial Q_{i\beta}^2}(0)Q_{i\beta}^2\right] \\ & + \sum_{i\neq j}\sum_{\beta}\frac{\partial^2 V_S}{\partial Q_{i\beta}\,\partial Q_{j\beta}}(0, t)Q_{i\beta}Q_{j\beta} \\ & + \sum_{i,j}\sum_{\beta\neq\gamma}\frac{\partial^2 V_S}{\partial Q_{i\beta}\,\partial Q_{j\gamma}}(0, t)Q_{i\beta}Q_{j\gamma} \end{aligned} \tag{3.33}$$

The terms on the right hand side of (3.33) have been divided into four groups. The diagonal matrix elements of those in the first two groups contribute to D, and the off-diagonal to V (in practice only the term linear in Q is significant in V, since the other corresponds to a two-quantum transition). The third group contributes only to R and the fourth to V. For a diatomic molecule (3.33) is simplified to[33]

$$\begin{aligned} D+R+V=\sum_i \Bigg[&\frac{\partial V_S}{\partial Q_i}(0,t)Q_i+\frac{1}{2}\frac{\partial^2 V_S}{\partial Q_i^2}(0,t)Q_i^2 \\ &+\frac{\vec{J}_i^2(t)}{2I_i(0)}\left(-\frac{2}{R_0}Q_i+\frac{3}{R_0^2}Q_i^2\right)\Bigg] \\ &+\sum_{i\neq j}\frac{\partial^2 V_S}{\partial Q_i\,\partial Q_j}(0,t)Q_iQ_j \end{aligned} \tag{3.34}$$

where R_0 is the equilibrium internuclear separation for the molecule.

Certain matrix elements of D, R, and V are required for the calculation of $\Phi(t)$ (see (3.14), (3.18), and (3.22)); they may be evaluated from (3.33) and (3.34). We give results only for the diatomic molecule case; the extension to polyatomics is straightforward. Consider the $\beta\rightarrow\gamma$ transition; let n_β, n_γ be the number of quanta in the two states. The transition can be a fundamental ($n_\beta=0$, $n_\gamma=1$), an overtone ($n_\beta=0$, $n_\gamma>1$), or a hot band ($n_\beta>0$). Evaluation of the anharmonic oscillator matrix elements using (3.30) then gives

$$\begin{aligned} D^i_{\gamma\gamma}(t)-D^i_{\beta\beta}(t)=(n_\gamma-n_\beta)\Bigg[&-\frac{\hbar f}{2\mu^2\omega_0^3}\frac{\partial V_S}{\partial Q_i}(0,t) \\ &+\frac{\hbar}{2\mu\omega_0}\frac{\partial^2 V_S}{\partial Q_i^2}(0,t)+\frac{\vec{J}^2(t)}{2I(0)}\left(\frac{\hbar f}{\mu^2\omega_0^3 R_0}+\frac{3\hbar}{I(0)\omega_0}\right)\Bigg] \end{aligned} \tag{3.35}$$

$$R_{i\beta j\beta}(t)=\delta_{n_\beta,1}\frac{\hbar}{2\mu\omega_0}\frac{\partial^2 V_S}{\partial Q_i\partial Q_j}(0,t) \tag{3.36}$$

$$V_{\beta\varepsilon}(t)=[\delta_{n_\varepsilon,n_\beta+1}(n_\beta+1)^{1/2}+\delta_{n_\varepsilon,n_\beta-1}n_\beta^{1/2}]\left(\frac{\hbar}{2\mu\omega_0}\right)^{1/2}\frac{\partial V_S}{\partial Q_i}(0,t) \tag{3.37}$$

A comparison of fundamentals and overtones shows that, although R does not contribute for overtones, D and V become larger, which should on balance lead to a broadening of spectral line widths along a progression of overtone bands.

Throughout this calculation anharmonicities have been assumed to be so large that the transfer of a single quantum from a mode containing several quanta to a nearby ground state molecule requires a substantial rotational or translational contribution to make up the energy difference.

In this case the resonant transfer contribution to an overtone band is small, and an isotopic dilution experiment would show little change in the linewidth. If the anharmonicities are small, however, there is a contribution from R to all the overtone bands of

$$R_{i\beta j1}(t)=(n_\beta)^{1/2}\frac{\hbar}{2\mu\omega_0}\frac{\partial^2 V_S}{\partial Q_i\,\partial Q_j}(0,t) \tag{3.38}$$

In this case isotopic dilution has a *larger* effect on the overtones than on the fundamentals. This does indeed seem to be observed in some recent experiments by Arndt and Yarwood[34,34a]

C. Infrared and Depolarized Raman Spectra

It was shown in Section II.A that infrared and depolarized Raman spectra are affected by vibrational dephasing, with the corresponding time correlation functions (for symmetric tops) being approximately given by

$$C_n(t)\sim\langle P_n(\cos\theta_i(t))Q_i(t)Q_i(0)\rangle \tag{3.39}$$

where $n=1$ for infrared and 2 for depolarized Raman spectra. To obtain information about rotational relaxation, the vibrational effects in (3.39) must somehow be treated. The crudest approximation is simply to neglect them completely, which is valid only if vibrational relaxation is much slower than reorientation. Bartoli and Litovitz[35] and Rakov[36a] have suggested a somewhat better approximation, in which the temperature dependence of the total linewidth is fit to the form

$$\delta_{\text{TOT}}=\delta_{\text{VIB}}+C\exp(-E_a/kT) \tag{3.40}$$

The vibrational part is assumed to be independent of temperature, whereas the rotational part has an exponential temperature dependence. However, since δ_{VIB} is often temperature dependent as well, this is not a very accurate approximation. Sometimes[36b] dephasing rates for a Raman-active mode are used for a different mode, an approximation that is certainly not valid, since each mode has its own phase relaxation mechanisms.

Bratos et al.[37] and Bartoli and Litovitz[35] have suggested a procedure that is widely used in studying parallel bands, where the isotropic Raman line is present. They neglect the correlation between rotational and vibrational degrees of freedom, so that $C_n(t)$ becomes a product of two separate correlation functions:

$$C_n(t)\approx\langle P_n(\cos\theta_i(t))\rangle\langle Q_i(t)Q_i(0)\rangle \tag{3.41}$$

If (3.41) is valid, then the rotational correlation function may be obtained experimentally by dividing $C_n(t)$ by the isotropic Raman correlation function $\langle Q_i(t)Q_i(0)\rangle$. The approximation of (3.41) neglects two types of vibration-rotation coupling: the dependence of the moments of inertia on the vibrational coordinates (see Section III.B) and the effect of orientation-dependent potentials (such as dipole-dipole coupling) on the vibrational dephasing process.

Two recent studies have challenged the separability assumption of (3.41). van Woerkom et al.[21] have carried out a careful analysis of the effect of resonant transfer on the infrared band shape using a dipolar interaction and avoiding the assumption of separability of vibrations and rotations. They use the cumulant expansion method outlined in Section III.A and obtain explicit results in the rotational diffusion limit. The functional form they obtain gives a good fit to the effect of isotopic dilution on the band shape of CH_2Cl_2.[25] The factorization assumption of (3.41) gives a rather different result. Lynden-Bell[26] has demonstrated a still more dramatic result: even in the absence of rotational relaxation or orientational correlations, the lineshapes for infrared absorption and polarized and depolarized Raman scattering differ. These two studies show that extreme caution must be used in attempting to extract information about rotational relaxation from spectra strongly perturbed by vibrational phase relaxation.

The most direct experimental evidence for the nonseparability of rotations and vibrations is the work of Amorim da Costa et al.[38] They compared the reorientation times obtained from Rayleigh scattering with those from Raman scattering using the separability assumption. For pure liquids the Raman bands were systematically broader by up to 150%; although this may be due to collective reorientation effects present in Rayleigh lineshapes but not in Raman, they found that the discrepancy persisted and in fact increased in dilute solutions of acetonitrile in CCl_4, where these collective effects are not present. Even more striking was the observation of large differences in reorientation times obtained from different Raman modes of the same molecule. The discrepancy between Rayleigh and Raman orientation times was largest in cases in which the isotropic Raman linewidth was broadest, suggesting that correlations between vibrations and rotations are important and cannot be neglected.

IV. THEORETICAL MODELS FOR VIBRATIONAL DEPHASING

The theory presented in Section III provides a framework for the description of vibrational dephasing; through a series of controlled and usually well-satisfied assumptions it relates the phase relaxation to the correlation functions of microscopic variables. The theory remains purely formal, however, until some approximate model is introduced for these correlation functions. In this section we outline a number of theories that have been applied to the problem of vibrational dephasing. The treatment is by no means chronological; in fact, we begin with the results of a very recent molecular dynamics simulation that was able to test a number of the assumptions made in the other theories.

A. Molecular Dynamics Simulation

The method of computer simulation via molecular dynamics has been applied to classical atomic fluids for some 20 years[39]; a wealth of information about collective and single particle motion in liquids has been obtained. More recently, simulations of diatomic fluids such as liquid nitrogen[40] have allowed the study of rotational relaxation in the classical approximation. Vibrational phase and energy relaxation have not been accessible to computer simulation until quite recently. There are two reasons for this: first, the fundamentally quantum nature of vibrations, even in liquids, makes a purely classical simulation more questionable than in the case of rotations; second, the vibrational period is often so much shorter than the characteristic rotational and translational time scales that a simulation of vibrating molecules would require an extremely short integration time step, leading to prohibitively lengthy calculations. This latter feature forced Riehl and Diestler,[41] in their one-dimensional simulation of vibrating diatomics, to choose an unphysically small vibration frequency (about $10\,\mathrm{cm}^{-1}$) to make the calculation feasible.[42] It appears therefore that the direct simulation of $\langle Q(t)Q(0)\rangle$ is not a very promising approach in most cases.

In a recent paper,[27] Oxtoby, Levesque, and Weis showed that the general theory presented in Section III can be used to avoid the problems that arise in a direct simulation. Equations such as (3.13), (3.18), and (3.22) relate the vibrational phase correlation function to matrix elements of D, R, and V, which are given in (3.35)–(3.37). The time dependence of these matrix elements arises through the dependence of the potential on the relative positions and orientations of the molecules; a fundamental simplication arises from the fact that the derivatives in (3.35) are all evaluated at the equilibrium position $Q=0$. Their statistical properties can therefore be evaluated through a simulation of rigid, nonvibrating, classical molecules.

The most general calculation that appears to be possible would combine an exact evaluation of Φ_1 (3.13) with a calculation of Φ_2 and Φ_3 to second order in a cumulant expansion [(3.16) and (3.21)]; this corresponds to an exact treatment of D and an approximate treatment of R and V. In a strongly interacting fluid, D gives rise to inhomogeneous broadening through the presence of a range of environments, whereas R and V will probably be smaller, and the approximate treatment should give good results. The evaluation of the time integral in (3.13) makes the calculation somewhat lengthier than a simple correlation function evaluation, but model calculations[43] on a strongly coupled fluid have demonstrated that their evaluation is feasible.

In the simulation of liquid nitrogen,[27] however, it was found that a much simpler calculation gave accurate results. Nitrogen is a "weakly interacting" fluid in the sense that the fluctuations in its vibrational frequency are small compared to the mean frequency. Thus a cumulant expansion in D as well as R and V may be carried out and truncated at second order. Since population relaxation is 12 orders of magnitude slower than dephasing,[44] V may be neglected, and the dephasing is then given by (3.19) in terms of the autocorrelation function of $\Delta\omega(t)$, which in turn is given by (3.20).

The simulation was carried out using an atom-atom Lennard-Jones potential

$$V_S = \sum_{i,j=1}^{N} \sum_{\alpha,\beta=1}^{2} \nu(r_{i\alpha j\beta}) \tag{4.1}$$

where

$$\nu(r) = 4\varepsilon\left[\left(\frac{\sigma}{r}\right)^{12} - \left(\frac{\sigma}{r}\right)^{6}\right] \tag{4.2}$$

and $r_{i\alpha j\beta}$ is the distance between the αth atom of molecule i and the βth atom of molecule j. In calculating derivatives of V_S with respect to the vibrational coordinate Q, ε and σ were held constant. This potential surface was chosen more for its simplicity than in the expectation that the results would be highly accurate. Further simulations are planned[43] using a multiparameter repulsive atom-atom potential plus dispersion and quadrupole terms.[45] The density and temperature were chosen to simulate a thermodynamic state near the boiling point of liquid nitrogen ($\rho/\sigma^3 = 0.62965$; $kT/\varepsilon = 2.03$ in reduced units). Details of the calculation are given in Ref. 27. In the first calculation, the vibration-rotation coupling term in $\vec{J}(t)$ from (3.34) was not included; its effect will be studied in future work.

The root mean square frequency fluctuation obtained from the calculation is

$$\langle(\Delta\omega)^2\rangle_c^{1/2} = 2.32\times 10^{11}\ \text{sec}^{-1} = 1.23\ \text{cm}^{-1} \tag{4.3}$$

The normalized autocorrelation function of $\Delta\omega(t)$, $C(t)$, is shown in Fig. 1; the correlation time τ_c, given by its integral, is

$$\tau_c = \int_0^\infty dt\, C(t) = 0.149\ \text{psec} \tag{4.4}$$

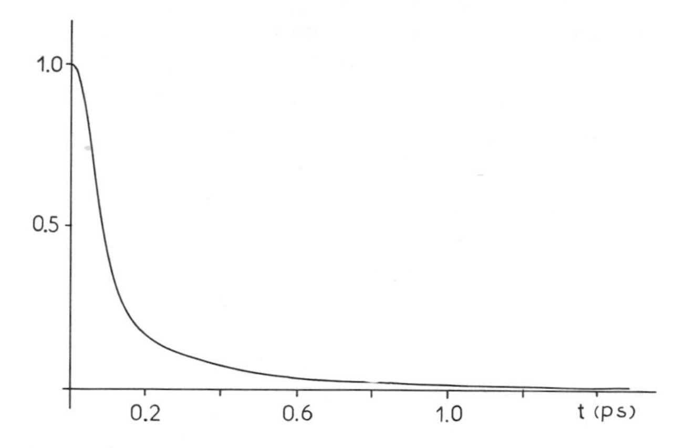

Fig. 1. Autocorrelation function of frequency fluctuations, $\langle\Delta\omega(t)\Delta\omega(0)\rangle_c$.

The spectrum is given [see (3.19)] by the Fourier transform of

$$\exp\left[-\int_0^t dt' \int_0^{t'} dt'' \langle(\Delta\omega)^2\rangle_c C(t''-t')\right] \tag{4.5}$$

Two limiting cases are of interest: if $\langle(\Delta\omega)^2\rangle_c^{1/2}\tau_c \gg 1$, then (4.5) corresponds to a Gaussian whose broadening is due entirely to a static distribution of frequency fluctuations. If $\langle(\Delta\omega)^2\rangle_c^{1/2}\tau_c \ll 1$, then (4.5) becomes $\exp(-\tau/\tau_v)$ where

$$\tau_V^{-1} = \langle(\Delta\omega)^2\rangle_c \tau_c \tag{4.6}$$

This Lorentzian spectrum arises then in the rapid modulation, or motional narrowing, limit. The results for liquid nitrogen give[27]

$$\langle(\Delta\omega)^2\rangle_c^{1/2}\tau_c = 0.035 \ll 1 \tag{4.7}$$

so that the motional narrowing limit obtains, the broadening is homogeneous, and the line should be Lorentzian, as is in fact was observed.[46,47] The calculated value of τ_V, 125 psec, agrees surprisingly well (considering the crudeness of the potential) with picosecond pulse studies[48] ($\tau_v = 150 \pm 16$ psec) and isotropic Raman line width measurements[46,47] ($\tau_v = 158 \pm 14$ psec).

Two additional calculations provided interesting insights. In the first, nitrogen was treated as harmonic by setting the anharmonicity parameter f [see (3.35)] to zero. This gave a τ_v of 3640 psec, a factor of 30 longer than that obtained when the vibrational anharmonicity was taken into account. This shows that model calculations which treat the vibrations as harmonic cannot be expected to give quantitatively accurate results. In a second calculations, the resonant transfer Hamiltonian R was suppressed, and the result $\tau_v = 115$ psec was found. This is *shorter* than the dephasing time found previously, showing that in this case resonant transfer leads to a *slower* dephasing rate. This result seems contrary to intuition, since one would expect that R, by allowing an additional dephasing mechanism, would lead to a more rapid dephasing; indeed, if R and D were uncorrelated, this would always be the case. However, there are cross terms between R and D in (3.19) which in this case lead to a lengthening of the dephasing time. The existence of such cross terms has been pointed out by others[26,28,29] although some calculations[21,31] neglect them. The simulation results suggest that there may be cases in which isotopic dilution leads to a *broadening* of spectral bands.

B. Harmonic Oscillator in a Heat Bath

The simplest model for vibrational phase and population relaxation is that of a harmonic oscillator weakly coupled to a heat bath; it has been studied by a number of authors. The first studies[49,50] took the coupling to the heat bath to be linear in the vibrational coordinate Q:

$$\mathcal{H}' = F(t)Q \tag{4.8}$$

For a harmonic oscillator, $\mathcal{H}'$ then has no diagonal matrix elements, so that, in the notation of Section III.A, the coupling Hamiltonian D is absent and only V is present. In this case only Φ_3 differs from unity, and τ_v and the population relaxation time T_1 are related through

$$\tau_v = 2T_1 \tag{4.9}$$

This result was derived through cumulant expansion[49] and projection operator[50] techniques. Equation 4.9 is in complete disagreement with experiments for diatomic liquids; τ_v for liquid nitrogen, for example, is 150 psec[48] whereas T_1 is longer than 56 *sec*,[44] a difference of 12 orders of magnitude.

Subsequent work has shown that relaxation of either of two assumptions (the harmonic nature of the oscillator or the purely linear coupling to the heat bath) is necessary to resolve the discrepancy. When either assumption is relaxed, the perturbation Hamiltonian D becomes nonzero; thus, as discussed in Section III.A, zero-frequency Fourier components can contribute to dephasing in addition to the very small high-frequency contributions. Madden and Lynden-Bell[51] included a Q^2 term in $\mathcal{H}'$ and used density matrix techniques developed for nuclear magnetic resonance problems to calculate the phase and population relaxation rates of different transitions in a harmonic oscillator; they demonstrated that this Q^2

term gave rise to a large zero-frequency contribution to τ_v but not to T_1. Oxtoby and Rice[52] and Knauss and Wilson[53] derived similar results using projection operator techniques, and Diestler[54] showed the importance of diagonal perturbations for a two-level system weakly coupled to a bath. As discussed in Section IV.A, for N_2 (liquid) the effect of anharmonicity with linear coupling is much more important than the bilinear coupling; both give rise to vibrationally elastic relaxation processes which are absent for a harmonic oscillator with linear coupling.

C. Static Environment Theories

Bratos et al.[37,55] developed an early theory for the infrared and Raman lineshapes of a diatomic impurity in a monatomic fluid. In this simple case, only the environmental fluctuation Hamiltonian D contributes to phase relaxation. Bratos et al.[37,55] carry out a cumulant expansion and argue that the cumulant averages may be replaced by ordinary averages; they avoid truncation of the expansion, which would be valid in the weak coupling limit. Finally, they assume that the slow modulation limit obtains, so that if $\Delta\omega(t)$ is the fluctuating frequency of the oscillator, the correlation time of $\Delta\omega(t)$ is long compared with $\langle(\Delta\omega(0))^2\rangle^{-1/2}$. This assumption implies that dephasing results from a collection of essentially static environments.

Molecular dynamics experiments[27] suggest, however, that the opposite limit of rapid modulation (or motional narrowing) is valid at least for liquid nitrogen and quite likely for other diatomics or diatomic/monatomic mixtures. The assumptions made in the earlier work[37,55] thus require reexamination. The correlation time of $\Delta\omega(t)$ is taken in Refs. 37 and 55 as the time required for an atom in the nearest neighbor shell of the diatomic to move out of that shell, or, in other words, the time for an atom to diffuse over an interatomic distance. This time is estimated to be 10^{-10} sec, three orders of magnitude longer than the correlation time obtained in the computer experiments.

The origin of this discrepancy has been analyzed in some of our recent work.[56] The frequency fluctuation of an oscillor arises from many different neighbors and thus has the form

$$C_M(t)=\left\langle\left[\sum_{j\neq i} F_{ij}(t)\right]\left[\sum_{k\neq i} F_{ik}(0)\right]\right\rangle \tag{4.10}$$

where F_{ij} is the frequency fluctuation of molecule i due to molecule j. C_M is to be contrasted with a binary approximation

$$C_B(t)=\left\langle\sum_{j\neq i} F_{ij}(t)F_{ij}(0)\right\rangle \tag{4.11}$$

The time scale of C_B is determined by the time during which two given molecules interact and thus is governed by slow, diffusive processes. C_M, on the other hand, relaxes on the much shorter time scale of the time between collisions, since, if an atom leaves the nearest neighbor shell, it is on the average replaced by another one. The time scale estimated in Ref. 55 applies to C_B but not to C_M. In several recent papers[26,57] the motional narrowing limit is assumed, but the correlation time is still estimated from C_B. The toal function C_M has both two and three body parts; each one separately includes a slow diffusive contribution, but in their sum this vanishes and the total correlation time is much shorter. The qualitative dependence of dephasing times on diffusion constant obtained in Refs. 26 and 57 therefore may not be accurate.

In a subsequent paper, Bratos and Chestier[33] included the vibration-rotation coupling effect, still within the static environment approximation. Levant[58] recently presented a numerical evaluation of the theory of Bratos et al. In addition to the static environment approximation, he took frequency perturbations arising from different neighbors to be statistically independent and used the low-density form of the pair distribution function with a Sutherland or Lennard-Jones plus hard sphere potential. He found that the isotropic Raman line narrowed with increasing temperature. Dijkman and van der Maas[59] studied the broadening of a Morse oscillator in a liquid, modeling the fluid as a static f.c.c. lattice unperturbed by the diatomic impurity. The interaction potential was taken to be a central dispersion term plus a $1/R^{12}$ atom-atom repulsive term. Broad, asymmetric lines were found.

D. Correlation Function Modeling

Rothschild[31,60] has demonstrated the usefulness of a simple two-parameter modeling of the correlation function for the frequency fluctuations $\Delta\omega(t)$ [see (3.19)]. His work is an application of Kubo's[22] general line shape theory. He suggests the exponential form

$$\chi(t) \equiv \langle \Delta\omega(t)\Delta\omega(0)\rangle_c = \langle(\Delta\omega)^2\rangle_c \exp(-t/\tau_c) \qquad (4.12)$$

Note, however, that it is the *cumulant* average of the mean square frequency shifts $\Delta\omega$ that appears in (4.12),

$$\langle(\Delta\omega)^2\rangle_c \equiv \langle(\Delta\omega)^2\rangle - \langle\Delta\omega\rangle^2 \qquad (4.13)$$

rather than simply the mean square, as used in Refs. 31 and 60. The assumption of an exponential decay for this normalized autocorrelation function is at least qualitatively reasonable for liquid nitrogen, as shown by the computer simulation results of Fig. 1. After this exponential form for $\chi(t)$ is inserted into (3.19), the vibrational coordinate correlation function becomes[60]

$$\Phi(t) = \langle Q(t)Q(0)\rangle \propto \exp\left[-\langle(\Delta\omega)^2\rangle_c\{\tau_c^2(\exp(-t/\tau_c)-1)+\tau_c t\}\right] \quad (4.14)$$

At short times ($t \ll \tau_c$)

$$\Phi(t) \to \exp\left[-\langle(\Delta\omega)^2\rangle_c t^2/2\right] = \exp\left[-\tfrac{1}{2}M_2 t^2\right] \quad (4.15)$$

where M_2, the second moment of the normalized spectrum, gives the mean square fluctuation in the frequency shifts $\Delta\omega$ experienced by the different molecules. At long times ($t \gg \tau_c$),

$$\Phi(t) \to \exp\left[-\langle(\Delta\omega)^2\rangle_c \tau_c t\right] \quad (4.16)$$

so that $\tau_v^{-1} = \langle(\Delta\omega)^2\rangle_c \tau_c$

This modeling of $\chi(t)$ is related to, but somewhat different from, the memory function modeling approach which has been applied to other problems.[61] The memory function $K(t)$ is defined through

$$\frac{\partial}{\partial t}\Phi(t) = -\int_0^t d\tau K(t-\tau)\Phi(\tau) \quad (4.17)$$

Rothschild has shown[60] that if the dephasing time is long compared to the correlation time of $\chi(t)$, τ_c, then $K(t)$ becomes equal to $\chi(t)$.

Rothschild has demonstrated the usefulness of the two-parameter form (4.12) both for fitting experimental spectra[60] and for predicting dephasing times.[31] One of the parameters can be obtained from M_2, the second frequency moment of the spectrum; if the spectrum is close to Lorentzian, however, this measurement is difficult, since the form of the spectrum in the far wings must be known (the second moment of a true Lorentzian is infinite). The combination $\langle(\Delta\omega)^2\rangle_c\tau_c$ can then be obtained from the long time behavior of $\Phi(t)$ or the low-frequency behavior of the spectrum. Applying this fitting procedure to the spectra of deuterochloroform, quinoline, tetravinyl tin, and isopropyl alcohol, Rothschild found correlation times τ_c ranging from 0.13 to 0.6 psec.[60] In a subsequent paper[31] he estimated the dephasing time of liquid nitrogen. τ_c was taken to be half the velocity correlation time, and $\langle(\Delta\omega)^2\rangle$ was estimated for a Lennard-Jones atom-atom potential using the known pair distribution function for several molecular orientations. Several corrections to this calculation are pointed out in Ref. 27; however, the overall qualitative agreement with the experimental dephasing time should still remain when they are taken into account.

E. Cell Model

Diestler and Manz[62] applied a cell model to the dephasing of liquid nitrogen and oxygen. They treated the coupling to the translational states of the fluid by replacing the actual potential felt by the diatomic by a "smeared-out" average potential arising from the nearest neighbor shell. Orientations were averaged over; thus the vibration was coupled only to its own translational motion in a Lennard-Jones Devonshire[63] cell. The cell potential obtained (from an atom-atom Morse potential) was then expanded in the vibrational coordinate of the diatomic, and the resulting radial Schrödinger equation was solved approximately. Taking the binding potential of the diatomic to be harmonic, Diestler and Manz obtained a dephasing time τ_v ($=2\tau_{ph}$) of about 3200 psec for N_2 near its boiling point and 1620 psec for O_2, much longer than the corresponding experimental values of 150 and 85 psec. To account for this discrepancy, the authors suggested that coupling of vibrations to rotations may be much more important than coupling to translations, as described by the cell model. However, another reason for the discrepancy may be the assumption of harmonic binding forces in the diatomic, which was shown in the computer simulation results presented in Section IV.A to give dephasing times too long by a factor of 30. Application of the cell model to N_2 and O_2, including their anharmonicities, would clearly be of interest.

F. Isolated Binary Collision Model

Fischer and Laubereau[64] have developed a binary collision model for dephasing in liquids. They studied the collinear collision of a diatomic AB with an atom C, using an exponential repulsive potential:

$$V(Q, R) = E \exp(\gamma_B Q/L - R/L) \tag{4.18}$$

where E is the relative kinetic energy, L is the range of the potential (estimated as the molecular diameter divided by 17.5), Q is the vibrational coordinate of the oscillator, R is the distance from atom C to the center of mass of A-B, and γ the mass ratio

$$\gamma_B = m_A/(m_A + m_B) \tag{4.19}$$

The change in $R(t)$ during a collision can be obtained for an exponential potential:

$$\exp[-R(t)/L] = \operatorname{sech}^2[(E/2\mu')^{1/2} t/L] \tag{4.20}$$

where

$$\mu' = m_C(m_A + m_B)/(m_A + m_B + m_C) \tag{4.21}$$

The vibrational frequency fluctuation $\Delta\omega(t)$ can be obtained by expanding $V(Q, R)$ in Q and taking the difference between its expectation values in the first excited and the ground vibrational state. For an anharmonic oscillator AB this gives

$$\begin{aligned}\Delta\omega(t) &= \frac{1}{\hbar}(V_{11}(t) - V_{00}(t)) \\ &= \frac{1}{2}\frac{E}{\mu\omega_0}\left(\frac{\gamma_B}{L}\right)^2\left(1-\frac{fL}{\mu\omega_0^2\gamma_B}\right)\operatorname{sech}^2\left[\left(\frac{E}{2\mu'}\right)^{1/2} t/L\right], \end{aligned}\tag{4.22}$$

where μ is the reduced mass of the oscillator and f is defined in (3.30). If the molecule C collides with the A end of the diatomic, the frequency shift $\Delta\omega(t)$ is the same except that γ_B is replaced by $\gamma_A = 1-\gamma_B$. If collisions occur at a rate τ_{BC}^{-1}, half with the A atom and half with B, and if these collisions are uncorrelated, then the correlation function of the fluctuating frequency is obtained by averaging over the time of minimum impact parameter in a series of collisions, giving

$$\begin{aligned}\left\langle \Delta\omega(u)\Delta\omega(\tau)\right\rangle &= \tau_{BC}^{-1}\int_{-\infty}^{\infty} dt\Delta\omega(t)\Delta\omega(t+\tau) \\ &= \frac{1}{2}\sum_{i=A,B}\frac{\gamma_i^4}{4\tau_{BC}\mu^2\omega_0^2L^4}\left[1-\frac{fL}{\mu\omega_0^2\gamma_i}\right]^2 \\ &\quad\times\int_{-\infty}^{\infty} dt\langle E^2\operatorname{sech}^2\left[\left(\frac{E}{2\mu' L^2}\right)^{1/2} t\right] \\ &\quad\otimes\operatorname{sech}^2\left[\left(\frac{E}{2\mu' L^2}\right)^{1/2}(t+\tau)\right]\rangle \end{aligned}\tag{4.23}$$

The average on the right side is over a Boltzmann distribution of energies E. τ_V is then obtained from (4.5):

$$\begin{aligned}\tau_v^{-1} &= \frac{1}{2}\int_{-\infty}^{\infty} d\tau\,\langle\Delta\omega(0)\Delta\omega(\tau)\rangle \\ &= \sum_{i=A,B}\frac{\gamma_i^4}{16\tau_{BC}\mu^2\omega_0^2L^4}\left[1-\frac{fL}{\mu\omega_0^2\gamma_i}\right]^2\left\langle\left|\int_{-\infty}^{\infty} dt\,E\operatorname{sech}^2\left[\left(\frac{E}{2\mu' L^2}\right)^{1/2} t\right]\right|^2\right\rangle \\ &= \frac{kT}{\omega_0^2L^2\tau_{BC}}\cdot\frac{1}{2}\sum_i\gamma_i^4\frac{\mu'}{\mu^2}\left[1-\frac{fL}{\mu\omega_0^2\gamma_i}\right]^2 = \frac{kT}{\omega_0^2L^2\tau_{BC}}\cdot\frac{\mathrm{B}}{\mathrm{A}} \end{aligned}\tag{4.24}$$

where

$$A \equiv \frac{2\mu^2}{\mu'(\gamma_A^4+\gamma_B^4)}\tag{4.25}$$

is a mass factor and

$$B = \frac{\gamma_A^4\left[1-\frac{fL}{\mu\omega_0^2\gamma_A}\right]^2 + \gamma_B\left[1-\frac{fL}{\mu\omega_0^2\gamma_B}\right]^2}{\gamma_A^4+\gamma_B^4} \tag{4.26}$$

is a measure of the effect of anharmonicity on τ_v, since $B \rightarrow 1$ as $f \rightarrow 0$. The dephasing time constant in a picosecond pulse experiment, τ_{ph}, is $\tau_v/2$ (see Section II.B). A comparison with (27) of Ref. 64 shows two corrections: an overall factor of $\frac{9}{4}$[65] and an anharmonicity correction given by the factor B in (4.24) (in the original work[64] the oscillator was assumed to be harmonic). τ_{BC}, the mean time between binary collisions, is estimated in Ref. 64 as $\rho d^2/6\eta$ (where ρ is the density, d the molecular diameter, and η the shear viscosity). Although the model was designed for the interaction of a diatomic with an atom, it can be approximately extended to other cases by making appropriate choices for the masses involved. A more extended discussion is given in Ref. 66.

A comparison with experiment requires knowledge of the anharmonic force constant f. For diatomics this is easily obtained from the vibration-rotation interaction [see (3.31)], but for polyatomic molecules it can only be found from multiparameter fits of high resolution spectra with isotopic substitution. Some of the few values of f which are thus available[32,66] are listed in Table 1. An examination of the last column of the table shows that anharmonicity is quite important in all the cases studied, giving a decrease in the dephasing time by as much as a factor of 20. The other parameters in the calculation are taken from Ref. 64 and from viscosity tables. Some of the mass parameters A differ from those used in Ref. 64; the reasons for the present choice are given in Ref. 66.

The results of the binary collision model are listed in Table II as τ_v(IBC) and may be compared with the experimental results given in the

TABLE I

	ω_0(cm^{-1})	L(10^{-9} cm)	A(amu)	η(cp)	ρ(g/cm^3)	τ_c(10^{-13} s)	$-f$ (10^{14} g/cm sec^2)	B
N_2	2326	2.20	56	0.158	0.808	1.26	17.8	20.3
O_2	1552	2.06	64	0.190	1.14	1.30	8.74	17.4
CS_2	656	2.65	107.8	0.363	1.263	1.25	1.21	6.65
CH_3I	525	2.69	7.97	0.50	2.279	1.68	1.08	6.22
$CHCl_3$	3020	2.92	0.0335	0.51	1.49	1.27	1.75	6.99
CH_3CH_2OH	2928	2.62	2.4	1.20	0.7893	0.23	1.67	6.24
CH_3CCl_3	2940	3.16	2.4	1.20	1.339	0.57	1.68	7.86

TABLE II

	$\omega_0(\text{cm}^{-1})$	τ_v(IBC) (psec)	τ_v(hyd) (psec)	τ_v(exp) (psec)
N_2	2326	50	121	150[a]
O_2	1552	23	57	85[b]
CS_2	656	8.9	21	21[c]
CH_3I	525	0.62	1.2	2.3[d]
$CHCl_3$	3020	0.12	0.50	1.35[e]
CH_3CH_2OH	2928	(0.25)	0.30	0.52[f]
CH_3CCl_3	2940	(0.25)	0.35	2.6[f]

[a] Ref. 48
[b] Refs. 46 and 47.
[c] Ref. 47.
[d] Ref. 67.
[e] Ref. 68.
[f] Ref. 69.

last column. The results for CH_3CH_2OH and CH_3CCl_3 are in parentheses, since these vibrations are not quasidiatomic; thus the choice of parameters is somewhat arbitrary.[66] The results of the binary collision model seem qualitatively reasonable for most of the molecules to which it has been applied, but they appear to underestimate the dephasing time. Several possible reasons for this include (1) the fact that only collinear collisions are treated, which will give more rapid dephasing than the frequent noncollinear ones; (2) the assumption of independent binary encounters (interference between separate collisions will result in a more isotropic environment and smaller frequency fluctuations, thus leading to longer dephasing times); (3) in the higher viscosity liquids the time between collisions is probably underestimated in this viscosity-based approximation. In addition, the attractive forces are neglected in the exponential repulsive potential approximation.

G. Hydrodynamic Models

Hydrodynamics has been successfully applied to a number of processes that take place on an atomic scale. Translational[70] and rotational[71] diffusion constants can be estimated to an accuracy of 10 to 20% by treating the diffusing atom or molecule as a rigid macroscopic body immersed in a viscous continuum. The velocity autocorrelation function of liquid argon is modeled quite well as that of a Brownian particle in a viscoelastic medium.[70] In these molecular scale calculations the use of slip rather than stick boundary conditions[71] appears to give the best results. The first application of hydrodynamics to a vibrational problem was the work of Keizer[72]; he studied the coupling between vibration and translation in an asymmetric oscillator and demonstrated that the relaxation through viscous drag of the momentum of the center of mass was coupled to the energy of the vibration. He was, however, concerned only with population relaxation and not with dephasing.

Metiu, Oxtoby, and Freed[73] studied dephasing for an oscillator in a viscous medium; the Langevin equation satisfied by the internal coordinate Q of the oscillator was taken to be

$$m\ddot{Q}(t)+m\omega_0^2 Q(t)+\int_{-\infty}^{t} ds\tilde{\zeta}(t-s)\dot{Q}(s)=f(t) \tag{4.27}$$

where m is the mass of each atom in the homonuclear diatomic; $\tilde{\zeta}(t)$ is the memory function that was approximated, using hydrodynamics, as the Fourier transform of the frequency-dependent friction coefficient on a Brownian sphere (at low frequencies, for slip boundary conditions, $\tilde{\zeta}(\omega)$ approaches $4\pi\eta R$ where η is the shear viscosity and R the Brownian particle radius, whereas at higher frequencies $\tilde{\zeta}(\omega)$ becomes complex). The correlation function of the random force $f(t)$ is related to $\tilde{\zeta}(\omega)$ through the fluctuation-dissipation theorem.[74] Equation 4.27 was solved for $\langle Q(t)Q(0)\rangle$, and the resulting linewidth was calculated to be

$$W=\text{Re }\tilde{\zeta}(\omega_0)/2m \tag{4.28}$$

The width therefore depended on a very high frequency Fourier component of the friction coefficient. In Ref. 73 a Maxwell relaxation form was taken for the frequency-dependent shear and bulk viscosities, and W was found to be larger than experiment; however, it is clear that the true $\tilde{\zeta}(\omega)$ must fall off much more rapidly than a simple power law at high frequency and will give a width orders of magnitude too *small*. The discrepancy with experiment found in Ref. 73 was attributed to a breakdown in the hydrodynamic description at high frequencies.

A closer examination of the problem[66] shows, however, that the difficulty lies not in the use of hydrodynamics but in the use of the linear Langevin equation, (4.27). As was demonstrated in Section III.A, pure dephasing involves only zero-frequency Fourier components of generalized forces, whereas population relaxation involves high-frequency components [see (3.22)]. Why then does the line shape involve the *high*-frequency friction coefficient? The answer is that (4.27) is the Langevin equation for a harmonic oscillator linearly coupled to a heat bath, and, as was shown in Section IV.B, this case is anomalous in that there is no "pure" dephasing—dephasing arises only through population relaxation. To describe true molecular dephasing, (4.27) must be generalized to include both nonlinear coupling to the bath and anharmonicity in the oscillator. Since either change would give rise to a nonlinear stochastic integrodifferential equation (and therefore would be virtually insoluble), a suitable approximate treatment was needed.

The approach suggested[66] is based on the observation in molecular dynamics simulations of liquid nitrogen that by far the most important contribution comes from *linear* coupling of an *anharmonic* oscillator to a bath. This term depends entirely [see (3.35)] on the time integral of the correlation function of the force on the bond of the oscillator:

$$\begin{aligned}\tau_v^{-1} &= \frac{1}{2}\int_{-\infty}^{\infty} dt \langle \Delta\omega(t)\Delta\omega(0)\rangle \\ &\approx \frac{1}{2}\left(\frac{f}{2\mu^2\omega_0^3}\right)^2 \int_{-\infty}^{\infty} dt \left\langle \frac{\partial V_S}{\partial Q_\alpha}(t)\frac{\partial V_S}{\partial Q_\alpha}(0)\right\rangle \\ &= \frac{1}{2}\left(\frac{f}{2\mu^2\omega_0^3}\right)^2 \int_{-\infty}^{\infty} dt \sum_{i,j=A,B} \left\langle \frac{\partial \underline{\xi}_i}{\partial Q_\alpha}\cdot \mathbf{F}^i(t)\mathbf{F}^j(0)\cdot\frac{\partial \underline{\xi}_j}{\partial Q_\alpha}\right\rangle \end{aligned} \tag{4.29}$$

Here i and j refer to the two atoms of the molecule; $\underline{\xi}_i$ is the displacement of each under the vibration Q_α. Assume now that the forces on the two atoms of the diatomic are uncorrelated, giving (after evaluating $\partial\underline{\xi}_i/\partial Q_\alpha$)

$$\tau_v^{-1} = \frac{1}{2}\left(\frac{f}{2\mu^2\omega_0^3}\right)^2 \sum_{i=A,B} (\gamma_i)^2 \int_{-\infty}^{\infty} dt \langle \mathbf{F}_n^i(t)\mathbf{F}_n^i(0)\rangle \tag{4.30}$$

where $\gamma_i \equiv m_j/(m_i+m_j)$ and $\mathbf{F}_n^i(t)$ is the component of the force on atom i pointing along the bond (rotation of the diatomic can be neglected for the time scales of interest here). For slip boundary conditions, the time integral of the correlation function of the force on a Brownian sphere is[74]

$$\int_{-\infty}^{\infty} dt \langle \mathbf{F}_x(t)\mathbf{F}_x(0)\rangle = 8\pi kT\eta R \tag{4.31}$$

For the diatomic we take the model (as in Ref. 73) of a cylinder with two half spheres at the end; the presence of *half* spheres reduces the time integral by a factor of 2 (as discussed in Ref. 73) giving

$$\tau_v^{-1} = \frac{\pi f^2 kT\eta}{2\mu^4\omega_0^6} \sum_{i=A,B} (\gamma_i)^2 R_i \tag{4.32}$$

Suppose we assume that

$$\frac{\partial^2 V_S}{\partial Q_\alpha^2} \approx \frac{1}{L}\frac{\partial V_S}{\partial Q_\alpha} \tag{4.33}$$

This is exact for an exponential potential and appears as well to be a good approximation for the Lennard-Jones potential simulation of liquid nitrogen.[43] Then the quadratic coupling to the heat bath can be treated as well, and the hydrodynamic theory gives the result

$$\tau_v^{-1} = \frac{\pi f^2 kT\eta}{2\mu^4\omega_0^6} \sum_{i=A,B} \gamma_i^2 R_i \left[1 - \frac{\mu\omega_0^2\gamma_i}{fL}\right]^2 \tag{4.34}$$

This can be compared with the result of the binary collision model of Section IV.F[64]:

$$\begin{aligned}\tau_v^{-1} &= \frac{\mu' f^2 kT}{2\mu^4\omega_0^6\tau_c} \sum_{i=A,B} \gamma_i^2 \left[1 - \frac{\mu\omega_0^2\gamma_i}{fL}\right]^2 \\ &= \frac{\pi f^2 kT\eta}{2\mu^4\omega_0^6} \left(\frac{2\mu'}{\frac{4\pi}{3}\rho(d/2)^3}\right) \sum_{i=A,B} \gamma_i^2 \frac{d}{2}\left[1 - \frac{\mu\omega_0^2\gamma_i}{fL}\right]^2 \end{aligned} \tag{4.35}$$

A comparison of (4.34) and (4.35) shows a close similarity between the two results; they differ in the presence of the factor $(2\mu'/(4\pi/3)\rho(d/2)^3)$, which is approximately the inverse packing fraction in the liquid, and by the replacement of R_i by $d/2$, where d is the molecular diameter. It is somewhat surprising that theories from the two extremes of hydrodynamics and isolated binary collisions should lead to such similar results; it is interesting to note the introduction of collective modes in the binary collision description through the use of a viscosity-dependent collision frequency.

The results of the hydrodynamic theory are listed as τ_v (hyd) in Table II; on the whole they are somewhat better than the binary collision results, although the difference is not great, since the final forms of the two theories are so similar. The only major discrepancy is for CH_3CCl_3; since its viscosity is the same as that of CH_3CH_2OH, the hydrodynamic theory gives a similar predicted dephasing time; in fact the dephasing is five times slower. The reasons for this particular discrepancy are not immediately obvious.

H. Resonant Transfer Effects

The theories outlined in Sections. B–G consider only the frequency fluctuations induced by the medium and thus apply only to $\Phi_1(t)$, the part of the dephasing arising from the operator D (see Section III.A.). In the pure fluid there is another potentially important contribution from resonant vibrational energy transfer between neighboring molecules. This contribution to $\Phi_2(t)$ (see Section III.A) arises from the operator R in the total Hamiltonian and can be described as line broadening caused by the presence of vibrational excitons. In the absence of resonant transfer, the vibrational coordinates Q_i and Q_j of two different molecules are uncorrelated, and the Raman spectrum depends only on the single molecule correlation function $\langle Q_i(t)Q_i(0)\rangle$. Resonant transfer introduces a dynamical coupling between the coordinates of different molecules; thus the spectrum depends on the two molecule correlation function $\langle Q_i(t) \sum_j Q_j(0)\rangle$, whose description is inevitably more complicated. Nevertheless, the theory outlined in Section III.A (based on Refs. 21 and 26) shows that within the weak coupling (or truncated cumulant expansion) approximation the calculation is straightforward and the result rather simple. In contrast, approaches[57] that attempt to calculate $\langle Q_i(t)Q_i(0)\rangle$ and $\langle Q_i(t)\sum_{j\pm i}Q_j(0)\rangle$ separately (when R is nonzero) become quite involved; as Wang[57] has pointed out, the two separate correlation functions have a much more complicated form than their sum.

A number of early theories[75,76] of resonant energy transfer were based on the assumption that the vibrational coupling could be treated as static (that is, neglecting motional narrowing). Kakimoto and Fujiyama,[75] for example, used an atom-atom Lennard-Jones potential in the static approximation,[77] whereas Tokuhiro and Rothschild[76] used an exponential repulsive potential to study the effect of isotopic dilution on several vibrational transitions of chloroform. Most simple liquids, however, fall within the motional narrowing regime; thus a theory for the dynamics of the resonant energy transfer must be developed.

Döge[78] treated the dynamics by using the correlation function modeling method discussed in Section IV.D; by assuming an exponential correlation function for the matrix elements of the resonant transfer Hamiltonian R, he found a vibrational phase correlation function which was Gaussian at short times and exponential at long times [see (4.14)]. van Woerkom et al.[21] carried out an extensive calculation of the effect of resonant transfer on the *infrared* lineshape, assuming dipole-dipole coupling and a rotational diffusion model. Wang[57] has applied a similar model to the description of the isotropic Raman lineshape. There appear to be some discrepancies between these different calculations. Three molecule contributions appear in the cumulant expansion of R [see (3.18)]; these are present in Refs. 26 and 57 (although only two-molecule terms are explicitly calculated), whereas they are not present in the calculation of Ref. 21 [see (32)]. The effect of translational diffusion also appears to be different in Refs. 21 and 57. The final result found by Wang (after a number of approximations) is that the resonant transfer contribution should be proportional to $\rho\eta/T$, where ρ, η, and T are the density, viscosity, and temperature of the liquid.

When both resonant transfer and environmentally induced frequency fluctuations are present (i.e., both D and R appear in the perturbation Hamiltonian), the description becomes more complicated because their effects cannot be separated. The separability assumption has been made in a number of papers[31,78]; for example, Döge[78] took $\Phi(t) = \Phi_1(t)\Phi_2(t)\Phi_3(t)$ [just as in (3.8)], but assumed that $\Phi_2(t)$ depended only on resonant transfer and not on environmental fluctuations. As stressed in Section III.A, although $\Phi_1(t)$ depends only on D, $\Phi_2(t)$ depends on both D and R. The cross correlations between D and R which arise in a cumulant expansion were present in the calculation of van Woerkom et al.[21] but were assumed to have a negligible effect on the lineshape. Lynden-Bell[26] was probably the first to exhibit them explicitly; she discussed general symmetry requirements for the effect of D and R on infrared and polarized or depolarized spectra for various types of interaction (dispersion, dipolar, and quadrupolar). She evaluated the pair contributions in (3.19) for a model in which translational diffusion occurs while rotations are slow. The molecular dynamics simulation of liquid nitrogen[27] discussed in Section IV.A showed the importance of cross correlations between D and R; they led to a narrowing of the predicted Raman line in contrast to the intuitive expectation of broadening when excitonic states arise. Knauss[28] studied both resonant transfer and environmental fluctuation effects using a time-dependent superposition approximation for four-point correlation functions, expressing the linewidth in terms of the van Hove time-dependent correlation function and deriving an approximate description for the concentration dependence of dephasing in binary mixtures. Wertheimer[29] generalized the binary collision model of Fischer and Laubereau,[64] discussed in Section IV.F, to include the effects of both anharmonicity and resonant energy transfer.

V. EXPERIMENTAL RESULTS

Isotropic Raman linewidths have been measured for a large number of molecular liquids, often simply for the purpose of separating vibrational from rotational relaxation in the depolarized lineshape. This section surveys some of the molecules in which dephasing has been investigated most systematically through temperature- or pressure-dependent studies, isotopic dilution, or comparative linewidth measurements for several different Raman-active transitions in a given molecule.

A. Nitrogen and Oxygen

The isotropic Raman linewidths of liquid nitrogen and oxygen were measured by Clements and Stoicheff[47] and by Scotto[46] and were found to be significantly smaller than the corresponding gas phase linewidths. The full widths at half height of the (Lorentzian) lines were found to be $0.067 \pm 0.006\ \text{cm}^{-1}$ for N_2 and $0.125 \pm 0.01\ \text{cm}^{-1}$ for O_2; for comparison, the envelope of the gas phase Q-branch of N_2 has a width of $2.6\ \text{cm}^{-1}$.[79] The broadening in the gas phase is due to vibration-rotation coupling, leading to a splitting of the Q-branch into a series of lines corresponding to the rotational states of the molecule. The reduction in linewidth by a factor or 40 in the liquid is due to motional narrowing, which sets in when the rotational lines begin to overlap in the dense gas. The liquid-state linewidth of N_2 corresponds to a τ_v of 158 ± 14 psec. Laubereau,[48] using the coherent picosecond pulse technique discussed in Section II.B, found a signal decay time of 75 ± 8 psec, which corresponds to a τ_v of 150 ± 16 psec. The agreement between the two techniques, which is expected for a motionally narrowed line, serves to validate the experimentally difficult picosecond dephasing measurements.

Clouter and Kiefte[80] investigated the Raman linewidth in nitrogen and oxygen along the coexistence curve from the triple point, through the normal boiling point, to the critical point. The linewidths increased rapidly as the critical point was approached, probably due to the rapidly decreasing density in that temperature range. As the temperature was lowered from the normal boiling point to the triple point, the linewidth of O_2 increased by a factor of 2, whereas that for N_2 remained approximately constant. Brueck[81] has studied the effect of vibration-rotation interaction in N_2, including only the squared angular momentum term in (3.34); he evaluated the angular momentum relaxation by using the J-diffusion model of Gordon[82] (according to which free rotation takes place between collisions that completely randomize the angular momentum). The mean time between collisions τ_j was estimated from a cell model and from a rough hard sphere model. He found good agreement with the experimental data,[80] especially at lower densities; in the normal liquid range the predicted line widths were too small by as much as a factor of 2. Brueck's work demonstrates the importance of vibration-rotation coupling for nitrogen even at liquid-state densities.

LeDuff[79] studied the Raman band shape of N_2 dissolved in inert solvents (SF_6, CCl_4, $CHCl_3$, and SO_2). The observed full widths at half height ranged from 1 to 1.3 cm^{-1}, indicating motional narrowing by a factor of more than 2 compared to the gas, but much less than in liquid nitrogen. This difference seems to indicate that N_2 (liquid) has a less open structure than the other solvents, so that rotational motion is more strongly perturbed. A subsequent study[83] investigated the solvent effect on the linewidths of H_2, D_2, and HF, as well as N_2.

Hesp et al.[83a] measured the dephasing rate of N_2 as a function of concentration in liquid mixtures of N_2 and Ar, using both picosecond and Raman lineshape techniques. The two methods were in agreement and showed a reduction in τ_v from 174 ps for pure N_2 to 70 ps in dilute Ar solutions. This reduction can be qualitatively understood in a hydrodynamic model since the viscosity of Ar is twice that of N_2. Hesp et al.[83a] attempted to calculate the mean square frequency fluctuations

$\langle \Delta\omega(0)^2 \rangle_c$ using the approach of Ref. 37, but their results are suspect since for pure N_2 they differ by a factor of 20 from the molecular dynamics simulation results of Ref. 27.

B. Methyl Iodide

Phase relaxation has been studied more extensively in methyl iodide than in any other molecule. Studies have probed temperature[67,84–88] and pressure[84] dependence, isotopic dilution effects,[85] and dilution in nonpolar solvents.[85,89] Comparative studies of the three polarized Raman lines[85,89] have been made.

Goldberg and Pershan[89] measured isotropic Raman half widths for all three polarized lines of both CH_3I and CD_3I at a single temperature. They commented on the fact that the ν_1 line (C-H or C-D stretch) is narrower by a factor of 2 in the deuterated compound and said that such a sizeable difference is difficult to account for. Such isotope effects can be explained on the basis of (3.35). Matrix elements of d and R are proportional to $\mu^{-2}\omega^{-3}$ and $\mu^{-1}\omega^{-1}$ where μ and ω are the reduced mass and frequency. Since $\omega \sim \mu^{-1/2}$, these matrix elements vary as $\mu^{-1/2}$. Since their square determines the linewidth (at least if cumulant truncation is valid) the width should vary as μ^{-1} and therefore be a factor of 2 smaller for a C-D stretch than a C-H stretch, as was indeed observed.

Wright et al[67] and others[85–88] carried out temperature-dependent studies at constant pressure of the line width of ν_3 (C-I stretch; they observed a narrowing of the linewidth (lengthening of dephasing times) as the temperature was raised in the liquid phase. Since a constant pressure experiment involves simultaneous changes in temperature and density, it was unclear which was the primary cause of the narrowing. Pressure-dependent studies by Jonas and co-workers[84] provided the answer: they found that for temperature changes at constant density the linewidth remained approximately constant, and it increased with increasing density at constant temperature. The constant pressure results are thus due entirely to the expansion of the liquid on heating. These qualitative trends can be understood with a number of different viscosity-dependent theories. Note that the binary collision model predicts a linewidth proportional to $\eta\rho^{-1}T$, the hydrodynamic theory ηT, and Lynden-Bell's theory[26] $\eta\rho T^{-1}$. Lynden-Bell and Tabisz[90] have fit the data of Ref. 84 to the form

$$\tau_v^{-1} = A'\eta\rho T^{-1} + C \tag{5.1}$$

The density dependence was obtained correctly; A' was found to increase with temperature, which the authors suggest may be due to a decrease in the distance of closest approach with increasing temperature.

Döge et al.[85] have recently carried out a comprehensive study of vibrational phase relaxation in methyl iodide. All three polarized bands were studied as a function of temperature at constant pressure, and dilution studies in CD_3I and the nonpolar solvent CS_2 were carried out. The results obtained were that ν_1 (the C-H stretch) phase relaxation was only weakly affected by temperature, isotopic dilution, and dilution in CS_2; ν_2 (symmetrical C-H bend) and ν_3 both showed narrowing with increased temperature and with dilution in CS_2, but only ν_2 showed significant narrowing on isotopic dilution. The authors interpreted the results in terms of three different primary broadening mechanisms: for ν_1 they suggested that the broadening is due to intramolecular vibrational energy exchange with nearby states, an effect that should be only weakly dependent on solvent and temperature. For ν_2, they proposed that resonant energy transfer due to dipole-dipole coupling dominates, thus giving a large isotopic dilution effect; additional evidence for this is the fact that the ν_2 mode has the highest intensity in the infrared spectrum, which is important because dipolar resonant transfer depends on the fourth power of the dipole matrix element μ_{10}. For ν_3, they suggested that environmentally induced frequency fluctuations arising from the permanent dipoles of the neighboring molecules is dominant, thus accounting for the large effect of dilution with CS_2.

C. Acetonitrile

Griffiths[91] measured the isotropic Raman linewidths for three bands (ν_1, ν_2, and ν_4) of acctonitrilc (CH_3CN) and its deuterated analog, CD_3CN. They ranged from 1.2–1.9 cm^{-1}, and the deuterated species showed slightly broader lines. This result for ν_1 (the C-H or C-D stretch) is surprising in light of the discussion of isotope effects on ν_1 for CH_3I in Section V.B. Jones et al.[87] also studied the Raman spectrum of CH_3CN, but concentrated primarily on rotational relaxation.

A study of acetonitrile by Breuillard-Alliot and Soussen-Jacob[92] shows that in certain favorable cases information about dephasing can be obtained from *infrared* spectra, although the interpretation is less direct than it is for the isotropic Raman line. These authors applied first the Rakov approach[35] discussed in Section III.C, in which the infrared linewidth is taken as the sum of a temperature-independent vibrational part and a rotational part characterized by an exponential temperature dependence; the vibrational effect was so large that this technique was not successful. They also attempted a reverse decomposition in which the rotational line width was estimated from the dielectric relaxation time and subtracted from the total line width; since for a pure fluid the dielectric relaxation depends on collective as well as single-particle reorientations, this approach may not be accurate. For dilute solutions of CH_3CN in nonpolar CCl_4, the far-infrared spectrum reflects only single-particle motions; under these experimental conditions a vibrational correlation function may be estimated (under the assumption that rotations and vibrations are uncoupled). The authors were able to do this for three of the four parallel bands of CH_3CN.

Whittenburg and Wang[93] studied the Rayleigh and Raman spectra of acetonitrile as a function of concentration in CCl_4. Their Raman studies were of the ν_2 line (C-N stretch) and gave a longer relaxation time for rotation than that of Griffiths,[91] based on the ν_1 line (C-H stretch). This discrepancy may be due to correlation between vibrations and rotations which are neglected when their effects on the linewidth are assumed to be additive. They observed a broadening of the line upon dilution with CCl_4; this is the same trend as that observed for CH_3I and can be explained with a binary collision or a hydrodynamic model. It would imply that dipolar interactions or resonant energy transfer are not important in liquid CH_3CN, since they would lead to a narrowing of the line on dilution[94]

Jonas and co-workers[95] studied the isotropic Raman lineshape for the ν_1 line (C-H stretch) as a function of both pressure and temperature. They observed that (*1*) at constant density a temperature increase led to line broadening; (*2*) at constant temperature a density increase also led to broadening; (*3*) at constant pressure, a temperature increase leads to broadening. Result (*3*) shows that the expansion upon heating in this case is not sufficient to overcome the pure temperature effect, in contrast to the result for CH_3I. A faster phase correlation decay rate was observed for CH_3CN than for CD_3CN, in agreement with the isotope effect predicted in Section V.B and in disagreement with the result of Griffiths[91] The line shape was fitted to the Kubo form discussed in Section IV.D; however, the second moment of the band was difficult to obtain. $\omega^2 I(\omega)$ showed a large peak 70 cm^{-1} from the band center which may have given a spuriously large contribution to the second moment. The correlation time for the frequency fluctuation, τ_c, obtained using this large second moment, is 0.04 psec, which seems too small. The linewidths were fitted very accurately using the original form[64] of the Fischer-Laubereau binary collision model, with an Enskog collision frequency and a temperature-dependent hard sphere diameter obtained from viscosity data. Since the anharmonicity of the C-H vibration is significant (see Table I) and was not included in the original binary collision model, the quantitative agreement is probably fortuitous, although the qualitative trends are correct. An isotopic dilution study showed no measurable effect on the linewidth, indicating that resonant transfer is unimportant. This is in agreement with the results of Döge et al.[85] for the C-H stretch of CH_3I. However, the conclusion drawn by the two authors was different. Döge concluded that intramolecular energy relaxation was the cause of phase relaxation in ν_1, whereas Jonas et al. concluded that sharply repulsive collisions were the cause (which would be independent of isotopic substitution and unrelated to the dipole moments of the solvent). At present, the experimental evidence does not appear sufficient to decide between the two alternatives.

Yarwood et al.[34a] recently studied the ν_1 (C-H stretch) and ν_3 (symmetric CH_3 deformation) Raman bands of acetonitrile as a function of temperature and in solution. They analyzed their spectra using the two-parameter Kubo form of Section IV.D. The linewidth of ν_1 showed only a very weak temperature dependence at constant pressure, but narrowed upon dilution in CCl_4; dilution in polar solvents had a smaller effect. The second moments calculated (from the low-frequency part of the band) were a factor of 20 smaller than those of Ref. 95, probably because in the latter work, as discussed above, there appears to be additional absorption on the high-frequency side of the band which is not directly related to the dephasing of ν. Yarwood et al. suggest that the dominant mechanism for phase relaxation in this mode comes from short-range, repulsive collisions, in addition to which there is a small effect arising from dipolar interaction. The ν_3 mode showed a much larger narrowing on dilution in CCl_4, indicating a stronger dipolar interaction, as expected on the basis of the larger infrared intensity of ν_3. Its overtone, $2\nu_3$, narrowed even more strongly upon dilution in CCl_4, as would be predicted from the dependence on overtone number given in (3.35). A comparison was made[34a] with infrared bandwidths of ν_1 and ν_3, and it was shown that the assumption of separability of vibration and rotation led to rotational relaxation times τ_{1R} from the two modes which were in disagreement by a factor of 2 to 3. This indicates once again that care must be taken in obtaining rotational correlation functions from either infrared or depolarized Raman spectra by simply dividing by the isotropic Raman correlation function.

D. Haloforms

Laulicht and Meirman[96] and Wright and Rogers[97] studied the Raman spectra of the ν_1 mode (C-D stretch) of deuterated chloroform ($CDCl_3$). They found that the isotropic linewidth was a sizeable fraction of the anisotropic and estimated the rotational correlation function on the assumption that vibrations and rotations are uncorrelated (see Section III.C). This is a somewhat risky assumption in cases in which the vibrational contribution is the major part of the total depolarized linewidth.[38] Brodbeck et al.[98] also studied the isotropic lineshape of ν_1 in $CHCl_3$ and $CHBr_3$, as well as their deuterated analogs. They found that the dephasing of the two different C-H oscillators was virtually identical (as was the dephasing of the C-D oscillators). Isotopic dilution studies indicated no significant resonant transfer effect (just as discussed in the last two sections for the C-H stretches of CH_3I and CH_3CN). An infrared study[99] showed the importance of vibrational relaxation for several modes of $CHCl_3$, especially ν_1. The ν_1 absorption line was found to exhibit an anomalous absorption on the high-frequency side, which is not observed in the Raman spectra.

Rothschild and co-workers[100] studied the Raman spectra of all three parallel modes of $CHCl_3$ and $CDCl_3$. Vibrational phase relaxation was fastest for the ν_1 mode; isotopic dilution studies showed no significant narrowing, in agreement with the results of Ref. 85. The ratio of the linewidths of the ν_1 modes in $CHCl_3$ and $CDCl_3$ was found to be $\sqrt{2}$, rather than the factor of 2 one would expect for a motionally narrowed line (see Section V.B). This may be due to a vibration-rotation interaction effect on the isotropic lineshape. Dipole-dipole resonant transfer was estimated to give only a small contribution to the other two parallel bands, ν_2 and ν_3. In a subsequent paper, Tokuhiro and Rothschild[76] used a static environment model to estimate the effect of the repulsive potential on resonant transfer. They predicted that the depolarized ν_4 band would show the largest effect; infrared linewdith experiments showed that the full width at half height decreased by $2\ cm^{-1}$ as $CHCl_3$ was diluted with $CDCl_3$.

Jonas and co-workers[68] carried out a pressure- and temperature-dependent study of the isotropic Raman lineshape of ν_1 in $CHCl_3$ and $CDCl_3$. They found that in both cases an increase in density at constant temperature led to a shorter phase relaxation time; an increase in temperature at constant density gave a shorter relaxation time for $CDCl_3$ but had little effect for $CHCl_3$. They evaluated second moments and fit their data to the Kubo relaxation form (Section IV.D). They found that the product $\langle(\Delta\omega)^2\rangle_c^{1/2}\,\tau_c$ ranged from 0.25 to 0.45 and that the motional narrowing limit gave a good description of the experimental linewidths. $\langle(\Delta\omega)^2\rangle_c$ was shown to increase with pressure at constant temperature and to decrease with temperature at constant pressure, as expected using intuitive arguments. τ_c decreased with a pressure increase at constant T, as expected, and increased with T at constant pressure due to a decrease in density. One anomalous result found is that τ_c differs more between $CHCl_3$ and $CDCl_3$ than does the second moment: since the effect of the isotopic substitution on the mass is so small, one would expect the dynamics (which govern τ_c) to change very little. The results were fit to the original form of the isolated binary collision model,[64] as were those for CH_3CN discussed in Section V.C. The molecules, however, were treated as colliding with structureless particles of the mass of H or D, which seems somewhat unrealistic.

A Raman study of liquid CHF_3[101] showed that the ν_3 (C-F stretch) mode was quite narrow, whereas that of the ν_1 (C-H stretch) was broad and changed only slightly with temperature over a range of 170 degrees in the liquid phase (at constant pressure). This may of course be due to a compensation of temperature effects by density changes. No isotopic dilution effect was observed.

E. Benzene

Several investigators have studied the isotropic Raman linewidth of the ν_2 (symmetric C-C stretch) mode in C_6H_6 and C_6D_6. Griffiths et al.[18] observed an unusual effect upon isotopic dilution. The linewidth of the ν_2 mode of C_6D_6 did not decrease with increasing dilution, whereas that of C_6H_6 did and approached that of C_6D_6 in the infinite dilution limit. If resonant transfer were the only process affected by isotopic dilution, the linewidths of benzene and its deuterated analog would have been affected in the same way. In the language of Section III.A, $\Phi_2(t)$ should be essentially the same for the two compounds. The experiment suggests that $\Phi_3(t)$ is changed as well, so that it is the *population* relaxation rate that is perturbed most by dilution. The authors[18] point out that benzene has several energy levels near ν_2, whereas benzene-d_6 does not, and suggest that intermolecular population relaxation is the cause of the increased linewidth in C_6H_6 compared to C_6D_6. It is not clear why intramolecular relaxation due to collisions is not observed (if it were present, the dilute C_6H_6 width would differ from the C_6D_6 width). These experiments show that caution is necessary in attributing isotopic dilution effects entirely to resonant transfer, since population relaxation is also affected. Wherever possible, studies of a Raman line under isotopic dilution should include both the mode on the original compound and that on the deuterated one. Neuman and Tabisz[102] studied the same line in benzene diluted in CCl_4; they found a linear decrease in linewidth as a function of concentration, which is in agreement with the hypothesis of Griffiths et al.[18] on relaxation in benzene. They also found that the linewidth decreased with increasing temperature (at constant pressure).

LeSar and Kopelman[103] carried out isotopic dilution studies of the linewidths of 10 different fundamentals of benzene and benzene-d_6. Some of them showed very large narrowing effects upon dilution, including one band (called ν_{11} by them) which narrowed from a full width at half height of 24 cm^{-1} to 7 cm^{-1}. The authors interpreted their results in terms of a solid-like exciton model for liquid benzene. Tanabe and Jonas[104] carried out pressure- and temperature-dependent studies of seven Raman lines in C_6D_6, including two totally symmetric bands and five degenerate bands. In the latter case the isotropic spectrum is absent; vibrational dephasing could only be estimated. The procedure used was as follows: the reorientation was modeled using Kubo[22] line shape theory, with the rotational diffusion constants obtained from NMR data and the second moments from the known moments of inertia about the principal axes. Assuming the total anisotropic lineshape to be a product of vibrational rotational parts, the vibrational relaxation time τ_v was obtained. τ_v decreased with increasing pressure at constant T for all the modes observed, whereas temperature dependence at constant pressure varied from one to another (probably because of competing density and temperature effect). A comparison was made with the binary collision model,[64] although further approximations had to be made to describe the complex molecular deformations in a quasidiatomic picture.

F. Other Molecules

Laubereau and co-workers[17,69,105] have applied coherent picosecond pulse techniques to the determination of phase relaxation times for a number of molecules besides nitrogen,[48] which was discussed in Section V.A. In each case their measurements were restricted to atmospheric pressure and to a single temperature. In this way dephasing times were obtained for CCl_4,[17,105] CH_3CH_2OH,[69,105] CH_3CCl_3,[69] and SnBr.[17]

Several other small molecules have been studied, Barral et al.[106] studied the Raman linewidth of liquid fluorine over a temperature range 75 to 110 K. They found an isotropic linewidth of $0.9\ cm^{-1}$ (much larger than those for nitrogen and oxygen[46,47], but their main emphasis was on rotational relaxation. Wang and Wright[107] studied the Raman line of HCl across the liquid-solid transition and found that the liquid phase linewidth was about 50% broader than the solid, indicating more rapid dephasing in the liquid phase. This may be due to the density change upon melting and the greater importance of repulsive forces in the liquid or to vibration-rotation coupling. Ouillon[108] observed motional narrowing of the isotropic Raman lines in some solvents and broadening in others.

Bartoli and Litovitz[36] measured Raman linewidths for a number of small molecules and also for C-Br stretches of a series of bromine-substituted hydrocarbons of up to eight carbons in length. Constant et al.[109] studied the C-Cl and C-Br stretches of t-butyl chloride and t-butyl bromide. The spectra of the pure liquids were broader than those of a 20% solution in CCl_4 and n-hexane and were insensitive to temperature changes from -40 to 50°C. Second moments of the isotropic spectrum were obtained. The vibrational correlation function of t-butyl chloride, for which $\tau_c\langle(\Delta\omega)^2\rangle^{1/2}\sim 0.5$ showed a nonexponential decay, whereas that for t-butyl bromide ($\tau_c\langle(\Delta\omega)^2\rangle^{1/2}\sim 0.3$) was exponential. Jones et al.[110] observed that the isotropic Raman linewidths for several bands of 1,2,5-thiadiazole decreased with increasing temperature at constant pressure.

The study of lineshapes of overtone bands should provide useful information about vibrational dephasing; however, many factors affect the interpretation of these spectra. The effect of resonant transfer on overtones was discussed at the end of Section III.B; depending on the anharmonicity of the vibration it may or may not give a significant contribution to the overtone linewidths. Environmentally induced frequency fluctuations have a complex effect as well; since the magnitude of the fluctuations is proportional to the overtone number [see (3.35)] the linewdith should increase as the square of the overtone number as long as the motional-narrowing limit holds; for the higher overtones the lines may become Gaussian and increase linearly with overtone number. However, even these conclusions must be modified by two other considerations: first, the cumulant expansion truncation breaks down at some point, and second, higher-order derivatives of the solvent potential with respect to vibrational coordinate become important for higher overtones. The linewidths of the second overtones studied by Arndt and Yarwood[34] showed no simple trends. As Madden and Wennerstrom[111] have pointed out, resonance Raman studies may be useful because they allow observation of long progressions of overtone bands.

References

1. D. J. Diestler, *Top. Appl. Phys.*, **15,** 169 (1976).
2. A. Laubereau and W. Kaiser, *Ann. Rev. Phys. Chem.*, **26,** 83 (1975); *Rev. Mod. Phys.*, **50,** 607 (1978).
3. K. Eisenthal, *Ann. Rev. Phys. Chem.*, **28,** 207 (1977).
4. W. A. Steele, *Adv. Chem. Phys.*, **34,** 1 (1976).
5. F. Legay, in C. B. Moore, ed., *Chemical and Biochemical Applications of Lasers*, Vol. 3.
6. See for example D. Robert and L. Galatry, *J. Chem. Phys.*, **64,** 2721 (1976).
7. J. de Bleijser, P. C. M. van Woerkom, and J. C. Leyte, *Chem. Phys.*, **13,** 387 (1976); **13,** 403 (1976).
8. S. Bratos, *J. Chem. Phys.*, **63,** 3499 (1975).
9. A. Abragam, *The Principles of Nuclear Magnetism*, Oxford University Press, London, 1961.
10. C. B. Harris, *Chem. Phys. Lett.*, **52,** 5 (1977).
11. A. H. Zewail, T. E. Orlowski, R. R. Shah, and K. E. Jones, *Chem. Phys. Lett.*, **49,** 520 (1977); A. H. Zewail and T. E. Orlowski, *Chem. Phys. Lett.*, **45,** 399 (1977).
12. T. J. Aartsma and D. A. Wiersma, *Chem. Phys. Lett.*, **42,** 520 (1976).
13. B. J. Berne and R. Pecora, *Dynamic Light Scattering*, Wiley, New York, 1976.
14. W. M. Gelbart, *Adv. Chem. Phys.*, **26,** 1 (1974).

15. G. Placzek, *Z. Physik*, **70,** 84 (1931).
16. R. G. Gordon, *Adv. Mag. Res.*, **3,** 1 (1968).
17. A. Laubereau, G. Wochner, and W. Kaiser, *Phys. Rev.*, **A13,** 2212 (1976).
18. J. E. Griffiths, M. Clerc, and P. Rentzepis, *J. Chem. Phys.*, **60,** 3824 (1974); **63,** 2262 (1975).
19. A. Laubereau, *J. Chem. Phys.*, **63,** 2260 (1975).
20. C. B. Harris, R. M. Shelby, and P. A. Cornelius, *Phys. Rev. Lett.*, **38,** 1415 (1977).
21. P. C. M. van Woerkom, J. de Bleyser, M. de Zwart, and J. C. Leyte, *Chem. Phys.*, **4,** 236 (1974).
22. R. Kubo, in D. ter Haar, ed., *Fluctuations, Relaxation, and Resonance in Magnetic Systems*, Plenum, New York, 1962.
23. N. H. March, W. H. Young, and S. Sampanthar, *The Many-Body Problem in Quantum Mechanics*, Cambridge University, Cambridge, 1967.
24. W. Feller, *An Introduction to Probability Theory and its Applications*, Wiley, New York, 1950.
25. P. C. M. van Woerkom, J. de Bleyser, and J. C. Leyte, *Chem. Phys. Lett.*, **20,** 592 (1973).
26. R. Lynden-Bell, *Mol. Phys.*, **33,** 907 (1977).
27. D. W. Oxtoby, D. Levesque, and J.-J. Weis, *J. Chem. Phys.*, **68,** 5528 (1978).
28. D. C. Knauss, *Mol. Phys.*, **36,** 413 (1978).
29. R. Wertheimer, *Mol. Phys.*, **35,** 257 (1978).
30. G. Herzberg, *Molecular Spectra and Molecular Structure*, Vol. 1, van Nostrand Reinhold, New York, 1945.
31. W. G. Rothschild, *J. Chem. Phys.*, **65,** 2958 (1976).
32. J. Overend, private communication. See, for example, M. Suzuki, *Bull. Chem. Soc. Jap.*, **48,** 1685 (1975).
33. S. Bratos and J. P. Chestier, *Phys. Rev.*, **A9,** 2136 (1974).
34. R. Arndt and J. Yarwood, *Chem. Phys. Lett.*, **45,** 155 (1977).
34a. J. Yarwood, R. Arndt, and G. Döge, *Chem. Phys.*, **25,** 387 (1977).
35. F. Bartoli and T. A. Litovitz, *J. Chem. Phys.*, **56,** 404 (1972).
36. A. V. Rakov, *Tr. Fiz. Inst. Akad. Nauk USSR*, **27,** 111 (1964); *Opt. Spectrosc.*, **7,** 128 (1959).
36a. A. E. Boldenskul and V. E. Pogorelov, *Opt. Spectrosc.*, **28,** 248 (1970).
37. S. Bratos and E. Marechal, *Phys. Rev.*, **A4,** 1078 (1971).
38. A. M. Amorim da Costa, M. A. Norman, and J. H. R. Clarke, *Mol. Phys.*, **29,** 191 (1975).
39. See, for example, A. Rahman, *Phys. Rev.*, **136,** A405 (1964); L. Verlet, *Phys. Rev.*, **159,** 98 (1967).
40. J. Barojas, D. Levesque, and B. Quentrec, *Phys. Rev.*, **A7,** 1092 (1973); P. S. Y. Cheung and J. G. Powles, *Mol. Phys.*, **30,** 921 (1975).
41. J. P. Riehl and D. J. Diestler, *J. Chem. Phys.*, **64,** 2593 (1976).
42. See, however, the simulations of population relaxation of highly energized bromine molecules in argon: D. L. Jolly, B. C. Freasier, and S. Nordholm, *Chem. Phys.*, **21,** 211 (1977); **23,** 135 (1977).
43. D. W. Oxtoby, D. Levesque, and J.-J. Weis, unpublished work.
44. S. R. J. Brueck and R. M. Osgood, *Chem. Phys. Lett.*, **39,** 568 (1976).
45. J. C. Raich and N. S. Gillis, *J. Chem. Phys.*, **66,** 846 (1977).
46. M. Scotto, *J. Chem. Phys.*, **49,** 5362 (1968).
47. W. Clements and B. P. Stoicheff, *Appl. Phys. Lett.*, **12,** 246 (1968).
48. A. Laubereau, *Chem. Phys. Lett.*, **27,** 600 (1974).

49. A. Nitzan and R. Silbey, *J. Chem. Phys.*, **60,** 4070 (1974).
50. D. J. Diestler and R. S. Wilson, *J. Chem. Phys.*, **62,** 1572 (1975).
51. P. A. Madden and R. M. Lynden-Bell, *Chem. Phys. Lett.*, **38,** 163 (1976).
52. D. W. Oxtoby and S. A. Rice, *Chem. Phys. Lett.*, **42,** 1 (1976).
53. D. C. Knauss and R. S. Wilson, *Chem. Phys.*, **19,** 341 (1977).
54. D. J. Diestler, *Chem. Phys. Lett.*, **39,** 39 (1976).
55. S. Bratos, J. Rios, and Y. Guissani, *J. Chem. Phys.*, **52,** 439 (1970);
56. D. W. Oxtoby, *Mol. Phys.*, **34,** 987 (1977).
57. C. H. Wang, *Mol. Phys.*, **33,** 207 (1977).
58. R. Levant, *Mol. Phys.*, **34,** 629 (1977).
59. F. G. Dijkman and J. H. van der Maas, *J. Chem. Phys.*, **66,** 3871 (1977).
60. W. G. Rothschild, *J. Chem. Phys.*, **65,** 455 (1976).
61. B. J. Berne and G. D. Harp, *Adv. Chem. Phys.*, **17,** 63 (1970).
62. D. J. Diestler and J. Manz, *Mol. Phys.*, **33,** 227 (1977).
64. J. E. Lennard-Jones and A. F. Devonshire, *Proc. R. Soc.*, **A163,** 53 (1937); **165,** 1 (1938).
64. S. F. Fischer and A. Laubereau, *Chem. Phys. Lett.*, **35,** 6 (1975).
65. In Ref. 52 it was incorrectly stated that this factor was $\frac{9}{8}$.
66. D. W. Oxtoby, *J. Chem. Phys.*, **70,** 0000 (1979).
67. R. B. Wright, M. Schwartz, and C. H. Wang, *J. Chem. Phys.*, **58,** 5125 (1973).
68. J. Schroeder, V. H. Schiemann, and J. Jonas, *Mol. Phys.*, **34,** 1501 (1977).
69. A. Laubereau, D. von der Linde, and W. Kaiser, *Phys. Rev. Lett.*, **28,** 1162 (1972).
70. R. Zwanzig and M. Bixon, *Phys. Rev.*, **A2,** 2005 (1970).
71. C. Hu and R. Zwanzig, *J. Chem. Phys.*, **60,** 4354 (1974).
72. J. Keizer, *J. Chem. Phys.*, **61,** 1717 (1974).
73. H. Metiu, D. W. Oxtoby, and K. F. Freed, *Phys. Rev.*, **A15,** 361 (1977).
74. T. S. Chow and J. J. Hermans, *Physica*, **65,** 156 (1973).
75. M. Kakimoto and T. Fujiyama, *Bull. Chem. Soc. Jap.*, **47,** 1883 (1974).
76. T. Tokuhiro and W. G. Rothschild, *J. Chem. Phys.*, **62,** 2150 (1975).
77. The expression in (5) of Ref. 75 is incorrect.
78. G. Döge, *Z. Naturforsch.*, **28a,** 919 (1973).
79. Y. Le Duff, *J. Chem. Phys.*, **59,** 1984 (1973).
80. M. J. Clouter and H. Kiefte, *J. Chem. Phys.*, **66,** 1736 (1977).
81. S. R. J. Brueck, *Chem. Phys. Lett.*, **50,** 516 (1977).
82. R. G. Gordon, *J. Chem. Phys.*, **44,** 1830 (1966).
83. Y. LeDuff and W. Holzer, *Chem. Phys. Lett.*, **24,** 212 (1974).
83a. H. M. M. Hesp, J. Langelaar, D. Bebelaar, and J. D. W. Van Voorst, *Phys. Rev. Lett.* **39,** 1376 (1977).
84. J. Hyde Campbell, J. F. Fisher, and J. Jonas, *J. Chem. Phys.*, **61,** 346 (1974).
85. G. Döge, R. Arndt, and A. Khuen, *Chem. Phys.*, **21,** 53 (1977).
86. M. Constant, M. Delhaye, and R. Fauquembergue, *Compt. Rend. Acad. Sci.*, **B271,** 1117 (1970).
87. D. R. Jones, H. C. Andersen, and R. Pecora, *Chem. Phys.*, **9,** 339 (1975).
88. G. D. Patterson and J. E. Griffiths, *J. Chem. Phys.*, **63,** 2406 (1975).
89. H. S. Goldberg and P. S. Pershan, *J. Chem. Phys.*, **58,** 3816 (1973).
90. R. M. Lynden-Bell and G. C. Tabisz, *Chem. Phys. Lett.*, **46,** 175 (1977).
91. J. E. Griffiths, *J. Chem. Phys.*, **59,** 751 (1973).
92. C. Breuillard-Alliot and J. Soussen-Jacob, *Mol. Phys.*, **28,** 905 (1974).
93. S. L. Whittenburg and C. H. Wang, *J. Chem. Phys.*, **66,** 4255 (1977).

94. For reasons that are unclear to us, the authors of Ref. 93 draw exactly the opposite conclusion from their data.
95. J. Schroeder, V. H. Schiemann, P. T. Sharko, and J. Jonas, *J. Chem. Phys.*, **66,** 3215 (1977).
96. I. Laulicht and S. Meirman, *J. Chem. Phys.*, **59,** 2521 (1973).
97. D. A. Wright and M. T. Rogers, *J. Chem. Phys.*, **63,** 909 (1975).
98. C. Brodbeck, I. Rossi, Nguyen-van-Thanh, and A. Ruoff, *Mol. Phys.*, **32,** 71 (1976).
99. J. Soussen-Jacob, E. Dervil, and J. Vincent-Geisse, *Mol. Phys.*, **28,** 935 (1974).
100. W. G. Rothschild, G. J. Rosasco, and R. C. Livingston, *J. Chem. Phys.*, **62,** 1253 (1975).
101. J. DeZwaan, D. W. Hess, and C. S. Johnson, Jr., *J. Chem. Phys.*, **63,** 422 (1975).
102. M. N. Neuman and G. C. Tabisz, *Chem. Phys.*, **15,** 195 (1976).
103. R. LeSar and R. Kopelman, *J. Chem. Phys.*, **66,** 5035 (1977).
104. K. Tanabe and J. Jonas, *J. Chem. Phys.*, **67,** 4222 (1977).
105. D. von der Linde, A. Laubereau, and W. Kaiser, *Phys. Rev. Lett.*, **26,** 954 (1971).
106. J. C. Barral, O. Hartmanshenn, and P. Rigny, *Chem. Phys. Lett.*, **26,** 79 (1974).
107. C. H. Wang and R. B. Wright, *Mol. Phys.*, **27,** 345 (1974).
108. R. Ouillon, *Chem. Phys. Lett.*, **35,** 63 (1975).
109. M. Constant, R. Fauquembergue, and P. Descheerder, *J. Chem. Phys.*, **64,** 667 (1976).
110. D. R. Jones, C. H. Wang, D. H. Christensen, and O. F. Nielson, *J. Chem. Phvs.*, **64,** 4475 (1976).
111. P. A. Madden and H. Wennerstrom, *Mol. Phys.*, **31,** 1103 (1976).

STRUCTURE AND DYNAMICS OF SIMPLE MICROCLUSTERS

M. R. HOARE

Institut für Physikalische Chemie
*Stuttgart University, Germany**

CONTENTS

* Permanent address: Department of Physics, Bedford College, Regent's Park, London NW1 4NS.

I. INTRODUCTION

A *microcluster* may be definied as an aggregate, whether of atoms, ions, molecules and so on, so small that an appreciable proportion of these units must be present in its surface at any given time. If we take the term "appreciable" to mean greater than say 10%, this sets an upper limit to size in the region of 10^4 to 10^5 units, a figure virtually independent of the details of geometrical structure. The same limit, in broad statistical mechanical terms, corresponds to the possibility of thermal fluctuations ($\sim N^{-\frac{1}{2}}$) at around the 1% level.

The general acceptance of the term *microcluster* in recent literature is welcome in that it expressly avoids the structural connotations of, for example, microcrystallite, microdroplet and the vagueness of colloid, nucleus, aerosol particle, grain, and so on, terms which in any case have come to be applied over a much broader size range. As used here, the term *microcluster* is also taken to imply a certain lifetime condition; although clusters may well be metastable with respect to fragmentation, we assume that the lifetime for this is considerably longer than the characteristic inverse frequencies for internal motions—in short that the cluster can be identified with a bounded region of phase space within which time averages over its natural motion are meaningful. In this respect we rule out the purely transient type of configuration sometimes referred to as a cluster without the implication of binding energy or stability.

The remarkable increase of interest in microcluster physics in the last decade has been marked by the confluence of several previously separated fields, each with its peculiar experimental and theoretical traditions. Thus on one hand, the long-standing need to understand the nucleation process in terms of atomistic rather than continuum models has merged with a host of new problems raised by the development of methods for the generation and observation of particles composed of no more than a few atoms. Ancient issues in crystal growth theory have taken on new life when considered in terms of computable interactions between a few neighbors; the occurrence of condensation in supersonic flow regimes—once a nuisance phenomenon—has been put to good use in cluster studies; greatly improved mass spectrometers developed for molecular fragmentation have found immediate application to cluster beams. The photographic process, for so long an almost isolated focus of interest in small, included metal particles, is now joined by a variety of systems in which very small metal aggregates can be prepared in an extended matrix.

From these interwoven themes a number of broad issues emerge. In the field of catalysis, certainly that with the widest economic implications, there has been a shift of interest from surfaces as a general adsorbing substrate with locally important electronic properties to the study of definite few-atom features of cluster type known to be present to a significant degree on the surfaces of many active preparations.

At quite another extreme, recent astrophysical theories have put increasing emphasis on the role of small interstellar grains as a factor in the generation of chemical species and the regulation of hydrogen equilibria in the galaxy.

These are, so to speak, the fields of direct application in the study of cluster properties. To them can be added a number of indirect ones in which the cluster approximates a unit in some larger structure for purposes of calculation. Thus, for example, cluster-like units, although of lesser interest in defining the properties of an ideal crystal, take on altogether more importance in the case of "amorphous" systems or when nonideal effects such as self-diffusion or vacancy energies are to be studied.

The content of theoretical work in these various fields shows a number of unifying features. A particular focus of interest has been the realization that important thresholds for the onset of effectively "macroscopic" properties may, somewhat against our preconceptions, actually lie in the microcluster size region. Thus "melting" transitions are surprisingly well-defined at the $N = 100$ level, and superconductivity and ferromagnetism have been shown to set in at sizes that seem surprisingly low for a phenomenon of long-range order. Another universal question is whether there may exist "magic numbers" for cluster stability or perhaps electronic properties.

In preparing this review we do not attempt to unify all these themes, preferring to concentrate on one more limited but best understood aspect—the essentially classical mechanics of *simple* microclusters. By simple we mean those microclusters composed of atoms interacting by two-body central forces. Even within these limitations we are selective, concentrating most on the classical stability problem and those aspects of dynamics which are best understood in terms of it. Actual applications of cluster mechanics, including the most important one of nucleation theory are given in less depth—in part to keep the present article to manageable proportions, in part because of the excellent review articles in existence elsewhere. The result, it is to be hoped, will be fairly self-contained, as well as a useful orientation in the field of cluster studies in general and a possible starting point for the more difficult areas of quantum mechanical and electronic properties.

Although we do not intend to review experimental work exhaustively, we begin with a short resume of five different fields in which microclusters have assumed a central importance. We then give a fairly extensive account of cluster statics and proceed by stages to dynamics and thermodynamics. The main part leads up to a discussion of the most important results obtained so far by computer simulation studies of molecular dynamic and Monte Carlo type, and we conclude with brief sections on nucleation and crystal growth.

Scales of Size and Number

The reader is strongly advised at this stage, and before using the cluster literature, to acquire a ready feeling for the linear, surface, and volume relationships of small structures as a function of the number of units N. For this the crudest of approximations is good to well within an order of magnitude.

Let r, R be the radius of a single unit and a whole cluster, respectively, and both assumed to be spherical. Assume that N units make up a sphere with the minimum possible surface/volume ratio. Then, neglecting entirely the packing fraction and geometrical details of the structure we can estimate $V=(4/3)\pi R^3 \cong Nv=(4/3)N\pi r^3$ where v and V are the volume of unit and cluster, respectively. Thus to this approximation $N \cong (R/r)^3$ and $R \cong N^{1/3}r$. Taking a standard radius $r=1$ Å$(=10^{-4}\mu)$ we find that, for $R=10$, 30, and 100 Å respectively, the cluster numbers are $N=10^3$, 10^4, and 10^6 to well within the order of magnitude.

To estimate the surface number N_s we again neglect packing fraction and simply assume that a surface area $4\pi R^2$ is covered by $N_s \pi r^2$ units. Then evidently $N_s=4(R/r)^2=4N^{2/3}$. Taking the surface occupancy ratio to be $N_s/N=4N^{-1/3}$ we obtain for the ratios $N_s/N=0.5$, 0.1, 0.04 the numbers $N=500$, 5×10^4, 10^6 which for the 1 Å packing unit gives $R=8$, 80, and 100 Å.

These estimates are surprisingly close to the effective values for nearly spherical packing geometries. For the icosahedral packings introduced in Section III. B, we know exactly that $N_s/N=I_n/I_{n-1}$ where I_n are the icosahedral numbers $I_n=1$, 13, 55, 147, 309, 561 . . . . For the $n=7$ icosahedron, $N_s/N=362/923=0.39$ . . . . We may compare this with $3/n=0.43$ . . . and $4/N^{1/3}=0.41$ . . . . (For $n=9$, $N=2057$ the three figures agree to better than 1%.)

II. EXPERIMENTAL FOREGROUND

There are, broadly speaking, three types of experiment in which microclusters may be generated and observed. These correspond to different degrees of physical isolation:

1. Freely translating in space
2. Supported upon a substrate
3. Embedded in a matrix

Not surprisingly it was the second of these which, with the coming of ultrahigh-voltage electron microscopy, led to the most direct observations of microcluster structure and thus to the buildup of interest in the subject in the 1960s. However, the generation of clusters in nozzle beams was achieved at almost the same time. This type of experiment, coupled with advances in mass spectrometry and electron diffraction, has accounted for much of our present knowledge of cluster properties under "free" conditions. Except in the special case of the photographic image, the study of embedded clusters has proceeded more slowly, but appears now as one of the most promising sources of spectroscopic information on the very smallest of metallic aggregates. These categories are, however, not entirely distinct; in the case of deposited particles, different degrees of intimacy with the substrate are possible, and there exist indefinite regions, such as the field of granular films, in which the matter observed achieves only a partial degree of cluster-like ordering.

As a prelude to our mainly theoretical discussion we selectively review some of the main types of experiment now available for the generation, detection, and analysis of microclusters in the laboratory and otherwise.

A. Micrography and Electron Diffraction

The earliest reports of cluster micrography in the 100 Å size region are those of Mihama and Yasuda,[1] Ino,[2] Kimoto and Nishida,[3] and Allpress and Sanders.[4] These workers examined a number of f.c.c. metals deposited on rocksalt or mica cleavage planes and found that in the smallest observable size range (50 to 500 Å), there was almost invariably a tendency to compound crystallinity rather than single crystal growth. Moreover, under certain conditions the particles observed tended to possess fivefold external symmetry, which careful examination of both micrograms and diffraction patterns showed as arising from polyhedra of iosahedral or pentagonal-bipyramidal type. Although the occurrence of such particles seemed to depend somewhat on the rate of deposition and the presence of inert gas, it was shown beyond doubt that the growth mechanism involved was the same whether particles formed in attachment to the substrate or were collected after precipitation from the gas phase. These findings have since been confirmed at greater resolution. In particular, Gillet and co-workers[5,6] found, using direct images, diffraction, and Moiré-fringe observation, that there is a transition from perfect pentagonal symmetry somewhat below 100 Å, to a faulted structure with clearance regions, which seems to presage the growth of normal crystalline units. Among other interesting effects that emerged is the observation by Kimoto and Nishida[7,8] that the b.c.c. metals chromium and iron have difficulty in taking up this lattice in the 100 Å range and instead tend to an A15 (β-tungsten) modification unknown in the bulk. Reformation to the normal b.c.c. structure eventually occurs on further growth.

It does not seem too much to expect that improvements in electron microscopy will eventually lead to the direct study of natural cluster images from the $N = 2$, 3, 4 cases upward. Unnatural images of individual atoms are, of course, already available in field-emmission microscopy. It is also claimed that nuclei of perhaps as few as two of three atoms can be made visible at surfaces by using them as development nuclei for the growth of larger particles of a different metal.[9]

B. Generation of Cluster Beams

Although the work just discussed lends support to the idea that isolated microclusters on atomically plane supports approximate the structural properties of the same objects in free space, there is an obvious need for measurements on clusters which are not only in free translation but also possibly at effective temperatures outside the normal solid range. Thanks to the development of condensing supersonic nozzles and parallel improvements in mass spectrometry, these experiments have proved to be less difficult than might have been imagined.

The existence of condensation in high-velocity expanding jets of moist air has been known for many years,[10] and more refined experiments have shown that careful control of the nozzle and thermodynamic parameters can lead to the selective production of quite high concentrations of bound clusters from the dimer upward to the $N = 100$ range or less selectively into the Rayleigh-scattering region and above.[11,12] Four types of generation have been used, each with certain experimental advantages: (*1*) simple effusion of saturated vapour, (*2*) free-jet expansions, (*3*) constrained nozzle expansions, and (*4*) nozzle expansions with carrier gas.

Foster, Leckenby, and Robbins[13] achieved the mass spectrometric discrimination of Na_n clusters effusing from an oven with sizes in the range $n = 2$ to 12. They also measured the ionization energies of the species and demonstrated that approximately 50% of the progression of this to the bulk work function is already achieved at Na_4. A spread of ionization energy with $n > 6$ was also found, presumably due to the occurrence of isomeric structures from this point on. No particular trends in the relative populations of Na_n in effusing vapor were noted in this work. However, in complete contrast to this, a recent study by Kimoto on lithium vapor[14] shows a marked depopulation of the *odd* clusters Li_3, Li_5, Li_7, and Li_9, compared to the even ones Li_2 to Li_8. Significantly, the absence of *odd* clusters does not continue above $n = 11$, and the cluster Li_{13} is prominent. Whether these results are simply kinetic in origin, with growth proceeding in units of the dimer Li_2 or whether there is an alternating stability condition of quantum mechanical origin between odd and even structures remains to be seen.

Earlier Becker, Bier, and Henkes[15] and Greene and Milne[16] carried out studies of rare gas and hydrogen clusters in nozzle expansion systems and were able to measure relative concentrations from the dimer up to the region $N=20$. The German work has been directed in part toward the possibility of producing charged D_{2n} clusters for acceleration and injection into thermonuclear machines, although little of this aspect has been heard for some years.[17] More recently, Gspann, Krieg, and Vollmar have reported the generation of cluster beams of both ^4He and ^3He and the measurement of scattering between each and Cs atoms.[18] However, the clusters involved are in the $N=10^6$ size range. It is worth noting that He clusters may be the only species unequivocally known to be liquid-like under the conditions of nozzle expansion.

Another remarkable series of experiments on free-jet cluster beams has been described by Farges and co-workers.[19–21] Argon clusters with size range to below $N=100$ were generated, and the beam was crossed with an electron diffraction unit. The resulting Debye-Scherrer ring patterns could be interpreted in terms of solid clusters with pentagonal symmetries in the lower size range, giving way to a normal f.c.c. structure at the level of some 1000 atoms upward. Although subject to inevitable statistical averaging, these results are the first to demonstrate solidity and noncrystallographic local order in clusters under free-space conditions. This work is being extended to clusters of H_2O and N_2. Other electron diffraction measurements on free jets have been reported by Stein.[22]

When measurements of cluster growth kinetics rather than individual particle properties are required, the free-jet method has the disadvantage that the thermal history of the condensing material is almost impossible to calculate. The use of constrained nozzle expansions are then an advantage and can be made to give results at very high supersaturation ratios which are free from both impurity and boundary-layer effects and for the which the full hydrodynamic evolution up to and during condensation is computable. Using this method, Wegener and co-workers[23,24] have carried out Raleigh scattering measurements on controlled expansions of H_2O using a He/Ne laser and have arrived at accurate measurements of the initial stages of growth. Although they only observed clusters in the $R>20$ Å size range, the results make possible the testing of nucleation theories based on atomistic cluster models with critical nuclei undoubtedly in the $N<100$ range. These measurements have recently been extended to the system of argon condensing in a large excess of helium carrier gas, yielding even more accurately predictable thermal conditions in the condensation region.[25]

C. Homogeneous Chemical Generation

The homogeneous liberation of smoke-forming particles during combustion and other rapid reactions is a subject with a long history. Recently Bauer et. al.[26,27] have refined methods for the shock-tube heating of metallo-organic vapors such that the evolution of particle size distribution behind a shock front is directly measurable by laser light scattering. The method has been applied successfully to iron microclusters (from $Fe(CO)_5$) and is suggested as a possible method of producing silica microclusters from silanes in attempts to simulate interstellar grains (see the next section). The unique beauty of this work is that it provides the first effective *calorimetry* of cluster formation: by observing the particle size distribution as a result of the growth process $Fe_n + Fe = Fe_{n+1}$ it is possible to arrive at a size-dependent enthalpy of condensation, ΔH_n. In the case of iron, a growth law of the form $\Delta H_n = \Delta H_\infty(1 - n^{-0.15})$ is measured.

D. Astrophysical Clusters

Among the lesser-known aspects of the physics of microclusters is their suspected role as interstellar grains. Interest in interstellar matter has increased considerably in recent years with the availability of satellite observations and the discovery of sufficient interstellar molecules to provide a rich cosmic chemistry. At present the main constituents of interstellar grains are believed to be silicates and graphite, formed in the atmospheres of M-type giants and carbon stars, respectively, and blown out by radiation pressure. An extensive review of their properties is given by Wickramsinghe and Nandy.[28] Although the average grain size in typical interstellar clouds has long been supposed to be some 0.1 μ for graphite and perhaps 0.01 μ for silicates, a recent tendency has been to revise these estimates downward, and it can reasonably be supposed that at least an appreciable tail of the grain size distribution exists well into the microcluster region as we define it. Some theoretical work actually predicts an increase in grain number density with density with decreasing radius according to the Oort-Van de Hulst law: $n(r) = \exp(-Ar^3)$, and this is thought to agree with measured interstellar extinctions in the visible and ultraviolet region.

A main reason for the shift of interest towards grains in the microcluster size region is the realization that certain processes of adsorption and desorption may owe their characters entirely to the smallness of the particles, irrespective of the detailed physical chemistry of their surfaces. Temperatures of interstellar grains vary somewhat according to size, material, and the ambient radiation field, but are thought never to exceed some 20 K and to be typically in the range 5 to 15 K. Under these conditions it is a major problem to explain how small molecules formed at a grain surface can become desorbed at all to contribute to interstellar extinction and unblock adsorption sites for further synthesis. The possibility of some outside influence in the form of soft cosmic ray or ultraviolet photodesorption cannot be ruled out, but this is an unwelcome hypothesis and may be untenable for processes in the interior of highly absorbing dense clouds. Two alternative hypotheses have been put forward to explain desorption without an external agent. Watson and Salpeter[29] believe that the energy of bond formation can, for nonchemisorbed species, couple sufficiently to outgoing translational modes to give a high probability of desorption. In opposition to this, Allen and Robinson[30] claim that a more likely mechanism is the *whole-grain* heating effect which would be appreciable in the microcluster size range and could well lead to desorption within the required time scales. They back this suggestion with detailed calculations using the McGinty equilibrium clusters (described in Section III.B.3) as a mechanical model. Adsorbed lifetimes for species such as OH, CO on solid N_2, ice, graphite, and quartz have appropriately low values, depending very sensitively on grain radius and rather less so on grain material. However, radii of appreciably less than 100 Å are required for the mechanism to be effective. With sizes well below this there is abundant energy available for the grain to disintergrate; thus the size distribution itself may be directly influenced by atomic-scale processes. We refer to the papers cited for further background and experimental data.

E. Matrix Isolation Methods

The method of matrix isolation, developed some years ago for the spectroscopic study of labile molecules and free radicals,[31] has been successfully adapted for the study of both metallic and ionic clusters. Vapor from a high-temperature Knudsen cell is allowed to impinge on a cooled substrate at about 10 K in the presence of a controllable excess of inert gas. Electronic spectra can then be taken in the visible or ultraviolet range in the case of metals and vibrational spectra in the infrared in the case of ion pairs.

When metals are deposited in this way, a number of features usually appear in the electronic spectra. There are peak shifts, which may be either to the red or the blue of the free-atom transitions, lines are in general broadened, and multiplet structure may appear in the case of degenerate states. All these symptoms may change in a complicated way as the gas/metal ratio is altered and upon warming up. Although in earlier work almost all perturbations were regarded as matrix effects, the view that the clustering of atoms could also play a part has gained ground and in recent work definite assignments of dimer, trimer, and even quadromer bands have begun to appear. Some notable studies are those by Andrews and Pimentel[32] (Li, Li_2), Brewer and Chang[33] (Pb, Pb_2) Francis and Webber[34] (Ca, Ca_2) Schultze, Becker, and Leutloff[35] (Ag, Ag_2, Ag_3), and Moskowitz and Hulse[35] (Cu, Cu_2, Cu_3, Cu_4). It does not seem likely that spectral information alone will lead to discrimination between clusters with N greater than 4 or 5, but quantum mechanical calculations of energy level distributions are now becoming available to assist in this process. The field remains extremely active and has attracted the interest of groups concerned with catalytic activity. As Moskovitz and Hulse wisely put it, nothing could be more directly indicative of the vast sensitivity of catalytic activity to the microdispersion of the metal than the subtle shifts of the electronic spectra seen in the matrix-supported state.

For the sake of completeness, we should at least recall the long history of work on silver atom clusters in their role as the photographic latent image. (See Ref. 37 for an up-to-date resume.) It has long been believed, on the grounds of quantum efficiency measurements in the region of reciprocity law failure, that the latent image required *two* silver atoms for physical stability and perhaps some few more to become an active development center. A carefully argued theory of Mitchell[38] centers on the claim that the critical image particle consists of three silver atoms tetrahedral to an Ag^+ ion. The nature of the latent image seems, however, to be far from settled, and a number of quite unorthodox theories still receive attention. One of these even denies a photolytic role to the incoming light, assuming that this simply leads to an aggregation effect among silver atoms already formed at the sensitization stage.

Matrix isolation of ionic species is a rather newer development which has led to interesting results in the infra red region. Again clustering is important, and interest in small stable ionic structures and their isomers appears to be growing.

F. Other Experimental Results

When electronic and nuclear properties of metal clusters are considered, a vast literature opens up that we can only point to in this summary. Properties that have been measured include the Mössbauer effect,[39] NMR,[40] superconductivity,[41] ESR,[42] superparamagnetism,[43] and much else. However the small particles in many of these studies are on the large side for microclusters, in some cases well into the micron range. A very good selection is found in the articles in Ref. 18. The best bibliography is probably that in Ref. 44.

III. CLUSTER STATICS: THE CLASSICAL STABILITY PROBLEM

No proper treatment of the dynamics of a microcluster system at finite kinetic energy can escape the fact that we deal with an N-body problem, and our expectations of such an exercise must be correspondingly modest. Although it is true that N need not be particularly large for the results to be interesting and that the Hamiltonian for the problem can be kept relatively simple, there is a compensating disadvantage that, even where the structure involved is solid-like, there will not usually be any counterpart of the translational symmetry that can often simplify calculations on crystalline systems. In any case, we should like, at least in a qualitative way, to obtain information as to the *whole* class of structures available to a small cluster under conditions in which only the total energy is given.

Let us concentrate at first on the internal motions of simple rotationless clusters. By simple we mean that the component units are either atoms or effectively structureless and identical; the internal dynamics are given in terms of $3N$-6 internal coordinates, which we symbolize $\mathbf{r}^N$ together with their conjugate momenta $\boldsymbol{\rho}^N$. For neutral systems at least we can assume separability and write for the classical Hamiltonian:

$$H(\boldsymbol{\rho}^N, \mathbf{r}^N) = \sum_i \frac{\boldsymbol{\rho}_i^2}{2m} + V(\mathbf{r}^N) \tag{1}$$

Here for the sake of definiteness we may assume a Cartesian system with origin in atom 1 and oriented in the plane of atoms 1, 2, and 3 such that the coordinates of the position vector $\mathbf{r}^N$ are

$$\begin{aligned}\mathbf{r}^N &\equiv \{\mathbf{r}_{2x}; \mathbf{r}_{3x}, \mathbf{r}_{3y}; \mathbf{r}_{4x}, \mathbf{r}_{4y}, \mathbf{r}_{4z} \ldots \mathbf{r}_{Nx}, \mathbf{r}_{Ny}, \mathbf{r}_{Nz}\}\\ &\equiv \{\hat{\mathbf{r}}_2; \hat{\mathbf{r}}_3; \mathbf{r}_4, \ldots \mathbf{r}_N\}\end{aligned}$$

We have marked the vectors $\hat{\mathbf{r}}_2$ and $\hat{\mathbf{r}}_3$ to show that they are only one- and two-dimensional, respectively. More formally we speak of $\mathbf{r}^N \in \mathscr{C}_{\hat{N}}$ where $\mathscr{C}_{\hat{N}}$ is the internal configuration space which is the product of Euclidean spaces R^n thus:

$$\mathscr{C}_{\hat{N}} = R_+^1 \otimes R_+^2 \otimes R_+^{N-3} \qquad R_+^1 \equiv (0, \infty)$$

It must be admitted at the outset that, although numerical solutions of the equations of motion based on $H(\boldsymbol{\rho}^N, \mathbf{q}^N)$ have taught us a lot about the natural motion of clusters in phase space, their analytical mechanics virtually begin and end with the examination of the potential energy function $V(\mathbf{r}^N)$ in the space $\mathscr{C}_{\hat{N}}$. Thus we may seek local minimum configurations $\mathbf{r}_0^N$ and go on to deduce vibrational frequencies from their radii of curvature, and we may work out moments of the mass distribution to make rigid rotor approximations—virtually all else is a matter of geometrical insight or brute numerical solution. Perhaps we should put it more constructively that the fascination of cluster mechanics lies in the interplay between these two elements—our geometrical insight can lead to definite mechanical hypotheses testable in computer simulations, whereas, on the other hand, computer runs occasionally lead to geometrical results that come as a surprise.

A.

1. *The Topography of $V(\mathbf{r}^N)$ for Simple Clusters*

The statics of a microcluster is in effect the study of the topography of the potential energy function $V(\mathbf{r}^N)$ in the internal configuration space $\mathscr{C}_{\hat{N}}$. Even under the simplest assumptions about the interaction energy, questions of extraordinary complexity arise, some idea of which is essential for the most qualitative approach to cluster dynamics.

Of the many cases of interest, the simplest is that of two-body central forces between the component atoms. Although this condition is certainly violated in many real systems—either through the occurrence of angle-dependent contributions or, more seriously, three-body terms—it provides a useful prototype as well as simplifying enormously the design of computer programs. Under these conditions we can write for the N-atom potential energy:

$$\begin{aligned} V(\mathbf{r}^N) &= \sum_{i=1}^{N-1} \sum_{j=i+1}^{N} v(|\mathbf{r}_i - \mathbf{r}_j|) \\ &= \sum_{i=1}^{N-1} \sum_{j=i+1}^{N} v\{[(\mathbf{r}_{ix} - \mathbf{r}_{jx})^2 + (\mathbf{r}_{iy} - \mathbf{r}_{jy})^2 + (\mathbf{r}_{iz} - \mathbf{r}_{jz})^2]^{1/2}\} \end{aligned} \tag{2}$$

in which $v(r)$ is the interatomic pair potential and $\mathbf{r}_i$, $\mathbf{r}_j$ refer to the three-space position vectors of pairs of atoms. The coordinate restrictions

$$r_{1x} = r_{1y} = r_{1z} = r_{2y} = r_{2z} = r_{3z} = 0$$

are implied throughout. Later we use the abbreviation $r_{i\alpha}$ for the three-space components of atom i with α standing for x, y, z. Although many types of function $v(r)$ may be used in physical models, it is necessary to apply some restrictions if $V(\mathbf{r}^N)$ is to have satisfactory behavior in *all* $\mathscr{C}_{\hat{N}}$ rather than merely some subspace. In general, we would like $v(r)$ to be continuous at least down to a hard core at $r_{\min}$ and to possess derivatives up to the second over the interval $(r_{\min}, \infty)$. More specifically, we are only interested in cases where $v(r)$ is a *well potential* satisfying the following (not of course independent) conditions:

1. $v(r) \to 0^-$ as $r \to \infty$.
2. $v(r) \to \infty$ for $r < r_{\min}$ and $r_{\min} \geqslant 0$.
3. $v'(r_0) = 0$ for a unique r_0 with $r_{\min} < r_0 < \infty$.
4. $v''(r_0) > 0$ and $v(r_0) < 0$.

Pair potentials of interest in cluster studies include the following:

1. $v(r) = (n-m)^{-1}[nr^{-m} - mr^{-n}]$ (Mie)
2. $v(r) = r^{-6} - 2r^{-12}$ (Lennard-Jones)
3. $v(r) = [1 - e^{\alpha(1-r)}]^2 - 1$ (Morse)
4. $v(r) = Ae^{-ar^2} - Be^{-br^2}$ (Gaussian)
5. $v^{\alpha\beta}(r) = z^\alpha z^\beta / r + A \exp(-r/\rho)$ (Born-Meyer)

In 1, 2, and 3 r has been scaled such that $r_0 = 1$ and v such that $v(r_0) = -1$. This is always possible when $v(r)$ is a two-parameter formula. Note that, although 1 and 2 satisfy all the conditions 1 to 4 above, 3 and 4 always violate condition 2 and 5 will in the case of ions of opposite charges $z^\alpha = -z^\beta$. When these are used, it may be necessary to introduce a cutoff at some artificial $r_{\min} > 0$ at which the potential is made infinite. Condition 1 is essential to the physical problem and enables us to define a reference state $V(\infty^N) = 0$ where ∞^N indicates the condition with all atoms infinitely separated. We note the obvious fact that the harmonic pair potential $v(r) = (r-1)^2 - 1$ violates this condition and requires a truncation at $r = 2$ and a hard core at some r in (0, 1) to satisfy conditions 1 and 2 above. Although a cluster can be defined with strictly harmonic interactions, its inability to dissociate and the vast overweighting of non-nearest-neighbor interactions for $N > 4$ makes it physically a most unrealistic object. Harmonic approximations can only be made in local sense about some particular point, usually a predetermined minimum. We discuss a number of alternative ways of doing this in the section on vibrational analysis (Section IV.B).

We now turn our attention to the central problem of cluster statics, that of characterizing the minimum locations of a particular potential energy function $V(\mathbf{r}^N)$ when it is composed of given pair potentials $v(r)$. The geometrical background to this is quite standard, but to gain insight into the nature of the stability problem it is useful to rehearse some of the details with a definite picture of a three-dimensional few-atom cluster in mind. A necessary condition for stability is that there exist *stationary points* of the function $V(\mathbf{r}^N)$ for which the partial derivatives vanish:

$$(\partial V(\mathbf{r}^N)/\partial r_{i\alpha}) = 0 \quad \text{for all} \quad i, \alpha = x, y, z \tag{3}$$

There are undoubtedly a vast number of such points when N exceeds about 6, and they can be expected to increase with N extremely rapidly. However, only a relatively small subset of stationary points corresponds to *unconditional minima*, the further requirements for which is that the *Hessian matrix* H with elements

$$H_{i,\alpha;j,\beta} = (\partial V(\mathbf{r}^N)/\partial \mathbf{r}_{i\alpha}\, \partial \mathbf{r}_{j\beta})$$

be *positive definite* at the point in question. The rest correspond to the great variety of *saddle points* with different combinations of local curvature. With $V(\mathbf{r}^N)$ bounded below, there exists a unique *absolute minimum* value for the potential energy which will in general be achieved at many points in $\mathscr{C}_{\tilde{N}}$. (Note that it is not a sufficient condition for local minimality that all the elements of H be themselves positive; there may always be "valleys" falling away from the stationary point at angles to the coordinate directions.) At any point $\mathbf{r}^N \epsilon \mathscr{C}_{\tilde{N}}$ the eigenvalues of H may be determined and give the principal radii of curvature at that point; the corresponding eigenvectors define principal axes and a set of normal coordinates $\{\xi^N(\mathbf{r})\}$ in terms of which the Hessian is diagonal.

The expression of these properties in structural terms, by which we mean statements about positional configurations $\mathbf{r}_i$ in three-space, is not altogether straighforward. We must first consider the multiplicity of points $\{\hat{\mathbf{r}}_1, \hat{\mathbf{r}}_2 \ldots \mathbf{r}_N\} \equiv \mathbf{r}^N$ which differ only by permutations of the indices. The (N-3)! structures that differ only in the labeling of atoms may be called *geometrically equivalent* (g.e.); others that differ essentially we may term *geometrically distinct* (g.d.). When using the term we shall also assume the class of geometrically equivalent structures to include those which are equivalent under proper symmetry operations in three-space. As for improper symmetry operations, we can also recognize that the situation is complicated by the existence of *enantiomorphic* configurations. Such pairs of structures, which can undoubtedly occur for $N \geqslant 6$, share the same potential energy while not being superimposable by normal rotations. Moreover, they share the same distance matrix $D_{ij} = |\mathbf{r}_i - \mathbf{r}_j|$ to within atom numberings.

Extending our three-space description, we say that a structure $\mathbf{r}^N \equiv \{\hat{\mathbf{r}}_1, \ldots, \mathbf{r}^N\}$ is *stable* if $V(\mathbf{r}^N)$ possesses a local minimum there, and *metastable* if $V(\mathbf{r}^N)$ is merely stationary. In the case of a minimum $\mathbf{r}_0^N$ the value $E_0 = -V(\mathbf{r}_0^N)$ is called the *binding energy* of the structure. We apply this term only at a stable minimum. The structure with the greatest binding energy over all is referred to as the *absolutely minimal* one (strictly the multiplicity of corresponding g.e. structures). In general, absolutely minimal structures cannot be positively identified, even though, for well potentials, the existence of such a point set in $\mathscr{C}_{\hat{N}}$ is guaranteed by the bound (5) below.

All other structures $\mathbf{r}^N \in \mathscr{C}_{\hat{N}}$ are called simply *unstable.* To every atom in an unstable structure can be associated a quantity $\boldsymbol{F}_i = \nabla_{\mathbf{r}_i} V(\mathbf{r}^N)$, a notional external force needed to hold the structure in the position specified. The whole unstable structure is in equilibrium under the N forces of this kind. There is a large subspace $\mathscr{C}_{\hat{N}}^{+} \subset \mathscr{C}_{\hat{N}}$ over which the hessian $H(\mathbf{r}^N)$ is *positive definite* even though $\mathbf{r}^N$ is not minimal. This corresponds to the great variety of structures that are in stable equilibrium under applied forces and yet yield real normal vibration frequencies, that is for which

$$[\omega_k(\mathbf{r}_0^N)]^2 = (A/m)(\partial^2 V(\xi^n)/\partial \xi_k^2)\big|_{\mathbf{r}=\mathbf{r}_0} > 0 \tag{4}$$

where A is a dimensionless quantity.

The normal-mode frequencies thus serve to classify points in $\mathscr{C}_{\hat{N}}$. This is of most importance with regard to the stationary points. At a *local minimum* all ω_k are real; at a *saddle point* at least one is complex; at so-called "monkey saddles" one or more is zero. With two-body well potentials, the existence of local maxima with all frequencies complex can be ruled out. Finally, we note that, as $\mathbf{r}^N \to \infty^N$ (all particles infinitely separated) all frequencies ω_k must tend through real values to zero.

When we consider constructive procedures for finding actual minimum or saddle-point structures in $\mathscr{C}_{\hat{N}}$ it must be admitted that very little of any generality can be said. Although the assumption of two-body well potentials is a severe limitation to possible forms of $V(\mathbf{r}^N)$, the only really quantitative statement it enables us to make is that there certainly exists a somewhat weak lower bound to the potential energy:

$$V(\mathbf{r}^N) \geqslant \tfrac{1}{2}N(N-1)v(r_0) \tag{5}$$

This expresses the idea that we can always imagine a simplex in $\frac{1}{2}N(N-1)$ dimensions in which all atoms are nearest neighbors and that the binding energy of this will always bound the values attainable with the function (2). In three-space the above statements hold with equality when $N=2$, 3, or 4 (the pair, the equilateral triangle, and the regular tetrahedron, respectively).

With these meagre insights, we are virtually forced back on geometrical experimentation. Nevertheless there prove to be a number of speculative guidelines which, although difficult to define mathematically, lead to definite progress in the discovery of minimal structures and supplement the many computational methods now available for searching in local regions of configuration space. Let us postpone the computational aspect until we have examined more closely what can be said in geometrical terms.

It seems clear, for example, that regions in configuration space giving low potential energy according to (2) tend to be those whose three-space structures possess a high proportion of interatomic distances $D_{ij} = |\mathbf{r}_i - \mathbf{r}_j|$ close to the pair distance $\mathbf{r}_0$; indeed, any stable structure can be expected to show an appreciable number of such terms. This brings out the connection, no doubt obvious from the start to a crystallographer, between minimal configurations in $\mathscr{C}_N$ and the packing constructions that can be made with hard spheres or sticks of unit length. Now although the use of packing models has led to important discoveries about "soft" cluster minima, they have a number of shortcomings we should stress at the outset. First, the effect of nearest-neighbor "contacts" may only be decisive in fixing a minimum when the pair potential is sufficiently "hard" and short-ranged that non-nearest-neighbor forces neither contribute much to $V(\mathbf{r}^N)$ nor cause significant compression of the structure due to the compliance in the repulsive part. Second, an "obvious" packing structure may in fact be metastable under a given $v(r)$, and its mode of collapse may not be at all obvious from a ball or spoke model. Finally, as anyone who has experimented with glued-spheres (or their counterpart in the computer) will have learned, packed structures of more than a few atoms which lack lattice symmetry tend to adopt ugly properties, with gaps and bistable sites at their surface. Nevertheless, the *packing property*, as we call this tendency, leads to important characterizations of minima with given N and also to insights into the relationships between minima for different numbers of atoms.

2. *Topological Properties of Minima*

It is useful to consider what information about the minimality of a structure might reside in the topological properties of its three-space graph. By graph we mean the abstract object composed of vertices and edges in which, for our purposes, we identify each atom in a configuration with a vertex and each nearest-neighbor relationship with an edge. To make this association we must first show that the nearest-neighbor property is unequivocally defined for all pairs (i, j) in the cluster. This is well known to be possible by a construction familiar in the theory of extended random packings (see, for example, Collins[45]). We construct the *Voronoi polyhedra* by considering the perpendicularly bisecting planes between all pairs of atoms, and partitioning space with the minimal polyhedra constructed from these planes. The nearest-neighbor relationship is then simply the property of sharing a common Voronoi face. In the case of finite structures, some of the Voronoi polyhedra are infinite, and we may use the property of possessing an infinite polyhedron to define rigorously the difference between *surface* and *interior* atoms. (Some precautions are needed where Voronoi faces meet at a point. It is important to exclude such points, thus denying the nearest-neighbor property to atoms separated by diagonals of an octahedron of a cube.)

Having determined the graph $\mathcal{G}(\mathbf{r}^N)$ for a structure we may tabulate its *adjacency matrix* $\mathbf{A}(\mathbf{r}^N)$, having the property that $A_{ij}=1$ if i, j are nearest neighbors and zero otherwise. In practical cases it is usually possible to find a cut-off distance which will convert a distance matrix $D_{ij}=|\mathbf{r}_i-\mathbf{r}_j|$ by putting $D_{ij}=1$ if $|\mathbf{r}_i-\mathbf{r}_j|<r_c$ and zero otherwise. However, with soft potentials and complex structures this method may be unreliable.

The matrix $A(\mathbf{r}^N)$ is in itself a useful bookkeeping device. We may use it to define at least three types of approximation of the true potential of (2). Consider, for example,

$$\tilde{V}_1(\mathbf{r}^N)=\sum_{i=1}^{n-1}\sum_{j=i+1}^{n} A_{ij}(\mathbf{r}^N)v(|\mathbf{r}_i-\mathbf{r}_j|)$$

or

$$\tilde{V}_2(\mathbf{r}^N)=v(r_0)\sum_{i=1}^{n-1}\sum_{j=i+1}^{n} A_{ij}(\mathbf{r}^N) \tag{6}$$

or

$$\tilde{V}_3(\mathbf{r}^N)=\sum_{i=1}^{n-1}\sum_{j=i+1}^{n}\{v(r_0)A_{ij}(\mathbf{r}^N) + \tfrac{1}{2}\chi(|\mathbf{r}_i-\mathbf{r}_j|, r_0, \delta)(\partial^2 V(\mathbf{r}^N)/\partial\mathbf{r}_i\,\partial\mathbf{r}_j)(|\mathbf{r}_i-\mathbf{r}_j|)^2\} \tag{7}$$

with $\chi(r_1 r_0, \delta)$ a cut-off function that is unity for r $r_0-\delta<r<r_0+\delta$ and zero otherwise. Each of these has been used extensively in model calculations. The first, V_1, counts only nearest-neighbor interactions but assigns their contributions realistically according to $v(r)$. The second, V_2, is cruder in that it counts an identical contribution $v(r_0)$, the pair energy, for each neighbor pair. Although V_2 is clearly no use for vibrational analysis, V_1, although also discontinuous, usually gives well-defined second derivatives about a minimum. The approximation V_3 in (7) makes this explicit in what we might justly christen the "Cowgum potential"—atoms in contact count one pair-energy unit but are also allowed to vibrate with infinitessimal amplitude about the point of contact. In applying this, all second derivatives are usually put equal to a single spring constant.

The greater importance of the graph representation of a structure lies, however, in the specification of relationships between clusters of different sizes. We define $\mathbf{r}^N$ to be a *substructure* of $\mathbf{r}^M$ if both $N<M$ and the graph $\mathcal{G}(\mathbf{r}^N)$ is a *subgraph* of $\mathcal{G}(\mathbf{r}^M)$. [$\mathcal{G}(\mathbf{r}^N)\subset\mathcal{G}(\mathbf{r}^M)$].* (Note that in this

* This is not orthodox notation but is well suited for the purpose.

context the statement that structures are *geometrically equivalent* translates as: their graphs are equivalent to within isomorphism. For general orientation and graph theoretic vocabulary see, for example, Ref. 46).

The substructure relationship is of particular importance when the structures concerned are minimal. We say that minimum $\mathbf{r}_0^N$ *supports* a second $\mathbf{r}_0^{N+1}$ if in fact the second is stable and $\mathscr{G}(\mathbf{r}_0^N) \subset \mathscr{G}(\mathbf{r}_0^{N+1})$. More generally we say that a sequence of minima $\mathbf{r}_0^N, \mathbf{r}_0^{N+1}, \ldots, \mathbf{r}_0^{N+k}$ is a *stable growth sequence* if $\mathscr{G}(\mathbf{r}_0^N) \subset \mathscr{G}(\mathbf{r}_0^{N+1}) \subset \ldots \subset \mathscr{G}(\mathbf{r}_0^{N+k})$. The structure $\mathbf{r}_0^N$ may be called the *seed structure* of the sequence, and the vertex corresponding to the complement of $\mathscr{G}(\mathbf{r}^N)$ in $\mathscr{G}(\mathbf{r}^{N+1})$ is called the *growth vertex*. The set of all geometrically distinct minima $\mathbf{r}^{N+1}$ supported by $\mathbf{r}^N$ is called its *immediate growth structures* (IGS), and one of these with greatest binding energy is the *minimum immediate growth structure* (MIGS). A sequence of structures in which every step is a MIGS with respect to the previous one is called a *minimal growth sequence* (MGS).

Three further characterizations complete our vocabulary. We say that a structure $\mathbf{r}_0^N \in \mathscr{C}_{\hat{N}}$ is *primitive* if all its atoms are surface atoms and that it is *compact* if every structure $\mathbf{r}_0^{N+1}$ for which $\mathscr{G}(\mathbf{r}_0^N) \subset \mathscr{G}(\mathbf{r}_0^{N+1})$ has its growth vertex corresponding to a surface atom. Finally, we define a structure as *crystalloid* if it contains at least a substructure whose graph is a subgraph of the infinite nearest-neighbor graph of a Bravais lattice and large enough to comprise at least one unit cell of the latter. If the whole structure $\mathbf{r}_0^N$ meets this condition, it is called *crystalline.*

The usefulness of this somewhat elaborate language lies first in framing definite questions about the set of all minimal structures available to N atoms for a given potential, and second in constructing algorithms for finding at least certain interesting subsets of the total.

Here we simply list a number of questions of different degrees of interest for the mechanics and growth theory of N-atom systems in the microcluster range. For example:

- Is there an upper limit to the possible size of stable *primitive* structures for simple pair potentials?
- Are all *primitive* structures also *compact*?
- Are there pair potentials for which *all* stable minima are *compact*?
- Are *crystalline* structures constructable by *stable growth sequences*? If so, are these sequences minimal?
- Given that a certain structure is supported by a particular substructure for a potential $v(r)$, what can be said about its stability under alternative potentials?

We return to these questions later and provide answers to some of them for particular pair potentials. Meanwhile, the reader might like to pause and guess the outcome—for, say, the Lennard-Jones systems—on the basis of native intuition and crystallographic prejudices.

3. *Numerical Minimum Searches*

The computer programs used to find actual minimum configurations for many-dimensional functions such as $V(\mathbf{r}^N)$ are all important in cluster studies, and the choice of an efficient one, apart from saving astronomical amounts of computing time, may be essential if values of N in the hundreds or thousands are to be considered at all. There is still widespread ignorance among physicists of the whole field of "optimization theory" and the very refined techniques now available for the analysis of functions of very many variables. Here we outline some of the principles involved, with a brief indication of the practical details.

The search for a minimum of a function in a space of high dimensionality is a vastly more complex undertaking than in a space of, say, three dimensions. Methods that work tolerably well in three-space, for example, the "sectioning" method, where minimization is done serially in one coordinate at a time, become enormously inefficient when attempted in a few hundred variables. Treatises on optimization theory[47] have spoken of the "curses of high dimensionality," one of which is the impossibility of moving efficiently in orthogonal zig-zag motions. Many methods have been devised to speed up the search process, and their efficiency depends in practice on a combination of the ultimate number of steps required and the storage and function-evaluation operations at each stage. Some algorithms are specific to functions in a configuration space and may be "pictured" physically—although they are none the more efficient for that. The force algorithm, for example, determines the net force on each atom in the structure and moves them serially in its direction and with a step proportional to the force's magnitude until some termination condition is satisfied. The related "steepest-descent" method moves the whole configuration point in the direction of grad $V(\mathbf{r}^N)$, either a predetermined step or until a conditional minimum in this direction is reached. Second-order methods in general involve determination of the Hessian H_{ij} at a point and stepping curvilinearly with the two-term Taylor expansion. This is inefficient if all the elements of H must be determined at each step, but there are shortcuts involving the redetermination of only those changing fastest. The method of termination of any of these methods depends on whether the full geometry of the minimum or simply the binding energy is required accurately. Since some cluster minima are quite flat in many of their dimensions, the movement of individual atoms rather than the energy itself must be used for termination if the geometrical structure is required. A useful property of cluster minima is that, when, as often occurs, the three-space structure is nearly spherical, the simple dilational transformation with respect to the center of mass is usually close to the steepest-descent (S-D) direction and can be alternated with other steps to save time in computing derivatives.

By general consensus, the best method for minimum searching with clusters is the "conjugate-gradient" procedure due to Fletcher and Powell.[48–50] In this a compromise is reached between stepping in the S-D direction and orthogonal "conjugate" directions determined from the Hessian and carried forward with a memory effect at each step. This was originally developed for studies of radiation damage and crystal defect equilibria, but is ideal for cluster purposes, provided care is taken to prevent premature termination. A version is available in the FORTRAN Scientific Subroutine Package.

Although the conjugate-gradient method is strongly recommedned, one final alternative is discussed because it has been widely used and raises important questions about the nature of the cluster potential energy surface. This is the molecular dynamic method in which the equations of motion of the cluster are simply solved either from the stationary initial condition given or from this with a little kinetic energy to "push it into a valley." In the natural motion the kinetic energy inevitably rises, going through a maximum before beginning to fluctuate. However, it is monitored and "quenched" at the maximum, whereupon the whole process may be repeated. In this way the system inevitable nears some minimum configuration. The method has been used on argon clusters by McGinty[51] and Farges et al.,[21] but it has several disadvantages, notably a tendency to take the system to minima not particularly close in configuration space to the starting point and differing topologically from it.

Usually when one speaks of relaxing a given structure, one means that a minimum is sought either with the same-neighbor topology or, if none exists, one in some sense "nearest" in configuration space. The notion of the nearest minimum to a given starting point is an important one for the understanding and description of the potential energy surface as well as in practical calculations. Let $\mathscr{C}_N^0$ be the subspace of the N-atom reduced configuration space over which $V(\mathbf{r}^N)$ takes finite values. Then it is clear that for "almost any" point $\mathbf{r}^N \epsilon \mathscr{C}_N^0$ there is a unique corresponding minimum $\mathbf{r}_0^N$, namely, that which is reached eventually on following the steepest-descent trajectory from $\mathbf{r}^N$. (The "almost any" here excludes only actual stationary points and the lower-dimensional sets of points lying on ridges leading precisely to a saddle.) The set of all points conjugate in this way to a given minimum is called its *catchment region.* The union of all catchment regions is $\mathscr{C}_N^0$ itself, and their intersections are only the lower-dimensional boundaries referred to above, that is, they partition $\mathscr{C}_N^0$ uniquely. These considerations lead to some obvious yet far-reaching distinctions. We say that a saddle point is *adjacent* to a minimum if it lies in the boundary set of its catchment region, and we speak of two minima as *neighboring* if they share a common boundary set.

Returning to our original point, we can now insist that the term *relaxation* of a structure be restricted to the process of finding the minimum of its catchment region—or one of the adjacent minima should it be a saddle point. The method used should then be an algorithm in which the trial points can never cross the boundary sets. The conjugate-gradient method appears to satisfy this condition.

B. The Natural History of Microcluster Structure

Our detailed approach to cluster mechanics must depend on which of several broad objectives we have in mind. If we require definite morphological information about shapes and multiplicities of structures and their relative energies, then we are bound to search for at least a representative sample of *true* minimum configurations for the assumed potential—which we can only hope will include the absolutely minimal structure. Once the pair potential is specified, these structures exist in their own right as geometrical entities, and the search for them can be undertaken independent of any thermodynamic or statistical mechanical applications.

If, on the other hand, we require only general predictions of thermodynamic behavior as, for example, the role of surface and volume effects, then it may be quite adequate to consider crude models that predict overall trends at the expense of geometrical reality. Although it requires caution, this approach has been used extensively in two of the areas that have contributed most of the stimulus to theoretical cluster mechanics—the prediction of nucleation rates for homogeneous condensation and the computation of phonon spectra for very small particles.

In practice, various compromises are inevitable. Since we cannot pretend to search configuration space globally, an inspired model structure is usually required as the starting guess for a local optimization search. In many cases (although by no means inevitably), this guess proves to lie in the catchment area of a true minimum which we can subsequently locate.

Here we emphasize the morphological point of view, which is more satisfying theoretically, although considerably more laborious to carry through. Having thus studied the "natural history" of simple clusters in some detail, we may then use it to draw conclusions about crystal growth mechanisms, nucleation rates, and thermodynamic functions as a separate exercise.

As a preliminary we classify the various methods for finding and confirming local minima under several headings.

1. Analogy with the crystalline state
2. Analogy with amorphous structures.
3. The use of aufbau algorithms
4. Structural design
5. Simulation of solidification from the liquid

Each of these has been used to advantage in different areas, and each is to an extent complimentary to the others. We describe them in order without particular attention to the historical sequence of the ideas involved.

1. *Analogies with the Crystalline State*

The most obvious way of modeling the geometry of an atomic or ionic microcluster is to dissect a small prism from an imaginary lattice of the bulk crystalline material and assume that it retains its identity (at least topologically) in the isolated state. The potential energies for different shapes and sizes of unit might then be determined either by evaluation of a realistic potential $V(\mathbf{r}^N)$ at the lattice points or, more crudely, by simple bookkeeping of nearest-neighbor contacts as in the approximation (6). Much of the early work on cluster gometry was done in this way. Detailed accounts of the various polyhedra obtainable from the common lattices are given by Van Harteveld and Hartog,[52] Romanovski,[53] and Nicholas[54] together with tabulations of the numbers of surface atoms N_s for different N. Rough estimates of binding energies were also made by simple contact counting, and from these the structures which maximize energy to this approximation were identified. As is to be expected, those structures which gave most inter-atom contacts were approximately spherical in shape and had surface planes with a high incidence of close packing (Fig. 1). Notice that these "spherical" microcrystallites bear some resemblance to the *Wulff polyhedra* which may be constructed for larger microcrystals by minimizing surface energy with respect to the choice of various low-index lattice planes as faces. The somewhat complex considerations underlying the Gibbs-Wulff theorem in crystal growth theory[55] are, however, out of place in the microcluster range because of the lack of uniformity of stress over the various facets and the large contributions from edge and corner atoms in the structure. Although some attempts have been made to restore the surface energy as a function of the vibrational characteristics of different planes and produce a modified Gibbs-Wulff construction,[56] this is destined to remain a very poor approx-

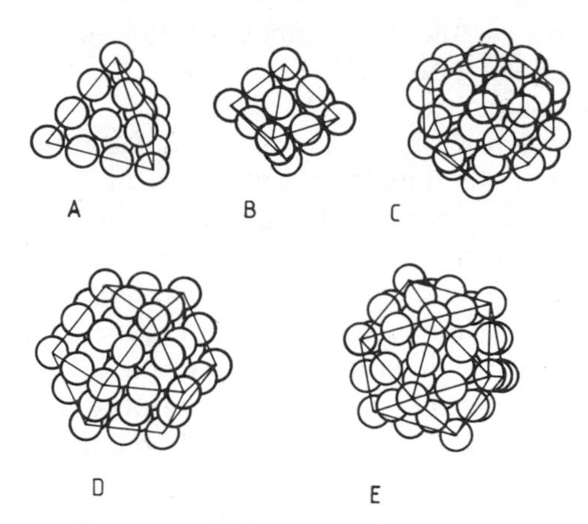

Fig. 1. Some spherical f.c.c. structures. A. $N = 20$ f.c.c. tetrahedron. B. $N = 19$ f.c.c. octahedron. C. $N = 43$ rhombicuboctahedron. D. $N = 55$ cuboctahedron. E. $N = 55$ Mackay icosahedron. The structure D is only precariously stable and deforms readily to E.

imation along with most others that force an explanation in terms of macroscopic quantities inappropriate to the few-atom problem.

The use of spherical f.c.c. microclusters to model rare-gas microclusters has, however, remained popular and has led to a number of thorough studies of thermodynamic properties.[56–58] These are discussed in detail in Section V. The spherical units have also been starting points for quantum-mechanical computations of electronic structure in metal microclusters.

A number of degrees of refinement are possible within the simple microcrystallite approach to cluster energies. Thus Allpress and Sanders first optimized potential within a chosen symmetry by allowing interplane distances to be chosen optimally and demonstrated for the first time the expansion of lattice constant to be expected at the surface.[59] Burton[58] attempted a complete optimization of a sequence of concentric f.c.c. spherical clusters (actually cuboctahedra and rhombicuboctahedra) with $N = 13, 19, 43, 55, 79, 87, 135 \ldots$, but was forced to limit relaxation to the radial direction in all but the first four structures. Binding energies and vibrational frequencies were obtained, the latter, however, only to $N = 55$. Subsequent work cast doubt on the accuracy of the numerical values but confirmed the most surprising qualitative result, namely, that the smallest $N = 13$ cuboctahedron is in fact structurally unstable and distorts with loss of octahedral symmetry in seeking a lower minimum. The nature of the distorted f.c.c. structure remained unclear until it was shown, using a different optimization method, to be in fact a regular icosahedron.[60] This was the first example of a close-packed unit proving to be metastable in isolation and led Hoare and Pal[61,62] to examine other structures of cuboctahedral type. The "dual" $N=13$ Hexagonal close-packed structure was likewise found to be metastable and to transform into the icosahedron. Optimization with a "sectioning" method at first indicated that a similar structural collapse might be possible with the two-shell cuboctahedron of 55 atoms also in Burton's series. A later reinvestigation with the conjugate-gradient method showed this structure to be marginally stable under the full interaction Lennard-Jones Potential.[63] It would now seem that the large cuboctahedra possess a precarious minimum, so near to a saddle point that a careful control of step length in optimization routines is needed to avoid stepping over the edge and missing it. (See Fig. 2)

The main lesson to be learned from these studies is that it is by no means self-evident that crystalline motifs represent the most economical groupings of a few atoms and may not even be structurally stable. With this realization, interest in them as true minima in the smallest size range was transferred to the non crystalline types of structure described in the next sections. This of course left wide open the question of how crystalline units actually *do* form by accretion of surface material on some type of nucleus—a problem to which we can only return after a considerable detour.

2. *Analogy with "Amorphous" Structures*

In direct contrast to the models just described, a number of studies have been made of clusters prepared from known simulations of amorphous packings. The field of amorphous packings is an old one (see, for example, Ref. (64) for general background) and relates to problems of

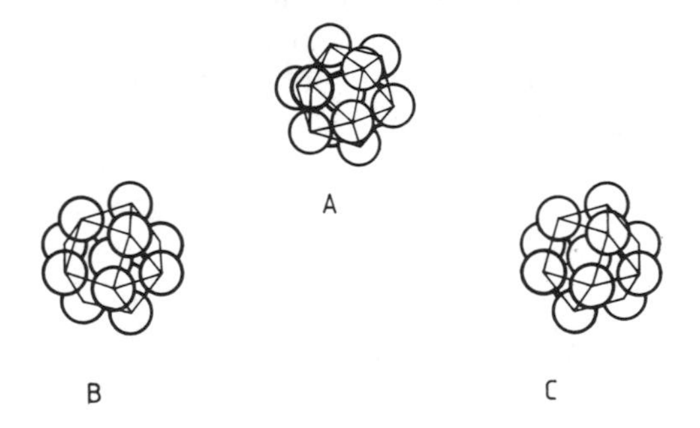

Fig. 2. Unstable clusters for $N=13$. The H.c.p. and f.c.c. structures B and C are metastable saddle points with respect to the $N=13$ icosahedron A.

great mathematical subtlety concerning possible bounds to attainable density in nonlattice packed structures. Although the mechanical and thermodynamic stability of an extended amorphous phase is properly a subject in itself, there is a clear connection with the theory of cluster structure inasmuch as any locally compact motif found to occur repeatedly in an extended packing, or otherwise designed, seems likely to represent a cluster minimum of high binding energy when treated in isolation.

Following the reverse of this train of thought, Boerdijk[65] studied a number of local packed structures designed to maximize local density even though not systematically extendable in three dimensions. A number of the Boerdijk sphere-packing structures are somewhat bizarre, for example, the spiral configuration of face-to-face $N=4$ tetrahedra, but others anticipate the growth series which we discuss from a different point of view in the next two sections. Moreover, Boerdijk enunciated an important principle—that the locally most compact structures tended to be those in which the degree of tetrahedriality in nearest-neighbor topology was a maximum. This tendency, which we noted in our graph-theoretic description, is clearly in opposition to the normal direction of crystallization, which requires a high proportion of *octahedral* nearest-neighbor relationships to occur. Boerdijk's principle seems a dominant factor in the lower size range, $N<50$, but to weaken in force as clusters become large enough for the tetrahedral space-filling deficit to build up. As shown later, it is indeed possible to include octahedra systematically within structures without initiating crystallization.

The most celebrated and influential random packing structure is of course that known as the Bernal model.[66–68] The Bernal coordinates, originally determined by direct measurements on a packed 7934 ball bearing structure have been made widely available and used as starting points for a number of investigations. The Voronoi polyhedron statistics for the model are well known and have been used in speculative accounts of the liquid and glassy state.[68] Writers on glass theory have often questioned what would happen if the Bernal structure were relaxed under a realistic potential—whether it would spontaneously crystallize or transform to an appreciably different amorphous structure, or retain its general character with only minor displacements.[69] Undoubtedly it would dislocate entirely if treated as a free giant cluster and simply allowed to run away in its natural motion. The question is more precisely whether the particular Bernal point in a configuration space of 23,796 dimensions can be transformed by a steepest-decent path on a Lennard-Jones energy surface to a structure with essentially the same nearest-neighbor topology. A relaxation of the whole structure can hardly be contemplated, even with the best programs available, but recently Barker, Finney, and Hoare[70] did carry out a relaxation of a 999-atom subunit under Lennard-Jones conditions. The program could not guarantee a true steepest-descent path, but contracted radially in between careful configurational movements with small steplength using the conjugate-gradient method. In the event, very little gross relative motion of atoms occurred, although there was a considerable gain in potential energy. What movement did occur was significantly in the direction of greater tetrahedriality, as could be inferred from the statistics of Voronoi polyhedron faces. Von Heimendahl obtained similar results.[71]

It is notoriously difficult to simulate in the computer the complex N-body mechanics that go with the shaking of beads or ball bearings, and it is doubtful whether the kind of jammed structure the Bernal model represents can be generated by step-by-step addition of atoms to a seed structure. However, Barker has recently described a 509-atom Lennard-Jones structure built in this way by condensation of atoms from random directions and careful optimization of potential energy between additions.[72] Although the density of the resulting aggregate is probably slightly less than the actual Bernal structure, the statistics of pair distances show remarkable similarity to the results of the relaxation experiment already described. The outcome is that it is possible to say with some confidence that there is no *geometrical* reason, as it were, why quite a large particle of glassy argon should not form and perhaps grow into a similar extended phase, if the deposition were carried out very close to 0 K. That this does little to solve the immense difficulties of preparing glassy argon under practical conditions need hardly be emphasized.

Since this work was carried out it has become clear that attempts to create noncrystalline clusters quasistatically in the computer have little to be said for them compared to the alternative of carrying out full molecular dynamic simulations at low temperature with gradual quenching of the kinetic energy until the motion becomes delocalized. Such simulations, although expensive, show that the phase point in natural motion at low liquid-like energies, has a remarkable ability to find interesting and characteristic low-lying minima of both locally structured and amorphous type. We describe these experiments separately in Section III.B.4.

3. *Aufbau Algorithms*

An alternative to the adoption of ready-made crystalline units as trial minima is to attempt to achieve *stable growth sequences* by selective addition of a series of atoms to a chosen *seed structure.* Various strategies are possible, depending on whether one is willing to optimize completely after each addition. McGinty first used an aufbau method based on the f.c.c. and h.c.p. lattices.[51] Two atoms were placed on neighboring sites, and next-neighbor positions were then tested exhaustively to determine the most favorable point for the addition of a third. With this assigned, the process was repeated to grow an extended sequence of crystalline clusters by atomic steps rather than a shell at a time. These clusters, which McGinty referred to as "nonequilibrium" type, were then relaxed to give equilibrium structures. Both force algorithms and MD methods were used, both of which indicated the $N=13$ and $N=55$ cuboctahedra to be unstable. When the latter was pushed slightly over the brink, it collapsed readily in a MD run to give an unspecified structure vibrating at some 1.5 K (Argon). McGinty went on to compute thermodynamic properties and nucleation rates, but did not attempt to identify the geometric form of the cluster minima he had created.

At about the same time, Hoare and Pal[61] carried out an extensive exploration of the *minimal growth sequences* derived by stepwise optimization from two fundamental seed structures, the $N=4$ tetrahedron and the $N=6$ octahedron. Although optimization under the Lennard-Jones potential was carried out to obtain true minima, it was possible to see the outcome of the first stages of this strategy without recourse to the computer, as indeed several previous workers had done. Consider first the tetrahedral sequence beginning at $N=4$. Additional atoms are positioned sucessively at pockets—positions tetrahedral to the facet giving the greatest gain of potential energy at each stage. There is no choice for $N=5$ and 6, all structures being obviously geometrically equivalent. However, at $N=7$ we are bound to form the pentagonal bipyramid, a very compact structure composed of a ring of five nearly regular tetrahedra. Thus we have an immediate instance of the ease with which fivefold motifs can form under conditions where compactness rather than crystalline symmetry dominates. When this algorithm is continued, an extremely interesting sequence unfolds. The faces of the pentagonal bipyramid become decorated with a ring of five more atoms, leaving a very favorable site vacant on the fivefold symmetry axis. When this is filled, the 12 exterior atoms are seen to have grouped themselves around the 13 in none other than the icosahderal configuration (Fig. 3). Thus the icosahedron is reached with the same inevitability as the $N=7$ pentagonal structure. In this description we of course ignored the slight deficit in packing the five tetrahedra into a ring on the reasonable assumption that this can be accomodated by a slight softness in the actual pair potential.

The train of thought has been followed by a considerable number of authors, the earliest we can cite being Werfelmeier.[73] Werfermeier, like

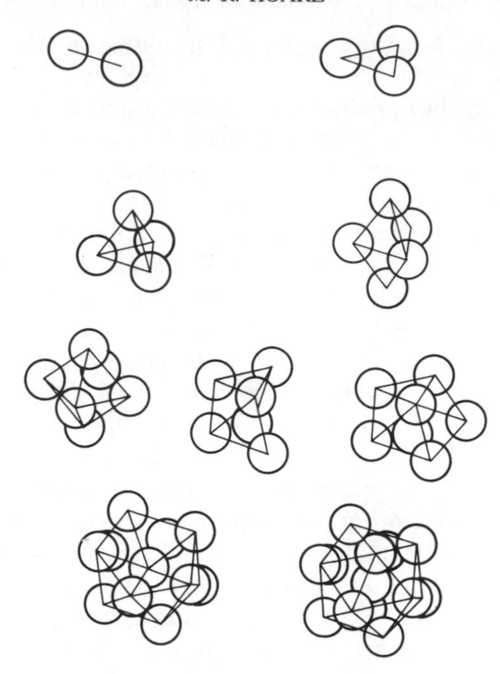

Fig. 3. The Werfelmeier growth sequence. Top to bottom sequence: Dimer, $N=3$ equilateral triangle, $N=4$ tetrahedron, $N=5$ bipyramid, $N=6$ octahedron, $N=6$ tritetrahedron, $N=7$ pentagonal bipyramid, $N=12$ icosahedron minus one, $N=13$ icosahedron.

Pauling, who repeated the exercise many years later,[74] was concerned primarily with predicting alpha particle structures in the nucleus. Similar growth sequences are described by Shternberg,[75] Bernal,[66] Komoda,[76] and others. It was almost certainly known in the nineteenth century to Kelvin and early crystallographers before him.

Hoare and Pal first followed the Werfelmeier sequence with full optimization under the Lennard-Jones potential at every stage and demonstrated beyond doubt that the $N=7$ and $N=13$ minima have perfect D_{5h} and icosahedral (I_h) symmetry, respectively. No trace of clearance regions such as would be expected with sufficiently hard potentials was found, and when these were introduced in a starting configuration they healed up immediately in the optimization process. In the case of the icosahedron, the 12 radial distances duly contracted to 0.969 Lennard-Jones pair distances as the 30 peripheral distances took up the length 1.013 needed to preserve the correct icosahedral ratio of 1.051 . . . . The binding energy was found to be $E_0 = 44.327$ pair units. Compare $E_0 = 42.0$ in the approximation (6).

Although the algorithm certainly "finds" the very stable structures from $N=7$ to $N=13$, it notably misses the $N=6$ octahedron, which might be expected to be the starting point of growth to the f.c.c. lattice. The high binding energy of the octahedron was confirmed (12.71 Lennard-Jones units compared to 12.30 for the $N=6$ triple tetrahedron), and, using it as a seed structure, the *minimal growth sequence* was built up by examining each *immediate growth structure* that could be made by positioning successive atoms at either tetrahedral or octahedral positions to the faces available. There were a number of surprises in the results of this exercise (Fig. 4). First it was discovered that the eight-atom structure with two atoms on adjacent facets of the octahedron undergoes distortion in both symmetry planes to give the interesting figure (Fig. 4*b*). This is the dodecadeltahedron of Bernal,[66] a polyhedron of D_{2h} symmetry of which every facet is an equilateral triangle and the two symmetry planes contain quadrilaterals with long sides of length 1.29 . . . times the equilateral edges (see also Cundy[77]). Surprisingly, this structure, which appeared spontaneously in our optimization program, had been conjectured

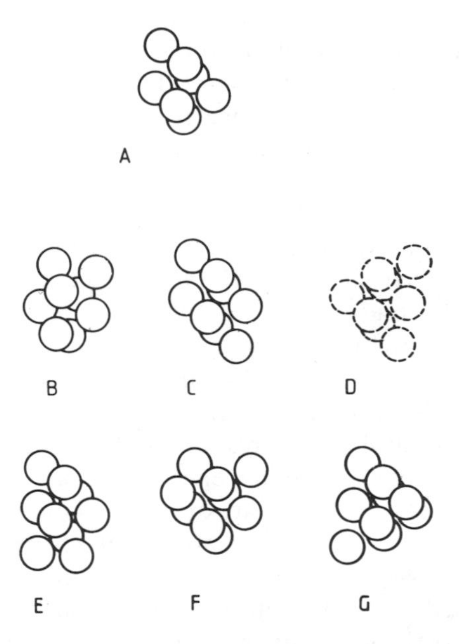

Fig. 4. Octahedral isomers for $N=7$ to 9. A. $N=7$ octahedron plus one. *B*. $N=8$ Dodecadeltahedron. C. $N=8$ *trans* isomer with regular octahedron. D. Metastable $N=8$ isomer which reforms to B. E. $N=9$ isomer with deltahedral distortion. F. $N=9$ isomer with regular octahedron. G. Regular $N=9$ isomer (truncated $N=10$ f.c.c. tetrahedron).

by Werfelmeier as the absolute minimum for eight atoms on the basis of sphere packing experiments alone. We now know that it is definitely *not* the absolute maximum for the Lennard-Jones potential, having binding energy $E_0 = 19.765$, compared to the pentagonal bipyramid plus one at $E_0 = 19.822$ pair units. The second surprise in the octahedral sequence is the instability of the isomer consisting of an octahedron with two additional atoms in the skewed positions (Fig. 4*d*). This moved spontaneously to the deltahedron just described. A third $N = 8$ isomer with atoms in *trans* positions about the octahedron (Fig. 4*c*) proved to be stable with only very minor distortion of the octahedron. Further study of the octahedral sequence as far as $N = 13$ (see also Section III.D) showed that in no case after the $N = 6$ octahedron itself could the minimal growth sequence on this seed compete in binding energy with the same sized isomers of tetrahedral type. Moreover, as the positioning of surface atoms was varied, the central octahedron moved in and out of the regular and deltahedral configurations. (Note that the graphs for the figures just described nicely distinguish the changes in neighbor relationship involved.)

We have gone into detail here first to give an idea of the geometrical subtleties that can occur with even the relatively low-dimensioned configuration spaces and must be multiplied enormously as the size of cluster increases, and second, to make the point that the growth of the bulk f.c.c. lattice for a Lennard-Jones substance is in no sense preconditioned by simple geometrical factors operating between a few atoms. In fact, whatever thermodynamic factors may supervene in the real crystallization process, there is no doubt that the statics of this tend to steer the growing units into noncrystallographic forms such as the icosahedron.

It is possible to trace the *minimal growth sequence* on the tetrahedral seed considerably beyond the $N = 13$ level provided that careful optimization of structures is carried out at each step. We refer to Ref. 61 for the geometrical details of this process and tabulations of the minimum potential energies discovered. Briefly, it was found possible to construct compact and economical structures of up to some 100 atoms which remained at an advantage in binding energy over simple spherical clusters such as those of Burton and McGinty over the whole range. The interiors of these clusters proved to have relatively undifferentiated tetrahedral nearest-neighbor topology, although certain motifs of distorted, interpenetrating icosahedra could be distinguished near the center, and there were marked distortions at the periphery due to a build up of the tetrahedral space-filling deficit. This type of structure, which in a sense represents the most uncompromising deviation possible from crystallinity, has become known as the polytetrahedral type.

4. *Structural Designs*

One positive aspect of the previous discussion could be said to offset the futility of searching configuration space globally, or of ever determining whether the lowest of a sequence of known minima is actually absolute. This is the possibility of disproving the absoluteness of any minimum by means of a counterexample or, more generally, using an energetic *variation principle* to improve on any structures already known. If, by whatever means, we can arrive at structure B which has manifestly greater binding energy than some other one A, then at least we can eliminate A from being absolutely minimal and draw definite physical conclusions as to B's advantage over A in thermodynamic processes—whatever additional minima C, D, and so on, may happen to exist. We should, of course, prefer to make the comparison in terms of free energy of formation at finite temperature, with or without quantum effects, but even in the absence of dynamical information, much can still be learned in this way.

In practical terms, the existence of a variation principle means that we can accept the conscious *design* of structures as a valid method, taking advantage of whatever intuition and geometrical cunning can be brought to bear. There is indeed a long tradition of this both in crystallography and the related folklore of sphere packings and space tesselations.[64,78,79]

Influenced by the experimental discovery of fivefold symmetries in fine particles, a number of designs were proposed for the construction of the small fivefold packing nuclei which were supposed to be the origin of the much larger observable particles. In particular, Bagley[80,81] published sphere packing designs of extendable pentagonal bipyramids with nearly close packed faces, and Fukano and Wayman[82] considered in greater detail how such structures might actually form by stepwise growth sequences. These were, however, pure hard-sphere packing experiments in which no energy factors other than the maximization of point contacts were taken into account, and softness in the potential was appealed to only as a means of distributing the strain in the inevitable clearance regions.

What proved to be a much more important family of structures was demonstrated by A. L. Mackay in 1962.[83] Mackay designed the series of icosahedral packings illustrated in Fig. 5, which, of all spherical structures, certainly come closest to the sphere and moreover achieve this with virtually close-packed surfaces. They can be described as being made up of 20 close-packed f.c.c. tetrahedra sharing a single vertex which is in turn the center of an innermost $N = 13$ icosahedron. To preserve overall icosahedral symmetry, a small distortion is required (ratio edge/radius = 1.051), but this strain, which is easily accommodated by a real potential, does not grow disproportionately with increasing shells, since new atoms appear regularly to neutralize it. Mackay pointed out that these structures can be formed by easy continuous deformations of the f.c.c. cuboctahedra of the same number of atoms (i.e., Fig. 1*d* to *e*), thus anticipating the possible instability of the $N=55$ case discussed in the preceding section. Simple geometrical considerations (see the appendix) give the icosahedral numbers as

$$I_n = \tfrac{1}{3}(10n^3 - 15n^2 + 11n - 3)$$
$$= 1, 13, 55, 147, 309, 561 \ldots$$

The Mackay icosahedra have been rediscovered on several occasions and frequently appear in experimental papers with the title "multiply twinned" structures. Far from being merely a crystallographic curiosity, they have now been shown beyond doubt to be the dominant motif in the growth process of rare-gas microclusters in the important size range of $N=50$ to $N=1000$. Conclusive evidence for this is now available from both dynamic computer simulations and the interpretation of electron diffraction patterns from argon cluster beams.[20,21]

In the optimization study already referred to, Hoare and Pal located precisely the $N=55$ Mackay minima for Lennard-Jones and Morse ($\alpha=3$) potentials.[60] The binding energies are $E_0/N=5.076$ and 7.534, respective (pair energy units). The surface bonds are slightly in compression for the Lennard-Jones structure and more so for the Morse structure as a result of the increased softness and range of this potential. A very slight convexity of the surface planes can be detected. Internally, a compression of the radial bonds is observed, increasing markedly toward the center and reaching 0.934 and 0.770 pair distances, respectively, for the innermost bonds under the two potentials. This is in effect the self-compression of a structure under its own rudimentary surface tension, which, for such a small cluster, quite overcomes the bulk tendency for surface to be slightly distended. However, the degree of compression observed, particularly with the Morse potential, is disturbing and points to the inadequacy to taking a potential derived indirectly from lattice measurements and applying it indiscriminately to cluster equilibria. (The use of an $\alpha=1.5$ Morse potential, regarded as quite normal for f.c.c. metals,[84,85] leads to quite absurd compressions approaching 50% in the central bonds at $N=55$.) The resolution of this problem would seem to lie in the introduction of some form of volume-dependent interaction when modeling metal clusters—a possibility well recognized in the computation of crystal defect statics, although not so far implemented in cluster studies. We return to aspects of the Mackay structures in the later sections on thermodynamics and crystal growth. In view of their central

Fig. 5. The first five Mackay icosahedra. $N = 13$, 55, 147, 309, and 561, respectively.

importance to simple cluster mechanics, we have in addition set out some of their geometrical and numerical properties in the appendix.

More recently Barker and Hoare have designed a number of more complicated structures of icosahedral symmetry which are close competitors for the absolute minimum role, particularly in the gaps between the "magic" icosahedral numbers I_n above. The most interesting and beautiful of these is perhaps the $N = 115$ structure (Fig. 6), which is *rhombicosidodecahedral* in its external form (See Toth[78] for geometrical terminology.) Originally arrived at in an attempt to construct the "icosahedron of icosahedra" by interposition of octahedra between $N = 13$ units, it was found on completion to be none other than a $N = 55$ Mackay structure with a further shell of atoms added *at stacking fault positions* with respect to the surface planes. Its surface, which is slightly in tension, can be decorated with further atoms to give other economical structures in the $N = 100$ to 150 size range. Another Lennard-Jones minimum that has been precisely determined is the Lennard-Jones rhombic triacontahedron (Fig. 7). The binding energy of this compares favorably with the nearby Mackay structure at $N = 149$, but, unlike the latter, it does not seem capable of systematic growth.

Much larger structures than those described have been constructed which retain overall icosahedral symmetry without the shell-by-shell

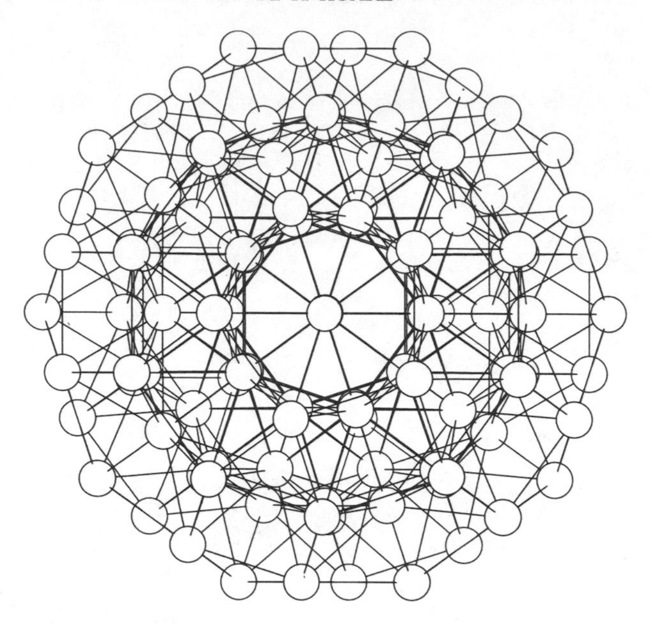

Fig. 6. The $N = 115$ rhombicosidodecahedral structure at the Lennard-Jones minimum. $E_0/N = 5.66$ pair-energy units. The square motifs on the surface have "bonds" slightly in tension; the triangles are slightly in compression relative to the pair distance.

growth and multiply twinned character of the Mackay series. These are in a sense further from the crystalline state and yet in a local manner highly ordered. Their existence seems to be of more importance for the theory of extended amorphous phases than that of isolated clusters, and we do not describe them further here. Details are found in Refs. 72, 86, and 87.

5. *Dynamic Minimum Searches*

We described earlier how dynamic methods could be used to find minima, the system simply being allowed to run away under its Newtonian equations of motion unitl a peak of kinetic energy is obtained, then cutting this off and repeating the process. Although guaranteed to produce some minimum or other rather quickly, this method has the disadvantage that the descent is almost totally haphazard and the final minimum is stumbled on, as it were, without particular reference to starting point or relative potential energy. A much more interesting variant of this can be carried out in which the kinetic energy is removed "adiabatically" by gradual increments over a long phase-space trajectory, the system thus being forced to cool progressively onto ever-lower total energy surfaces. Experiments under these conditions indicate that the

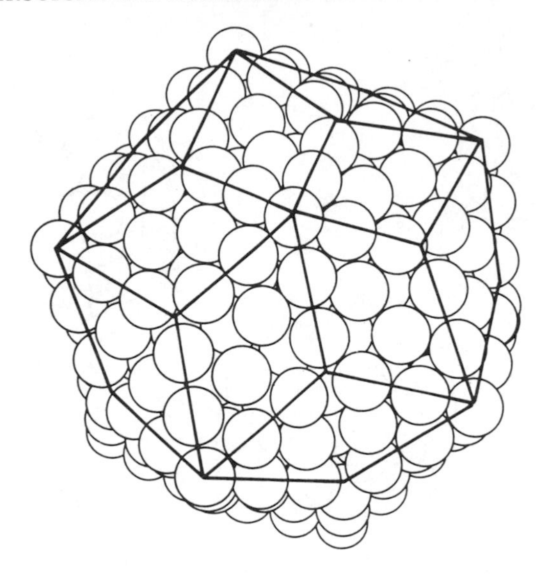

Fig. 7. Rhombic triacontahedral structure of 471 atoms at Lennard-Jones minimum. There are 13 internal icosahedra. $E/N = 6.47$ units.

eventual outcome of such a run, when the phase point becomes trapped in a region of localized motion, is the discovery of one of a set of minima very close to the absolute one. Very slow cooling and hence long, expensive runs appear to be necessary for satisfactory results; in fact, only one group among the many currently interested in computer simulations appears to have used it with success.

The work of Farges, de Feraudy, Raoult, and Torchet[20–22] provides some of the most dramatic results ever achieved in a few-body computer experiment. We describe it here, rather than in the later section on molecular dynamics, because it is immediately relevant to the previous discussion on growth minima in the important $N > 50$ range. In the terms just described, these authors cooled argon clusters from an initially liquid condition until the last vestiges of delocalized motion were seen to have disappeared, whereupon the average atomic positions over a sufficiently long trajectory were determined. (This proves virtually equivalent to a true minimization, provided the region of harmonic vibrations has been reached.) Strong regularities in the pair distribution statistics were observed, and when graphical renderings of the quenched structures were printed out it was clear that a form of crystallization had taken place. The drawings, reproduced here by permission, speak for themselves (Fig. 8). Although at first sight some of the structures appear rather featureless,

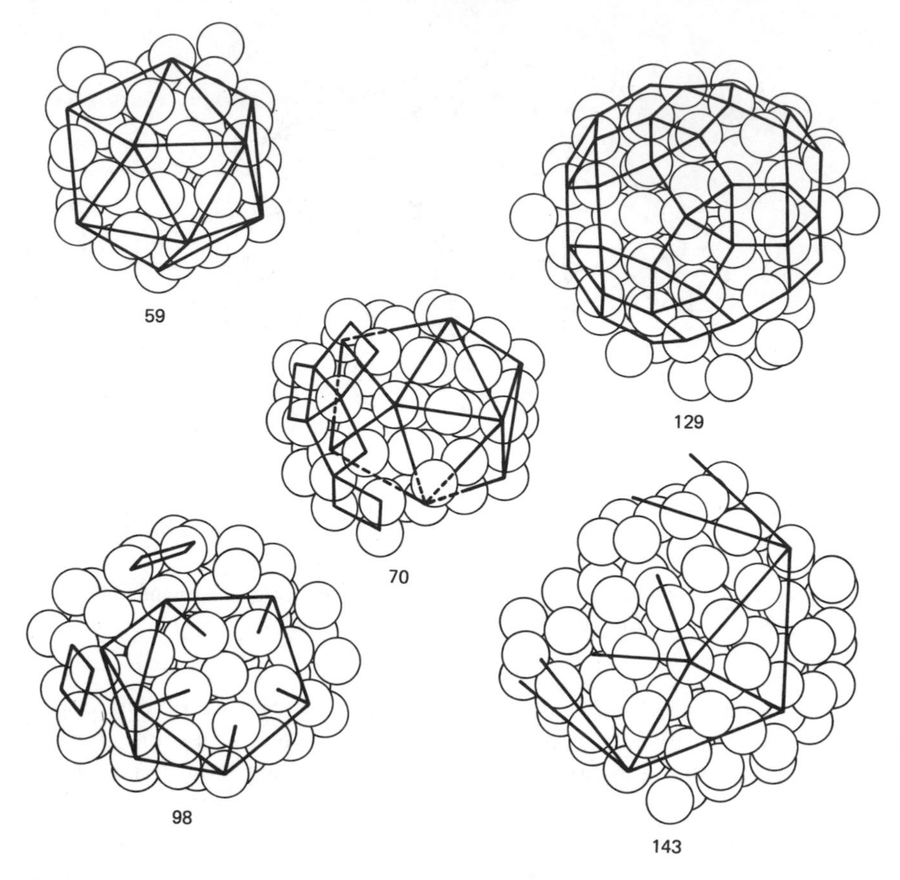

Fig. 8. Structures produced by molecular dynamics cooling of Lennard-Jones liquid droplets. Various motifs are visible, all related to the $N = 55$ icosahedron. (Results of J. Farges, M. F. de Feraudy, B. Raoult, and G. Torchet. Laboratoire de Diffraction Electronique, University of Paris-Sud, Orsay. Reproduced by permission.)

careful examination reveals that each possesses a core region of characteristic icosahedral symmetry, decorated by additional atoms in a more or less unsystematic way. In the diagrams the cores are indicated by polyhedral constructions. (Although it was impossible to do multiple runs for statistical comparison, there seems little reason to doubt that the cases drawn are truly representative.)

In the successive emergence of the several polyhedra—first the Mackay 55-icosahedron, then, by decoration of its surface in stacking fault positions, the 115-rhombicosidodecahedron, and finally the 147-icosahedron—we recognize the very structures which had earlier been designed and tested in static packing studies. Seeing them form "spontaneously" from computed liquid drops, we cannot reasonably doubt their importance in the formation of real microclusters, at least under conditions in which a liquid microdrop cools by contact with a homogeneous carrier gas or by evaporation.

C. Radial Distribution Statistics

We have decided not to take up space here with diagrams of the very interesting radial distribution statistics and interference functions that can be obtained from the geometrical coordinates of various clusters discussed in this section. These have been studied primarily in the experimental groups concerned with cluster beam data, and a considerable expertise has been built up for inverting Debye-Scherrer diffraction data to give structural particulars. A meticulous account of this process is found in Ref. 20.

The obviously noncrystalline character of structures such as the polytetrahedra and the icosahedral forms translates into definite recognizable features of the radial distribution $g(r)$ curves, even when the significance of these is limited by the finite size of the sample. Most notable is the splitting of the second peak of the radial distribution known to be a characteristic of amorphous packings in the bulk and the subject of much discussion in the physics of amorphous metals.[88] The origin of this feature is usually understood in terms of the disappearance of characteristic *octahedral* distances and their replacement by tetrahedral arrangements.[89] The resemblance between microcluster statistical geometry and that of metallic amorphous systems has led to suggestions that the latter may possess more highly ordered local structure than the simple Bernal-type packing would imply. Briant and Burton have emphasized the role of the 13-icosahedron as a dominant motif[90,91]; this author has argued in favor of much larger structures in the form of definite "packing amorphons" such as the 115-rhombicosidodecahedron (Fig. 6).[36] This type of medium-range structure would lead to a model of greater explanatory power, particularly in relation to the glass transition.[87,92]

D. Statistical Morphology at 0 K

From the time of the earliest studies in cluster physics, the problem of the multiplicity of diverse N-minima has led to a number of uncertainties—which most workers have been content to put aside in the knowledge that it would be far from easy to use the full details of isomeric minima even if they were to become available. In qualitative terms, the effect of this neglect may be described quite clearly. If we insist on describing the whole multiplicity of cluster types at given N by a single representative structure, then at the very least we neglect an important factor in the entropy of the condensed phase relative to the vapor and thus inevitably tend to *under*estimate the tendency for the condensation to occur. In certain cases the consequences of this "best single cluster" (BSC) approximation may be numerically insignificant—but without at least some exploration of the problem we can hardly assert this with confidence.

Although in some earlier papers descriptions of alternative minima were given,[61,93] the only systematic attempt to enumerate *all* existing minima for given values of $N>6$ is that recently completed by J. Mc Innes.[94–96] Using the vocabulary developed in Section III.A.2, we can describe the basis of this project quite concisely. Mc Innes set out to both enumerate and identify that subset of all N minima ($N=6,7,8\ldots$) which are generated by *Stable growth sequences* from all possible *primitive seed structures.* The method took the form of a growth algorithm in which *all immediate growth structures* from a given seed were accounted and carried forward—not merely the minimal ones (MIGS) to which our discussion in the previous sections was restricted. The computing procedure could be divided into two main tasks: finding a supposedly exhaustive set of seed structures and carrying out growth algorithms on each of these. We describe the second aspect first. The essentials of the algorithm used are as follows:

1. Select a seed structure of N_0 atoms. (An obvious example would be the regular tetrahedron with $N_0=4$.)
2. Place a further atom at all possible growth sites in turn (in practice the "pockets" at its surface) and eliminate those choices which must lead only to geometrically equivalent isomers.
3. Optimize all the distinct structures of the previous step under the given potential. Check again for geometrical equivalences and record (*a*) new stable (N_0+1) minima supported by the N_0 structure and, (*b*) any new structures arising by rearrangements.
4. Return to step (2) with each of the new structures thus obtained.
5. And so on as computing time allows.

This procedure must then be repeated with as many primitive seed structures as deemed necessary. The selection of the latter could not be made systematic, but, since the possible choices would seem to be very limited, we can with some confidence presume to have found them all. In the Mc Innes study the set of Bernal canonical polyhedra were chosen as the possible seeds,[66] but, since several of these reverted to others in the scheme either on initial optimization or after one or two growth steps, they could be eliminated. The remaining structures were (*1*) tetrahedron, (*2*) octahedron, (*3*) Archemedean antiprism with two axial caps, (*4*) trigonal biprism with three half-octahedral caps.

The original references should be consulted for details of the implementation on the computer and of various tricks that speed the discrimination of geometrically non-equivalent isomers. In spite of the latter, the whole procedure remains laborious and expensive and at present does not seem worth carrying out beyond the level of about $N=13$. A considerable number of distinct minima have nevertheless been found for the Lennard-Jones potential and $6<N<13$, and from this data base a further variety of information, much of it of quite unfamiliar type, can be extracted. A subsidiary study dealt with the Morse ($\alpha=3$) potential. For economy reasons however, the full algorithm was not rerun; rather the original Lennard-Jones minimum configurations were taken and each examined as starting configurations for optimization under the Morse potential. We attempt only a brief survey of the main results on geometry and binding energy—other properties related to vibration and thermodynamics are discussed in Sections IV.A and V.B.

Multiplicities and Pair Potential Dependence. The total count of distinct Lennard-Jones minima (not including enantiomorphs) discovered between $N=6$ and $N=13$ grew as follows: 2, 4, 8, 18, 57, 145, 366, 988 .... These figures are broken down in Table I to show the contributions from tetrahedral, octahedral, and other seeds and the numbers of enantiomorphic minima occurring. Some selections of computer-generated drawings of cluster reconstructions for each minimum are also given in Fig. 9 to 11, these being composed in order of decreasing binding energy. A crude extrapolation of the function $g(N)$, giving numbers of distinct isomers, leads to a fit of the form $g(N)\cong \exp[-2.5176+0.3572N+0.0286N^2]$ from which $g(14)=3279$ and $g(15)=10753$. Although the very steep rise in isomer counts with size can hardly be doubted, we can only take the latter figures as a vague indication.

The finding of 988 minima for as few as 13 atoms is at first surprising, but is easily comprehended when the tree-like interrelationship of structure and substructure within the growth sequences is allowed for. Only very few cases of structural collapse were observed with the Lennard-Jones potential, and these were all in the octahedral growth pattern. There are, however, numerous cases of distortion, such as the opening and closing of octahedral motifs changing into dodecadeltahedra. These and many other qualitative features may be discerned on careful study of the drawings. Further structural information is given in the figure captions.

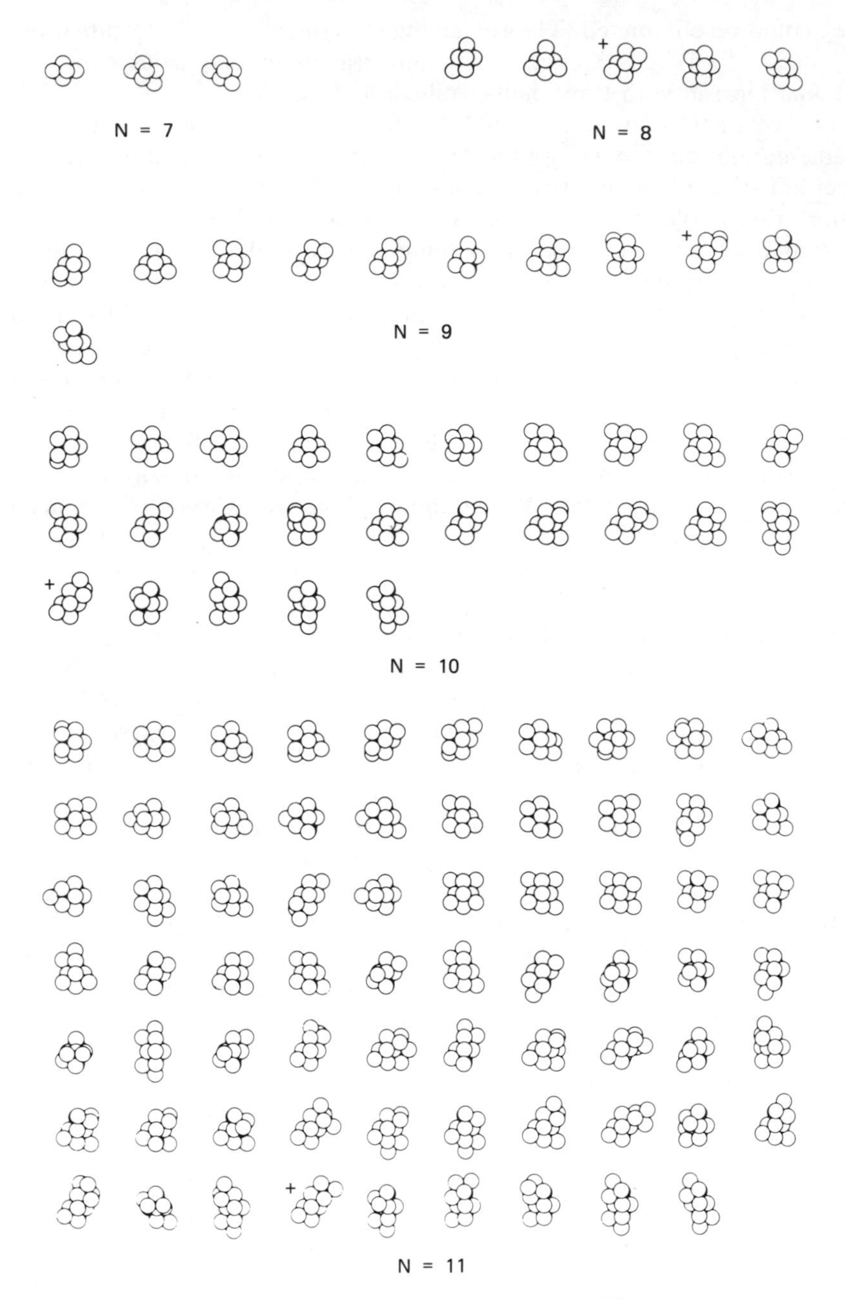

Fig. 9. Tetrahedral Lennard-Jones isomers for $N = 7$ to 11. The structures are accurate Lennard-Jones minima and are reproduced in order of decreasing binding energy, left to right and downward. The Boerdijk spiral configurations are marked with a cross.

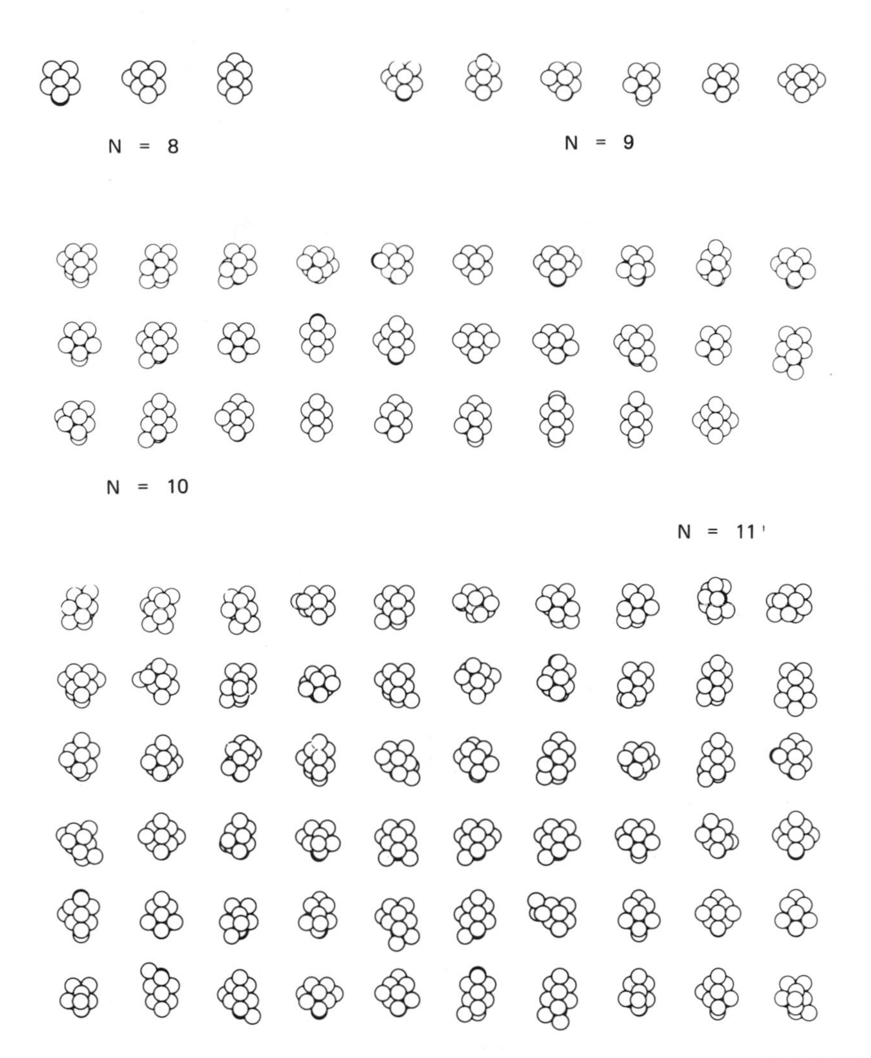

Fig. 10. Octahedral Lennard-Jones isomers for $N = 8$ to 11. Note particularly the deltahedral distortion present in some but not others.

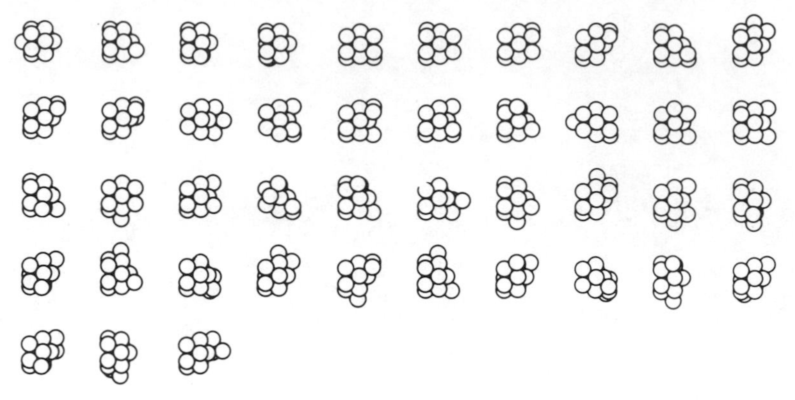

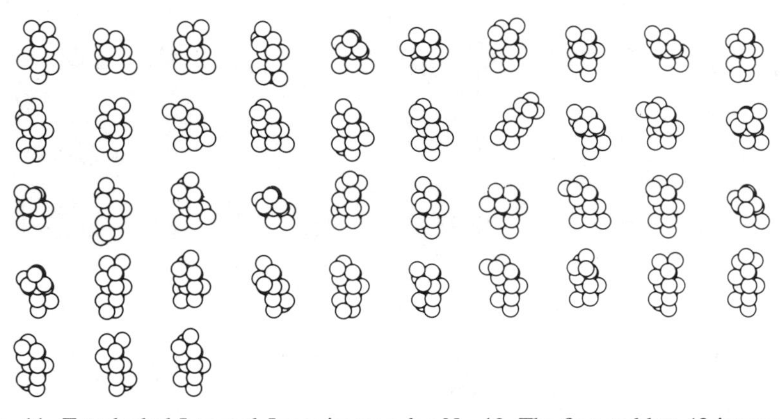

Fig. 11. Tetrahedral Lennard-Jones isomers for $N = 13$. The first and last 43 isomers from the total of 483 are selected, again in order of decreasing binding energy. Note the $N = 13$ icosahedron top left and the Boerdijk spiral, which proves to have by no means the least possible binding energy.

TABLE I
Multiplicity of Lennard-Jones Isomers by Types

N	6	7	8	9	10	11	12	13
Tetrahedral	1	3	5	11	25	69	171	483
Octahedral	1	1	3	6	29	60	143	338
Others	0	0	0	1	3	16	52	167
Total	2	4	8	18	57	145	366	988
Enantiomorphs	(1)	(1)	(1)	(3)	(11)	(19)	(47)	(131)

The most striking single result of the McInnes computations is undoubtedly the sensitivity of the results to the form of pair potential. When the set of Lennard-Jones minima for $N=13$ were tested for stability under the Morse ($\alpha=3$) potential, the vast majority (952 of 988) collapsed and reformed at one of the few stable minima remaining. A similar pattern is repeated at the smaller sizes, as set out in Table II. Again, this result is less astonishing when we see it as the consequence of the tree structure relating successive generations of isomers. It needs only one structural motif at $N=7$ or 8 to become unstable for all derived branches of the tree of minima to be removed. We studied the structures in Fig. 9 and 10 to determine precisely which Lennard-Jones isomers in fact survive under the Morse potential. Close examination of these then revealed the origin of the instability. It is *the opening up of isolated tritetrahedral motifs to give octahedra* rather than (as might conceivably have been the case) the reverse. Thus it would appear that the tetrahedral minima, although in a slight majority, are actually somewhat more precarious than their octahedral counterparts. Some tetrahedral minima do evidently survive, but only as substructures, shored up by further atoms which appear to hold the central unit in position. These findings confirm, in details previously unsuspected, the intuitive feeling that hard potentials give amorphous structures more readily than soft ones. The instability of tetrahedral motifs is, however, a rather different process from the jamming of atoms together in a packed phase.

TABLE II
Total known Isomers for Lennard-Jones and Morse ($\alpha=3$) Potentials

N	6	7	8	9	10	11	12	13
$g(N)_{\text{L-J}}$	2	4	8	18	57	145	366	988
$g(N)_{\text{Morse}}$	1	3	5	8	16	24	22	36

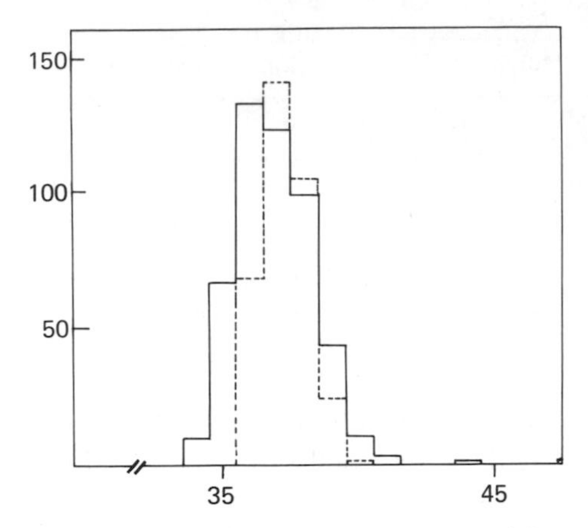

Fig. 12. Statistics of binding energy for the 988 known Lennard-Jones minima for $N=13$. The solid line is for the total sample, the broken one for the octahedral subset. The energy scale is in units of the Lennard-Jones pair energy.

Statistics of Binding Energy. The knowledge of sufficient minima for a statistical analysis of binding energy, rotational parameters,and so on opens up possibilities for entirely new kinds of discussion—one might almost say a new subject of "statistical statics." The histogram of numbers over binding energy (Fig. 12) for $N=13$ shows a slight asymmetry with a pronounced foot at the high energy side leading to a detached point which is the icosahedral minimum. This is well separated from its closest competitor ($E_0=44.3$ compared to $E_0=41.47$). The relatively small variance of binding energy reflects the fact that, in spite of the wide fluctuations in overall shape, none of the structures is large enough for these to bring into play either overall differences of coordination number or second-neighbor contributions.

When analyzed by types, the sample shows a narrowing in the variance of the octahedral subset energies, occurring predominantly in the middle of the range. For the $N=13$ set, the first 14 minima in order of decreasing binding energy are tetrahedral; thus PT structures seem likely to dominate the thermodynamic equilibrium even when the apparent absolute minimum structure no longer does so alone.

Many other statistical quantities can be derived from the files of minimum coordinates. In applications to catalysis it is of interest to know the frequency of occurrence of certain characteristic neighbor distances and perhaps also of certain few-atom geometrical features that might be suspected to have interesting electronic properties. It is also a matter of some importance to know whether clusters such as those depicted can retain their identity when deposited on substrates and, if so, whether they tend to adopt preferred orientations. These are, however, difficult questions only indirectly related to the results we have set out for the free, unperturbed clusters.

We cannot leave the McInnes clusters without posing the one entirely unequivocal question: Do the 988 known Lennard-Jones isomers for $N=13$, augmented by all their geometrically equivalent renumberings and enantiomorphs, actually exhaust the totality of minima which exist in the space $\mathscr{C}_{\tilde{N}}$ under the potential (2)? Formally there can be no question of a decisive proof; informally we can neither convincingly affirm nor deny. Certainly the partition of all minima into those supported by stepwise growth sequences and those requiring cooperative constructions is an interesting and physically suggestive one. We know from the model building studies in the previous sections that larger structures (e.g., the Mackay $N=55$ icosahedron) cannot be assembled stepwise from a primitive seed structure by a minimal growth sequence, and the possibility of something similar in the lower size range cannot be ruled out altogether. A more likely flaw in the counting algorithm is that we have not in fact discovered all possible seed structures in the relevant size range. Although there is nothing logically sacrosanct about the Bernal canonical polyhedra, it can only be emphasized that neither his ingenuity nor ours seems to have led to any additions to the list. Moreover, we can assume with some confidence that any unknown seed structures will occur, if at all, in the $N=12$ region and above, not at the earlier stage, which would lead to numerous branches of supported minima.

With this somewhat partial summing up, we simply admit to a certain confidence in having found "nearly all" the Lennard-Jones minima to $N=13$ (while at least avoiding the philosophical booby trap of saying that there are "probably 988").

A good many open questions remain in the subject of minimum enumerations, but it is wise to keep in mind the somewhat limited range of physical importance some of them have. It would not be of much use, for example, to know the full energy histogram for *all* the minima at, say, $N=50$, even if this could somehow be deduced. Of much greater interest would be the high-energy tail and details of the spacing of the first few isomers above the absolute minimum. Burton[97,98] has concentrated on this aspect of the isomer problem for certain of the spherical f.c.c. isomers where it is relatively easy to count the structures resulting from the displacement of one, two, and three atoms and so on at a time.

The last comment we make concerns the possible use of graph theory as an aid in isomer enumeration. Since we can associate with each minimum a unique nonplanar graph and since there exists some hope of enumerating such graphs systematically at least to low orders, we might suppose that graph theory could supply upper bounds to the number of possible minima for given N if not actual trial structures for computation. In practice the scope for this is extremely limited, because a graph can easily include totally nonphysical adjacency conditions (e.g., three atoms all tetrahedral to the same triangle) which are very troublesome to recognize and eliminate. However, some current interests in applied graph theory, for example, the enumeration of lattice "animals," have elements in common with the isomer problem and may provide a connection to be exploited even if only to obtain inexact estimates.[99] Further details of the graph-theoretic aspect of cluster enumeration are given in Ref. 95.

IV. STATISTICAL DYNAMICS

As emphasized earlier, there is a two-way interplay between the static and dynamic aspects of microcluster mechanics. We can, as illustrated in the preceding section, use dynamic calculations to probe the N-atom potential energy surface, and at the same time our qualitative knowledge of such surfaces can lead to useful insights about the dynamics in particular energy regimes. We begin this chapter with a brief account of some of the more important qualitative aspects of cluster motion, which at the same time serve to classify the regions in which it is possible to make some progress by analytical means. A number of conjectures are made, some of an obvious nature, others less so and inviting confirmation by computer simulations. As always, we must be thankful for whatever simplifying features can be grasped in the general complexity of N-body dynamics.

A. Qualitative Aspects of Dynamics

Conventionally we represent the natural motion of an N-atom system by the trajectory of a point in *phase space*, the $6N$-dimensional space $\mathscr{F}_N = \mathscr{C}_N \otimes \mathscr{P}_N$ composed of the product of a full configuration space $\mathscr{C}_N$ and a similar space $\mathscr{P}_N$ of momentum variables $\{p_{1x}, p_{1y}, p_{1z} \ldots p_{Nx}, p_{Ny}, p_{Nz}\}$. By working in center of mass coordinates and restricting to rotationless systems, we can simplify to the reduced phase space $\mathscr{F}_{\hat{N}} = \mathscr{C}_{\hat{N}} \otimes \mathscr{P}_{\hat{N}}$ in $6N-12$ dimensions. When only velocity-independent potentials are involved, it is natural, and indeed instinctive, to project mentally the notion of the phase point onto the configurational part of $\mathscr{F}_{\hat{N}}$ alone and "see" it as moving about the surface $V(\mathbf{r}^N)$ in a somewhat haphazard manner. In these terms certain essentials can be perceived. The motion of lowest energy available to a cluster consists of a bounded orbit within the catchment area of the absolute minimum configuration $V(\mathbf{r}^N_{\min})$. These motions are arbitrarily close to harmonic as kinetic energy approaches zero, corresponding to elliptical torii in phase space. Similar harmonic motions occur if excited in the neighborhood of any local minimum of the cluster. As the kinetic energy, and hence the amplitude of motion about a minimum, increase, a number of possible thresholds may be reached. A degree of anharmonicity sets in and may lead to fluctuation of the nearest-neighbor topology at a certain amplitude. This in itself does not have direct dynamic significance, but usually indicates the approach to a saddle point in the immediate surroundings. Possession of a certain minimum *activation* energy may be presumed to allow the phase point to pass the saddle into another catchment area where, depending on the topography on the other side, it may move into motion about a minimum of similar depth to the first or remain excited at relatively high potential and low kinetic energy for an appreciable period. Some clusters move in this way between geometrically equivalent minima symmetrical about the barrier. However, even in this apparently simple regime, all is not quite as it seems to be. It is conceivable that the possession of kinetic energy in excess of the height of the saddle above the minimum is *not* a sufficient condition for the phase point to reach it—we must assume the system to be *ergodic*, at least within the relevant energy limits. The problematics of this would lead us too far afield, although we return briefly to the topic in the next section.

For now let us be content with the conjecture that the onset of delocalized motion at low energies is associated with the lowest of all adjacent saddle points and the two minima that neighbor it. Study of simple cases suggests that this saddle point is necessarily *simple*, having only one negative principal radius of curvature. (Two of the minima for $N=7$ are reproduced in Fig. 13 in such a way that the intermediate saddle points are not difficult to visualize. The heights of some of these have been determined numerically and are reported elsewhere.[100]

Our next conjecture is somewhat less obvious. If any two minima are neighboring in the above sense, we believe the passage between them to be low dimensional in the collective coordinates defined with respect to either. This requires careful explanation. First we note that, in classfying types of saddle point motion, there is a distinction between simple motions, such as the riding of one surface atom across its neighbors to a nearby site, and the more complicated situation in which an internal atom forces its way through a constriction, or perhaps a number of atoms readjust cooperatively. We can speak qualitatively of the dimension of a saddle in $\mathscr{C}_{\hat{N}}$ as the rough number of coordinates which change effectively in a short trajectory of steepest descent passing through it. In somewhat the same way the easiest path between a minimum and a neighboring saddle can be given a dimensionality—but it is clear that many "easy" saddles are of dimensionality comparable to N. (Consider, for example, the passage between two $N=13$ icosahedra via the cuboctahedron saddle point.) Our conjecture is that, if the motion is reformulated in the normal coordinates $\{\xi_i\}$ about either of the minima, then it takes place substantially in the direction of one particular normal mode. Geometrically this means that, when a saddle is adjacent to a particular minimum, there is in general a line of low curvature between them, and high curvature with steep rise of potential in all orthogonal directions. Moreover, the passage is direct, without bends in the approach valley (a condition depending on the general smoothness of the central force potential and by no means satisfied on more complicated potential surfaces such as those occurring in unimolecular chemical reactions).

We would add to this a third conjecture to the effect that the greater the relative depth of a minimum with respect to its neighboring saddle(s) the greater the separation of curvature between the easiest passage and the orthogonal directions. Absolute minima can thus be expected to be extreme in this respect also; conversely, high-lying minima tend to be precarious, lying in broad, shallow regions on high plateaux. Here it is important to stress that the softmode postulated to lead toward a saddle is necessarily a geometrical one and its frequency may need to be distinguished from others whose low value is primarily a result of the mass factor A_i for the mode concerned. The simpler form of low dimensionality, that in $\mathbf{r}^N$, may nevertheless also occur. Note further that the above description modifies naturally to admit multiple saddles and consequently degenerate directions of low curvature resulting from symmetry in the structure considered.

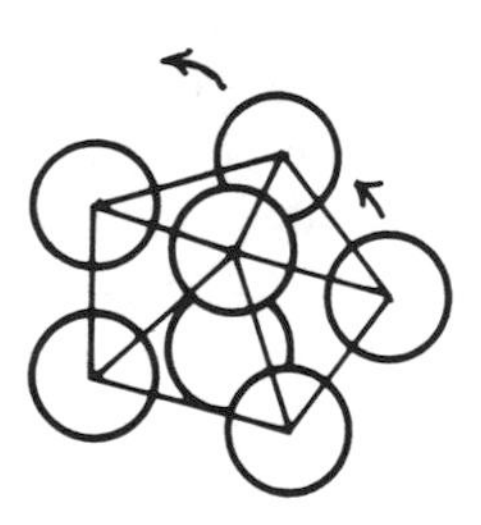

Fig. 13. Illustration of a saddle-point motion for $N=7$. By the cooperative motion of two atoms the pentagonal bipyramid transforms to the isomer composed of four face-to-face tetrahedra in staggered configuration. (The structure on the right is not a relaxed minimum as drawn.)

With these conjectures we leave the energy regime in which motion is effectively localized and consider the liquid-like state in which delocalized motion predominates. Several key qualitative questions arise. We may wonder how sharp is the transition from solid-like to liquid-like motion and whether even quite small clusters are able to mimic a sharp melting transition to a significant degree. It is also of importance to know whether the onset of liquid-like motion leads immediately to appreciable evaporation, or whether such states can in fact persist with almost negligible tendency to decay. How well defined, indeed, is the distinction between solid-like and liquid-like states, given that the phase point may well fluctuate up to visit a high-lying minimum region and remain temporarily trapped there before returning abruptly to the liquid region of phase space?

Clearly these questions are all one aspect and another of the general problem of describing how phase space is opened-up for the growth of entropy with increasing total energy E. Our difficulty in answering them is further compounded by the fact that, except at very low energies—on the order of a single pair energy—all clusters are dynamically metastable with respect to evaporation of one or more atoms. Subject to the usual caveat on ergodicity, the phase point can always be expected eventually to reach one of the many dissociation channels available to it, whereupon the motion effectively factorizes into that of a free particle of one or more atoms and a smaller cluster of lower energy. Since the dissociation energy is on the order of three pair energy units, the remaining cluster, although in all probability metastable, is considerably cooled by loss of a single atom. It is believed that the observation of predominantly solid-like argon clusters in nozzle-beam experiments is primarily the result of a fast cooling by successive evaporations from original liquid-like nuclei rather than a direct consequence of the mechanism of formation.[20] Similar considerations of the effect of single atom events—adsorption/desorption or chemical reaction—on the temperature and state of microcluster arise in the theory of processes at the surface of interstellar grains.[30]

At the present qualitative level we can still formulate a number of key questions about the dissociation region. It is usually assumed that, of the many decomposition channels available to an N-cluster, those leading to the loss of a single atom are overwhelmingly more effective than others for the ejection of dimers; whereas the probability of complete fission to fragments of comparable size is usually taken to be completely negligible at energies where single-atom evaporation is fast. If true, as seems likely on general grounds, this effect would be ascribable to both energetic and entropic factors, the great excess of phase space in single-atom outgoing channels combined with the lower energy requirement for escape. (A little-known paper by Cohen[101] discusses this effect in some detail, comparing the tendency of atomic and molecular microclusters to decay by single-particle channels, with the radically different behavior of atomic nuclei for which the preferred mode of decay frequently involves compound (alpha) particles. The difference, he suggests, is accounted for on a purely phase-space grounds, once the Pauli exclusion principle is correctly allowed for in the nuclear case.)

In spite of these complications, the difficulties introduced by dynamic metastability are less acute than it might seem. It is, after all, nothing new to compute thermodynamic and mechanical quantities for metastable phases and chemical compounds in the knowledge that their validity is conditioned by some time scale outside which either time or ensemble averages become meaningless. This, as we emphasized in the opening paragraphs, is in effect a precondition for our freedom to speak of a microcluster as a well-defined entity in the first place.

More serious perhaps is the almost complete lack of any practical theory of the lifetimes of microclusters with respect either to internal transformations or dissociation. This is understandable inasmuch as nucleation theory has been able, through various simplifying assumptions, to bypass the problem of cluster lifetimes. On the other hand, it is surprising in view of the fact that there was an early, fully acknowledged connection between the Weisskopf compound nucleus theory for beta decay and the Frenkel liquid drop theory of the nucleation process.[102,103] The work of Slater, in particular the so-called Slater New Theory,[104] is likewise relevant to the cluster lifetime problem and represents one of the few sources in print to give both a clear and penetrating account of the subtleties of the few-body problem for bound systems of classical particles at dissociative energies. We have made some progress toward a rudimentary lifetime theory using both the original Weisskopf formulation its near relative the Eyring theory of unimolecular processes.[105] Here we simply draw attention to the obvious parallels between cluster mechanics and molecular rate-process theory, extending through our description above and encompassing many analogous concepts such as activation energy, reaction coordinate, entropy of activation, and so on.

With these preliminaries, we return to considering what can be done in the realm of practical computations. These come under several headings, the main distinction falling between analytical treatments—virtually confined to simple vibrational and rotational analysis—and computer simulations of Monte-Carlo (MC) and molecular-dynamics (MD) types. The remainder of this section is concerned mainly with recent progress in the vibrational problem; a detailed account of the results of computer simulations follows in Section VI.

B. Classical Vibrational Analysis

Once a minimal structure has been determined by whatever method, it is a relatively simple matter with modern computers to obtain vibrational frequency spectra at least up to sizes on the order of $N = 150$. We note that, even with the full potential function (2) and a pair potential such as the Lennard-Jones, the second partial derivatives and hence the force-constant matrix elements, are not difficult to write explicitly when a Cartesian system is used. There is little or no advantage with large computers in using internal bond-coordinates or in eliminating the six zero-frequency modes (see especially Gwinn[106]).

The first explicit vibrational analysis of microcluster models was carried out by Reed,[107] who obtained some limited frequencies for a few simple, conjectured Lennard-Jones minima up to $N = 10$ and used them to estimate thermodynamic properties. Some years later there followed a number of studies based on the spherical family of f.c.c. microcrystallites and using in effect what we earlier called the cowgum potential. Thus Nishioka, Shawyer, Bienenstock, and Pound[56] considered microcrystallites of sizes $N = 13$, 19, 43, 55, 79, 87, 135, 141, 177, 201, 225, and 249 and, by careful choice of shape, computed normal mode frequencies on the assumption of a single harmonic spring constant for the bulk and another for surface planes. Overall distributions of normal-mode frequencies were obtained and used in estimates of cluster free energies. Abraham and Dave,[57] at about the same time, developed an elaborate form of the Einstein model for microcrystallites, taking into account the effect of coordination number on the Einstein frequencies for atoms in a consistent way, thus simulating in an approximate way the increase in entropy due to large-amplitude vibrations of surface and edge atoms. The use of a relatively crude model gave the advantage that much larger clusters than otherwise could be calculated, leading to useful predictions up into the N-500 range.[108]

Shortly afterward Burton[58] and McGinty[51] made the important advance of analyzing microcrystallite clusters both with all-neighbor interactions and an effective optimization under the Lennard-Jones potential. The results of optimization have been referred to already: their vibrational analyses were likewise the first to apply to realistic structures and force fields.

Later Hoare and Pal[61,62] were able to carry out systematic determinations of spectra for the series of polytetrahedral structures, comparing these and their derived thermodynamic data with the Burton and McGinty results. A few of the Hoare-Pal frequency spectra are illustrated in Fig. 14. At small cluster sizes not much can be said about the distribution of frequencies, and crude histograms must be interpreted

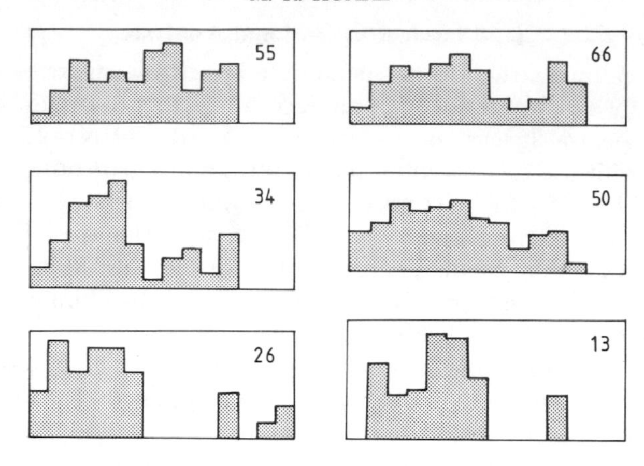

Fig. 14. Normal mode frequency histograms for various Lennard-Jones structures. The structures illustrated are $N = 55$ icosahedron, polytetrahedral forms for 66, 50, 34, and 26 atoms, the $N = 13$ icosahedron. Results were obtained in the nearest-neighbor approximation.[121] Later work shows that the low-frequency peak for $N = 13$ is reduced considerably when all-atom interactions are taken into account.

with caution. Various tendencies are at work in determining mode distributions. Higher frequencies than the pair frequency occur because of the multiplicity of bonds being stretched in uncorrelated motions of near neighbors, and also because longer range interactions compress the innermost regions, with consequent stiffening-up of the structure. At the same time, however, an increase in the number of atoms brings a corresponding growth in the incidence of lower frequencies due to torsional modes and breathing motions of spherical structures. At no point in the size ranges so far investigated does it seem reasonable to identify modes with any particular geometrical effect, such as surface or edge motions; where high symmetry is involved the degeneracies do not markedly affect the frequency histograms, which are at most vaguely bimodal. No resemblance to the phonon distributions for bulk f.c.c. argon can be expected, and none is found, unless it be a very slight tendency for the separation into two branches above about $N = 50$. The only systematic feature that might possibly correlate with detailed structure is the high frequency invariably found when an icosahedron is present. These findings are qualitatively confirmed in direct computer simulations.

One can expect that the low-frequency contributions are most likely to vary with structure and with details of the potential function. The spectra in Fig. 14 show some difference in spectra obtained for the $N = 13$ icosahedron first using only nearest-neighbor interactions and those with all interactions allowed. Neglect of longer-range contributions seems to result in a spurious low-frequency component, which could have detectable effect on low-temperature thermodynamic properties. Since several conflicting sets of frequencies now exist in the literature,[109] there is some need for a definitive comparison of vibrational analyses with and without the nearest-neighbor approximation in energy minimization and/or vibrational analysis.

Transition to Large Structures. With particles towards the upper size limit of the microcluster range, the transition from small-system to surface-dominated regime to eventual bulk behavior is a complicated study in itself. A vast body of work exists on the Weyl problem in which the normal vibrations of small, continuous elastic prisms are studied as a function of size and shape (see for example, Baltes[110]). Although detailed computational studies are still somewhat limited, there can be little doubt that microcluster vibrations will progressively imitate Debye-type elastic behavior in the low-frequency region as their size increases, but in the intermediate range a number of competing effects can be distinguished. Baltes, *loc. cit* summarizes much of the recent work on the Weyl problem in the context of small-particle heat capacities and phonon spectra, his work bringing up to date the classical article of Maradudin, Montroll, and Weiss.[111] In addition to collecting a variety of results derived by mode counting in different shaped domains, he recalls the early Planck-Schaffer modification of the Debye heat capacity law which applies when particles are so small that only the lowest possible frequency can be appreciably excited. Suppose that the available modes are ordered by frequency as $\omega_1 < \omega_2 < \ldots < \omega_n$ and that Θ_D is the normal, bulk Debye temperature. Then, in addition to the normal requirement for the validity of the Debye τ^3-law, $T \ll \Theta_D$, the condition that only ω_1 is appreciably excited implies further that

$$T \gg \Theta_0 = \hbar\omega_2 / k_B = \Theta_D/(3N')^{1/3}$$

where N' now indicates the number of atoms required for the operation of the T^3 law. Three regimes for the vibrational heat capacity can then be distinguished:

1. A single-mode regime
2. A Debye regime
3. A Dulong-Petit regime

From the condition just given it follows that regime (1) is observed to reasonably high temperatures with particles in the microcluster size range.

If we take reasonably high to be, say $\Theta_D/5$ it follows that $N' \cong 50$. In this regime the Planck-Schaefer heat capacity law is obeyed:

$$C_V(\tau) \sim T^2 \exp(-\hbar\omega_1/k_BT) \tag{8}$$

Vibrational Spectra and Isomerism. There has been very little systematic study of the relationships between spectra for different minima of the same number of atoms. Bonnissent and Mutaftschiev were the first to attempt this with the few isomeric structures of Lennard-Jones $N=6,7$, and 8.[93] The only investigation dealing with many and widely varied structures is that of Hoare and McInnes based on the Lennard-Jones minimal sets to $N=13$ described in the previous section.

As with our treatment of the energy minima themselves, there is something of a problem of data presentation—we have the complete vibrational spectra for 988 distinct $N=13$ minima, 33 frequencies in each; what now? One might argue that as far as thermodynamic quantities are concerned, the necessary averages will take care of themselves, the spread of frequencies coming naturally into play as the temperature rises. On the other hand, it is a pity to pass over the real content of mechanical information which residues in the spectra themselves and can give insight into the qualitative change of properties with size and the nature of the contrast between cluster and bulk properties.

When we look at tables of frequencies alongside the drawings of minima in Figs. 9 to 11, some broad trends suggest themselves. Compactness and high binding energy tend to go with high frequencies and a possibility of degeneracy; extended structures with exposed atoms and low binding energy tend to give low frequencies and little degeneracy. We quote a brief table of frequencies for Lennard-Jones isomers to exemplify these points and a make a number of finer distinctions.

Table III shows several sets of normal mode frequencies for different Lennard-Jones isomers in the $N=2$ to $N=13$ size range. They were computed by J. McInnes[95] using the Gwinn[106] method, taking all neighbor interactions into account. The figures quoted are scaled and can be converted to real-time frequencies for the rare gases on multiplying by the factors listed below. The first four entries give the dimer to the bipyramid, which are unique, and the next two are the only alternative isomers for $N=6$, the octahedron and the tritetrahedron. There follow two entries for $N=7$ and $N=13$ which give the isomers highest and lowest in binding energy, respectively. In these instances the stablest structures are of course the pentagonal bipyramid and icosahedron.

These few figures are sufficient for a number of the most important features of the spectra to be perceived. We note in particular (*1*) the

TABLE III
Vibrational Frequencies of Rare-gas Lennard-Jones Isomers

N	
2	1.91
3	1.65(2) 2.34
4	1.35(2) 1.91(3) 2.70
5	1.06(2) 1.59 1.87(2) 1.94(2) 2.52 2.79
6 (oct)	1.34(2) 1.41(3) 2.00(3) 2.45(3) 2.79
6 (tritet)	0.73 1.00 1.45 1.47 1.60 1.83 1.98 2.00 2.15 2.38 2.75 2.87
7 (pentag)	1.05(2) 1.47(2) 1.50 1.60(2) 1.97 2.19(2) 2.22(2) 2.81(2) 2.84
7 (highest)	0.66 0.81 1.10 1.43 1.56 1.64 1.66 1.85 1.98(2) 2.17 2.45 2.57 2.83 2.97
13 (Icosa)	1.15(5) 1.56(4) 1.70(5) 1.83(3) 2.15(3) 2.16(4) 2.49(5) 2.69 4.41(3)
13 (highest)	0.37 0.40 0.44 0.63 0.78 0.84 1.03 1.17 1.23 1.35 1.43 1.56 1.62 1.66 1.69 1.82 1.86 1.90 1.95 1.97 2.04 2.08 2.25 2.39 2.43 2.49 2.66 2.71 2.85 2.96 3.04 3.29 3.54

Ne 3.93×10^{11} Ar 4.13×10^{11} Kr 3.22×10^{11}
Xe 2.57×10^{11}

emergence of a high frequency, 2.8, as soon as complete tetrahedral units are present and its persistence in the next few structures, (*2*) the first occurrence of an appreciable soft mode in the $N=6$ tritetrahedron ($\omega=0.731$) (this probably associated with passage to the octahedral isomer), (*3*) the sparse and highly degenerate spectra associated with the D_{5h} and I_h symmetries at $N=7$ and $N=13$ and their lack of any of the soft modes seen in the extended isomer.

It must be remembered that, even at $N=13$, there cannot be any vestige of truly bulk behavior and that, were analysis possible on much larger clusters, it would certainly show a number of new effects such as the gradual emergence of distinct surface modes and low-frequency acoustic modes, tending, for compact structures at least, to imitate the harmonics of the free elastic sphere.

The transition from regime (1) to regime (2) is of particular interest for the larger microclusters with $10^2<N<10^6$. Here the heat capacity is most conveniently written

$$C_V(T)=\frac{d}{dT}\int d\omega D(\omega)\hbar\omega[\exp(\hbar\omega/k_BT)-1]^{-1}$$

where $D(\omega)$ is a mode density function that can include surface and shape contributions and even to an extent anharmonic effects. Further discussion of the determination of $D(\omega)$ is given later. We note that, if $D(\omega)$ is construed as a comb of delta functions at normal mode frequencies, the above quasi-continuum approach reverts to the ordinary quantum heat capacity formula (see Section V.A).

C. Anharmonic Theories and Quantum Corrections

Among the most interesting of recent developments in cluster dynamics is undoubtedly the application of the *self-consistent phonon* method by Etters, Kanney, Gillis, and Kaelberer.[112] A related, although less general application of the same is found in a series of papers by Matsubara.[113–115] This technique, although perhaps of doubtful validity in the strongly nonlinear region associated with structural transformations, is a valuable corrective to the assumption, unquestioned in most earlier work, that most of the truth about small-amplitude thermally excited motions is contained in the semiclassical analysis we have just outlined.

The key to the self-consistent phonon (SCP) method, which is in line of descent from the Grüneisen method for bulk anharmonic crystals, lies in its referring the vibrational motion of a cluster not to the local minimum condition $(\partial V(\mathbf{r}^N)/\partial \mathbf{r}_i)=0$ but to a point in configuration space defined by the vanishing of the ensemble average force:

$$\langle \partial V(\mathbf{r}^N)/\mathbf{r}_i \rangle = 0 \tag{10}$$

The method used to determine the coordinates that satisfy (10) takes account of both the thermal expansion of the cluster at a given temperature and the zero-point motion and its associated expansion. Under strictly harmonic conditions no difference in results can be expected, since the cluster has no effective thermal expansion; however, when realistic potentials are introduced, condition (10) can be expected to lead to a temperature-dependent average displacement which, at small amplitudes, has a stiffening effect on the spectrum.

We follow closely the description in Ref. 112. (A particularly clear account of the one-dimensional form of the method is given by Matsubara and Kamiza[114]).

First, the true Hamiltonian operator for the problem is rewritten in terms of atomic displacement coordinates $\mathbf{u}_i$ relative to the minimum in question. Thus

$$\mathcal{H} = \sum_{i=1}^{N} \left(-\frac{1}{2m} \nabla_i^2 \right) + \frac{1}{2} \sum_i \sum_{j>i} v(\mathbf{r}_i^0 - \mathbf{r}_j^0 + \mathbf{u}_i - \mathbf{u}_j) \tag{11}$$

where $\mathbf{r}_i^0$ and so on are the atomic positions in the classical minimum. A trial Hamiltonian is then constructed in the form

$$\mathcal{H}_h = \sum_{i=1}^{N} \left(-\frac{1}{2m}\nabla_i^2\right) + \frac{1}{4}\sum_i \sum_{j>i} (\mathbf{u}_i - \mathbf{u}_j) \cdot \mathbf{\Phi}_{ij} \cdot (\mathbf{u}_i - \mathbf{u}_j) \tag{12}$$

where $\mathbf{\Phi}_{ij}$ is a tensor quantity to be determined self-consistently. The next step is to consider a trial density matrix

$$\rho_h = \exp(-\beta\mathcal{H}_h)/Tr\exp(-\beta\mathcal{H}_h) \tag{13}$$

corresponding to a trial free energy F_t where

$$F_t = Tr[\rho_h(\mathcal{H} + \beta^{-1}\ln\rho_h)] = \langle\mathcal{H} + \beta^{-1}\ln\rho_h\rangle \tag{14}$$

the last bracket identifying this as a thermal average at $\beta = (k_B T)^{-1}$ over all states of the harmonic Hamiltonian. Interpreting this we find that

$$F_t = F_h + \tfrac{1}{2}\sum_i \sum_{j>i} \langle v(\mathbf{r}_i^0 - \mathbf{r}_j^0 + \mathbf{u}_i - \mathbf{u}_j)\rangle - \tfrac{1}{4}\sum_i \sum_{j>i} \mathbf{D}_{ij} : \mathbf{\Phi}_{ij} \tag{15}$$

where

$$F_h = \langle\mathcal{H}_h + \beta^{-1}\ln\rho_h\rangle \tag{16}$$

and D_{ij} is the thermal mean-square displacement between atoms i and j:

$$\mathbf{D}_{ij} = \langle(\mathbf{u}_i - \mathbf{u}_j)^2\rangle \tag{17}$$

It then emerges that the minimization of free energy with respect to the choice of $\mathbf{\Phi}_{ij}$ and $\mathbf{D}_{ij}$ implies the variational conditions

$$(\delta F_t/\delta\mathbf{\Phi}_{ij}) = (\delta F_h/\delta\mathbf{\Phi}_{ij}) - \tfrac{1}{4}\mathbf{D}_{ij} = 0 \tag{18}$$

$$(\delta F_t/\delta\mathbf{D}_{ij}) = \tfrac{1}{4}\langle\nabla_i\nabla_j v(\mathbf{r}_i^0 - \mathbf{r}_j^0 + \mathbf{u}_i - \mathbf{u}_j)\rangle - \tfrac{1}{4}\mathbf{\Phi}_{ij} \tag{19}$$

Thus the choice of correct $\mathbf{\Phi}_{ij}$ requires evaluation of the thermal average of the Hessian matrix for the pair potential over all displacements.

The actual minimization of the free energy according to (18) and (19) is carried out by diagonalizing the quadratic form on the right-hand side of (12) and forming a Gaussian average over the potential function. The resulting algebraic equations are then solved self-consistently, giving a set of mean displacements and their normal mode frequencies. We refer to the original papers for details. Certain simplifications of the method are possible; for example, Matsubara and Kamiya assume an Einstein approximation and take the high-temperature form for the free energy. They also show the advantages that result from the use of a Morse or a Gaussian potential.

The net result of this formalism is that an equilibrium condition is obtained which in effect takes account of both zero-point and thermal expansion of the cluster and provides the appropriately shifted, temperature-dependent normal-mode frequencies. The disadvantage of the method is its general laboriousness—perhaps an order of magnitude greater than for ordinary normal-mode analysis—and the fact that a good guess for the starting configuration is needed. The method favored by Etters et al. uses their own Monte-Carlo program to obtain classical minima (Section III.B.5) and then refines these by readjusting all bond lengths until the SCP mean force on every atom is effectively zero for the given temperature.

Etters, Kanney, Gillis, and Kaelberer computed spectra and bond lengths for a handful of clusters up to the $N=13$ isocahedron and went on to evaluate thermodynamic quantities. Bond lengths in each case showed the expected thermal expansion, which was, however, not simply increasing as in the bulk crystal and showed rather an abrupt increase at a characteristic temperature (8 to 12 K for argon in range $N=3$ to 13). Over a similar temperature range the frequencies showed somewhat peculiar trends, the most marked of which was the fall in the highest frequency for the $N=13$ icosahedron at the temperature mentioned above. Evidently, for highly compact clusters at least, there is a rather sudden onset of nonlinearity—presumably to be associated with the temperature at which vibrational amplitudes on average reach the point of inflection in the Lennard-Jones pair potential. We comment on the thermodynamic results derived from these vibrational analyses in the next section.

The use of the SCP method by Matsubara and co-workers has involved the Einstein approximation throughout with drastic simplifications of the assumed pair potential and a high-temperature form for the free energy. However, the results are more far reaching in that they lead to definite predictions of melting point and superconducting transition temperature.[115] These authors show that the self-consistent equations for small particles fail to have a solution at some characteristic temperature which they identify with the melting point. This temperature indeed shows the same type of dependence on radius as experimental results for several metals down to some 50 Å, but the onset of melting point depression is predicted to occur at radii some 10 times too low. A rescaling of the theory to eliminate this disagreement can be partially justified.

A different approach to nonlinear cluster mechanics, more closely allied to computer simulation methods lies in the determination of the mode density function $D(\omega)$ described earlier. Computations by this method include those of Dickey and Pascin[116] and Kristensen, Jensen, and Cotterill.[117] See also Esbjørn et al.[118]

The key to the derivation of $D(\omega)$ from computer simulations of cluster natural motion lies in its identity with the *power spectrum* for atomic motions and hence its relation to the velocity autocorrelation function through the Wiener-Kinchine theorem. Thus on defining the normalized velocity autocorrelation function $\gamma(t)$ as:

$$\gamma(t)=\left\langle \sum_i \bar{v}_i(t)\bar{v}_i(0) \Big/ \sum_i v_i(0)^2 \right\rangle \tag{20}$$

with the brackets indicating an ensemble average, we know that

$$D(\omega)=\int_0^\infty \gamma(t)\cos(\omega t)\,dt \tag{21}$$

In evaluating this it is the lower frequencies that carry the greatest inaccuracy because of the limited time available in runs for $\gamma(t)$, and this in turn affects low temperature predictions of thermodynamic properties.

Kristensen and co-workers determined $D(\omega)$ for clusters of 55, 135 and 429 atoms, the former being at low temperature the Mackay icosahedron. A range of temperatures was considered between about 0.1 and 0.4 (ε/k_B), with ε Lennard-Jones pair energy, covering the motion from harmonic vibrations through the anharmonic regime to the liquid-like state. At low temperatures the results, judged from the $N=55$ case, broadly resemble the normal-mode histograms in Fig. 14. As temperature increases, the curves broaden, and above a fairly well-defined temperature, zero frequency diffusive modes begin to appear. When the pair distribution curves for these frequencies are examined, it may be deduced that the dominant contribution to the diffusive modes comes from motions in the surface of the cluster.

Dickey and Pascin[116] have carried out similar determinations of $D(\omega)$ on small f.c.c. crystallites, although only in the solid range. They suggest tentative interpretations for some of the low-temperature peaks in terms of normal surface modes and contributions of edge atoms. By studying units of increasing size, they are able to demonstrate the eventual dominance of bulk over surface modes.

We return to the thermodynamic aspects of this work in the next section and to a more critical account of possible errors in computer simulations in Section VI.

D. Rotational Mechanics

There is little to be said about the rotational mechanics of clusters that is comparable in importance to the vibrational properties just discussed. Once the coordinates of a minimum configuration are known, it is a straightforward matter to obtain the inertia tensor and hence the principal moments of inertia I_A, I_B, I_c. These, together with the symmetry factor σ, are required for the rigid rotor partition function $Z_{rot} = (\pi^{1/2}/\sigma)(8\pi^2 k_B T/h^2)^{3/2} (I_A I_B I_C)^{1/2}$. The determination of σ is a simple matter when the structure corresponds to an identifiable polyhedron with a particular point group, but it is not a simple matter to examine hundreds of isomers in coordinate tabulations to find σ for each.

McGinty, in his paper on the molecular dynamics of clusters,[119] determines the time average of the factor $(I_A I_B I_C)^{1/2}$ during the natural motion at different temperatures and thus takes partial account of vibration-rotation interaction. A proper treatment of this is as difficult as in the case of molecules if not more so, but the effect appears to be small. Certain minima in more extended cluster configurations can undoubtedly be destabilized by rotation, and it is a simple exercise to estimate the angular velocity required to dissociate a peripheral atom from a long cluster. However, rotational factors are as a rule of minor importance in statistical mechanical terms once the size of the cluster exceeds about $N = 10$.

V. THERMODYNAMICS

Although a number of specific problems—notably the prediction of nucleation rates and crystal growth mechanisms—have tended to dominate thermodynamic calculations on microclusters, this enterprise can be seen in somewhat larger terms as part of our need to understand better the general characteristics of small-system behavior, especially as concerns the competing role of surface and volume contributions. Where specific problems are concerned, the calculations required may overlap in many particulars, although the rules of the game depend somewhat on the objective in mind. Thus, for example, in seeking to predict mucleation rates, it seems reasonably legitimate to consider artificial cluster structures which, although possibly nonexistent in nature, nevertheless account for the qualitative trends in free energy that underlie the condensation process. On the other hand, should we seek definite information on the morphology and stability of particular aggregates, we are begging the question if we assume anything more specific at the outset than the interatomic force field.

In work carried out so far, thermodynamic and mechanical studies of microcluster properties have tended to remain somewhat separate. This is in large part a matter of necessity, since, although we can within limits carry out a search for potential energy minima, the analogous search for free energy minima, even at a single defined temperature, is almost unthinkable. The most we can hope to do is to examine individual structures, determined in effect at 0 K, and compute their *entropic* behavior as a separate exercise as best we can.

The quickest route to the thermodynamics of clusters is through the well-known formulae for the harmonic oscillator/rigid rotor (HO/RR) approximation. This assumes the cluster to be in the free state; if it is supposed fixed to a substrate the three rotational degrees of freedom must be removed and replaced by vibrational ones for the cluster-surface motion. Most computations so far have been restricted to the case of free clusters, the objects considered usually being the supposedly most stable configurations known. This leads to what we have called the *best single configuration* (BSC) approximation. In the few cases in which a whole ensemble of isomers is available it becomes possible to attempt a *multiconfiguration* (MC) treatment of the system, allowing each minimum to become thermally excited and to contribute to equilibrium according to its respective free energy. This has been possible under certain restricting approximations with the McInnes set of isomeric structures.[94,96]

We now discuss the HO/RR approximation as applied to simple microclusters and present a very small sample of results for the free energy of polytetrahedral and multiply twinned structures. There follows a discussion of the improvements possible in a multiconfiguration approach and some pointers to progress beyond the harmonic region. Section VI, on computer simulation methods, then forms a natural sequel.

A. The Harmonic-oscillator Rigid-rotor Approximation

Once we pass beyond the detailed characterization of structurally stable microclusters, the calculation of low-temperature thermodynamic properties for a particular case has much in common with the standard textbook treatments for molecules. Only when delocalized motion sets in must we abandon this useful parallel.

To clarify these points and to establish notation it will do no harm to recapitulate some of the well-known formulae in their new setting. The starting point is as usual the Canonical ensemble partition function for a particular subsystem, in this case the structure defined by a particular minimum $\mathbf{r}_0^N$. We denote this $\mathscr{Z}^\alpha(N, T)$, using the superscript α to indicate the chosen isomer. We may wish to consider stationary, rotationless clusters from the start; if not, we are virtually bound to assume vibration and rotation separable and write

$$\mathscr{Z}^\alpha(N, T) = Z_{\text{trans}}(N, T) Z_{\text{rot}}^\alpha(N, T) Z_{\text{vib}}^\alpha(N, T) \exp(-E_0^\alpha/k_B t) \qquad (22)$$

Here we have been careful to fix the zero of energy at E_0^α above the minimum value on the understanding that the vibrational component is to be reckoned with respect to the minimum itself.

The HO/RR approximation is then defined in the usual way through the partition functions:

$$Z_{\text{trans}}(N, T) = \left(\frac{2\pi m N k_B T}{h_2}\right)^{3/2} V \tag{23}$$

$$Z_{\text{rot}}^{\alpha}(N, T) = \frac{\pi^{1/2}}{\sigma^{\alpha}} \left(\frac{8\pi^2 k_B T}{h^2}\right)^{3/2} (I_A^{\alpha} I_B^{\alpha} I_C^{\alpha})^{1/2} \tag{24}$$

$$Z_{\text{vib}}^{\alpha}(N, T) = \prod_{i=0}^{3N-6} \exp(-h\nu_i^{\alpha}/k_B T)[1 - \exp(-h\nu_i^{\alpha}/k_B T]^{-1} \tag{25}$$

in which

m = the mass of individual atoms
V = the volume of the container
ν_i^{α} = the normal-mode frequencies about the given minimum
σ^{α} = the symmetry factor for the cluster
$I_A^{\alpha}, I_B^{\alpha}, I_C^{\alpha}$ = the principal moments of inertia.

The main thermodynamic properties of interest are

Internal Energy $U^{\alpha}(N, T)$

$$U^{\alpha}(N, T) = E_0^{\alpha}(N) + E_0^{\alpha}(N) + E^{\alpha}(N, T) = k_B T^2 (\partial \ln \mathscr{Z}/\partial T)$$

Here ε_0^{α} is the zero-point energy and $E^{\alpha}(N, T)$ the thermal contribution. Assuming the HO/RR approximation

$$\varepsilon_0^{\alpha} = \tfrac{1}{2} \sum_{i=0}^{3N-6} h\nu_i^{\alpha} \tag{27}$$

and

$$E^{\alpha}(N, T) = \sum_{i=1}^{3N-6} \frac{h\nu_i}{[\exp(h\nu_i^{\alpha}/k_B T) - 1]} + 3k_B T \tag{28}$$

Heat Capacity $C_V^{\alpha}(N, T)$

$$C_V^{\alpha}(N, T) = 2k_B T \left(\frac{\partial \ln Z^{\alpha}(N. T)}{\partial T}\right)' + k_B T^2 \left(\frac{\partial^2 \ln Z^{\alpha}(N, T)}{\partial T^2}\right)$$

$$= k_B \sum_{i=1}^{3N-6} \left[\frac{(h\nu_i^{\alpha}/k_B T)^2 \exp(h\nu_i^{\alpha}/k_B T)}{(\exp(h\nu_i^{\alpha}/k_B T) - 1)^2}\right] \tag{29}$$

the summation being over all normal modes.

Entropy $S^{\alpha}_{\text{vib}}(N, T)$

$$S^{\alpha}_{\text{vib}}(N, T) = k_B T\left(\frac{\partial \ln Z^{\alpha}_{\text{vib}}(N, T)}{\partial T}\right) + k_B \ln Z_{\text{vib}}(N, T)$$

$$= k_B \sum_{i=1}^{3N-6} \left\{\left[\frac{(h\nu^{\alpha}_i/k_B T)}{\exp(h\nu^{\alpha}_i/k_B T) - 1}\right] - \ln\left[1 - \exp(-h\nu^{\alpha}_i/k_B T)\right]\right\} \quad (30)$$

Helmholz Free Energy $F^{\alpha}_{\text{vib}}(N, T)$

$$F^{\alpha}_{\text{vib}}(N, T) = -k_B T \ln Z^{\alpha}(N, T)$$

$$= E^{\alpha}_0(N) + \varepsilon^{\alpha}_0(N) + k_B T \sum_{i=1}^{3N-6} \ln\left[1 - \exp\left(-\frac{h\nu^{\alpha}_i}{k_B T}\right)\right] \quad (31)$$

From the latter we can write one of the most important secondary expressions, that for the relative concentrations $n(N)$ of two different clusters of the same number of atoms in thermal equilibrium. Indicating the two by the superscripts α and β, we find

$$\left(\frac{n^{\beta}(N)}{n^{\alpha}(N)}\right) = e^{-\Delta E_0/k_B T} \cdot e^{-\Delta \varepsilon_0/k_B T}$$

$$\times \left(\frac{\sigma^{\alpha}}{\alpha^{\beta}}\right)\left(\frac{I^{\beta}_A I^{\beta}_B I^{\beta}_C}{I^{\alpha}_A I^{\alpha}_B I^{\alpha}_C}\right)^{1/2} \times \prod_{i=1}^{3N-6} \left[\frac{1 - \exp(h\nu^{\alpha}_i/k_B T)}{1 - \exp(h\nu^{\beta}_i/k_B T)}\right]$$

$$(\Delta E = E^{\beta}_0 - E^{\alpha}_0 \text{ etc.}) \quad (32)$$

This is a particularly transparent representation of the factors that go into the relative *thermodynamic* stability of two structures. We see immediately how either of the last two *entropic* factors can fall out whenever the two structures considered are sufficiently close either in vibrational spectra or in mass distribution and symmetry. Under such conditions the equilibrium is determined entirely by the Boltzmann factor in the potential energy.

An important formula related to the above gives the concentration of N clusters of given type in an equilibrium ensemble relative to the monomer concentration. By elementary considerations of the chemical potential we arrive at what is in effect a version of the law of mass-action:

$$n(N) = (n(1)/Z(1, T))^N Z(N, T) \quad (33)$$

(see Ref. 108, Section 4.4.) On entering the partition functions for monomer and N-mer, respectively, we obtain an explicit statistical-thermodynamic formula for the number density $n(N)$. This is one of the starting points for the atomistic theory of nucleation, which is discussed further in Section VII.

Some of the approximations mentioned earlier can be applied in the above formula and may lead to drastic simplifications. In the Einstein models of Abraham and Dave[57] and Nishioka, Shawyer, Bienenstock, and Pound[56] the number of frequencies involved is reduced to a small number with the option of various refinements. At the crudest level, all frequencies are simply put equal; alternatively, subsets may be assigned more or less realistically to the contributions of surfaces, edges, vertices, and so on. Even the classical form of the Einstein model can be used, as by Andres,[120] who simply assigned $3N-6$ equal frequency modes at the pair frequency and in the high-temperature limit. Although totally unrealistic, this mode has its uses as a far-fetched limiting case against which other models can be tested.

Results in the HO/RR Approximation. A considerable number of studies have been published dealing with one aspect or another of microcluster thermodynamics in the harmonic approximation. In particular we cite the following: Abraham and Dave,[57] Nishioka, Shawyer, Bienenstock, and Pound[56] Burton,[58,98] McGinty,[51] Bonnissent and Mutaftschiev,[93] Hoare and Pal,[121] and Barker, Hoare, and Pal.[122] We do not attempt to review all this work here, but describe by way of a sample two extracts of the results of Barker, Hoare, and Pal, together with some tentative conclusions they have drawn.

Figure 15 shows curves of the heat capacity of some polytetrahedral (PT) argon clusters as a function of temperature in the low-temperature regime. By plotting $C_v(3N-6)k_BT^3$ we reduce per degree of freedom and at the same time observe the deviation from the Debye T^3 law. Even with the smallest clusters this deviation is slight except below 5 K, and the tendency with the larger ones is only to a very slight positive anomaly, considerably less than the T^2 law for small prisms or the Planck-Schaefer law would indicate. Only clusters of less than 10 atoms show considerable freezing out at the 1 K level.

In Fig. 16 we collect results for the vibrational free energy of a selection of clusters in the $N<60$ size range, including both polytetrahedral types and various f.c.c and multiply twinned forms. This confirms the thermodynamic advantage of the PT structures over all others shown, with the notable exception of the Mackay 55-icosahedron.

These results have since been extended to the higher range of about $N \cong 150$ without appreciable difference in the trends illustrated.[122] Some disagreement has been reported with our heat capacity results for the $N=13$ case, the origin of which may lie in the use of a nearest-neighbor approximation in the Hoare-Pal study. Recalculations with full interactions do in fact reduce the low-frequency components somewhat and thus lead to slight reduction in the low-temperature heat capacity.[122]

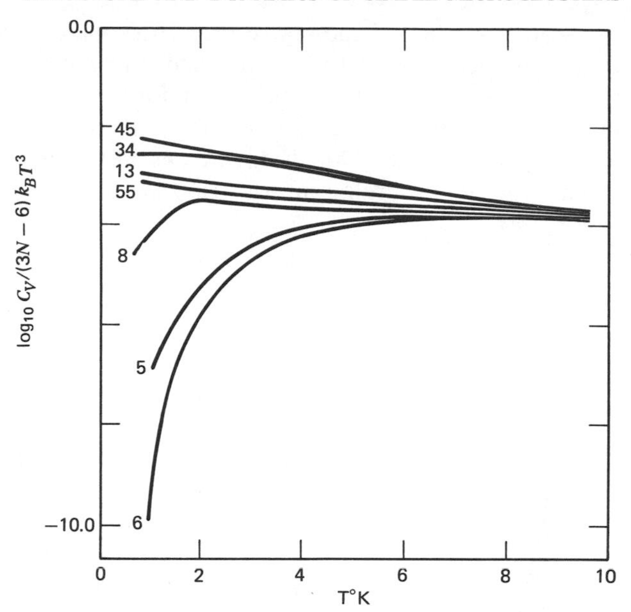

Fig. 15. The Debye T^3 law for various Lennard-Jones structures. The curves are scaled for argon.[121]

As general conclusions we note the following:

1. In the range $N<40$ there can be little doubt that the PT clusters possess a clear thermodynamic advantage over all microcrystalline forms of the same size, the free energy superiority paralleling that found previously for the potential energy alone. Thus in the size range considered—and very likely under all conditions of solid-like behavior—the predominant factor in stability is *energetic* rather than *entropic*. As the crucial $N=55$ point is approached, incomplete substructures of the Mackay icosahedron begin to compete with pure PT forms, and reconstructive transitions to the former become likely.
2. Apart from a small inflection at $N=13$ for the icosahedron, the dependence of zero-point energy E_0/N, internal energy $E(N, T)$, and Helmholtz free-energy $F(N, T)$ are all virtually monotonic in the cluster size when reduced per atom. Thus earlier suggestions that "magic numbers" may exist for certain full-shell spherical f.c.c. structures[58] cannot be substantiated. Certainly the numbers $N=13$, 55, 115, and 147 are "magic" in the sense that they mark crucial geometrical features, but these do not cause sudden shifts in any of the thermodynamic functions, because they emerge and assemble gradually with increasing N.

3. Although thorough studies of the above properties under a variety of interatomic potentials have yet to be published, there is no reason, as far as can be judged from isolated determinations, to doubt that the same trends will prevail when alternative force laws are introduced. However, this is a point on which further confirmation would be desirable, particularly in relation to aspects of crystal growth morphology.

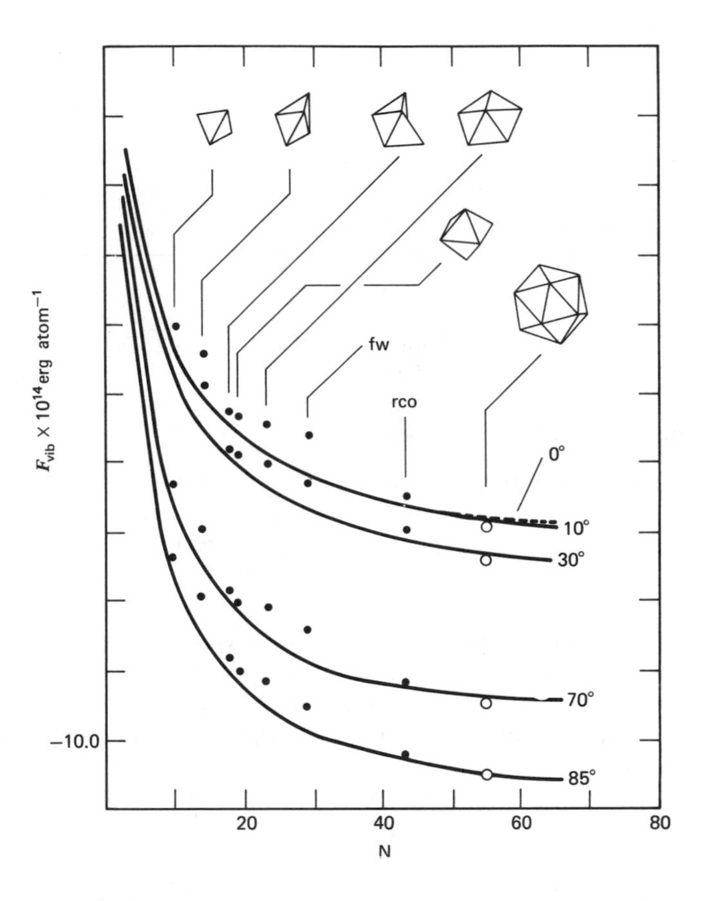

Fig. 16. Vibrational free energy per atom for Lennard-Jones argon clusters in the harmonic, nearest-neighbor approximation.[62] The curves represent values for the most stable isomers known of PT type. The black points give various f.c.c. structures of multiply twinned and spherical form. FW indicates the 29-atom cluster proposed as a growth unit by Fukano and Wayman[82] RCO indicates the $N = 43$ structure in Fig. 1C.

B. Multiconfiguration Thermodynamics

Although the introduction of multiple isomeric clusters into the previous thermodynamic picture is a step in the direction of reality, the difficulties introduced into the theory of cluster equilibria are by no means limited to those of finding the required sets of cluster minima themselves.

In qualitative terms we can see the situation as somewhat parallel to the hypothetical case of a molecule with a vast number of excited electronic states, each with its vibrational and rotational substructure. Thus at low enough temperatures only the ground-state isomer need be considered and the BSC approximation is adequate; as temperature increases, however, the higher isomers begin to be excited thermally, and each makes its contribution to vibrational and rotational entropy. Clearly the energy gap between the ground-state isomer and the next one or two minima available is a crucial factor in the way the equilibrium develops and in the justifiability of dropping higher states in any calculation.

The theoretical problem for clusters is, if anything, somewhat worse than this, because for some potential energy surfaces the energy required to excite a few higher isomers may already be greater than that needed to melt at least some of the lower ones. We then have a situation in which there is a fluctuating coexistence between liquid-like microdrops and transient solid-like species with characteristic energy-dependent lifetimes. Although computer experiments seem to indicate that excurisons to solid-like "islands" in phase space are rare, this aspect is a warning against overreliance on harmonic models.

We have already described the geometrical and energetic features of the isomer set for $N=13$ determined by Hoare and McInnes.[94–96] The conversion of these data into thermodynamic functions is not without additional difficulties. As a first step we write a multi-configuration partition function:

$$\mathscr{Z}(N, T)=\sum_{\{\alpha\}} Z^{\alpha}(N, T) \tag{32}$$

with summation implied as over all isomers. Since only the translational component is configuration independent, the most we can simplify is to write

$$\mathscr{F}(N, T)=Z_{\text{trans}}(N, T)\sum_{\{\alpha\}} Z^{\alpha}_{\text{rot}}(N, T)Z^{\alpha}_{\text{vib}}(N, T)e^{-E^{\alpha}_{0}(N)/k_{B}T} \tag{33}$$

In practice the rotational factors present an additional problem. Although the principal moments of inertia I_A, I_B, I_C are easy to determine, it is no small task to identify the symmetry factors for several hundred tabulated coordinate sets. Since only gross differences in mass distribution are significant and there is an appreciable cancellation effect, it seems reasonable to put all rotational P. F's equal to that for the ground-state isomer. When this is done we have the slightly more manageable expression:

$$\mathscr{Z}(N, T) = Z_{\text{trans}}(N, T) Z^0_{\text{rot}}(N, T) \sum_{\{\alpha\}} Z^{\alpha}_{\text{vib}}(N, T) e^{-E^{\alpha}_0/k_B T} \tag{34}$$

Using this, Hoare and McInnes have carried out extensive comparisons of single-configuration and multiple-configuration thermodynamic functions in the range $N=6$ to $N=13$.[96] The results confirm that marked deviations from the BSC approximation can occur and do so increasingly with rising temperature, particularly under conditions in which the gap between ground-state isomer and the next ones available is relatively narrow. Thus the $N=13$ icosahedron gives a comparatively good BSC approximation at least to $T^*=0.6$, and likewise the $N=6$ and $N=7$ clusters conform well because of the small number of alternatives available and the narrow spacing between their binding energies. However, distinctly greater deviations in the free energy function are observed at the $N=10$ and $N=11$ level, as would be expected.

An alternative approach to the estimation of configurational contributions to free energy is as follows. We assume, even more crudely than in (34), that vibrational as well as rotational partition functions are sufficiently insensitive to isomer geometry for us to write

$$\mathscr{Z}(N, T) = Z_{\text{trans}}(N, T) Z_{\text{rot}}(N, T) Z^0_{\text{vib}}(N, T) Z_{\text{config}}(N, T) \tag{35}$$

with

$$Z_{\text{config}}(N, T) = \sum_{\{\alpha\}} e^{-E^{\alpha}_0/k_B T} \tag{36}$$

a configurational partition function. If we assume Z_{config} alone to dominate isomer equilibria, then we can tabulate sets of configurational thermodynamic properties in their own right. This has been done by Hoare and McInnes, with results broadly comparable with those for the complete calculations just described. We note here that the idea of neglecting vibrational factors in favor of configurational ones is very much in the spirit of recent theories of glass thermodynamics.[123,124]

C. Thermodynamics in the Anharmonic and Liquid Regimes

At the time of writing there are only very limited thermodynamic results available from either the SCP method or the quasi-continuum computations discussed earlier. The well-known problem of obtaining entropy from dynamical trajectories still appears to introduce errors comparable to effects of interest. Although entropy differences can in theory be obtained by differentiation of the caloric equation of state curve $E(T)$ and then integrating the resulting estimates of $C_v(T)$, the error involved appears to be considerable, and there is wide disagreement between different authors. Alternative methods of estimating entropy from the mode distribution function $D(\omega)$ have been considered by Esbjørn et al.[118] and applied to clusters in the work of Kristensen et al.[117] Agreement in the two alternative calculations is somewhat poor, and more lavish use of computing time will probably be required before a final verdict on their accuracy is likely to be possible.

Results of the SCP method for entropy and free energy are likewise in a very preliminary stage. Etters, Kanney, Gillis, and Kaelberer[112] present entropy and free energy data for Lennard-Jones argon clusters up to $N = 15$ at 0 K. One of the remarkable features of their results is the very limited stability of the $N = 14$ cluster consisting of the 13-icosahedron plus one. Due, it would seem, to zero-point vibrational effects, the 14th atom appears to be barely bound to the inner icosahedron. This is an effect that may well be observable in mass-spectrometric studies of cluster beams, although it has not been mentioned in any yet published (e.g., Ref. 14).

For the sake of coherence, we have anticipated some of the results of computer simulation studies in this section. In the next we give a somewhat more critical account of the scope of this method in its application to microcluster systems.

VI. COMPUTER SIMULATIONS

Among the many examples of computer simulation as a tool for elucidating thermodynamic and transport properties, its use in microcluster dynamics can be said to occupy something of a special place. There can be few cases in which the relation of the simulated construct to the physical reality is logically as direct or in which, at least in laboratory physics, the absence of well-defined experiments gives such point to the computed results. Although, just as in simulations of the bulk phase, it is easy to run up against the limits of computing time and budgets, the results obtained for clusters lose none of their validity for being computed at low N, and devices such as the periodic boundary conditions needed to disguise the finite nature of bulk simulations can be dispensed with altogether. In short, although computer studies of the bulk liquid and solid phases have, on the whole, confirmed and quantified an intuitive picture of matter at the atomic level, in the case of clusters quite new and sometimes counterintuitive results have been brought to light by this means. It is not surprising that computer experiments have been carried out enthusiastically in many quarters and have played a crucial part in the growth of microcluster mechanics.

We attempt here a critical account of the main results of microcluster simulations to date, including both molecular dynamics (MD) and Monte Carlo (MC) methods and linking these where possible with the content of our previous sections. The result, it is to be hoped, will balance the long section on statics and support our contention that in the microcluster field geometrical insight and numerical mathematics come together more successfully than in almost any other branch of physics.

Monte Carlo and Molecular Dynamics Methods. We assume that the reader is familiar with the two main simulation methods, at least in their usual form of application to the study of bulk phases. Excellent descriptions are to be found in Refs. 125 and 126. The main sources we draw on in this section are the works of McGinty,[119] Kristensen, Jensen, and Cotterill,[117] Briant and Burton,[127] Farges, Feraudy, Raoult, and Torchet[20,21] (MD method), and Lee, Barker, and Abraham,[128] Etters, Danilowitz, and Kaelberer[109,129] (MC method).

Both MC and MD methods have been used for some time in cluster studies with similar degree of success; both make comparable demands on computing time and are subject to errors of similar magnitude, the origins of which lie more in the physics of the system than the numerical analysis. Both lead to certain difficulties of interpretation which are not really resolvable within the scope of the results.

In each case conclusions are drawn from a sample of phase space points that can be converted to either time or ensemble averages for geometrical characteristics, and thermodynamic and transport properties. In the MD method the samples consist of time trajectories for the natural motion of the system as determined by the solution of Newton's equations of motion. The total energy of the system being conserved, this corresponds to a *microcanonical ensemble* moving on a fixed energy surface. The *temperature* of the system must be assigned secondarily by forming a time average of the kinetic energy. In the MC method the sample consists of a Markov chain of configurations, so constructed as to satisfy a Boltzmann distribution in potential energy with given temperature. Thus the average involved corresponds to a *canonical ensemble.* We know from fluctuation theory that the differences introduced by nonconservation of energy even in quite small clusters are almost negligible and in any case unimportant provided we derive thermodynamic quantities in the appropriate manner. Only the question of ergodicity raises doubts as to the equivalence of the two methods, and this we can safely ignore. In practice the distinction between the two methods is not even as precise as it seems; MD trajectories have a tendency to "wander" due to numerical error and cease to be traceable to their initial condition. Monte Carlo runs, as usually carried out, involve the movement of only one atom at a time and hence lead to an appreciable "memory" effect not unlike that in a deterministic trajectory.

At first sight the dynamics of a single cluster is more easily computed than in the bulk MD or MC methods. The complication of periodic boundary conditions is absent and with it any form of artificially imposed symmetry; moreover, each choice of N yields results interesting in their own right and free of any necessity to imagine a thermodynamic limit $N \to \infty$. On the other hand, a price must be paid for this. Atoms evaporate from the surface at higher energies, and there may be difficulty in controlling angular momentum. Just as in the bulk simulations, the choice of initial conditions is a major problem—with a realistic interaction potential it is virtually impossible to start a run with predetermined temperature or total energy. Thus in practice a considerable amount of computing time must be devoted to equilibration before reliable trajectory samples can be taken.

The second problem, allowing for evaporation or fission of the cluster, is handled in various ways. Either the sample trajectory can be terminated as soon as one or more atoms escape with net kinetic energy, or the cluster must be surrounded with a containing surface that reflects back outgoing atoms with proper conservation of momentum and energy. The first alternative is impractical on account of the equilibration problem; the second leads to difficulties of interpretation and in comparing results of different calculations. (If the barrier is too close it imposes an unnatural external pressure; if it is too far away, a considerable part of the computed trajectory really belongs in the $(N-1)$ atom phase space.) Pessimistically, it could be argued that these difficulties reflect an inherent wrongness in attempting to compute time or ensemble averages in the dissociation region—there may be no genuine alternative to taking many more atoms in a large box and observing the total fragmentation equilibrium, a seemingly impossible task.

Fortunately it has been found that much interesting physics is discoverable in the low-energy region where evaporation is no problem.

With these preliminaries in mind, we now attempt a digest of the work cited earlier. This takes the form of a brief listing, with additional comments, of the main quantities that can be extracted from MD/MC data, then a resume of the more important results so far obtained. References are given for only a few items, where a particularly original move is mentioned. The headings in brackets are in effect suggestions of quantities that have not been given in published papers but that could be extracted at no extra cost from new computations or taped records.

1. *Geometrical quantities*
   Pair distribution function $g(r)$.
   Density as function of radial coordinate $\rho(r)$
   Radial distribution of potential energy $V(r)$
   Temperature dependence of effective radius $r_{\text{eff}}(T)$
   Mean trace of inertia tensor $(I_A I_B I_C)^{1/2}$ [119]
   Interference functions
   Static

$$I(k)=1+\frac{2}{N}\sum_{i,j}\frac{\sin(2\pi k\,|\mathbf{r}_i-\mathbf{r}_j|)}{2\pi k\,|\mathbf{r}_i-\mathbf{r}_j|}$$

   Dynamic

$$I(s)=\frac{(Z-F)^2}{s^4}\int_0^\infty\left(1+4\pi r g(r)\frac{\sin sr}{s}\right)dr+\frac{S}{s^4}$$

   (Ref. 21)
   (Z, atomic number; F, S, elastic and inelastic cross-sections.)
   [Distance and adjacency matrices **D**, **A**. Voronoi polyhedron statistics.]

2. *Dynamic quantities*
   Diffusion coefficient for atoms

$$D_N=\frac{1}{6}\frac{d}{dt}\langle|\mathbf{r}_k(t+t_0)-\mathbf{r}_k(t)|^2\rangle$$

   (Time t_0 to be in linear region)
   Decomposition of D_N into D_{radial} and D_{angular} [127]
   Activation energy for diffusion: $-k_B d\ln D_N/d(1/T)$
   Frequency spectra $D(\omega)$ [117]
   Temperature (rotating clusters)

$$T=\left(\frac{3N}{3N-6}\right)\frac{2}{3k_B}\left\langle\frac{1}{N}\sum_{i=1}^{N}\tfrac{1}{2}mv_i^2\right\rangle$$

   Pressure (through the virial of Clausius):

$$P=\rho T+\frac{2\rho}{N}\sum_i\sum_j V(|\mathbf{r}_i-\mathbf{r}_j|)$$

   Velocity autocorrelation function $\gamma(t)$

$$\gamma(t)=\left\langle\sum_i \bar{v}_i(t)\bar{v}_i(0)\Big/\sum_i v_i(0)^2\right\rangle$$

(Fourier inversion gives the power spectrum $D(\omega)$ above) Kinetics of melting
[Cluster lifetimes, rate constants, activation energies, and entropies of activation for dissociation and structural change.]

3. *Thermodynamic quantities*
   Thermodynamic functions: U, H, S, F, G.
   (The essential sequence for calculation is
   Kinetic energy $\rightarrow T \rightarrow U \rightarrow C_V \rightarrow S \rightarrow F$.)
   Heat capacities. Caloric equation of state $U(T)$.
   Rotational free energy.
   (From averaged partition function:

$$\langle Z_{\text{rot}}(N, T)\rangle = \pi^{1/2}(8\pi^2 k_B T/h^2)^{3/2}\langle(I_A I_B I_C)^{1/2}\rangle)$$

4. *Pseudothermodynamic quantities*
   Melting temperature $T_m(N)$, entropy of melting ΔS_m
   Surface energy

$$E^s(N, T) = E^c(N, T) - NE^b(T) + 6k_B T$$

(Definition of Burton.[127] E^s, E^c, E^b internal energies of surface, whole cluster, bulk per atom, respectively.)

Summarized Results on Simulated Cluster Dynamics. At the time of writing there exist three detailed studies of clusters by molecular dynamics[117,119,127] and two by the Monte Carlo method.[128,129] Although it would have been desirable to combine the results of all these into a single picture, it must be admitted that there are too many differences outstanding at the moment for this to be possible. Moreover, there are unresolved problems of interpretation which lead us to approach the Monte Carlo Data with caution.

The key work on MD simulation is by Briant and Burton.[127] This follows the pioneering studies of McGinty[119] and Kristensen, Jensen, and Cotterill,[117] who established the general procedures and first encountered some of the more insidious sources of error that beset the subject. Briant and Burton have claimed, and, it must be admitted, convincingly, that all previous MD work was carried out without due attention to the problem of equilibration. Although this would not have invalidated all the results obtained at higher temperatures, it has an especially serious effect on the melting transition region, in which much of the interest lies. In view of these uncertainties, we feel justified in concentrating on the work of Briant and Burton except in a few particulars.

These authors studied the dynamics of clusters of 2, 3, 4, 5, 6, 7, 13, 33, 55, and 100 atoms. The initial cold configurations taken were all based on the 55-icosahedron, substructures, or superstructures on this. The clusters were warmed up by the very slow addition of kinetic energy and allowed to equilibrate at each chosen temperature for from 3000 to some 10,000 steplengths before taking an accurate time average of kinetic and potential energy. By repeating this at various values of total energy, an energy-temperature diagram could be plotted for each cluster (the caloric equation of state). Because the clusters used were either absolute or close to absolute minima, they retained their identity without transitions well into the anharmonic region. Provided the equilibration procedure was properly observed, there then occurred a relatively sharp delocalization which, for the clusters of seven atoms and above, was marked by a definite fall in temperature. The effect is illustrated schematically by the curve ABCD in Fig. 17. In the transition region a slight tendency to differential melting beginning at the surface could be seen, although the temperature range of this was narrow. The sigmoid feature in the $E(T)$ curve is reminiscent of a Van der Walls loop, but is probably of the different kind shown by Hill to be a natural characteristic of small systems.[130] In a few runs the liquid-like clusters were recooled with

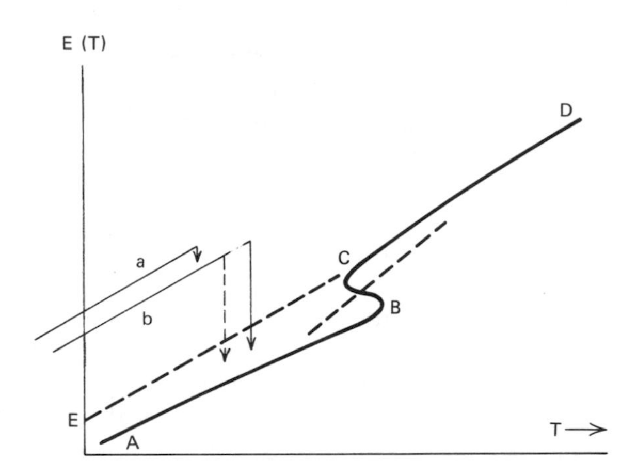

Fig. 17. Schematic diagram of the caloric equation of state as obtained in molecular dynamic and Monte Carlo simulations. The curve ABCD represents the resolved melting-transition region found by Briant and Burton,[127] from for example, the $N=55$ icosahedron. Without very slow cooling precautions the system will retrace the dashed curve DCE to an amorphous structure. Curves a and b indicate the transition behavior of small, nonabsolutely minimal isomers discovered in MC calculations by Etters, Danilowicz, and Kaelberer.[109] The dotted line indicates the averaged curve obtained in MC calculations and in MD ones when equilibration is not achieved.

somewhat inconclusive results. Up to 13 atoms, the heating curves were easily retraced and the original structures recovered. With $N=55$, only a glassy structure of higher energy than the Mackay-icosahedron appeared, although this proved to have a single 13-icosahedron embedded in it (DCE). Much has since been made of this single icosahedron.[90,91] As we know from the work of Farges et al. discussed earlier, sufficiently slow cooling leads to the crystallization out of both the 55-structure and its larger relatives. Undoubtedly these would have appeared had Briant and Burton cooled as carefully as they heated. We refer to the original paper for details of the many other quantities measured during the same runs and listed in our earlier tabulation.

The pioneer work on Monte Carlo simulation of clusters is that of Lee, Barker, and Abraham.[128] Their method made skillful use of an artificial containing volume both to prevent evaporation of the clusters and also as a device for obtaining the Helmholtz free energy of the system. We do not go into details of this here. Less well known is the recent series of papers by Etters, Kaelberer, and others,[109,129,131] who have used the MC method to study both the melting transition and the possibility of transitions between different isomeric structures. The inability of the MC method to discriminate between different branches of the $E(T)$ curve near a transition is a well-known effect,[125] but seems to have been rediscovered in cluster studies and to have generated much discussion. In fact, as the transition region approaches, the sampling algorithm draws points from both liquid- and solid-like regions so that an intermediate line is traced looking more like the result of a second-order transition. Once clear of the transition region in either direction, only one branch is sampled and the MC and MD results agree again.

In their most recent work, Etters, Danilowitz, and Kaelberer[109] show how the phase space for different isomers in the $N=7$ and $N=8$ cases may be explored by MC procedures in such a way as to show the passage of higher isomers into the structure of apparently lowest possible potential energy. Thus a temperature is reached beyond which the $E(T)$ curve terminates and sampling registers either the ground-state isomer or an intermediate one. In fact, there is evidence that systems started in a higher isomer configuration will cascade through intervening ones in order before arriving at the lowest (see Fig. 17). In this way isomers can be discovered without the use of geometrical intuition, and the Etters-Danilowitz-Kaelberer program faithfully produces the isomers for $N=6$ to 8 for Lennard-Jones argon with near perfect agreement of the binding energies with the static ones of Hoare and Pal.[61] The extent to which MC sequences can be said to mimic an actual phase space trajectory is not easily definable, although it seems that the process of dropping from one

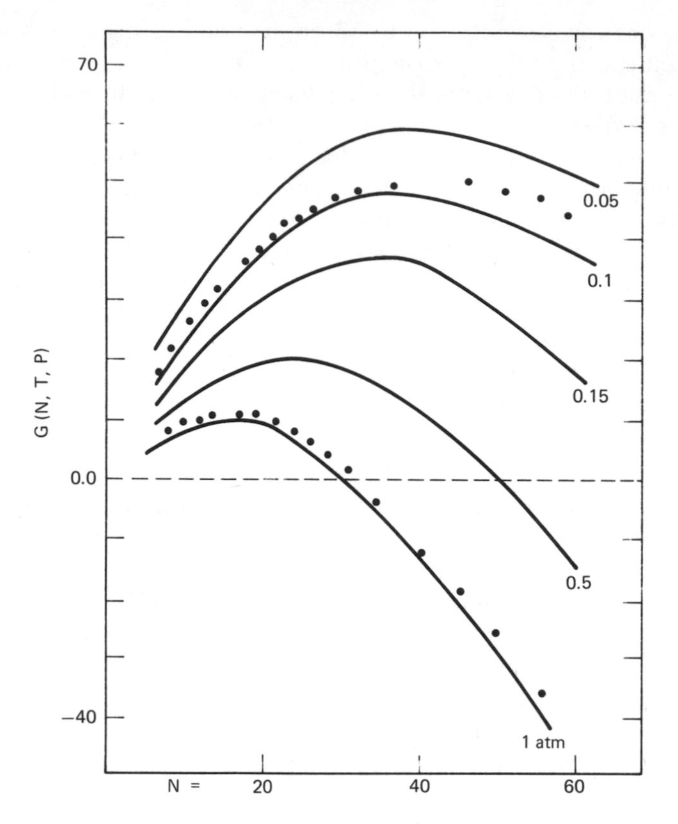

Fig. 18. Gibbs free energy of formation $G(N, T, P)$ for BSC minimal Lennard-Jones argon clusters as a function of pressure and size N. The maximum in the curve represents the critical nucleus. The temperature is 70.1 K throughout, and the energy scale is in units of 10^{21} J. The black points represent the corresponding values for McGinty's relaxed f.c.c. clusters described in Section III, B.3.

isomer catchment region to another is a close parallel to the natural motion of a phase point finding its way over a saddle. If this is so, then the claim made by Kaelberer, Etters, and Raich elsewhere[131] that they have discovered definite transition temperatures for interconversion of isomers is somewhat misleading. It may be that the temperature at which transition is easily observed corresponds to a saddle point activation energy, but it seems more likely that, as with any activated process, the initial state has an energy-dependent lifetime which decreases rapidly above a threshold. Briant and Burton have actually plotted examples of the fluctuating residence times between liquid and solid clusters as a function of energy in the transition region.[127]

We end this section on a slightly critical note. There has been much talk of nonergodic effects in several of the papers just discussed—sometimes one almost has the impression that failure of the system to visit some supposedly interesting part of phase space within the space of a usual computer run is grounds for immediate suspicion that the potential is nonergodic. There is, in fact, an impressive literature dealing with the ergodic problem for few-atom systems under potentials such as the Lennard-Jones, and interrelating theoretical work on the Kolmogorov-Arnold-Moser theorem with extensive computational experiments. We hesitated to introduce this topic here because the results, although fascinating, are extremely remote from the practical concerns of the rest of this article. However, if nothing else, these papers are a most valuable corrective to the casual use of ergodicity hypotheses in cluster simulations. The articles of Galgani, Scotti, and co-workers are a useful introduction.[132–135]

VII. NUCLEATION AND CRYSTAL GROWTH

We explained in the introduction our intention not to treat the role of cluster mechanics in nucleation theory in any depth, even though not to do so is perhaps to disguise the dominant influence of the latter throughout the whole development of microcluster studies. So much has been written in recent years on nucleation that for a full background we may safely refer the reader to the best of the recent review treatments, those by Abraham[108] and Burton.[136,137]

What follows here is intended only as a brief account of the special problems that arise when nucleation predictions are attempted on a truly *ab initio*, atomistic basis, that is, when, as in spirit of the whole of this article, all information on macroscopic bulk and surface properties is renounced and replaced by deductions from the intermolecular force field.

A. Atomistic Nucleation Theories

The usual object of fundamental theories of homogeneous, gas phase nucleation is the computation of nucleation rate as a function of temperature and pressure and hence the determination of critical supersaturation ratios for comparison with experiment. The route to this is invariably some form of the Gibbs-Volmer prescription in which an intermediate objective is the calculation of a critical nuclear size N^* at which the free energy of formation of clusters from the supersaturated gas is a maximum. Such nuclei, which can only form by fluctuations, are then assumed to grow freely through accretion of gaseous atoms at their surface. For this, a unit sticking coefficient is usually assumed.

The classical computation of critical nuclear size involves the use of the bulk surface tension for the substance—however dubious this may be at sizes of less than $N = 100$—and the implicit assumption of liquid, spherical nuclei. At first sight the conversion of this to a truly atomistic theory would seem to be simply a matter of using a different route to the free energy of formation, applying the approximate statistical mechanical formulae such as (31), and avoiding thereby any explicit reference to surface and bulk contributions. Although an approach of this kind is possible, the attempt to stick religiously to *ab inito* principles proves to raise as many new problems as it solves old ones. Thus, whereas in the Gibbs method the critical nucleus is a well-defined liquid drop of a certain radius, in a cluster theory the proper analog of this must be a whole set of isomeric clusters (not necessarily of the same N) sharing only the characteristic that their rate of growth under the given conditions equals on average their rate of decay. Even with the knowledge of small cluster isomerism described in a previous section, such a calculation is a hopeless task, and in any case of doubtful use given the many further approximations involved.

The easiest way out of these difficulties is, as we have seen, to compute the free energy of formation $\Delta G_f(N, T, P)$ for a single representative cluster for each size N, this being selected as the lowest known potential energy minimum. This *best single cluster* approximation was first carried through by McGinty[51] using the relaxed f.c.c. cluster set described in Section III.B, and later by Hoare and Pal using the best-known polytetrahedral and icosahedral set[121] for Lennard-Jones argon. The McGinty and Hoare-Pal results for argon are presented together in Fig. 18 where it can be seen that they differ little in spite of the considerable geometrical differences in the structures assumed. Although differences in the free energy are detectable, their effect on the derived value of N^* is virtually negligible. Although the absolute values of nucleation rates obtained in this way cannot be very accurate, the fact that true Gibbsian behavior (i.e., a maximum in ΔG_f versus N) is obtained at all is something of an achievement for a purely *ab initio* theory. Unfortunately, the extensive calculations of Abraham and co-workers[108] using microcrystalline models have not been given in a form in which they are comparable to ours. However, we doubt that there would be very severe disagreement given the general insensitivity of nucleation models to details of structure.

The assumptions of harmonic oscillator and rigid rotor behavior underlying the above calculations are not altogether unavoidable. McGinty has subsequently obtained molecular dynamic estimates of $\Delta G_f(N, T, P)$ in the temperature region where anharmonic behavior should be prominent and has shown that the required corrections are less than might be supposed.[119] Concerning the all-important BSC approximation, there has been some limited discussion of the magnitude of possible errors but no clear comparison in the range of the above data.[138,139] Calculations based on the McInnes isomer set have been carried out and yield appreciable differences when $N^* < 13$ between BSC and MC values for rare gases.

We await with great interest the first interpretable results on nucleation rates for argon condensing in a nozzle beam with helium carrier gas.[140]

B. Crystal Growth Mechanisms

Less has been written on the fundamentals of crystal growth in the microcluster region than on nucleation theory. The requirements of crystal growth theories are in some respects more stringent, in that definite morphological predictions are required; on the other hand, the absence of many of the conceptual difficulties referred to just now makes it a rather more attractive subject for cluster computations.

Although there is a long tradition of atomistic theories of crystal growth thermodynamics,[141] most of this has been concerned with the surfaces of large crystals and their various imperfections. For the most part the theory of the initial stages of growth at the microcluster level is a neglected one, it having long been assumed that not much can be said about growth until the growing object is somehow large enough to recognize itself as a crystal and take up the Gibbs-Wulff geometry.

We can add considerably to this in describing the probable intermediate stages in the formation of true microcrystals, through the packing experiments and computer simulations discussed earlier. Although these results point convincingly to the essential role of icosahedral intermediates in the growth of argon microcrystals in the gas phase (and the nozzle beam diffraction experiments confirm this), a number of more general questions remain to be taken up.

In the first place we should stress the not quite obvious fact that (even ideally) there can be no such thing as a deposition process at 0 K. An atom cannot simply be brought up from infinity and moved around, looking for the most favorable site on which to deposit itself. Even a supposedly cold atom would at least accelerate in under the attractive potential and settle with some transient dissipation of kinetic energy around the site. Thus in a very essential way finite temperature, thermal fluctuations, and the free energy function are at the root of all growth processes—however successfully we may be able to approximate the latter quantity in terms of potential energy alone.

Next we should point to the fundamental difference between the genesis of a crystalline or crystalloid microcluster from a liquid drop cooled by evaporation or in a carrier gas—and the corresponding process under conditions of atom-by-atom deposition at a seed structure. In the former case there is little difficulty in imagining a substructure such as the 55-icosahedron forming first by solidification of its 13-icosahedron core, then by cooperative rearrangement of the second and third shells about it. For the same structure to form from the vapor atom by atom a definite reconstructive rearrangement would be required to put the second shell in place. The first icosahedron could certainly form by the Werfelmaier sequence, but the second could not, for the simple reason that atoms would have to sit waiting balanced on a tetrahedral edge for their supporting partners to arrive. On the basis of our free energy computations as illustrated in Fig. 16, it seems hard to believe that, for Lennard-Jones systems, a reconstructive transformation could occur much below the $N=50$ level at which the 55-icosahedron is almost complete. We have serious doubts whether on very cold substrates such structures would assemble by vapor deposition if two-body forces alone were active.

A similar mystery attends the growth of the less stable multiply twinned pentagonal bipyramidal crystallites which, as we have seen, tend to form with some readiness in deposits of metal smokes (Section II.A). A number of authors, in particular Ino,[2] Allpress and Sanders,[59] and Fukano and Wayman,[82] seem to have assumed that the pentagonal nuclei for these structures simply assemble themselves, each atom moving unerringly to its required place. It would seem to us that the only situation in crystal growth where atoms might appear to "know where they are supposed to go" is the when the final state is of appreciably lower free energy than all others and where sufficient thermal fluctuations are present for each atom to wander between alternative positions before it finally settles. Even then the surfeit of kinetic energy must be removed before the structure will become permanent. The data in Fig. 16 show beyond doubt that the free energies for small multiply twinned structures of $N=10$, 14, 18, and 22 are quite unable to compete with those of tetrahedral/icosahedral type in the same size range. (In the McInnes isomer enumeration the tetrahedral f.c.c. pile for $N=10$ is actually the *least* rich in binding energy of 29 alternative isomers!) Thus there can be no free energy driving force to favor such a growth route as there undoubtedly is for the 55-icosahedron.

The easy way out of this conflict between theory and observation is to find fault with the Lennard-Jones potential. Once the necessity of two-body central forces is abandoned, it is not difficult to imagine the occurrence of several-atom forces which, through some peculiarity of surface densities of states, actually favor the formation of, say, a protruding tetrahedral vertex. If resolved electron micrographs of metal clusters in the $N=20$ range become available and show primitive multiply twinned units rather than polytetrahedra, we shall have plain visual evidence for the inappropriateness of two-body central force fields. Further discussion of crystal growth mechanisms involving pentagonal and icosahedral nuclei is found in Ref. 62.

A further question of some difficulty concerns the eventual growth of "normal" f.c.c. argon microcrystallites under the ideal conditions of nozzle beam condensation. Farges and his group have assembled considerable electron diffraction data to show that approximately in the $N = 1000$ range icosahedral scattering patterns cease to predominate and there is a progressive trend toward an ordinary f.c.c. structure. There are several possible explanations of this transition effect. Farges believes it likely that in some way one facet of a large Mackay icosahedron begins to grow at the expense of the others, giving rise eventually to a giant f.c.c. tetrahedron or a truncated form tending to the appropriate Gibbs-Wulff polyhedron. This process may have something in common with Gillet's suggestion[5] that, beyond a critical size, clearance regions can open up, with eventual loss of the fivefold symmetry on further growth.

VIII. CONCLUSION

In concluding what we have tried to make as balanced an account as possible of this subsection of a fast-developing field, we cannot but feel conscious of all that, for a variety of logical and editorial reasons, has been left out. If there is a simple rationale for stopping at this point it is that the systems we have considered exclusively can serve as prototypes and first approximations to more complex ones, as well as being in a more literal way the starting points for a variety of difficult calculations. Thus at present most quantum mechanical computations seem bound to start with the assumption of one or another of the classical minimum geometries discussed here. A great many quantum studies of small microclusters are now in print, in which the 13-icosahedron and the spherical f.c.c. forms are prominent. For recent entry points into this literature we refer to Gillet,[142] Ogawa,[143] and Baetzold[144] and the various papers in Ref. 18.

At the level of quantum mechanics and electronic properties, a number of quite new issues emerge which, in the case of metal microclusters, lead to a host of measurable quantities, some of which are listed in Section I. Among the particularly live issues at the time of writing is the question of *odd-even* differences in electronic and stability properties at very small N,[145] and whether planar and even linear structures may be minimal at greater binding energies than tetrahedral ones in certain ranges of N.[146] In this particular problem, as in more general terms, it must be admitted that virtually all quantum mechanical calculations to date have been with some form of semiempirical method such as CNDO or extended Hückel. Although these may certainly give useful qualitative indications, the accuracy of quantitative predictions and even the correctness of energy level sequences remains open to considerable doubt.

With somewhat more regret, we have omitted any account of the recent work on ionic and molecular clusters. Clusters of the water molecule are of obvious interest and have been the subject of several studies.[147] Deuterium clusters may emerge at any time as fuel for thermonuclear machines.[18] We have also carried out preliminary studies of the classical minima for the nitrogen system under the Lennard-Jones atom-atom potential.[148] Ionic clusters are now beginning to be studied extensively in matrix isolation, and the main characteristics of their minima, and vibrational modes are now understood in simple cases.[149–153]

In preparing this article literature has been surveyed up to approximately February 1978.

APPENDIX

The Mackay Icosahedral Clusters. The formula for the icosahedral numbers I_n giving the number of atoms in the complete Mackay icosahedra illustrated in Fig. 5 can be obtained as follows. We sum the numbers T_1 to T_7 accounted for thus:

T_1	(central atom)	1
T_2	(vertex atoms)	12
T_3	(inner face)	$15(n-2)(n-3)$
T_4	(inner edge)	$12(n-2)$
T_5	(outer face)	$10(n-2)(n-3)$
T_6	(outer edge)	$30(n-2)$
T_7	(internal atoms)	$(10/3)(n-2)(n-3)(n-4)$

Thus for the total N as a function of the number of shells

$$\begin{aligned} N(n) &= 13+42(n-2)+25(n-2(n-3)+(10/3)(n-2)(n-3)(n-4) \\ &= (1/3)[10n^3-15n^2+11n-3] \end{aligned}$$

The number in the surface $N_s(n)$ is then

$$\begin{aligned} N_s(n) = \Delta_n N(n) &= 12+30(n-2)+10(n-2)(n-3) \\ &= 10n^2-20n+12 \end{aligned}$$

As n becomes large we see that the tetrahedral numbers T_7 dominate over all others and $N \sim (10/3)\,n^3$. At the same time the surface is dominated by the triangle numbers T_5, so that $N_s \sim 10\,n^2$. Thus we obtain the asymptotic surface/volume ratio $N_s/N \sim 3\,n^{-1}$. The numerical values in Table IV show that this ratio is almost indistinguishable from the "crudest possible estimate" of $4\,N^{-1/3}$ derived in Section I. Both these

TABLE IV
Actual and Asymptotic Surface Ratios for Icosahedral Mackay Clusters

n	N	N_s	N_s/N	$3/n$	$4/N^{1/3}$
1	1				
2	13	12	0.92	1.50	1.70
3	55	42	0.76	1.00	1.05
4	147	92	0.63	0.75	0.76
5	309	162	0.52	0.60	0.59
6	561	252	0.45	0.50	0.48
7	923	362	0.39	0.43	0.41
8	1415	492	0.35	0.38	0.36
9	2057	642	0.31	0.31	0.31
10	2869	812	0.28	0.28	0.28

figures approximate the true N_s/N quite well for $n > \sim 6$. It will be noted that the asymptotic behavior $N_s/N \sim 3\,n^{-1}$ is true for *any* packing system whatever when the volume grows as a third-order polynomial in n and a complete covering shell is added at each step.

References

1. K. Mihama and Y. Yashda, *J. Phys. Soc. Jap.*, **21,** 1166 (1966).
2. S. Ino, *J. Phys. Soc. Jap.*, **21,** 346 (1966).
3. K. Kimoto and I. Nishida, *J. Phys. Soc. Jap.*, **22,** 940 (1967).
4. J. G. Allpress and J. V. Sanders, *Surf. Sci.*, **7,** 1 (1967).
5. E. Gillet and M. Gillet, *Thin Solid Films*, **15,** 249 (1973); **29,** 217 (1975).
6. M. Gillet, *J. Cryst. Growth.*, **36,** 23a (1976).
7. K. Kimoto and I. Nishida, *Thin Solid Films*, **17,** 49 (1973).
8. K. Kimoto and I. Nishida, *J. Phys. Soc. Jap.*, **22,** 744 (1967).
9. C. Chapon, C. Henry, and B. Mutaftschiev, *J. Cryst. Growth.*, **33,** 291 (1976).
10. K. Oswatitsch, *Z. Angew. Math. Mech.*, **22,** 1 (1942).
11. B. Raoult and J. Farges, *Rev. Sci. Inst.*, **44,** 430 (1973).
12. P. P. Wegener and A. A. Pouring, *Phys. Fluids*, **7,** 352 (1964).
13. P. J. Foster, R. E. Leckenby, and E. J. Robbins, *J. Phys.*, **B2,** 478 (1969).
14. K. Kimoto and I. Nichida, *J. Phys. Soc. Jap.*, **42,** 2071 (1977).
15. E. W. Becker, K. Bier, and W. Henkes, *Z. Phys.*, **146,** 333 (1956).
16. F. T. Greene and T. A. Milne, *J. Chem. Phys.*, **47,** 4095 (1967).
17. E. W. Becker, R. Klingelhöfer, and P. Lohse, *Z. Naturforsch*, **15a,** 644 (1960).
18. J. Gspann, G. Krieg and H. Vollmar, *International Meeting on Small Particles and Inorganic Clusters*, Lyon-Villeurbanne, September 1976. *Journal de Physique Colloque* C-2, 1977, p. 171.
19. J. Farges, B. Raoult, and G. Torchet, *J. Chem. Phys.*, **59,** 3454 (1973).
20. J. Farges, *Structure des agregats de symmetrie pentagonale formes lors de la detente en jet d'argon gazeux.* Thesis, University of Paris-Sud, 1977.

21. J. Farges, M. F. deFeraudy, B. Raoult, and G. Torchet, *International Meeting on Small Particles and Inorganic Clusters*, Lyon-Villeurbanne, September 1976. *Journal de Physique Colloque* C-2, 1977, p. 47.
22. G. S. Stein, in *Proceedings Ninth International Symposium on Rarefied Gas Dynamics.* American Institute of Aeronautics and Astronautics, New York, 1974 p. F9.1. See also Ref. 18, p. 53.
23. G. P. Stein and P. P. Wegener, *J. Chem. Phys.*, **46,** 3685 (1967).
24. G. P. Stein, *J. Chem. Phys.*, **51,** 938 (1969).
25. B. J. C. Wu, P. P. Wegener and G. D. Stein, *J. Chem. Phys.*, **69,** 1776 (1978).
26. S. H. Bauer and H. J. Freund, *J. Phys. Chem.*, **81,** 994 (1977).
27. S. H. Bauer and D. J. Frurip, *J. Phys. Chem.*, **81,** 1001, 1007, 1015 (1977).
28. N. C. Wickramsinghe and K. Nandy, *Rep. Prog. Phys.*, **35,** 157 (1972).
29. W. D. Watson and E. E. Salpeter, *Astrophys. J.*, **174,** 321 (1972).
30. M. Allen and G. Wilse Robinson, *Astrophys. J.*, **195,** 81 (1975).
31. B. Meyer, *Low Temperature Spectroscopy*, Elsevier, New York, 1971.
32. L. Andrews and G. C. Pimentel, *J. Chem. Phys.*, **47,** 2905 (1967).
33. L. Brewer and C. Chang, *J. Chem. Phys.*, **56,** 1728 (1972).
34. J. E. Francis and S. E. Weber, *J. Chem. Phys.*, **56,** 5879 (1972).
35. W. Schultze, H. U. Becker and D. Leutloff, Ref. 18, p. 7.
36. M. Moskowitz and J. E. Hulse, *J. Chem. Phys.*, 4271 (1977).
37. T. H. James, ed., *The Theory of the Photographic Process*, McMillan, New York, 1977.
38. J. F. Hamilton, in T. H. James, ed., *The Theory of the Photographic Process*, McMillan, New York, 1977, Chapter 4.
39. S. W. Marshall and R. M. Willenzich, *Phys. Rev. Lett.*, **16,** 219 (1966).
40. J. Charrolin, C. Froidevaux, C. Taupin, and J. M. Winter, *Solid State Comm.*, **4,** 357 (1966).
41. S. Matsuo, H. Miyata, and S. Noguchi, *Jap. J. Appl. Phys.*, **13,** 351 (1974).
42. K. Saiki, T. Fujita, Y. Shimiza, S. Sakoh, and N. Wada, *J. Phys. Soc. Jap.*, **32,** 447 (1972).
43. S. Takajo, S. Kobayashi, and W. Sazaki, *J. Phys. Soc. Jap.*, **35,** 712 (1973).
44. Kotaibutsuri, *Solid State Physics*, special issue on fine particles. Agne, Tokyo, December 1975. Text in Japanese.
45. R. Collins, *Proc. Phys. Soc.*, **83,** 553 (1964); **86,** 199 (1965).
46. W. Mayeda, *Graph Theory*, Wiley, New York, 1972.
47. D. J. Wilde and C. S. Brightler, *Foundations of Optimization Theory*, Prentis-Hall, Englewood Cliffs, New Jersey, 1967.
48. R. Fletcher and M. J. D. Powell, *Comp. J.*, **6,** 163 (1963).
49. R. Fletcher and C. M. Reeves, *Comp. J.*, **7,** 149 (1964).
50. R. Fletcher, *Comp. J.*, **13,** 317 (1970).
51. D. J. McGinty, *J. Chem. Phys.*, **55,** 580 (1971).
52. R. Van Hardeveld and F. Hartog, *Surf. Sci.*, **15,** 189 (1969).
53. W. Romanowski, *Surf. Sci.*, **18,** 373 (1969).
54. J. F. Nicholas, *Aust. J. Phys.*, **21,** 21 (1968).
55. J. P. Hirth and G. M. Pound, *Progress in Materials Science*, vol. 2, London, 1963. See also Ref. 141.
56. K. Nishioka, R. Shawyer, R. Bienenstock, and G. M. Pound, *J. Chem. Phys.*, **55,** 5082 (1971).
57. F. F. Abraham and J. V. Dave, *J. Chem. Phys.*, **55,** 1587; 4817 (1971).
58. J. J. Burton, *J. Chem. Phys.*, **52,** 345 (1970).
59. J. G. Allpress and J. V. Sanders, *Aust. J. Phys.*, **23,** 23 (1970).

60. M. R. Hoare and P. Pal., *Nature*, **230,** 5 (1972); **236,** 35 (1972).
61. M. R. Hoare and P. Pal, *Adv. Phys.*, **20,** 161 (1971).
62. M. R. Hoare and P. Pal, *J. Cryst. Growth*, **17,** 77 (1972).
63. J. Andrew Barker and M. R. Hoare, unpublished.
64. H. S. M. Coxeter, *Introduction to Geometry*, Wiley, New York, 1971, chapter 22.
65. A. H. Boerdijk, *Phillips Res. Rep.*, **7,** 303 (1952).
66. J. D. Bernal, *Nature*, **183,** 141 (1959); **185,** 68 (1960).
67. J. D. Bernal, *Proc. R. Soc.* (*London*), **280A,** 299 (1964).
68. J. L. Finney, *Proc. R. Soc.* (*London*), **319A,** 479 (1970).
69. M. H. Cohen and D. Turnbull, *Nature*, **189,** 131 (1964); **203,** 964 (1964).
70. J. Andrew Barker, J. L. Finney, and M. R. Hoare, *Nature*, **257,** 120 (1975).
71. L. Von Heimendahl, *J. Phys. F: Metal Phys.*, **5,** L141 (1975).
72. J. Andrew Barker, Ref. 18. p. 37.
73. W. Werfelmeier, *Z. Phys.* **107,** 332 (1937).
74. L. Pauling, *Science*, **150,** 297 (1965).
75. A. A. Shternberg, *Rost. Kristallov*, **5a,** 179 (1968). English translation: *Growth of Crystals*, **8,** 256. Consultants Bureau, New York, 1969.
76. T. Komoda, *Jap. J. Appl. Phys.*, **7,** 27 (1968).
77. H. M. Cundy, *Math. Gaz.*, **36,** 263 (1952).
78. L. Fejes-Toth, *Regular Figures*, McMillan, New York, 1964.
79. L. Fejes-Toth, *Lagerungen in der Ebene, auf der Kugel und im Raum*, Springer, Berlin, 1953.
80. B. G. Bagley, *J. Cryst. Growth*, **6,** 323 (1970).
81. B. G. Bagley, *Nature*, **225,** 1040 (1970).
82. Y. Fukano and C. M. Wayman, *J. Appl. Phys.*, **40,** 1656 (1969).
83. A. L. Mackay, *Acta. Cryst.*, **15,** 916 (1962).
84. L. A. Girofalco and V. G. Weizer, *Phys. Rev.*, **114,** 687 (1959).
85. D. Weaire, M. F. Ashby, J. Logan, and M. J. Weiss, *Acta Met.*, **19,** 779 (1971).
86. M. R. Hoare, *Ann. N. Y. Acad. Sci.*, **279,** 186 (1976).
87. M. R. Hoare, Proceedings Symposium *The Structure of Non-crystalline Materials*, Society of Glass Technology, Cambridge, 1976. Taylor and Francis, London, 1977, p. 175.
88. H. S. Chen, Reference 87, p. 79.
89. D. E. Polk, *Scripta. Met.*, **4,** 117 (1970).
90. C. L. Briant, *Faraday Disc. Chem. Soc.*, **61,** 25 (1976).
91. C. L. Briant and J. J. Burton, *Physica Status Solidi*, **82b,** 393 (1977).
92. J. Andrew Barker and M. R. Hoare (To be published).
93. A. Bonnissent and B. Mutaftschiev, *J. Chem. Phys.*, **58,** 3727 (1973).
94. M. R. Hoare and J. McInnes, *Faraday Disc. Chem. Soc.*, **61,** 12 (1976).
95. J. McInnes, *Statistical mechanics and homogeneous nucleation of atomic microclusters*, Thesis, University of London, 1976.
96. M. R. Hoare and J. McInnes. (To be published).
97. J. J. Burton, *Chem. Phys. Lett.*, **17,** 199 (1972).
98. J. J. Burton, *J. Chem. Soc. Faraday Trans*, II, **69,** 540 (1973).
99. F. Harary, *Graph Theory and Theoretical Physics*, Academic, New York, 1967.
100. M. R. Hoare and P. Pal (To be published).
101. B. L. Cohen, *Phys. Rev.*, **120,** 925 (1960).
102. V. Weisskopf, *Phys. Rev.*, **52,** 295 (1937).
103. I. Frenkel, *Sov. Phys.*, **9,** 533 (1936).
104. N. B. Slater, *Theory of Unimolecular Reactions*, Meuthen, London, 1959, Chapter 9.

105. H. M. Rosenstock, M. B. Wallenstein, A. L. Wahrhaftig, and H. Eyring. *Proc. Natl. Acad. Sci. U.S.A.*, **38,** 667 (1952).
106. W. D. Gwinn, *J. Chem. Phys.*, **55,** 477 (1971).
107. S. G. Reed, *J. Chem. Phys.*, **20,** 208 (1952).
108. F. F. Abraham, *Homogeneous Nucleation Theory* (supplement No 1 to *Advances in Theoretical Chemistry*), Academic, New York, 1974.
109. R. D. Etters, R. Danilowicz, and J. Kaelberer, *J. Chem. Phys.*, **67,** 4145 (1977).
110. H. P. Baltes, Ref. 18, p. 151.
111. *Theory of lattice dynamics in the harmonic approximation*, A. A. Maradudin, E. W. Montroll, G. H. Weiss and I. P. Ipatova, Solid State Physics, Supplement 3, Academic Press, New York (1971).
112. R. D. Etters, L. Kanney, N. S. Gillis, and J. Kaelberer, *Phys. Rev.*, **B15,** 4056 (1977).
113. T. Matsubara, Y. Iwase, and A. Momokita, *Prog. Theoret. Phys.*, **58,** 1102 (1977).
114. T. Matsubara and K. Kamiza, *Prog. Theoret. Phys.*, **58,** 767 (1977).
115. A. Nakanishi and T. Matsubara, *J. Phys. Soc. Jap.*, **39,** 1415 (1975).
116. J. M. Dickey and A. Pascin, *Phys. Rev.*, **B1,** 851 (1970).
117. W. Darmgaard Kristensen, E. J. Jensen, and R. M. J. Cotterill, *J. Chem. Phys.*, **60,** 4161 (1974).
118. P. O. Esbjørn, E. J. Jensen, W. D. Kristensen, J. W. Martin, and L. B. Pedersen, *J. Comp. Phys.*, **12,** 289 (1973).
119. D. J. McGinty, *J. Chem. Phys.*, **58,** 4733 (1973).
120. R. P. Andres, *Ind. Eng. Chem.*, **57,** 24 (1965).
121. M. R. Hoare and P. Pal, *Adv. Phys.*, **24,** 645 (1975).
122. J. Andrew Barker, M. R. Hoare, and P. Pal (To be published).
123. M. Goldstein, *J. Chem. Phys.*, **39,** 3369 (1963).
124. M. Goldstein, *J. Chem. Phys.*, **51,** 3728 (1969).
125. J. P. Valleau and G. M. Torrie, in B. Berne, ed., *Statistical Mechanics, Part A. Equilibrium Techniques*, vol. 5, *Modern Theoretical Chemistry*, Plenum, New York, 1977, Chapter 5.
126. J. Kusick and B. Berne, in *Statistical Mechanics. Part B. Time-Dependent Processes*, vol. 6, *Modern Theoretical Chemistry*, Plenum, New York, 1977, Chapter 2.
127. C. L. Briant and J. J. Burton, *J. Chem. Phys.*, **63,** 2045 (1975).
128. J. K. Lee, J. A. Barker, and F. F. Abraham, *J. Chem. Phys.*, **58,** 3166 (1973).
129. R. D. Etters and J. Kaelberer, *Phys. Rev.*, **A11,** 1068 (1975).
130. T. L. Hill, *J. Chem. Phys.*, **23,** 812 (1955).
131. J. B. Kaelberer, R. D. Etters, and J. C. Raich, *Chem. Phys. Lett.*, **41,** 580 (1976).
132. L. Galgani and A. Scotti, *Phys. Rev. Lett.*, **28,** 1173 (1972).
133. P. Boccherini, A. Scotti, B. Bearzi, and A. Loinger, *Phys. Rev.*, **A2,** 2013 (1970).
134. L. Galagni and A. Scotti, *Nuovo Cimento*, **2,** 189 (1972).
135. M. Casartelli, E. Diana, L. Galgani, and A. Scotti, *Phys. Rev.*, **A13,** 1921 (1976).
136. J. J. Burton in A. C. Zettlemoyer, ed., *Nucleation*, Elsevier, Amsterdam, 1977.
137. J. J. Burton, in Ref. 125, Chapter 6.
138. D. J. McGinty, *Chem. Phys. Lett.*, **13,** 525 (1972).
139. J. J. Burton, *Chem. Phys. Lett.*, **17,** 199 (1972).
140. P. P. Wegener, in progress, Private communication.
141. R. Lacmann, *Springer Tracts in Modern Physics* (*Ergebnisse der exakten Wissenschaften*), **44,** 1 (1960).
142. M. Gillet, *Surf. Sci.*, **67,** 139 (1977).
143. T. Ogawa, *Z. Physik*, **B28,** 73 (1977).
144. R. C. Baetzold, *J. Chem. Phys.*, **68,** 555 (1978).

145. R. Kubo, Ref. 18, p. 69.
146. R. C. Baetzold, Ref. 18. p. 175.
147. P. L. M. Plummer and B. N. Hale, *J. Chem. Phys.*, **56,** 4329 (1972).
148. J. Andrew Barker and M. R. Hoare (To be published).
149. T. P. Martin, *Phys. Rev.*, **B7,** 3906 (1973).
150. T. P. Martin, *Phys. Rev.*, **B15,** 4071 (1977).
151. T. P. Martin, *J. Chem. Phys.*, **67,** 5702 (1977).
152. D. O. Welch, O. W. Lazareth, G. J. Dienes, and R. D. Hatcher, *J. Chem. Phys.*, **68,** 2159 (1978).
153. T. P. Martin, *J. Chem. Phys.*, **69,** 2036 (1978).

A KINETIC APPROACH TO HOMOGENEOUS NUCLEATION THEORY

JOSEPH L. KATZ AND MARC D. DONOHUE

Department of Chemical Engineering
Clarkson College of Technology
Potsdam, NY 13676

CONTENTS

I. INTRODUCTION

There are some phase transitions which appear to occur without hinderance whenever the experimental conditions change such that a new phase becomes more stable, for example, the evaporation of a pure liquid or the melting of a pure crystal. For some substances, these transitions occur so reproducibly that they are often used as thermometric fixed points. However, there are other phase transitions for which this is not the case, for example, the condensation of a gas or the crystallization of a melt. Thus one often encounters terms like *supersaturated* gases or *supercooled* melts, and mention is made of the need to *nucleate* these substances. In reality the difference between phase changes which occur at equilibrium and those which do not occur at equilibrium is only apparent and results from the presence of free surfaces. If one eliminates all free surfaces, for example, by surrounding a liquid with another whose boiling point is much higher, the liquid can be superheated very readily. One finds that at atmospheric pressure, small droplets of all substances thus far studied can be superheated to about 90% of their critical temperature before they explosively vaporize.[1] Experiments have also shown that it is possible to superheat crystals when they are poorly wet by their melts.[2]

Phase transitions (except in experiments specifically designed to study homogeneous nucleation) usually occur in an irreproducible fashion as a result of the variety and widely varying effectiveness of the heterogeneous nuclei that are ever present in all systems, thus making it difficult to characterize accurately the crucial experimental parameters. There is, however, one case that is difficult to study experimentally but is accurately characterizable. This case is *homogeneous nucleation*, that is, nucleation in the absence of walls, dust, ions, bubbles, microcrystals, and so on.

Homogeneous nucleation is more than just the theoreticians' desire to simplify interesting problems until they can be solved. There are a number of "real" world situations in which it is probable that homogeneous nucleation is the dominant mechanism for phase transition, for example, condensation in supersonic nozzles,[3] explosions that occur when a cold liquid contacts a much hotter one,[4] formation of heavily microcrystallized ceramics.[5] Furthermore, it is indeed possible to devise experimental techniques to study homogeneous nucleation.[6–10] In recent years these techniques have advanced significantly. Today it is generally accepted that one can measure, to an accuracy of a few percent or better, the supersaturations required to cause a gas to condense.[11–13] Similarily, for superheated liquids, to an accuracy of about 0.2%, one knows the temperatures to which liquids can be superheated.[1,19] Thus one only need devise an acceptable theory, and it can be tested.

II. NUCLEATION IN DILUTE SYSTEMS

A. New Theory

In a dilute solution, whether gaseous (e.g., water vapor in air) or condensed, an instantaneous snapshot would show that nearly all solute molecules exist independently or in small clusters containing two, three, or possibly four molecules; larger clusters of solute molecules would be extremely rare.† Furthermore, it is usually the case that the concentration of independent molecules is very much larger than the concentration of all other clusters combined. The growth of clusters therefore occurs almost entirely by the addition of single molecules.‡ At *equilibrium*, since

† For simplicity of presentation, we assume that all clusters are compact and have the same shape; however, it is a straightforward matter to generalize our development to include one or more shape parameters.

‡ There are substances for which this assumption is not true (e.g., carboxylic acids, metal vapors). Although classical nucleation theory has been extended to include such substances,[15] the reformulation presented here has not yet been sufficiently generalized to encompass these substances.

every forward process (i.e., the addition of single molecules) must be matched by its corresponding reverse process, the backward process must occur almost entirely by the loss of individual molecules. Furthermore, since the backward process is entirely determined by the properties of the cluster (e.g., size, temperature, shape) and by its interactions with the surrounding medium, in sufficiently *dilute* solution the mechanisms by which molecules evaporate from a cluster are independent of solute concentration; therefore the decay of clusters in a supersaturated state also occurs almost entirely by the loss of single molecules. Thus the net rate J at which clusters containing i solute molecules (at time t) become clusters containing $i+1$ molecules is given by the equation

$$J(i, t) = f(i, t)n(i, t) - b(i+1)n(i+1, t) \tag{2.1}$$

where $n(i, t)$ is the concentration of i-sized clusters. The forward rate $f(i, t)$ and the backward rate $b(i)$ are, respectively, the rates that molecules arrive at and depart from the surface of a nucleus containing i atoms. Given $f(i, t)$, $b(i)$, and the concentrations $n(i, t)$ of clusters, the nucleation rate could, in principle, be calculated. However, the only term that is known, even approximately, is $f(i, t)$; it is equal to the product $\beta a(i)$ where β is the rate at which solute molecules impinge and condense on the surface per unit area at time t, and $a(i)$ is the surface area of a cluster containing i molecules. For an ideal gas

$$\beta = \frac{\alpha P(t)}{\sqrt{2\pi mkT}} \tag{2.2}$$

where the pressure, $P(t)$, can be a function of time, and α is the condensation coefficient.

Consider a system at equilibrium (e.g., a vapor in contact with its own liquid or a solution in contact with solute crystals). The nucleation rate is identically zero for clusters of all sizes. Equation (2.1) can therefore be used to determine the backward rate in terms of equilibrium concentration distribution $n_e(i)$† and the forward rate (at equilibrium). Thus

$$0 = f_e(i)n_e(i) - b_e(i+1)n_e(i+1) \tag{2.3}$$

where the subscript e denotes the true equilibrium state at the same temperature as the supersaturated state of interest.

To determine the backward rate, we return to the argument made above: For dilute systems (i.e., systems in which the number density of solute molecules is much smaller than the number density of solvent molecules) the interactions between a molecule leaving a nucleus and

† This is *not* the same distribution as the *constrained* equilibrium distribution all previous authors have used.

other solute molecules is negligible compared to its interactions with the solvent molecules. Therefore the backward or "evaporation" rate $b(i)$, although a complicated and unknown function of temperature, is independent of the concentration of the nucleating species, that is, the solute. Thus $b(i+1)$ is equal to $b_e(i+1)$. Substituting (2.3) into (2.1),

$$J(i,t)=f(i,t)n_e(i)\left[\frac{n(i,t)}{n_e(i)}-\frac{n(i+1,t)}{n_e(i+1)}\cdot\frac{f_e(i)}{f(i,t)}\right] \tag{2.4}$$

Since the forward rate $f(i)$ is the product of the arrival rate β and the area of the cluster, the ratio of the forward rates $f(i,t)/f_e(i)$ simplifies to the ratio of arrival rates β/β_e. (For an ideal gas β/β_e reduces to the supersaturation.) Rearranging (2.4) and dividing both sides by $(\beta/\beta_e)^i$,

$$\frac{J(i,t)}{f(i,t)n_e(i)(\beta/\beta_e)^i}=\frac{n(i,t)}{n_e(i)(\beta/\beta_e)^i}-\frac{n(i+1,t)}{n_e(i)(\beta/\beta_e)^{i+1}} \tag{2.5}$$

Note that the terms on the right-hand side differ only in the value of the index i. Summing from $i=1$ to some sufficiently large size, $i=\bar{i}-1$, successive terms cancel and one obtains

$$\sum_{i=1}^{\bar{i}-1}\frac{J(i,t)}{f(i,t)n_e(i)(\beta/\beta_e)^i}=\frac{n(1)}{n_e(1)(\beta/\beta_e)}-\frac{n(\bar{i})}{n_e(\bar{i})(\beta/\beta_e)^{\bar{i}}} \tag{2.6}$$

For most systems, a steady state is very rapidly achieved. Since the rate of change of concentration of clusters containing i molecules is given by

$$\frac{dn(i,t)}{dt}=J(i-1,t)-J(i,t) \tag{2.7}$$

at steady state, J becomes a constant for all sizes. In addition, as shown later, the numerator of last term on the right-hand side of (2.6) is a very slowly decreasing function of i, but the denominator is an ever increasing function. Therefore for sufficiently large $\bar{i}$, the last term on the right-hand side becomes negligible compared to the first. Factoring J out of the summation and rearranging gives

$$J=\frac{n(1)}{n_e(1)(\beta/\beta_e)}\Bigg/\sum_{i=1}^{\bar{i}-1}[f(i)n_e(i)(\beta/\beta_e)^i]^{-1} \tag{2.8}$$

Thus, as originally shown by Wiedersich and Katz,[16–18] the problem of predicting the rate of nucleation has been reduced to two simpler problems—a kinetic problem, that of knowing the forward rate and, more important, the arrival rate ratio β/β_e, and an equilibrium problem, that of determining the concentration of clusters in equilibrium with their bulk condensed phase. Given these quantities, the summation in (2.8) can be evaluated numerically rather easily. However, by making two very accurate mathematical approximations,† an analytical solution is possible.

The first is the conversion of the summation to an integral. Since the i dependence of the forward rate $f(i)$ and that of the term in the arrival rate ratio are known, evaluation of the integral does not require a more detailed specification of their meaning. However, we also require some knowledge of the i dependence of $n_e(i)$.

The concentrations $n_e(i)$ can be described by *equilibrium* thermodynamics, since they are characteristic of a true equilibrium state. For a fluid at equilibrium (in this case in both phase and thermal equilibrium), it is generally accepted that one has a Boltzmann distribution in the concentrations of clusters. Therefore

$$n_e(i) = \mathcal{N} \exp[-i\Delta\mu(i)/kT] \tag{2.9}$$

where $\mathcal{N}$ is a normalization constant and $\Delta\mu(i)$ is the difference in chemical potential between a molecule in a cluster of size i and the bulk condensed phase. Unfortunately, $\Delta\mu(i)$ is not rigorously known. However, although the final answer does depend on $\Delta\mu(i)$, to proceed further we only need to know that $\Delta\mu(i)$ decreases with increasing size. Then the product $f(i)n_e(i)(\beta/\beta_e)^i$ exhibits a maximum, and the integral is convergent.

The second approximation is to write the integrand as an exponential, expand the argument of the exponential about its maximum value in a Taylor series, and truncate the series after the quadratic term, that is,

$$J = \frac{n(1)}{n_e(1)(\beta/\beta_e)} \Big/ \int_{i=1}^{i} \exp[-H(i)]\, di \tag{2.10}$$

where

$$H(i) = -i\Delta\mu(i)kT + i \ln(\beta/\beta_e) + \ln f(i) + \ln N \tag{2.11}$$

and

$$H(i) \approx H(i^*) + (i - i^*)^2 H''(i^*)/2 \tag{2.12}$$

where i^* is chosen such that the first derivative is zero. A transformation of variables gives this as a standard integral (i.e., the error function) and for commonly encountered values of i^*, (2.10) becomes

$$J = \frac{n(1)}{n_e(1)(\beta/\beta_e)} \left(\frac{H''(i^*)}{2\pi}\right)^{1/2} \exp[H(i^*)] \tag{2.13}$$

$H(i^*)$ depends on the arrival rate ratio β/β_e and $\Delta\mu(i)$. As knowledge

† The accuracy of these approximations is thoroughly discussed in a paper by E. R. Cohen.[19]

in the fields of kinetic theory and equilibrium statistical mechanics progresses, it will be a straightforward matter to incorporate improved molecular theories into this result.

Nevertheless, one can make some approximations and proceed further. Assuming that the only difference between the bulk condensed phase and a cluster is in the surface to volume ratio, $\Delta\mu(i)$ can be written in terms of the surface free energy σ†

$$\Delta\mu(i) = \sigma a(i)/i \tag{2.14}$$

The surface area $a(i)$ can be written as $a_0 i^{2/3}$ where a_0 is a constant that depends on the geometry and the volume per molecule v (e.g., for spherical molecules $a_0 = 4\pi(3v/4\pi)^{2/3}$). The distribution becomes

$$n_e(i) = \mathcal{N} \exp[-\sigma a_0 i^{2/3}/kT] \tag{2.15}$$

Using (2.15), (2.13) becomes

$$J = \frac{n(1)}{n_e(1)(\beta/\beta_e)} \mathcal{N}\beta a_0 \left[\frac{a_0\sigma}{9\pi kT} - \frac{1}{3i^{*2/3}}\right]^{1/2} \exp\left[-\frac{a_0\sigma i^{*2/3}}{3kT} - \frac{2}{3}\right] \tag{2.16}$$

where i^* is obtained by solution of the equation

$$\ln \beta/\beta_e = \frac{2}{3}\left[\frac{a_0\sigma i^{*2/3}}{kT} - 1\right] \Big/ i^* \tag{2.17}$$

A further simplification can be made since $a_0[i^*]^{2/3}/kT$ is almost always much greater than 1. Then (2.17) can be solved and substituted into (2.16), giving

$$J = \frac{n(1)}{n_e(1)\beta/\beta_e} \mathcal{N}\beta a_0 \left[\frac{a_0\sigma}{9\pi kT}\right]^{1/2} \exp\left[\frac{-4}{27}\left(\frac{a_0\sigma}{kT}\right)^3 (\ln \beta/\beta_e)^{-2}\right] \tag{2.18}$$

Although the rate of nucleation depends slightly on the preexponential fugacity ratio, $n(1)/n_e(1)$, one sees that the real driving force for nucleation arises from the ratio of arrival rates β/β_e in the exponential. Given a thermodynamic expression for the fugacity ratio and a kinetic expression for the arrival rate ratio, one can calculate the rate of nucleation.

The most simple expressions for these quantities occur when we have an ideal gas. Then the fugacity ratio becomes the partial pressure ratio P/P_e. When ideal gas kinetic theory is valid, one can use (2.2) for the

† In the context of the constrained equilibrium distribution there has been great controversy about the validity of equations similar to (2.14). Lothe and Pound[19] have argued that rotational and vibrational contributions also need to be included in this equation. Others,[20] including an author (JLK),[21] have argued that to a large extent, they already are. This controversy is not the subject of this paper. Nevertheless, by eliminating the concept of supersaturation, it should be of some help in clarifying the issue, since to the extent that modern statistical mechanics or simulation methods (e.g., Monte-Carlo and molecular dynamics) can answer the *equilibrium* question, the controversy will be resolved.

arrival rate ratio and obtain

$$\beta/\beta_e = P/P_e$$

Substituting this result into (2.18), the fugacity ratio and the arrival rate ratio cancel in the preexponential, and one obtains

$$J = \frac{\mathcal{N}\alpha P a_0}{\sqrt{2\pi mkT}}\left[\frac{a_0\sigma}{9\pi kT}\right]^{1/2} \exp\left[\frac{-4a_0^3\sigma^3}{27(kT)^3(\ln P/Pe)^2}\right] \tag{2.19}$$

Although this result is identical to classical nucleation theory, the similarity is fortuitous. The appearance of the supersaturation P/Pe in the expression for the rate of nucleation arises from a simplification of the arrival rate ratio β/β_e and not from thermodynamic arguments.

B. Comparison with Previous Versions of Nucleation Theory

Classical nucleation theory, as it is commonly presented today,[23,24] follows the form in Frenkel's text[25] and is attributed by him to Zeldovich[26] and Becker and Döring.[27] In this form, the rate of nucleation is calculated by relating the actual concentrations of clusters $n(i)$ to a hypothetical state that is *constrained* to be in equilibrium at the same temperature and supersaturation. Today this theory is popularly accepted and is used in numerous texts.[24] It is also the basis for a number of theories describing nucleation in nonideal systems (e.g., associated vapors,[15] binary[28] and multicomponent mixtures,[29] diffusion limited nucleation,[30] etc.).

The problem with previous versions of nucleation theory is that they rely on this hypothetical and unattainable constrained equilibrium distribution. This has led to a great deal of confusion between properties that can be obtained by thermodynamic procedures and inherently kinetic properties. Evidence of this is present everywhere in the nucleation literature, and one often finds words like "metastable," "constrained to be in equilibrium," and so on to describe this hypothetical state. Unfortunately, some authors have not realized that this is an artifice and have confused the *constrained* equilibrium distribution with the actual distribution that would occur in a real system. This misconception is so firmly held that they have actually suggested that experimentalists in this field should try to observe the minimum in the constrained equilibrium distribution. Some of those who have realized the hypothetical nature of this state have tried to justify its existence in principle by calling on an all-powerful Maxwell's demon. Maxwell's demon was attributed with the power to break up all clusters beyond a certain size and to return the molecules as monomers (individual molecules) to the system. However, all this process succeeds in doing is creating a system in which there is a finite *steady state* rate of nucleation and in which the distribution of clusters is characteristic of the steady-state nucleation and not the constrained equilibrium distribution. The attainment of a constrained equilibrium distribution is beyond even the powers of Maxwell's demon. Nevertheless, if one is to obtain such a distribution from thermodynamics, one needs a prescription for some method capable in principle of creating such a hypothetical equilibrium distribution. No such method exists.

Fortunately, as we have shown, it is neither necessary nor useful to the development of nucleation theory to hypothesize the existence of a constrained equilibrium distribution. A true equilibrium distribution does exist for the saturated state. The concentrations of clusters in this state can, at least in principle, be measured, and the rates of nucleation can be predicted using them.

In addition to the elimination of the constrained equilibrium distribution, the most striking difference between the theory presented here and all other versions is that the nucleation rate is shown to be determined by the ratio of arrival rates, a kinetic property, and not by the thermodynamically defined supersaturation. Although it is true that for sufficiently ideal systems (e.g., an ideal gas and arrival rates given by $\beta = \alpha P/\sqrt{2\pi mkT}$) these two ($\beta/\beta_e$ and P/P_e) are equivalent, for less ideal systems they are not. This difference makes the extension of theory to nonideal systems more transparent and eliminates a great deal of the confusion regarding the true meaning of the inherently kinetic terms.

III. NUCLEATION IN DENSE SYSTEMS

In the previous section on dilute solutions, we used the approximation that the backward or "evaporation" rate is independent of solute supersaturation. This is an excellent assumption at very low concentrations, because the interactions between solute molecules leaving a cluster and other solute molecules are negligible. At high concentrations this assumption can no longer be valid, since molecules leaving a cluster do interact with other nearby solute molecules. For a crystal nucleating from its melt, these interactions totally dominate (there are no solvent molecules).

Katz and Spaepen[31] have shown that by making what appears to be a most reasonable assumption, this problem can be overcome. This assumption concerns the variation of the evaporation rate with concentration at constant temperature. They assumed that the effect of solute density on the probability of a molecule leaving a cluster is the same for clusters of all sizes, that is,

$$\frac{b(i) \text{ at } T, C}{b(i) \text{ at } T, C'} = \frac{b(j) \text{ at } T, C}{b(j) \text{ at } T, C'} \quad \text{for all } i \text{ and } j \tag{3.1}$$

To use this result, it is necessary to choose the concentration C' and the size j so that $b(i)$ at the system temperature T and composition C can be related to known quantities. The most readily obtainable values are for the equilibrium condition ($C' = C_{sat}$ at system temperature) and for a macroscopic bulk phase ($j = \infty$). Then (3.1) can be written

$$b(i) = b_e(i)b(\infty)/b_e(\infty) \equiv b_e(i)m(T, C) \tag{3.2}$$

where m is in general a function of temperature and concentration (or pressure), but is independent of i. The quantities $b_e(i)$ and $m(T, C)$ can both, in principle, be determined experimentally. Using the equilibrium condition, (2.3), $b_e(i)$ can be determined from measurements of the concentrations $n_e(i)$ together with an appropriate expression for $f_e(i)$ from kinetic theory. The quantity $m(T, C)$ can be determined from the rate of growth of one macroscopic phase at the expense of another at supersaturated conditions.†

For the dilute solution case discussed previously, m is equal to unity. In concentrated solution, m is not unity, but it is a straightforward matter to repeat the derivation and obtain

$$J = \frac{n(1)}{n_e(1)} \Big/ \sum_{i=1}^{\bar{i}-1} \left[f(i)n_e(i)\left(\frac{\beta}{m\beta_e}\right)^{i-1} \right]^{-1} \tag{3.3}$$

This result is identical to the dilute solution result except that we have $\beta/m\beta_e$ instead of the ratio of arrival rates β/β_e.

A complication arises in condensed systems. This concerns the minimum size of a nucleus. The large density difference between a gas phase and a nucleating condensed phase makes the identification of the condensed-phase nuclei straightforward; any cluster containing two or more solute molecules is considered to be a potential nucleus. However, for nucleation in condensed systems (e.g., melt → crystal, very concentrated solution → crystal, crystal → crystal) where the density difference between phases is small, this criterion is no longer useful. Therefore other structural characteristics which make these two condensed phases distinct from each other must be used to identify the nucleus.

For a crystal-crystal transformation, the difference in crystallography between the two phases is the obvious choice for identification of a phase transition. Specifically, an assembly of neighboring molecules can be identified as the nucleus of a new phase if the lines connecting the centers of the molecules form polyhedra that are characteristic of the new phase. For example, a f.c.c. nucleus that forms in a b.c.c. matrix can be identified by outlining those parts of the system which contain a specific array of

† This is discussed in more detail in the next section.

regular octahedra and tetrahedra typical of the f.c.c. structure and quite distinct from the array of distorted octahedra that make up the b.c.c. structure.

The identification of crystalline nuclei in a melt is similar; the only difference is that the melt has even less characteristic structure than a b.c.c. crystal. It has been shown that an excellent model for melts of noble and late transition metals is the dense random packing of hard spheres.[32] The polyhedra that make up the structure of such a fluid are a random mixture of tetrahedra, with a few octahedra and occasional deltahedra. The preponderance of tetrahedra, however, makes this structure distinctly different from the more ordered crystal states. This is illustrated by considering the ratio of tetrahedra to octahedra: for random packing this ratio is 15 to 1; for a f.c.c. structure it is only 2 to 1. Therefore, since the octahedra are a necessary element of a f.c.c. structure and only incidental to the dense random packing, this suggests that a nucleus of a f.c.c. crystal can be identified as any part of the system that is made up of properly aligned octahedra.

The identification of crystalline nuclei in a concentrated solution is quite similar. Even though little is known about the structure of these solutions, the identification of regular polyhedra is, in principle, always possible.

This procedure for identification of nuclei in dense systems makes it clear that it is impossible to identify nuclei containing fewer than a certain number of molecules; there is no way to decide whether a pair or triplet of molecules in the dense system is actually a crystal nucleus. The minimum number of molecules is determined by the types of polyhedra that characterize the nucleating phase. For the example of f.c.c. crystals, the minimum identifiable nucleus size is one octahedron and therefore requires at least six atoms to identify the nucleus. The few octahedra that are always present in the dense random packing of hard spheres can perhaps be considered as incipient nuclei for crystallization from the melt.

Accounting for the minimum size of an identifiable nucleus in the calculation of the nucleation rate is straightforward. Since we consider only clusters larger than some minimal size, i_0, we need only sum over the size range i_0 to $\bar{i}$ in (3.3). The result is that

$$J = \frac{n(i_0)}{n_e(i_0)} \Bigg/ \sum_{i=i_0}^{\bar{i}-1} [f(i)n_e(i)(\beta/m\beta_e)^{i-1_0}]^{-1} \tag{3.4}$$

where $n_e(i_0)$ is the concentration of clusters of size i_0 in the true equilibrium state.

A second complication arises when we consider the nucleation of a solid from its melt. This concerns the existence of a melt in equilibrium at the system temperature. In (2.3) we assumed that there will always be an equilibrium state between the two phases at the system temperature. There is no difficulty in creating a real equilibrium state for crystallization from solution; the saturated solution involves a lower concentration and can always be obtained by dilution of the supersaturated solution. However, for crystallization from the melt, the equilibrium concentration must be obtained by decreasing the pressure (at constant temperature), and at the lower pressure the liquid may not exist. This is illustrated in Fig. 1, a pressure-temperature diagram for a pure substance. Shown as solid lines are the crystal-melt, melt-vapor, and crystal-vapor equilibrium states. The dashed line represents the maximum tension that can be exerted on the liquid and not cause cavitation. Only for nucleation temperatures above the triple point does the equilibrium state at $P_e(T)$ exist. However, since it is possible to have liquids and crystals at negative pressures (i.e., tensions), the crystal-melt equilibrium line can be extrapolated to negative pressures (and in principle measured), as shown by the dotted line. Thus the equilibrium distribution of crystal nuclei in the melt is, in principle, experimentally accessible at all pressures above the cavitation pressure. However, when the extrapolated crystal-melt equilibrium line falls below the cavitation limit, it is no longer possible, even in principle, to relate the cluster concentrations to an equilibrium state. States whose temperatures are typical of these three different cases are shown in Fig. 1.

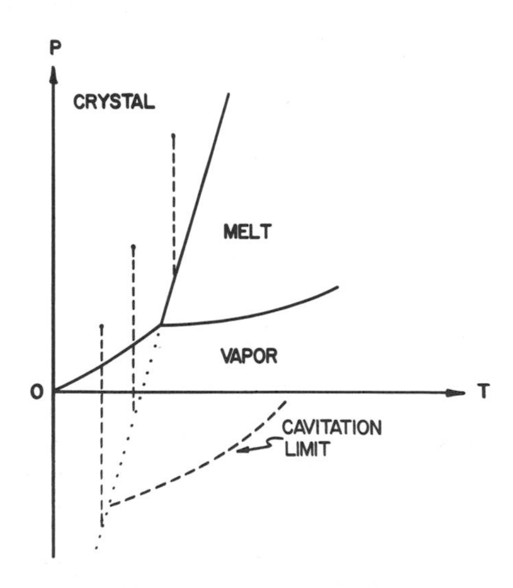

Fig. 1. Pressure-temperature diagram for a pure substance showing three different undercooled liquid states and their corresponding equilibrium reference states.

IV. A GENERAL APPROACH TO NUCLEATION THEORY

In the previous sections, we used the steady-state assumption and eliminated the backward rate (by writing it in terms of the forward rate) at the very beginning of the derivation. This approach has the advantage of simplicity but does not lend itself to generalization for multiple and/or time-dependent backward and forward processes. However, it is possible to postpone the use of the steady-state assumption and the elimination of the backward rate to a quite late stage in the derivation.†

This is accomplished by defining a recursive relation[17]

$$Z(i+1)=Z(i)\frac{b(i+1)}{f(i+1)} \quad \text{for} \quad i>i_0 \tag{4.1}$$

and

$$Z(i_0)\equiv 1 \tag{4.2}$$

when i_0 is the minimum number of molecules required to have an identifiable nucleus. This relation is easily solved:

$$Z(i)=\prod_{j=i_0+1}^{i} b(j)/f(j) \tag{4.3}$$

By multiplying (2.1) by $Z(i)$ we obtain a form in which the two terms on the right-hand side differ only in their indices:

$$J(i,t)Z(i)=f(i)n(i)Z(i)-f(i+1)n(i+1)Z(i+1) \tag{4.4}$$

On summing, successive terms cancel and we obtain

$$\sum_{i=i_0}^{\bar{i}-1} J(i,t)Z(i)=f(i_0)n(i_0)-f(\bar{i})n(\bar{i})Z(\bar{i}) \tag{4.5}$$

The upper limit $(\bar{i}-1)$ is any sufficiently large number, since, as we now show, the last term in this equation is negligible if $\bar{i}$ is large enough. Using (4.3) one obtains for the ratio of the two terms on the right-hand side

$$\frac{f(\bar{i})n(\bar{i})}{f(i_0)n(i_0)}\prod_{j=i_0+1}^{\bar{i}}\frac{b(j)}{f(j)}=\frac{n(\bar{i})}{n(i_0)}\prod_{j=i_0+1}^{\bar{i}}R(j) \tag{4.6}$$

† For simplicity of notation, we have omitted the time dependence of $f(i)$, $b(i)$, and $n(i)$. In the general case, $f(i)$ is a function of time only because the supersaturation may be a function of time. However, for $b(i)$ and $n(i)$ there are two different time scales that must be considered. The first concerns the thermal and structural equilibrium of the clusters. This should be attained very rapidly for most systems but cannot in general be treated without explicit time dependent expressions for $b(i)$ and $n(i)$. The second time scale concerns the time dependence of the supersaturation. This effect is usually much slower than the first. Then the backward rate and the concentrations of clusters can be related to their equilibrium values, even though they may be functions of time.

where

$$R(j) \equiv \frac{b(j)}{f(j-1)} \tag{4.7}$$

is the ratio of departure and arrival rates that determine $J(i, t)$.
For two bulk phases in equilibrium, the arrival and departure rates must be equal (i.e., $R_e(\infty) = 1$). However, as the supersaturation is increased, one phase always grows at the expense of the other. For this to occur, the forward rate must be larger than the backward rate (i.e., $R(\infty) < 1$). Since the backward rate (per unit area) will, in general, increase with decreasing size and the forward rate (per unit area) is essentially constant, there is, at any supersaturation, a critical size i^* where the backward and forward rates are equal (i.e., $R(i^*) = 1$). For all i greater than i^*, $R(i)$ is less than 1, whereas for all i less than i^*, $R(i)$ is greater than 1. Since there are only a finite number of i less than i^*, but an arbitrarily large number of i greater than i^*, it is always possible to make the product of $R(j)$s arbitrarily small.

Neglecting the last term, (4.5) becomes

$$\sum_{i=i_0}^{\bar{i}} J(i, t) Z(i) = f(i_0) n(i_0) \tag{4.5a}$$

This result, combined with (2.7) and an appropriate model for the backward and forward rates, provides the complete time-dependent rate of nucleation.

In most cases, the steady-state rate of nucleation is established in a matter of microseconds. Then $J(i, t)$, as shown previously, is a constant and can be factored out of the summation to obtain

$$J = f(i_0) n(i_0) \Big/ \sum_{i=i_0}^{\bar{i}-1} Z(i) \tag{4.8}$$

where $Z(i)$ is the product given in (4.3).

A. Nucleation by a Single Mechanism

When there is only one mechanism whereby molecules arrive at or depart from a surface, $R(i)$ can be calculated by considering the backward and forward rates associated with the equilibrium distribution of nuclei $n_e(i)$ which exists at saturated conditions (i.e., at the equilibrium concentration or pressure for the system temperature). Since this is a true equilibrium state, the arguments leading to (2.3) apply. Combining (2.3) and (3.2) with the definition of $R(i)$ gives

$$R(i) \equiv \frac{b(i)}{f(i-1)} = \frac{b(i)}{f_e(i-1)} \frac{f_e(i-1)}{f(i-1)} = m \frac{n_e(i-1)}{n_e(i)} \frac{f_e(i-1)}{f(i-1)} \tag{4.9}$$

and

$$Z(i) = \frac{f(i_0)}{b(i)} \prod_{i=i_0+1}^{i} R(i) = \frac{f(i_0)}{b(i)} \prod_{i=i_0+1}^{i} m \frac{f_e(i-1)}{f(i-1)} \frac{n_e(i-1)}{n_e(i)} \tag{4.10}$$

Equation 4.8 can now be solved, resulting in

$$J = f(i_0)n(i_0)\left[1 + \sum_{i=i_0+1}^{i-1} \frac{f(i_0)}{f(i)} \frac{n_e(i_0)m^{i-i_0}}{n_e(i)} \prod_{i=i_0+1}^{i} \frac{f_e(j-1)}{f(j-1)}\right]^{-1} \tag{4.11}$$

A further simplification occurs when f/f_e is independent of the nucleus size. This is usually the case and occurs when the arrival rate $f(i)$ is just the product of the impingement rate of molecules per unit area and the surface area of the nucleus. This is equivalent to assuming that the ratio of condensation coefficients is independent of concentration. Then (4.11) can be written

$$J = \frac{n(i_0)}{n_e(i_0)}\left[\sum_{i=i_0}^{\bar{i}-1} \frac{1}{f(i)n_e(i)(\beta/m\beta_e)^{i-i_0}}\right]^{-1} \tag{4.12}$$

The factor $(\beta/m\beta_e)^{i-i_0}$ can, in principle, be obtained from the growth velocity of one bulk phase at the expense of another. The growth velocity u is given by

$$u = (\beta - \gamma)v_0 \tag{4.13}$$

where v_0 is the volume of molecule, and γ is the rate at which molecules leave or evaporate per unit area, $\gamma = b(i)\ a(i)$. Therefore

$$m = \frac{b(\infty)}{b_e(\infty)} = \frac{\gamma}{\gamma_e} = \frac{\beta - uv_0}{\beta_e} \tag{4.14}$$

and

$$\frac{\beta}{m\beta_e} = \left[1 - \frac{u}{\beta v_0}\right]^{-1} \tag{4.15}$$

since the growth velocity u is zero at equilibrium. Measurement of u, in conjunction with a knowledge of β from kinetic theory, makes this factor experimentally available.

The term $n_e(i)$ is the distribution of concentrations of nuclei in the system at true equilibrium; in principle it can be measured. Even if no such measurements are available, it can be estimated using any of a number of molecular theories applicable to equilibrium thermodynamics.

Given an appropriate expression for $n_e(i)$, (4.12) can be evaluated directly by performing the summation numerically. However, as shown before, an analytical expression can be obtained by converting the sum into an integral, expanding in a Taylor series about the maximum term, and truncating. One obtains equations identical to (2.13), (2.16), and (2.17) except for the substitution of $\beta/m\beta_e$ for β/β_e and $n(i_0)/n_e(i_0)$ for $n(1)/n_e(1)$.

There are two limitations to the development presented above. First, we have assumed that there is no change in the *mechanism* for growth or decay when supersaturation changes. This assumption may not always be valid, but it is essential to the development presented thus far (e.g., for a saturated solution at a given T, the solute molecules may exist as dimers). Second, in some cases it is difficult to obtain values for $m(T, C)$. Although one can relate m to the growth velocity of macroscopic crystals, this relationship is valid only if the mechanism of growth does not change as the cluster increases in size from a critical-sized cluster to a macroscopic crystal. However, even if this difficulty does arise, one can still use nucleation rate data at one supersaturation to determine m and then predict the supersaturation dependence of the nucleation rate.

Note that in this derivation we have assumed that the solvent does not co-condense or co-precipitate with the solute. For such cases, this approach must be generalized to include the effects of the binary nucleation.

B. Nucleation with Multiple Mechanisms

The utility of the general derivation leading to (4.5) becomes apparent when one tries to examine time-dependent nucleation phenomena or when there are multiple mechanisms for the addition and loss of molecules from a cluster. One very important example of such a nucleation process occurs in the claddings of fuel rods in nuclear reactors. Energetic neutrons knock atoms from their lattice sites, creating high concentrations of vacancies and of interstitial atoms. The vacancies can cluster and grow to become voids, typically 50 to 100 Å in diameter. The interstitial atoms also nucleate and form interstitial loops, that is, an extra plane of atoms growing radially. These structural defects lead to swelling and warping of the fuel rods.

The description of this nucleation phenomenon is more complicated because the vacancies and the interstitials are matter and antimatter to each other. Not only can they combine directly with each other to produce a nothing (except for the release of energy), but each can also combine with clusters of the opposite species to make the cluster smaller by one unit. Furthermore, the clusters are capable of evaporating not only the self specie, but also the opposite specie.

In the limit that vacancies and interstitials arrive at a void at the same rate, no growth occurs. Attempts to model this phenomenon using previous versions of nucleation theory led to a great deal of confusion. In fact, it was the attempts[33] to solve this complex nucleation problem that led Wiedersich and Katz[16] to develop the general formalism presented here.

Consider the nucleation of voids in a metal containing excess concentrations of both vacancies and interstitials. The forward and backward rates are now the sum of two terms:

$$f(x) = [\beta^v + \gamma^i(x)]a(x)\dagger \tag{4.16}$$

$$b(x) = [\beta^i + \gamma^v(x)]a(x) \tag{4.17}$$

where β^v is the condensation rate (per unit area) of vacancies, $a(x)$ is the surface area of a vacancy cluster, $\gamma^i(x)$ is the rate (per unit area) at which a cluster composed of x vacancies emits an interstitial atom, β^i is the condensation rate (per unit area), and $\gamma^v(x)$ is the rate of emission of vacancies from a cluster containing x vacancies.

For the determination of the emission terms, detailed balance can be used again in an even more restricted sense. For the metal at the same temperature but with equilibrium concentrations of both vacancies and interstitials, not only is it true that $J(x)=0$ and therefore $f_e(x)n_e(x) = b_e(x+1)n_e(x+1)$, but also that the fluxes in cluster space caused by vacancies and interstitials must separately be equal to zero (i.e., at equilibrium the vacancy condensation is balanced by the vacancy emission). Thus

$$\beta_e^v a(x)n_e(x) = \gamma_e^v(x+1)a(x+1)n_e(x+1) \tag{4.18}$$

and

$$\beta_e^i a(x+1)n_e(x+1) = \gamma_e^i(x)a(x)n_e(x) \tag{4.19}$$

Solving (4.18) and (4.19) for $\gamma^v(x)$ and $\gamma^i(x)$, substituting into (4.16) and (4.17), and using (4.3) for $Z(x)$, we obtain

$$Z(x) = \prod_{j=i_0+1}^{\bar{x}} \frac{b(j)}{f(j)} = \prod_{j=i_0+1}^{\bar{x}} \frac{\beta^i a(j) + \beta_e^v a(j-1)n_e(j-1)/n_e(j)}{\beta^v a(j) + \beta_e^i a(j+1)n_e(j+1)/n_e(j)} \tag{4.20}$$

In this case $Z(x)$ does not simplify by cancellation of terms as it did in (4.12). However, the rate of nucleation can still be obtained by using (4.20) with (4.8), since it is a straightforward matter to evaluate the summation numerically.[18,33]

† We have changed the symbol for the cluster size from i to x to avoid confusion with the superscript i which here denotes interstitial.

V. FUTURE WORK

The theory presented here results from the first clear understanding of nucleation kinetics.[16–18,30] It is, however, not yet complete. To predict the equilibrium concentration of clusters $n_e(i)$, more accurate molecular theories need to be developed and included. Also very important is the incorporation of a more sophisticated kinetic theory to predict more accurately the ratio of arrival rates. The power of this treatment should make these improvements straightforward.

In addition, there are many extensions of this theory that should be made. These include extension to mixtures, associated substances, boiling nucleation and heterogeneous nucleation. Each of these extensions will require major reworking of the theory.

Acknowledgments

The ideas presented in this article grew out of work on void nucleation with Dr. Hartmut Wiedersich.[16,33] The extension to dense systems was made in collaboration with Dr. Franz Spaepen.[31] Credit for the original thoughts presented here must be fully shared with them. Financial support that made collaboration with them possible was received from the Department of Energy via its contract EY-76-C-02-2183 (Nucleation of Voids) and a fellowship from the John Simon Guggenheim Memorial Foundation.

References

1. M. Blander and J. L. Katz, *AIChE J.*, **21,** 833 (1975).
2a. D. P. Woodruff, *The Solid-Liquid Interface*, Cambridge University Press, New York, 1973, p. 171.
2b. G. W. Sears, *J. Chem. Phys. Solids*, **2,** 37 (1957).
3. P. P. Wegener, *Acta Mech.*, **21,** 65 (1975).
4. R. C. Reid, *American Scientist*, **64,** 146 (1976).
5. A. I. Berezhnoi, *Glass-Ceramics and Photo-Sitalls*, Plenum Press, New York, 1970, p. 67.
6. C. T. R. Wilson, *Philos. Trans. R. Soc. (London)*, **189,** 265 (1897).
7. A. Langsdorf, *Rev. Sci. Inst.*, **10,** 91 (1939).
8. K. Oswatitsch, *Z. Angew. Math. Mech.*, **22,** 1 (1942); *Jahrb. Dtsch. Luftfahrtforsch.*, **1,** 703 (1941).
9. F. B. Kenrick, C. S. Gilbert, and K. L. Wismer, *J. Phys. Chem.*, **28,** 1297 (1924).
10. H. Wakeshima and K. Takata, *J. Phys. Soc. Jap.*, **13,** 1398 (1958).
11a. J. L. Katz, *J. Chem. Phys.*, **52,** 4733 (1970).
11b. J. L. Katz, C. J. Scoppa II, N. G. Kumar, and P. Mirabel, *J. Chem. Phys.*, **62,** 448 (1975).
11c. J. L. Katz, P. Mirabel, C. J. Scoppa II, and T. L. Virkler, *J. Chem. Phys.*, **65,** 382 (1976).
11d. P. Mirabel and J. L. Katz, *J. Chem. Phys.*, **67,** 1697 (1977).
12. R. C. Miller, R. J. Anderson, and J. L. Kassner, Jr. in M. Kerker, ed., *Colloid and Interface Science*, vol. II, Academic, New York, 1976, pp. 1–21.
13a. B. J. C. Wu, P. P. Wegener, and G. D. Stein, *J. Chem. Phys.*, **68,** 308 (1968).
13b. P. P. Wegener and B. J. C. Wu, *Adv. Colloid Interface Sci.*, **7,** 325 (1977) and references therein.

14a. T. J. Jarvis, M. D. Donohue, and J. L. Katz, *J. Colloid Interface. Sci.*, **50,** 359 (1975).
14b. V. P. Skripov, *Metastable Liquids*, translated from the Russian by R. Kondor, Wiley, New York, 1974.
14c. R. E. Apfel, *Nature Phys. Sci.*, **238,** 63 (1972).
14d. J. G. Eberhard, W. Kremsner, and M. Blander, *J. Colloid Interface Sci.*, **50,** 369 (1975).
15a. J. L. Katz, H. Saltsburg, and H. Reiss, *J. Colloid Interface. Sci.*, **21,** 560 (1966).
15b. J. L. Katz and M. Blander, *J. Colloid Interface. Sci.*, **42,** 496 (1973).
16. H. Wiedersich, in R. S. Nelson, ed., *The Physics of Irradiation-Produced Voids*, AERER 7934, 1975, p. 147.
17. J. L. Katz and H. Wiedersich, *J. Colloid Interface Sci.*, **61,** 351 (1977).
18. H. Wiedersich and J. L. Katz, *Advances in Colloid and Interface Science.*, **10,** 33 (1979). A shortened version of this is presented in the *Proceedings of the Workshop on Correlation of Neutron and Charged Particle Damage*, CONF-760673, June 1976, p. 21.
19. E. R. Cohen, *J. Stat. Phys.*, **2,** 147 (1970).
20a. J. Lothe and G. M. Pound, *J. Chem. Phys.*, **36,** 2080 (1962).
20b. J. Lothe and G. M. Pound, *J. Chem. Phys.*, **48,** 1849 (1968).
20c. F. F. Abraham and G. M. Pound, *J. Chem. Phys.*, **48,** 732 (1968).
21a. H. Reiss, *J. Stat. Phys.*, **2,** 83 (1970).
21b. F. H. Stillinger, *J. Chem. Phys.*, **48,** 1430 (1968).
22a. H. Reiss and J. L. Katz, *J. Chem. Phys.*, **46,** 2496 (1967).
22b. H. Reiss, J. L. Katz, and E. R. Cohen, *J. Chem. Phys.*, **48,** 5553 (1968).
22c. M. Blander and J. L. Katz, *J. Stat. Phys.*, **4,** 55 (1972).
23a. R. L. Gerlach, *J. Chem. Phys.*, **51,** 2186 (1969).
23b. M. V. Buikov and V. P. Bakhanov, *J. Colloid and Interface Sci.*, **28,** 388 (1966).
23c. V. K. La Mer, *Ind. Eng. Chem.*, **44,** 1270 (1952).
23d. A. J. Barnard, *Proc. R. Soc. (London)* **A220,** 132 (1953).
23e. J. E. McDonald, *Am. J. Phys.*, **30,** 870 (1962); *Am J. Phys.*, **31,** 31 (1963).
24a. F. F. Abraham, *Homogeneous Nucleation Theory*, Academic, New York, 1974.
24b. A. S. Michaels, (ed, *Nucleation Phenomena*, American Chemical Society Publications, Washington, D.C., 1966.
24c. A. E. Nielsen, *Kinetics of Precipitation*, Pergamon, New York, 1964.
24d. J. Heicklen, *Colloid Formation and Growth*, Academic, New York, 1976.
24e. A. G. Walton, *The Formation and Properties of Precipitates*, Interscience, New York, 1967.
24f. C. A. Knight, *The Freezing of Supercooled Liquids*, van Nostrand, Princeton, N. J., 1967.
24g. R. F. Strickland-Constable, *Kinetics and Mechanisms of Crystallization*, Academic, New York, 1968.
25. J. Frenkel, *Kinetic Theory of Liquids*, Dover, New York, 1955.
26a. Ya. B. Zeldovich, *Acta. Physicochim. (U.S.S.R.)*, **18,** 1 (1943).
26b. Ya. B. Zeldovich, *J. Exp. Theort. Phys. (Russ.)* **12,** 525 (1942).
27. R. Becker and W. Döring, *Ann. Phys. (Leipzig)*, **24,** 719 (1935).
28a. H. Reiss, *J. Chem. Phys.*, **18,** 840 (1950).
28b. B. S. Holden and J. L. Katz, *AIChE J.*, **24,** 260 (1978).
29. J. O. Hirschfelder, *J. Chem. Phys.*, **61,** 2690 (1974).
30a. Yu. Kagan, *Russ. J. Phys. Chem.*, **34,** 42 (1960) (English translation) of *Zh. Fiz. Khim.*, **34,** 92 (1960).
30b. M. Blander, D. Hengstenberg, and J. L. Katz, *J. Phys. Chem.*, **75,** 3613 (1971).

31. J. L. Katz and F. Spaepen, *Phil. Mag.*, **B37,** 137 (1978).
32a. G. S. Cargill III, in H. Ehrenreich, F. Seitz, and D. Turnbull, ed., *Solid State Physics*, Academic, vol. 30, New York, 1975 p. 227.
32b. J. L. Finney, *Proc. R. Soc.*, **319A,** 479 (1970).
32c. J. D. Bernal, *Proc. R. Soc.*, **280A,** 299 (1964).
33a. J. L. Katz and H. Wiedersich, *J. Chem. Phys.*, **55,** 1414 (1971).
33b. J. L. Katz and H. Wiedersich, *J. Nucl. Materials*, **46,** 41 (1973).
33c. H. Wiedersich and J. L. Katz, *Proceedings of the Conference on Defects and Defect Clusters in B.C.C. Metals and their Alloys, AIME Nuclear Metallurgy Series*, **18,** 530 (1973).
33d. H. Wiedersich, J. J. Burton, and J. L. Katz, *J. Nucl. Materials*, **51,** 287 (1974).

DYNAMICS OF CRYSTAL GROWTH

JOHN D. WEEKS AND GEORGE H. GILMER

Bell Laboratories
Murray Hill, New Jersey 07974

CONTENTS

I. INTRODUCTION

In this article we discuss from a microscopic point of view theories for the transfer of material from a fluid to a crystalline phase. We emphasize the processes occurring in the interfacial region, where the atoms or molecules from the fluid assume the ordered arrangement of the crystal lattice. Our purpose is to evaluate the different factors that affect crystal growth, including the structure of the interface, impurities, defects in the crystal lattice, and the mobility of atoms at the interface. Morphological stability of the crystal[1] and other aspects of crystal growth associated with bulk transport are not discussed.

Recent progress in understanding fundamental aspects of interface kinetics has come primarily from studies of a kinetic Ising (or lattice gas) model[2] of the crystal-vapor interface. This model gives an atomic-scale representation of the growing crystal, including the atomic processes of condensation, evaporation, and surface migration, and can describe the structure of different crystal faces as well as the effects of lattice imperfections such as screw dislocations and impurities.[3–6] The kinetic Ising model has a relatively simple mathematical description and yet it is capable of treating all the above phenomena which are known to have important influences on the crystal growth kinetics. Thus it provides a convenient starting point for further theoretical analysis and is also well suited for computer simulation "experiments." Monte Carlo (MC) simulations can provide an arbitrarily accurate treatment of the model with all the complexities mentioned above, limited only by the amount of computer time available, whereas the theoretical methods are usually applied to more idealized situations.

Using this model, we describe several recent theoretical advances that have been made in our understanding of both the static (equilibrium) and dynamic (growth) behavior of low-index impurity-free faces of a perfect crystal. The computer simulations are also reviewed. These results provide an important test of the accuracy of the analytic methods in their regime of applicability and indicate the changes that arise in more complicated situations.

Section II gives a brief mathematical description of the model for the simplest case of the (001) solid-vapor interface of a perfect impurity free simple cubic (SC) crystal. Some limitations inherent in the model are pointed out. In Section III we review several simple theories of crystal growth for this case. These theories are derived from an exact kinetic equation which describes the impingement, evaporation, and surface migration of atoms on the crystal surface. However, the surface structure is treated by mean field approximations. These methods prove most accurate at high temperatures or high impingement rates where continuous (nonnucleated) growth occurs. At low temperatures and for small driving forces, crystal growth must proceed by a nucleation mechanism. Section IV contains a discussion of a version of the classical nucleation theory of Becker and Doering[7] applied to nucleation on a crystal surface. This serves as a basis for the development of a more detailed atomistic nucleation theory. Both approaches are compared to a MC simulation specifically designed to test the ideas of nucleation theory. The growth rates of low index faces are then related to the nucleation and spreading of clusters at the crystal surface.

None of these theories adequately describes the transition region from nucleated to continuous growth. In Section V we discuss the change in the equilibrium structure of the crystal-vapor interface that occurs at the *roughening temperature* T_R. Below T_R the surface is basically flat with only small clusters of adatoms or surface vacancies present, whereas above T_R arbitrarily large clusters can be found and the interface extends over many lattice planes. It is this change in the structure of the interface that then allows for different modes of growth above and below T_R. The static properties of the roughening transition are analyzed and related to phase transitions in several other model systems.

In Section VI dynamic properties of the roughening transition are analyzed using methods that have been successfully applied to study dynamic critical phenomena. A renormalization group analysis relates the change in growth rates above and below T_R to the changes in equilibrium correlations between different parts of the interface that occur at the roughening temperature.

In Section VII we discuss the relation of linear defects in the crystal lattice to crystal growth rates. The classical theory of spiral growth about screw dislocations is briefly reviewed and compared with recent MC calculations. For an understanding of the more complex situations where both spiral growth and two-dimensional (2d) nucleation occur, we rely mainly on the MC results. We also compare the kinetics of growth with that of evaporation and suggest reasons for the asymmetry observed in experimental data. Monte Carlo data on a crystal containing a small columnar hole exhibits a similar asymmetry.

In Section VIII we consider two aspects crystal growth in the presence of impurities: (1) impurity enhancement of 2d nucleation growth kinetics and (2) the trapping of weakly bonded impurities at the surface of a growing crystal. Again, most of the quantitative results are obtained by simulation. Final comments and conclusions are found in Section IX.

II. THE KINETIC SOS MODEL

We consider here the simple case of a (001) solid-vapor interface of an impurity-free SC crystal. More general situations can be easily described. Consider the ordinary lattice gas (Ising model) in which each site in a SC lattice is either vacant or occupied by a single atom whose interaction energy with another atom in a nearest neighbor site is ϕ. Longer-ranged interactions are ignored. If we further require that every occupied site be directly above another occupied site (thus excluding "overhangs") we obtain the *solid-on-solid* (SOS) model.[4] The SOS system can be equally well described as an array of interacting columns of varying integer heights. The surface configuration is represented by a square array of integers which specifies the number of atoms in each column perpendicular to the (001) plane, that is, the height of each column. Growth or evaporation of the crystal involves the "surface atoms" at the tops of their columns.

This SOS model is a generalization of the familiar terrace-ledge-kink model[8] in the sense that it permits clusters of adatoms, surface vacancies, and irregular step structures (see Fig. 1). [Note that the addition or removal of an atom from a *kink site* (denoted K in Fig. 1) leaves another kink site present. These repeatable step sites play an important role in the theory of crystal growth.] The SOS model is a special case of the lattice gas model, since it is required that each atom have another one directly below it, that is, overhangs are excluded. At low temperatures the SOS model is an accurate approximation to the lattice gas because the surface remains quite flat and overhangs are energetically unfavorable. The SOS model is simpler to study than the usual lattice gas model because a 2d array of height variables is required to specify a configuration rather than the 3d array of occupation numbers needed for the lattice gas.

The energy of a particular configuration (i.e., a particular choice of the set of heights $\{h_i\}$ for all the columns in the system) is determined by

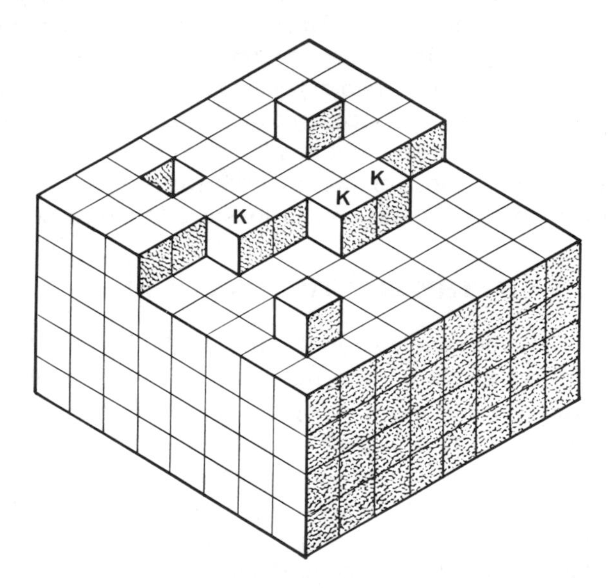

Fig. 1. Atoms on a (001) face of a SC crystal. Surface atoms may have up to four lateral neighbors. An atom in a kink site, indicated by a K in the figure, has two lateral neighbors.

counting the number of nearest neighbor bonds. Thus

$$E_{SOS}(\{h_i\}) = E_0 - \frac{\phi}{2} \sum_{j,\delta} \min(h_j, h_{j+\delta}) - \phi \sum_j h_j \tag{2.1}$$

Here E_0 is the energy of the crystal when $h_i = 0$ for all columns of atoms. The second term on the right-hand side accounts for the lateral bonds; the number of such bonds between column j and one of its nearest neighbors at $j+\delta$ is equal to the smaller of the numbers h_j and $h_{j+\delta}$. The summation in j includes all columns, and δ includes all nearest neighbors of column j. The factor of $\frac{1}{2}$ corrects for double counting. The third term accounts for the vertical bonds within a given column.

The equilibrium properties of an open system are determined by the grand canonical partition function. This is a summation over all sets $\{h_i\}$, that is,

$$\Xi = \sum_{\{h_i\}} \exp[-\beta E_{SOS}(\{h_i\}) + \beta\mu N(\{h_i\})] \tag{2.2}$$

where

$$N(\{h_i\}) = N_0 + \sum_j h_j \tag{2.3}$$

N_0 is the number of atoms in the configuration with all $h_i = 0$ and $\beta = (kT)^{-1}$ with T the absolute temperature and k Boltzmann's constant. (Throughout this article references to low or high temperatures imply values of the dimensionless temperature kT/ϕ much less than or greater than unity.) The value of the chemical potential for two-phase equilibrium is $\mu = \mu_{eq} = -3\phi$.[9] If we insert this in (2.2) and replace E and N using (2.1) and (2.3), we obtain

$$\Xi = \exp(-\beta E_0 + \beta\mu_{eq} N_0) \sum_{\{h_i\}} \exp[-(\beta J/2) \sum_{j,\delta} |h_j - h_{j+\delta}|] \tag{2.4}$$

Here $J \equiv \phi/2$, and we have made use of the identity

$$|h_j - h_{j+\delta}| = h_j + h_{j+\delta} - 2\min(h_j, h_{j+\delta}) \tag{2.5}$$

The partition function in (2.4) contains only differences between the height variables h_j and $h_{j+\delta}$. Configurations that differ by a vertical translation of an integral number of lattice spacings have identical probabilities. This is consistent with equilibrium between the vapor and crystal phases and confirms the value of the chemical potential $\mu_{eq} = -3\phi$.

Equation 2.4 can be obtained directly in the case of the equivalent spin system in zero field. The energy of the lateral pairs of overturned spins is

(neglecting an additive constant)

$$E_{\text{Ising}} = \frac{J}{2} \sum_{j,\delta} |h_j - h_{j+\delta}| \tag{2.6}$$

The total number of spins is fixed, and the canonical partition function is equivalent to (2.4). For simplicity, we often refer to the expression in (2.6) as the energy of the SOS system, although it includes chemical potential terms when the lattice gas interpretation is used.

Although no one has been able to evaluate (2.4) exactly using analytic methods, much progress has been made recently by studying related models with slightly different interaction energies.[10–12] The MC method has also been used to obtain very accurate thermodynamic and structural data for this system.[9,13,14] This work is reviewed in Section V.

Dynamics is introduced into the model by creating or annihilating atoms at random positions on the surface. This simulates the molecular exchange between the solid and vapor phases. The rate of creation (deposition) of atoms per site at the surface, denoted k^+, is assumed to be independent of the neighboring surface configurations. Physically we are envisioning the random impingement and attachment of atoms from the vapor phase. The deposition rate is proportional to the pressure in a pure vapor system, and in general we write

$$k^+ = k_{eq} \exp(\beta \Delta\mu) \tag{2.7}$$

where $\Delta\mu$ is the deviation of the chemical potential from its equilibrium value and k_{eq} is the deposition rate at equilibrium. Its value is calculated in Section III.

Although it is reasonable to assume that the deposition rate is independent of the neighboring surface configurations, the annihilation (evaporation) rate of a surface atom depends very critically on the number of nearest-neighbor bonds that must be broken in the evaporation process. We assume here that the evaporation rate of an atom with m lateral neighbors ($0 \leq m \leq 4$ in a cubic lattice) is

$$k_m = \nu e^{-m\beta\phi} \tag{2.8}$$

where ν is the evaporation rate of an isolated adatom at the surface. Thus the more neighbors an atom has, the slower is its evaporation rate. Note the very strong temperature dependence of the evaporation rate caused by the activated process of breaking bonds. The net evaporation rate, and hence the total growth rate of the crystal, depends on the detailed structure, in particular, the amount of clustering found at the interface.

Another important process in the kinetics of crystal growth is surface migration or diffusion, in which a surface atom hops from one site to an unoccupied site on the surface. Generally the rate of migration is greater than that for evaporation, since the migrating atoms may remain within the region of the attractive surface interactions. Radioactive tracer methods and measurements on the relaxation of perturbations in the structure of metal-vapor interfaces have demonstrated the importance of mass transport along the surface.[15] It is easy to see that the qualitative effect of surface diffusion is to increase the growth rate, since atoms which impinge on sites with few bonds to the crystal can jump to more favorable positions and hence are less likely to evaporate. For simplicity in the mathematical formalism that follows, we consider a restricted form of surface migration in which a surface atom on a given layer can hop only to unoccupied sites in the *same* layer. Generally this is the most important process. This restriction is not needed in the MC simulations.

The above picture models most directly growth from the vapor phase of materials with short-ranged intermolecular forces, for example vapor-deposited growth of semiconductor crystals. Other kinetic models with different impingement and evaporation probabilities may be more appropriate for melt and solution growth cases. (The energy parameter ϕ must of course also be chosen differently for these cases.) However, the equilibrium structure of the interface and (as shown in Section V) many features of the dynamic behavior of the system near the roughening transition are independent of the kinetics assumed. In what follows, we concentrate on the simple vapor growth case.

We emphasize that the above represents a probabilistic *model* for the dynamics of crystal growth. An exact description of the incorporation of atoms from the vapor into the crystal lattice would involve the solution of Newton's equations of motion for the system. The impingement from the vapor, the loss of the initial kinetic energy, and the final attachment to the lattice are idealized in our model, as is the description of the evaporation process. However, the model does give a consistent and physically reasonable description of the cooperative interactions among clusters of atoms that are crucial to the crystal growth process. A more fundamental treatment based on Newton's equations seems at present prohibitively difficult.

In comparing our results to those of real systems, one must keep in mind the limitations of the model we use. For example, the lattice structure must be chosen *a priori*, and hence this model cannot be used to investigate the formation of dislocations or extended lattice defects. However, if a lattice containing defects is specified initially, their effect on the kinetics can be measured. The assumption that atoms occupy sites in a perfect lattice is also an oversimplification. However, recent work reviewed in Sections V and VI suggests that this restriction is not important in studies of the roughening transition.

In all our discussions we assume that it is possible to achieve equilibrium conditions where the crystalline and fluid phases coexist with one another. If the temperature is below the triple point temperature, the disordered phase is the vapor, whereas it is a dense fluid above the triple point. The energy parameter ϕ must be changed to try to describe these two very different situations. The SOS model itself gives no indication of where this change should occur or what its magnitude should be. When the temperature is increased still further, it is often very difficult in the laboratory to achieve pressures large enough to maintain two-phase equilibrium and prevent the crystal from melting. Thus at high temperatures the model may describe an experimentally unrealizable situation. Further, the no-overhang restriction of the SOS model becomes increasingly unrealistic at very high temperatures. Hence the high temperature limit of the model may not be applicable to the more common crystalline materials. The predictions of the model are most realistic at low and moderate temperatures, and it is in this regime that we concentrate our efforts in the sections that follow. Properly interpreted, the kinetic SOS model provides a useful compromise between mathematical simplicity and physical reality. In this article we hope to demonstrate its utility as a versatile model of the crystal growth process.

III. SIMPLE THEORIES FOR THE DYNAMICS OF CRYSTAL GROWTH

In this section we use the kinetic SOS model to derive several simple theories for the dynamics of crystal growth. We are mainly interested in calculating the crystal growth rate as a function of the imposed driving force and the temperature for the simple case of growth on a (001) face of a SC crystal that is free of impurities and dislocations. Crystal growth in more complicated situations is discussed in Sections VII and VIII.

We first write down a set of kinetic equations which in principle provide an exact description of growth in the kinetic SOS model.[16] Although an exact solution of these equations is not possible, some approximate solutions based on mean field theory are easy to find. These prove most accurate at high temperatures and high deposition rates and provide a starting point for further developments in the continuous growth regime.

A. Exact Kinetic Equations

Let $C_n(t)$ be the fraction of sites in the nth layer parallel to some (001) reference plane which is occupied by atoms at time t. Because of the exclusion of overhangs, the fraction of surface atoms in the nth layer (i.e., atoms at the tops of their columns with the site directly above unoccupied) is given by

$$P_n = C_n - C_{n+1} \tag{3.1}$$

In other words, P_n is the probability of finding a particular column whose height extends to layer n. We suppress the argument t when no confusion will result.

By definition,

$$C_n = \sum_{k=n}^{\infty} P_k \qquad \sum_{n=-\infty}^{\infty} P_n = 1 \tag{3.2}$$

Finally, let $P_{n;m}$ be the probability that a surface atom in the nth layer has m lateral neighbors ($0 \leq m \leq 4$). Clearly,

$$P_n = \sum_{m=0}^{4} P_{n;m} \tag{3.3}$$

Then $f_{n;m}$, the fraction of surface atoms in the nth layer with m lateral neighbors (bonds), is given by

$$f_{n;m} \equiv P_{n;m}/P_n \tag{3.4}$$

It is easy to write an exact equation for the kinetic SOS model describing the rate of change of the $C_n(t)$ using (3.1) to (3.4) along with (2.8):

$$\frac{dC_n(t)}{dt} = k^+[C_{n-1}(t) - C_n(t)] - \nu \sum_{m=0}^{4} e^{-m\beta\phi} P_{n;m}(t) \tag{3.5}$$

$$= k^+ P_{n-1}(t) - k(n, t) P_n(t), \tag{3.6a}$$

$$= k^+[C_{n-1}(t) - C_n(t)] - k(n, t)[C_n(t) - C_{n+1}(t)] \tag{3.6b}$$

Here:

$$k(n, t) = \nu \sum_{m=0}^{4} e^{-m\beta\phi} f_{n;m}(t) \tag{3.7}$$

represents the *effective evaporation rate* of surface atoms in layer n at time t.

The first term on the right in (3.5) is the product of the deposition rate k^+ with the fraction of sites in layer n that are available for deposition, that is sites that are occupied in layer $n-1$ but not already occupied in layer n. The second term gives the rate at which surface atoms in layer n evaporate; this depends on the number of nearest neighbors of each atom as determined by the $P_{n;m}$ functions.

The total crystal growth rate R is the difference between the rate of deposition in all layers R^+ and the rate of evaporation from all the layers R^-:

$$R = R^+ - R^- \tag{3.8}$$

where, using (3.5) to (3.7),

$$R^+ = k^+ \sum_{n=-\infty}^{\infty} P_{n-1} = k^+ \tag{3.9}$$

and

$$R^- = \sum_{n=-\infty}^{\infty} k(n, t) P_n \tag{3.10}$$

Equation 3.5 is not a closed equation for the $C_n(t)$, since it also involves the more complicated functions $P_{n;m}(t)$ or equivalently $f_{n;m}(t)$. These functions describe the amount of clustering present during growth and hence contain the essential physics of the growth process. The amount of clustering varies dramatically with temperature and deposition rate (see Fig. 2). Rather, (3.5) is the first member of a hierarchy of equations relating lower-order distribution functions to higher-order functions.

Note in particular that with our restricted version of surface diffusion (no change in surface level on diffusion) all effects of surface diffusion are contained implicitly in the $f_{n;m}$ functions. Surface diffusion increases the growth rate by increasing the fraction of atoms having many neighbors (i.e., atoms in a cluster) which consequently are much less likely to evaporate.

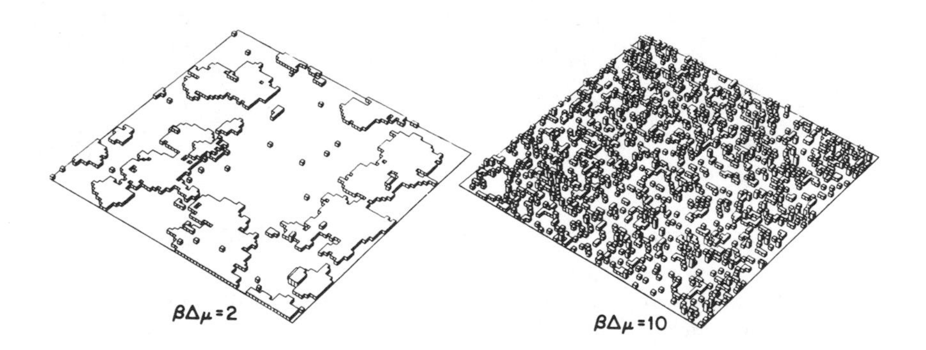

Fig. 2. Typical MC interface configurations after deposition of 25% of a monolayer on a flat (001) face of the crystal. The temperature is identical for the two cases, but $\beta\Delta\mu = 2$ for the interface on the left and $\beta\Delta\mu = 10$ on the right.

As rewritten in (3.6), all these complications are hidden in the definition of the effective evaporation rate $k(n, t)$ in (3.7). We discuss several theories of crystal growth which make some approximations based on physical intuition or mathematical convenience as to the form of the $k(n, t)$. These closures implicitly describe the amount of clustering present and allow (3.6) to be solved to give approximations to the growth profile $C_n(t)$ and the growth rate R. This basic approach is most accurate at high temperatures or high driving force where the amount of clustering can be estimated using the assumption of a random distribution of adatoms. At very low temperatures and low driving force, the more explicit cluster formation picture of classical nucleation theory is required. This is reviewed in Section IV together with recent work which puts the theory on a more fundamental basis and tests by MC simulations many of its assumptions. At higher temperatures approaching the roughening temperature, the nucleation picture breaks down. A theory describing dynamics near the roughening transition is given in Section VI.

B. Approximate Evaluation of the Evaporation Rate

1. *Wilson-Frenkel Theory*

The simplest possible assumption about the evaporation rate during growth is to assert that it is the same as that from a surface at equilibrium. This equilibrium rate can be easily calculated. A layer can, in principle, grow (or be removed) by successive creations (annihilations) of atoms only at kink sites (see Fig. 1). The *equilibrium* deposition rate $k_{eq}^{+} \equiv k_{eq}$ must then equal the kink site evaporation rate $\nu e^{-2\beta\phi}$. Further, this must equal the total equilibrium evaporation rate R_{eq}^{-} if the growth rate R in (3.8) is to be zero. Thus

$$R_{eq}^{-} = k_{eq} = \nu e^{-2\beta\phi} \tag{3.11}$$

The simple *Wilson-Frenkel* expression[17] for the growth rate is then, using (3.8) and (2.7),

$$R_{\mathrm{WF}} = k^{+} - R_{eq}^{-} = k_{eq}(e^{\beta\Delta\mu} - 1) \tag{3.12}$$

Equation 3.12 in general greatly overestimates the growth rate, since it assumes the maximum possible (equilibrium) clustering at the surface. Thus it represents an upper bound to the actual growth rate. It should be accurate at very high temperatures when essentially all atoms evaporate at the same rate and perhaps also in the case of very rapid surface diffusion. However, there is no indication of where it can be trusted and where it might fail. For small enough $\Delta\mu$, the WF growth rate is linearly proportional to the driving force (i.e., linear growth is predicted), so (3.12) fails completely in the nucleation regime.

2. *Temkin Mean Field Theory*

More ambitious approaches make use of the detailed equations of motion, (3.5) to (3.7). For this method to be useful, one must first assume a form for the effective evaporation rate $k(n, t)$. The Temkin mean field theory[18] approximates $k(n, t)$ as the evaporation rate of an atom with a "typical" number of nearest neighbors. In our model an atom with m neighbors has an evaporation rate $\nu e^{-m\beta\phi}$ [see (2.8)]. The Temkin theory asserts

$$k^T(n, t) = \nu e^{-\langle m\rangle_n \beta\phi} \tag{3.13}$$

where $\langle m\rangle_n$ is the average number of neighbors of an atom in layer n. This in turn is approximated by the average number of neighbors in a *random distribution* of the atoms. If the probability of finding an atom at a given site in the nth layer is C_n, then $4C_n$ gives the average occupancy of the four nearest-neighbor sites in a random distribution. Thus the Temkin approximation for $k(n, t)$ is

$$k^T(n, t) = \nu e^{-4\beta\phi C_n(t)} \tag{3.14}$$

The Temkin approximation for $k(n, t)$ depends only on the average occupancy of a layer, as is characteristic of mean field approaches. Such an expression would in fact be exact for a system with infinitely long-ranged and infinitely weak forces in the lateral directions. Every atom then interacts equally with every other atom in that layer and has an evaporation rate determined by the average occupancy as given in (3.14). This limit is an artificial one, far removed from the nearest-neighbor interactions originally assumed.

Furthermore, since clustering has no effect on the energy or evaporation rate of the atoms, growth cannot proceed by a nucleation mechanism. As a result, the mean field equations always give a region of "metastable states" where the predicted growth rate is rigorously zero, although there is a finite driving force applied to the system. These metastable states are artifacts of the mean field approximation. In fact, because nucleation can occur (even if it is very improbable), there will always be some growth in response to a finite driving force, even well below the roughening temperature. The critical value of the driving force required to produce growth in mean field theory is much greater than that found by the MC calculations. (In the latter case there is always some growth in response to a finite driving force, but we can define a region where the growth rate is negligibly small.) Thus mean field theory does not give even a qualitatively accurate description of growth below T_R.

Above T_R the width of the metastable state regime becomes very much smaller (but never disappears entirely), and mean field theory offers a quantitatively accurate theory over a rather wide range of nonzero driving forces. It also correctly describes the approach at very high temperatures to the Wilson-Frenkel limiting growth law.

Strictly speaking, however, because there are metastable states even above T_R, the mean field theory fails to provide the limiting linear growth law that should result from the application of a very small (infinitesimal) driving force. It is also incapable of describing the effect of surface diffusion on the growth rate. Since surface diffusion does not change the average layer concentration in our model, the mean field expression, (3.14), predicts the same growth rate for a system with and without surface diffusion. This must be looked on as a rather serious failing of the mean field approach.

Figure 3 compares the numerical integration of (3.6) using the mean field expression, (3.14), with the MC results (with no surface diffusion allowed) at a temperature just above T_R. Note that the mean field results exhibit a small metastable state region, whereas the MC calculations show linear growth for small $\Delta\mu$. Aside from this region, the two curves are in good agreement. The results become even better at higher temperatures. At lower temperatures the metastable state region greatly increases in width and the mean field theory is very inaccurate.

3. *Two Rate Model*

Next we review a simple and completely analytic theory of crystal growth introduced by Weeks, Gilmer, and Jackson[16] that does predict a

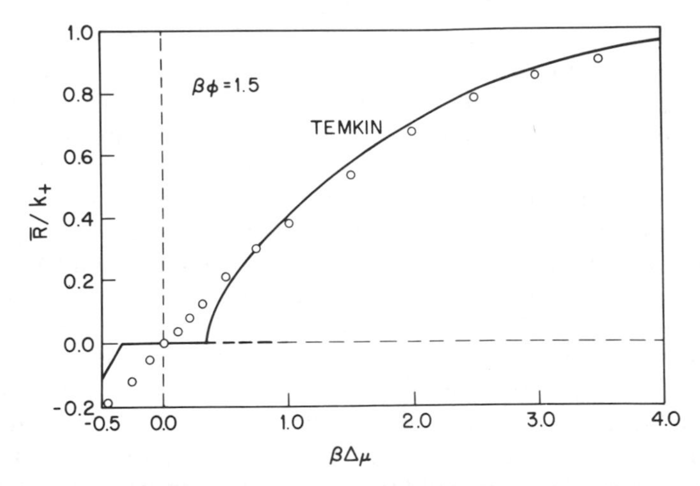

Fig. 3. Comparison of growth rates calculated by the MC method, circles, and the Temkin mean field model, solid line. The Temkin model has metastable states in the region $-0.33 < \beta\Delta\mu < 0.33$. Here $\beta\phi = 1.5$.

linear growth rate at small driving force. Provided a single empirical parameter is chosen properly, the theory gives results in quantitative agreement with MC calculations above T_R. Unfortunately, this approach also gives linear growth below T_R, but the predicted growth rate is very small and so differs little in magnitude from the correct value. The qualitative effects of surface diffusion can also be understood in this approach. A reader primarily interested in the results of this method could skip the derivation that follows and begin with (3.43).

We begin the analysis by making several simplifications and approximations in the basic equations, (3.5) to (3.7). One simplification is to study only the steady-state solution. After an initial transient period (typically after a few layers have grown), MC simulations have shown that the average growth rate $\bar{R}$ reaches a steady-state value (see Section IV). It is reasonable to suppose that when this occurs, the structure of a given layer n at time t can be related to that of layer $n-1$ at time $t-\tau$, where τ is the time required to deposit a monolayer of material on the surface. Thus we expect

$$C_n(t) = C_{n-1}(t-\tau) = C_{n+1}(t+\tau) \tag{3.15}$$

and

$$k(n, \tau) = k(n-1, t-\tau) = k(n+1, t+\tau) \tag{3.16}$$

Here we assume that at steady-state growth, the state of each successive layer repeats the conditions of the preceding layer at an earlier time. [Note from (3.7) that for (3.16) to hold we must assume the $f_{n;m}$ repeat themselves in the same manner.]

Strictly speaking, for a crystal of infinite cross-sectional area which is observed over an infinite period of time, this very plausible assumption is incorrect. Because of fluctuations in the growth condition of widely separated regions of a given layer (e.g., variations in the rate of formation and separation of critical nuclei), the density profile of an infinite crystal gradually becomes more and more diffuse. (This is true even at equilibrium if the temperature is greater than T_R. See Sections V and VI.)

This effect can be seen most clearly in the limit where evaporation is completely neglected. It is easy then to solve exactly for the $C_n(t)$ and the $f_{n;m}(t)$ and show that, despite the constant growth rate in this limit, the $C_n(t)$ never rigorously obey the steady-state conditions (3.15).[16] However, the rate of change of the interface width is very small compared to the growth rate of the crystal. Thus (3.15) and (3.16), which relate conditions on adjacent layers, hold to a very good approximation even in this rather artificial limit. Nevertheless, the steady-state ($t \to \infty$) profile is infinitely diffuse.

Under more realistic conditions, evaporation considerably inhibits the increasing diffuseness of the profile. In fact, a finite crystal then develops a well-defined steady-state growth profile whose width should depend only very weakly on the size of its cross-sectional area once it is much greater than the size of a critical nucleus. By assuming steady-state conditions as in (3.15) and (3.16), we are implicitly considering the local profile and effective evaporation rate of such a finite system. In the practical case where one studies steady-state growth on a finite crystal, (3.15) and (3.16) should be exact.

Assuming (3.15) and (3.16) hold, we can define continuous functions of the variable $x \equiv n - \bar{R}t$ such that

$$C_n(t) \equiv c(n - \bar{R}t) \equiv c(x) \tag{3.17}$$

and

$$k(n, t) \equiv k(n - \bar{R}t) \equiv k(x) \tag{3.18}$$

Here, $\bar{R}$, the average growth rate, is given by

$$\bar{R} = 1/\tau \tag{3.19}$$

where τ is the time required to deposit a monolayer on the surface. Equation 3.6 can then be rewritten with the help of (3.17) and (3.18) as

$$-\bar{R}\frac{dc(x)}{dx} = k^+[c(x-1) - c(x)] - k(x)[c(x) - c(x+1)] \tag{3.20}$$

At high deposition rates or high temperatures, the local profile $c(x)$ is a slowly varying function of x. That is, for fixed time t, $C_n(t)$ changes only a small amount from $C_n(t)$ to $C_{n\pm1}(t)$ or, for fixed n, $C_n(t)$ is smoothly varying in time from t to $t \pm \tau$. Then a Taylor series expansion of $c(x \pm 1)$ about $c(x)$ should be a good approximation:

$$c(x \pm 1) = c(x) \pm \frac{dc(x)}{dx} + \frac{1}{2}\frac{d^2c(x)}{dx^2} \pm \ldots \tag{3.21}$$

We assume here that this expansion is valid under all conditions. We keep terms through second order in the expansion (3.21) and substitute them into (3.20); expansion at least through second order is necessary to distinguish between the derivative on the left-hand side of (3.20) and the finite differences on the right. This gives the differential equation

$$-\bar{R}\frac{dc(x)}{dx} = k^+\left[-\frac{dc(x)}{dx} + \frac{1}{2}\frac{d^2c(x)}{dx^2}\right] + k(x)\left[\frac{dc(x)}{dx} + \frac{1}{2}\frac{d^2c(x)}{dx^2}\right] \tag{3.22}$$

Defining

$$\rho(x) \equiv -dc(x)/dx \tag{3.23}$$

and integrating, we get

$$\rho(x) = \rho(0) \exp\left[\int_0^x \frac{2(k^+ - k(y) - \bar{R})}{k^+ + k(y)} dy\right] \tag{3.24}$$

The origin of the coordinate system in x has not yet been fixed. We choose it such that

$$c(0) = \tfrac{1}{2} \tag{3.25}$$

Then, since $c(-\infty) = 1$ and $c(\infty) = 0$, we have

$$c(x) = 1 - \int_{-\infty}^{x} \rho(y) dy = \int_{x}^{\infty} \rho(y)\, dy \tag{3.26}$$

where $\rho(x)$ is given by (3.24), and (3.25) requires

$$\int_{-\infty}^{0} \rho(x)\, dx = \int_{0}^{\infty} \rho(x)\, dx \tag{3.27}$$

Equations 3.24 to 3.26 give the average growth rate $\bar{R}$ and profile $c(x)$ in the steady-state continuum approximation as a function of the effective evaporation rate $k(x)$. The latter of course is unknown. If we use the Temkin expression, (3.14), for $k(x)$, we obtain a complicated nonlinear integral equation. This choice is only an approximation, but it does express the basic physical fact that the effective evaporation rate is less when surface atoms have many neighbors (on the average) than when they have few.

We consider here an extremely simple approximation for $k(x)$ which still maintains this essential physical feature but also allows (3.24) to (3.26) to be solved analytically:

$$\begin{aligned} k(x) &= k_s; \qquad x \leq 0 \\ &= k_f; \qquad x > 0 \end{aligned} \tag{3.28}$$

Here k_s and k_f are constants (which may depend on T, k^+, and the rate of surface migration) whose values will be chosen later. Using (3.25), (3.28) asserts that the effective evaporation rate is some "slow" constant value k_s for layers that are more than half-filled (these presumably have more nearest neighbors) and some "fast" constant value k_f for layers that are less than half-filled (these presumably have fewer nearest neighbors). Clearly $k_f > k_s$. We refer to this approach as the two rate model.

One possible (but rather extreme) choice for these parameters is

$$\begin{aligned} k_s &= \nu e^{-4\beta\phi} \\ k_f &= \nu \end{aligned} \tag{3.29}$$

Here we approximate the effective evaporation rate by that of isolated adatoms if the layer is less than half-filled and that of a completely filled layer if it is more than half-filled. More realistic choices are discussed later. We consider now the qualitative conclusions one can make independent of the particular choice of these parameters.

Substituting (3.28) into (3.24), we find

$$\rho(x)=\rho(0)\exp\left\{\frac{2(k^+-k_s-\bar{R})x}{k_s+k^+}\right\},\qquad x<0 \tag{3.30}$$

$$\rho(x)=\rho(0)\exp\left\{\frac{2(k^+-k_f-\bar{R})x}{k_f+k^+}\right\},\qquad x>0 \tag{3.31}$$

Equation 3.27 then implies

$$\frac{2(k^+-k_s-\bar{R})}{k_s+k^+}=\frac{-2(k^+-k_f-\bar{R})}{k_f+k^+} \tag{3.32}$$

thus determining the average growth rate $\bar{R}$ as

$$\bar{R}=\frac{(k^+)^2-(k_sk_f)}{k^++[(k_s+k_f)/2]} \tag{3.33}$$

Then the interface profile $c(x)$ from (3.25) and (3.26) is

$$\begin{aligned} c(x)&=\tfrac{1}{2}e^{-x/L}, \qquad x>0\\ &=1-\tfrac{1}{2}e^{x/L}, \qquad x<0 \end{aligned} \tag{3.34}$$

where L is

$$L=\frac{k^++[(k_s+k_f)/2]}{k_f-k_s} \tag{3.35}$$

L gives a measure of the interface width.

Equation 3.33 shows that $\bar{R}=0$ when $k^+=(k_sk_f)^{1/2}$. Since there is no growth at equilibrium, we fix the product $(k_sk_f)^{1/2}$ as

$$(k_sk_f)^{1/2}=k_{eq}=\nu e^{-2\beta\phi} \tag{3.36}$$

Note that the choices for k_s and k_f in (3.29) obey (3.36). Having required (3.36) to hold, there is only one parameter, say,

$$\bar{k}\equiv(k_s+k_f)/2 \tag{3.37}$$

left free to vary.

Recalling the WF growth rate, (3.12), and using (3.37), the average growth rate in (3.33) can be written as

$$\bar{R}=\frac{(k^+)-k_{eq}^2}{\bar{k}+k^+}=R_{\mathrm{WF}}\left[\frac{k_{eq}+k^+}{\bar{k}+k^+}\right] \tag{3.38}$$

The interface width L in (3.35) can also be written, using (3.31) and

(3.37), as

$$L=\frac{\bar{k}+k^+}{\sqrt{\bar{k}^2-k_{eq}^2}} \tag{3.39}$$

Equation 3.38 shows that the growth rate tends to the WF result for two limiting cases. The first is at very high deposition rates where $k^+ \gg \bar{k}$. The second is at very high temperatures where $\beta\phi \ll 1$. Then, $k_s \to k_f$, since the effects of the bonds on the evaporation rate become unimportant. From (3.36) to (3.38) we see $\bar{R} \to R_{\mathrm{WF}}$.

Note that the interface width given by (3.39) becomes increasingly large as either the deposition rate or the temperature is increased, in agreement with MC calculations and physical intuition.

Since k_f must be positive, we can define a new parameter $\bar{c}$ such that

$$k_f = \nu e^{-4\beta\phi\bar{c}} \tag{3.40}$$

In analogy with the Temkin expression, (3.14), we physically interpret $\bar{c}$ as some "effective" probability per surface atom of finding a neighboring site occupied when averaged over the region $x>0$. Using (3.36) and (3.37), we then have

$$\bar{k}/k_{eq} = \cosh\left[\alpha\beta\phi\right] \tag{3.41}$$

where

$$\alpha \equiv 2(1-2\bar{c}) \tag{3.42}$$

Using (2.3), (3.33), and (3.39), the growth rate can be written

$$\frac{\bar{R}}{k^+}=\frac{2\sinh(\beta\Delta\mu)}{\exp(\beta\Delta\mu)+\cosh(\alpha\beta\phi)} \tag{3.43}$$

and the interface width is

$$L=\frac{\exp(\beta\Delta\mu)+\cosh(\alpha\beta\phi)}{\sinh(\alpha\beta\phi)} \tag{3.44}$$

On physical grounds the parameter $\bar{k}$ and hence $\bar{c}$ or α should be a function of the temperature and the deposition and surface diffusion rates. Here we take a semiempirical approach and choose α to best fit experimental data. We rather arbitrarily choose α such that the growth rate given in (3.43) agrees with the MC value for $\beta\phi=4$ when $\bar{R}/k^+=\frac{1}{2}$. This choice is $\alpha=1.3$ or $\bar{c}=0.175$. Figure 4 compares the growth rates given by (3.43) with $\alpha=1.3$ to the MC calculations for several different values of $\beta\phi$. Note that only the point at $\beta\phi=4$ and $\bar{R}/k^+=0.5$ has been used to fit the parameter α. Figure 5 repeats the most physically relevant part of the data on an expanded scale. No surface diffusion was allowed in the MC results.

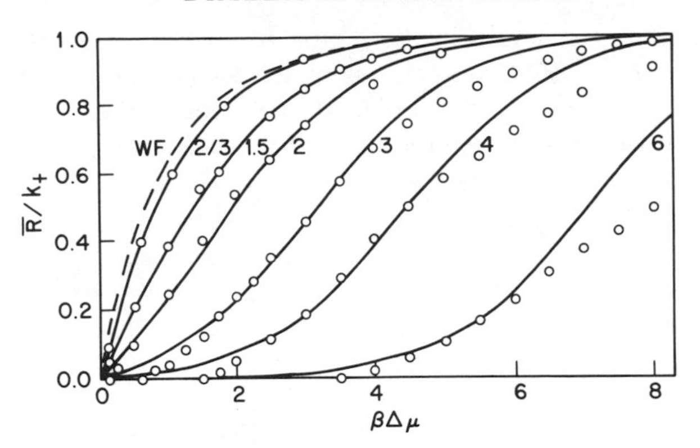

Fig. 4. The growth rate as calculated by the two-rate model, solid curves, is compared to MC data, circles, for a wide range of parameters. The number adjacent to the circles indicates the value of $\beta\phi$, and the dashed curve is the Wilson-Frenkel rate.

There is good qualitative agreement between theory and computer "experiment" over a very wide range of choices of temperature and deposition rate. In the region of high temperature and high deposition rates, where the original assumptions leading to the model are valid, the theory provides quantitatively accurate results.

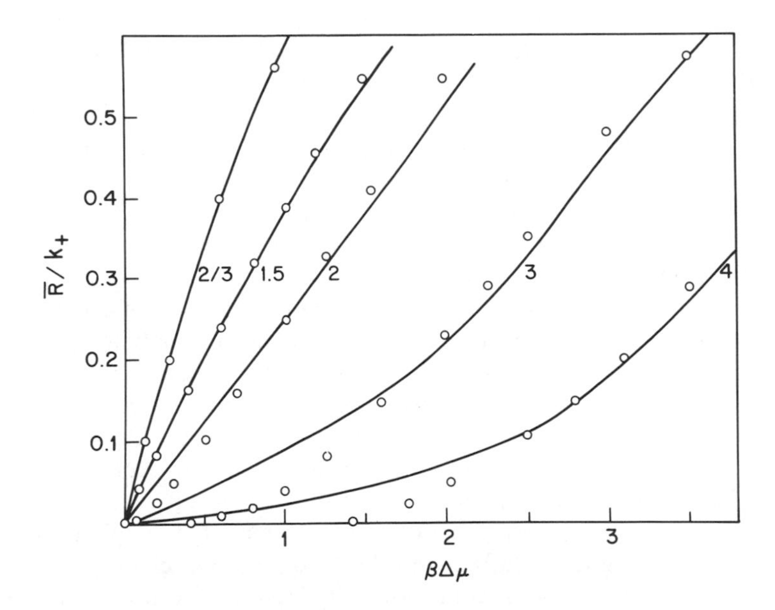

Fig. 5. An expanded plot of some of the data shown in Fig. 4.

The theory is least accurate in the region of low temperatures and low deposition rates, precisely that region in which the continuum approximation and our crude estimates for $k(x)$ are most in error. Figure 5 concentrates on this region. Equation 3.43 predicts that as $\beta\Delta\mu \to 0$ or $k^+ \to k_{eq}$

$$\bar{R}/k^+ \propto \beta\Delta\mu \tag{3.45}$$

In the low temperature, low deposition rate region, growth occurs by a nucleation mechanism, and the expected dependence is (see Section IV)

$$\bar{R}/k^+ \propto \exp(-c/\beta\Delta\mu) \tag{3.46}$$

For small $\beta\Delta\mu$, (3.43) predicts growth rates whose magnitudes are very large relative to those of (3.46).

However, the actual growth rate is in any case very small in this region, and (3.43) is not far off in absolute magnitude from the MC results. Furthermore, the theory gives the correct temperature dependence of the growth rate.

The above results were derived assuming that there was no surface diffusion. The qualitative effects of surface diffusion show up in the two rate model in the choice of the parameter $\bar{c}$. Because clustering is enhanced with surface diffusion present, a larger value of $\bar{c}$ or, equivalently, a smaller value of α would be appropriate. Using (3.43), we see that the growth rate of a crystal with surface diffusion (and smaller α) is equal to that of a system without surface diffusion but at a smaller value of $\beta\phi$. Within the two rate model, then, the effect of surface diffusion is to shift the growth rate curves in Fig. 4 toward the WF result. This conclusion is in qualitative agreement with MC simulations at temperatures above T_R or for high driving forces.

Thus in general the two rate model has proven very successful. We have made the simplest possible assumption about the effective evaporation rate $k(x)$ which still maintains the essential physical feature that atoms with many nearest neighbors evaporate more slowly than do atoms with few neighbors. This assumption allowed us to derive analytic expressions for the growth rate and interface profile in the continuum steady-state approximation. These expressions give a qualitatively correct description of the growth rate and density profile over a very wide range of deposition rates and temperatures. Above the roughening temperature, the theory is in quantitative agreement with the MC results.

Below T_R the predicted growth rate is very small for small driving forces. However, the theory fails to predict a qualitative change in the growth rate at the roughening transition. The relative accuracy of the theory at low temperatures arises from a fortuitous cancellation of errors between the continuum approximation and two rate approximation for $k(x)$, and not to a correct description of the fundamental physics in the nucleation regime. Furthermore, the subtleties of the roughening transition cannot be dealt with by such simple methods as these. The next sections discuss the more complicated theories needed in these cases.

IV. NUCLEATION THEORY OF CRYSTAL GROWTH

A. Classical Theory of Heterogeneous Nucleation

When the temperature is far below the roughening point, none of the approximations described above is very accurate at small values of the driving force. It is necessary to include the formation of large clusters of atoms that play an essential part in the 2d nucleation process. In fact, deposition on a low index face in this regime can be described in terms of the random nucleation of 2d clusters of atoms that expand and merge with one another to form complete layers. As we shall see, a good approximation for the crystal growth rate can be obtained if accurate values of the 2d nucleation rate are available.

We first employ the classical Becker-Doering theory[7] to calculate the 2d nucleation rates on the SC(001) face. This theory provides a basis for the development of a more detailed atomistic nucleation theory. Both theories are then compared directly with a MC simulation that is specifically designed to compute nucleation rates.

Suppose that a crystal surface initially in equilibrium with its vapor at low temperatures is subjected to a finite driving force. At first most of the clusters on the surface contain only a few atoms because of the small vapor pressure corresponding to the low temperature and driving force. The small clusters are more likely to lose atoms by evaporation than to expand, since the atoms at their periphery are only weakly bonded to the crystal. However, there may be a few clusters present that are large and compact. These are more likely to grow to larger sizes. These "supercritical" clusters can be generated only by an improbable series of impingement and evaporation events, but once they are formed, they usually grow until they cover the entire crystal face or merge with other clusters at the same level.

We derive the rate of nucleation of supercritical clusters in terms of the net rates of promotion of clusters along the sequence of increasing sizes until they become stable against decomposition. At the beginning of the sequence, the net rate at which empty sites are promoted to single adatoms is

$$I_0 = k^+ - n_1 \nu \tag{4.1}$$

where n_1 is the average number of adatoms per site. Note that k^+ is an accurate expression for the rate of creation of adatoms when only a small number of the surface sites are located adjacent to adatoms or clusters. Again, this is valid for small values of the driving force. (In general k^+ represents an upper bound to this rate.)

Generalizing this procedure, we define the net rate of promotion of the n_i clusters containing i atoms to size $i+1$ as

$$I_i = \gamma_i^+ n_i - \gamma_{i+1}^- n_{i+1} \tag{4.2}$$

where γ_i^+ is the average rate of attachment of atoms to a cluster containing i atoms, and γ_{i+1}^- is the average rate of detachment from a cluster of $i+1$ atoms. We neglect the merger of two or more clusters in these equations, since the density of clusters is small.

The crux of nucleation theory is the evaluation of γ_i^+ and γ_{i+1}^-. A cluster of i atoms may exhibit a large number of different configurations, since the only requirement is that each atom be joined to another atom in the cluster by nearest-neighbor bonds. Thus the γ's are averaged over clusters with a large variation in condensation and evaporation rates. Furthermore, the densities of the various cluster configurations depend on both the temperature and the driving force. Several procedures have been proposed to solve this problem. One method that is accurate at very low temperature and high driving force is to include only the lowest energy configurations in the computation.[19,20] Exact values for the γ's can then be calculated, but the nucleation rates obtained always represent a lower limit to the actual rate.

The classical theory[7] makes use of an artificially constrained equilibrium to evaluate γ^+ and γ^-. We imagine that a constraint is imposed on the system that prevents clusters above a certain finite size from growing. The system then reaches a new (constrained) equilibrium state where the steady-state nucleation rate is zero, even in the presence of a large driving force. The cluster concentrations take on values $n_i^{(o)}$ corresponding to a state in which the net rates of creation and annihilation of all cluster sizes are equal. Thus, according to (4.2),

$$\gamma_i^+ n_i^{(o)} = \gamma_{i+1}^- n_{i+1}^{(o)} \tag{4.3}$$

This equation also follows directly from the principle of microscopic reversibility. Equation 4.3 is then used to eliminate γ_{i+1}^- in (4.2), and the result is

$$I_i = \gamma_i^+ n_i^{(o)} \left[\frac{n_i}{n_i^{(o)}} - \frac{n_{i+1}}{n_{i+1}^{(o)}} \right] \tag{4.4}$$

This expression is particularly useful for the evaluation of the nucleation rate under steady-state conditions, where $I_i = I$ for all i. Then (4.4) can be solved for I, provided n_j for some particular j is known or can be approximated. Very accurate expressions for the nucleation rate are obtained if we set $n_j = 0$, where j is sufficiently large that the corresponding cluster is not likely to decompose. The larger clusters usually merge with one another and are, in a sense, removed from the system in the form of new layers of the crystal. Dividing both sides of (4.4) by $\gamma_i^+ n_i^{(o)}$ and then summing over i, we obtain

$$I \sum_{i=0}^{j-1} \frac{1}{\gamma_i^+ n_i^{(o)}} = \frac{n_0}{n_0^{(o)}} \cong 1 \tag{4.5}$$

where $n_j = 0$ is assumed. Note that n_0 and $n_0^{(o)}$ are the fractions of empty sites during nucleation and at equilibrium, and both are approximately unity under the assumed conditions.

The values of γ_i^+ and $n_i^{(o)}$ in (4.5) can be experimentally estimated from the macroscopic diffusion rates and surface free energies in the case of 3d nucleation.[21] However, the properties of small surface clusters are very different from those in bulk material; furthermore, in the case of 2d nucleation, the macroscopic properties are not usually available. Therefore we derive expressions for $n_i^{(o)}$ on the SC(001) face from a theoretical estimate of the excess free energy associated with the presence of a cluster of i atoms,

$$\Delta G_i \cong (\phi/2) \cdot 4\sqrt{i} - i\Delta\mu \tag{4.6}$$

Here $\phi/2$ is the energy of a broken bond, and the number of broken bonds at the edge of a cluster is approximated by $4\sqrt{i}$, the exact value for square clusters. The last term in (4.6) represents the free energy gained in the transfer of atoms from the fluid to the crystal. If nucleation involves large clusters at high temperatures, the configurational entropy of the cluster should also be included in (4.6). This can be done by replacing the $T = 0$ edge energy $\phi/2$ by the step edge free energy (see Ref. 22 for a theory of the equilibrium properties of steps at finite temperatures). This is useful in cases in which the critical clusters are very large and the edge free energy per unit length is approximately equal to that of an infinite step. For the present calculation at low temperatures, the edge energy differs negligibly from the edge free energy, and the simple expression in (4.6) can be used. Accordingly,

$$n_i^{(o)} \cong \exp\left[\beta i \Delta\mu - 2\beta\phi\sqrt{i}\right] \tag{4.7}$$

The exponential has its minimum value when $i = i^* \cong (\phi/\Delta\mu)^2$, and this defines the *critical cluster size*. Clusters with $i \gg i^*$ have large equilibrium concentrations, since the free energy $i\Delta\mu$ gained by the transfer of atoms from the fluid to the crystal is much greater than the edge energy, $2\phi\sqrt{i}$.

The nucleation rate is derived from (4.5) and (4.7),

$$\frac{I}{k^+} \cong \left[\frac{1}{4}\sum_{i=0}^{j-1} \exp\left[-\beta i \Delta\mu + 2\beta\phi\sqrt{i}\right] i^{-1/2}\right]^{-1} \tag{4.8}$$

where the approximate expression $\gamma_i^+ = 4k^+\sqrt{i}$ is used. This value for γ^+ is exact in the case of direct impingement on the sites at the peripheries of square clusters. Here we have excluded surface diffusion. However, adatoms may also contribute to γ_i^+ when their mobility is large. Later we consider some of the effects of surface diffusion on the nucleation process. It is apparent that mobile adatoms may cause a dramatic increase in γ^+ and hence in I. Note that the sum in (4.8) is dominated by the terms in the vicinity of i^*, and the result is not at all sensitive to the value chosen for j provided that $j \gg i^*$.

An analytical expression for I is obtained if we replace the sum in (4.8) by an integral and change variables to $y = \sqrt{i}$:

$$\frac{I}{k^+} \cong \left[\frac{1}{2}\int_0^{\infty} e^{-\beta\Delta\mu y^2 + 2\beta\phi y}\, dy\right]^{-1}$$

or

$$\frac{I}{k^+} \cong \left[\frac{1}{2}\exp\left[\beta\phi^2/\Delta\mu\right](\beta\Delta\mu)^{-1/2}\int_{-\phi\sqrt{\beta/\Delta\mu}}^{\infty} e^{-\xi^2}\, d\xi\right]^{-1} \tag{4.9}$$

If $\Delta\mu \ll \beta\phi^2$, we have the simple result

$$I/k^+ = \frac{\sqrt{\pi\beta\Delta\mu}}{2}\exp\left[-\beta\phi^2/\Delta\mu\right] \tag{4.10}$$

and even when $\Delta\mu \to \infty$ this expression is half that of (4.9).

The nucleation rate in (4.10) is the result of a number of approximations, and for this reason it is important to test its validity before deriving expressions for the crystal growth rate. Furthermore, a test of this result also provides information on the validity of the classical approach to nucleation. Several authors have measured heterogeneous nucleation rates by the MC method.[23,24] Here we report some of our recent MC data and an atomistic nucleation theory that corresponds precisely to the conditions of the simulation model.[25] By utilizing the control over the atomic processes afforded by the MC method, we measured the nucleation rates directly without the complicating effects caused by the merger of large clusters.

We simulate nucleation on the SC(001) face using exactly the same transition probabilities employed for the growth rate calculations mentioned earlier. However, in this case a current list is maintained that contains all clusters of two or more atoms, together with the locations of the atoms that are a part of each cluster. Whenever the number of atoms in a cluster exceeds twice the number in the critical cluster, it is removed from the surface. That is, the heights of the columns corresponding to atoms in that large cluster are each reduced by one atomic spacing. In this way, only the clusters that have an important part in nucleation are present at any time. The nucleation rate is proportional to the rate at which clusters are removed. As one would expect, the rate of removal was essentially unchanged in one simulation in which the clusters were allowed to grow to a larger size.

The MC nucleation data are plotted as symbols in Fig. 6 as a function

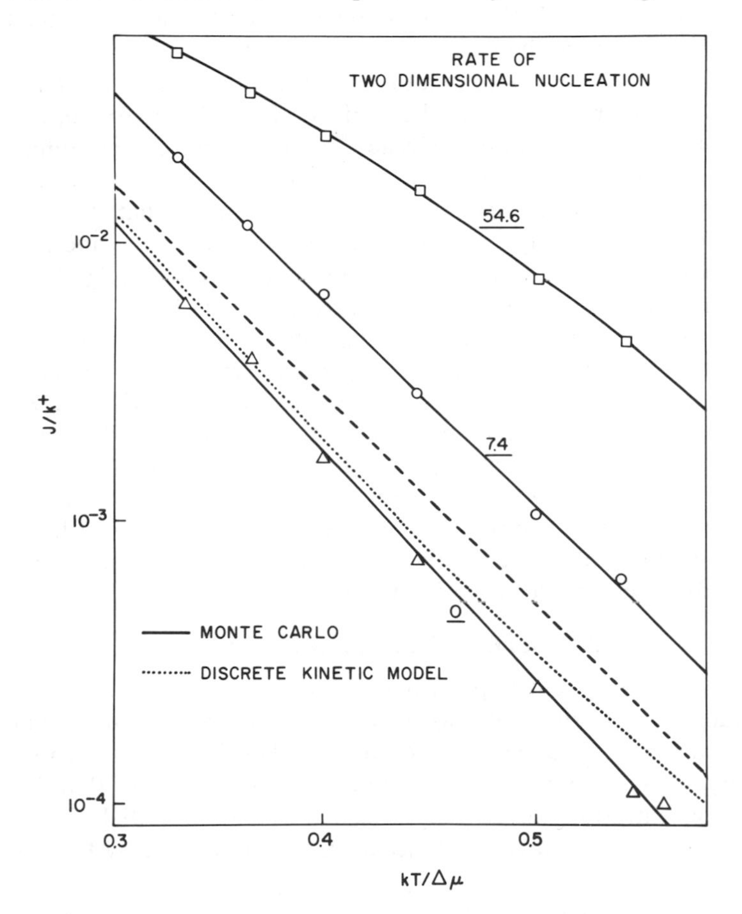

Fig. 6. Comparison of MC data on 2d nucleation rates with theory. The symbols with the solid curves through them are MC data. The numbers adjacent to these curves are the ratio of the migration to evaporation rates. The dashed curve is the classical nucleation rate (4.10). The dotted curve is the nucleation rate predicted by the atomistic nucleation theory.

of $1/\beta\Delta\mu$. Several values of the surface mobility are included, but the curve labeled with 0 corresponds to immobile adatoms as was assumed in the theory discussed here. The temperature $kT/\phi=0.25$ is less than half of the roughening temperature, and entropy effects should be small, since the edge free energy of a step is only slightly less than the edge energy. The nucleation rate predicted by (4.10) is indicated by the dashed curve; this result is somewhat higher than the MC data corresponding to zero mobility. In view of the approximate (upper bound) value of γ^+ employed in the derivation of (4.10), this discrepancy is not surprising. The expression in (4.10) is nearly linear in this plot. The fact that the MC data are approximately linear with the same slope is a good indication that the classical theory contains the essential physics in this case.

The primary effect of mobile adatoms is to increase the rate at which atoms arrive at empty sites on the surface. In addition to impinging directly from the vapor, atoms already on the surface may jump to neighboring vacant sites. If the surface were in equilibrium with the vapor, the argument of microscopic reversibility implies that the rate of impingement on an empty site would be increased by the factor $1+K_m/K^-$, where K_m/K^- is the ratio of the frequencies of migration to evaporation. The MC data labeled 7.4 corresponds to $K_m/K^-=7.4$, and indeed at small $\Delta\mu$ values they are approximately 8.4 larger than the zero mobility data. At the larger $\Delta\mu$ values, kinetic effects are apparently important, and a smaller increase is observed. The data with $K_m/K^-=54.6$ exhibits a correspondingly large nucleation rate only at the smallest values of $\Delta\mu$. At large $\Delta\mu$ values, a much smaller increase is measured. When $\Delta\mu=3kT$, about 30% of the atoms that impinge on the surface are removed with the supercritical clusters. This rapid removal rate depletes the adatom population far below the value corresponding to equilibrium, and hence the microscopic reversibility argument is not appropriate.

B. Atomistic Nucleation Theory

A qualitatively different approach to nucleation is possible when the critical nucleus is very small, as is usually the case in most vapor deposition and molecular beam experiments. Then it is feasible to solve numerically, without approximation, detailed kinetic equations that determine the concentrations of all of the clusters near the critial size. The time dependence of the adatom concentration is governed by the equation

$$\frac{dn_1}{dt}=k^+ + 2k_1^- n_2 - (4k^+ + k_0^-)n_1 \tag{4.11}$$

The first two terms on the right-hand side account for the formation of adatoms by direct impingement and by the disintegration of dimers, respectively. The other terms represent the depletion of the adatom population by a direct impingement on one of the four neighboring sites to produce a dimer (surface migration is excluded), and by evaporation. Similarly, the rate of change of the dimer population is

$$\frac{dn_2}{dt}=4k^+n_1+2k_1^-n_3-(6k^++2k_1^-)n_2 \tag{4.12}$$

The number and complexity of the equations increases when larger clusters are considered. In the case of the three-atom clusters, it is necessary to distinguish between the different possible configurations, as illustrated in Fig. 7.

We have programmed the equations for all possible configurations of clusters containing less than seven atoms. We count as the nucleation rate the rate of formation of seven-atom clusters, and these are not allowed to disintegrate. The equations are integrated with respect to time by the Runge Kutta method starting at $t=0$ with $n_i(0)=0$. When the $n_i(t)$ reach stationary values corresponding to the steady state, the nucleation rate is determined.

The results of this calculation are indicated in Fig. 6 by the dotted curve. This curve is within the statistical error of the MC data at all points except for the smallest values of $\Delta\mu$. Here the critical cluster size is in the vicinity of seven atoms, and it is not valid to assume that none of the seven atom clusters will disintegrate. This is probably the best agreement

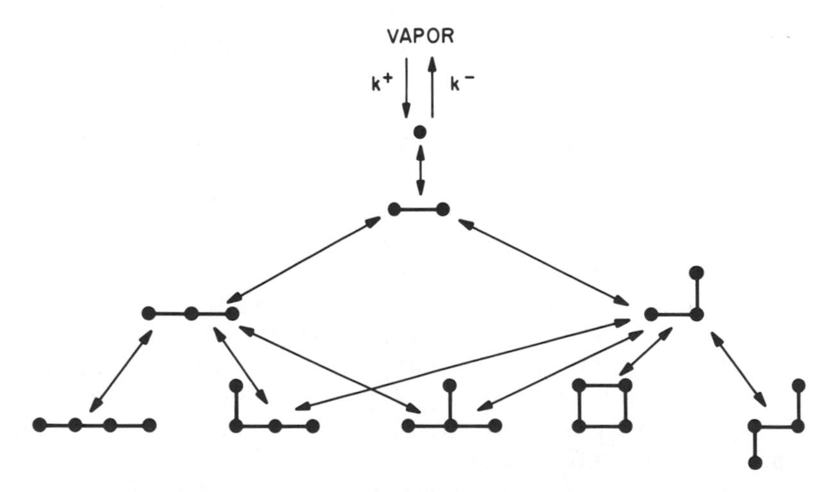

Fig. 7. Schematic diagram of the cluster "reactions" involved in nucleation on a SC(001) face. All configurations of clusters with four or less atoms are included.

between a nucleation theory and "experiment" that has been obtained. Although the theory is designed specifically to correspond to this particular MC model, the excellent agreement justifies two basic assumptions of nucleation theories in general: (*a*) the predominance of single atom additions and removals in the kinetics of cluster growth and decay and (*b*) the dominant effect of clusters close to the critical nucleus in size in determining the nucleation rate.

C. Interface Motion by 2d Nucleation

The crystal growth rate is completely determined by the nucleation rate I in certain limiting cases. If the time between nucleation events is very large, then each such event results in the growth of a complete layer of atoms,[26] and

$$R = aIN_0 \qquad (IN_0 \ll v/L) \tag{4.13}$$

Here a is the layer spacing; N_0 is the number of sites on the close-packed face, v is the average speed of expansion of the edge of a cluster, and L is the largest linear dimension of the face. The condition on I requires that the average time between nucleation events should be much longer than the time necessary for a cluster to expand and cover the entire face. Hence it is unlikely that two or more clusters would contribute to the growth of the same layer. In this limiting case, the growth rate is proportional to the surface area through the term N_0. In practice, this relation is normally applicable only to the kinetics of minute crystals or to growth rates that are so small that they are difficult to measure. Electrocrystallization provides a notable exception, since the rate of growth is proportional to the current to the electrodes and very small currents can be detected. Budevski et al.[27] applied voltage pulses to generate isolated clusters on silver crystals in small capillaries. They were then able to monitor the rate of expansion of the cluster by measuring the subsequent current flow at a potential that was too small to induce appreciable nucleation.

In most cases of practical importance, a very large number of clusters nucleate at each level, and the growth rate is independent of the surface area, but instead is determined by the rate at which the clusters nucleate and merge with one another. Kolmogoroff was the first to obtain an exact analysis of a restricted version of this problem.[28] His result is limited to the first layer deposited onto an initially flat substrate, but it also provides an estimate of the multilevel growth rate and illustrates most of the essential features of this problem.

Kolmogoroff's analysis can be applied to the nucleation of circular clusters that expand with a constant radial velocity v. Also, we assume a constant nucleation rate I starting at $t=0$. (The nucleation induction time is assumed to be negligible in comparison with the time required to deposit a layer.) Then we can calculate the probability $x_1(t)$ that an arbitrary point in the first layer is covered by a cluster. First consider the probability q_i that a nucleus formed during the interval $(t_i, t_i+\Delta t_i)$ would cover the point before time t $(t_i<t)$. The density of clusters nucleated during Δt_i is $\rho I \Delta t_i$, where ρ is the number of sites per unit area of the surface. Only those nucleated within a radius $v(t-t_i)$ of the point will reach it before time t, and hence

$$q_i = \pi v^2 (t-t_i)^2 \rho I \Delta t_i \tag{4.14}$$

The probability that the point remains uncovered at time t is the product of the probabilities that it did not get covered by nucleation events during all of the subintervals Δt_i in t, that is,

$$1-x_1(t)=\prod_i (1-q_i) \tag{4.15}$$

We may use the relation $1-q_i \cong \exp(-q_i)$ in (4.15) and (4.14) to derive the exact result (in the limit as $\Delta t_i \to 0$)

$$1-x_1(t)=\exp\left[-\pi I \rho v^2 \int_0^t (t-t')^2\, dt'\right] \tag{4.16}$$

Performing the integration using the dimensionless time variable $\tau = t(I\rho v^2)^{1/3}$, we have

$$x_1(\tau)=1-\exp\left[-\frac{\pi}{3}\tau^3\right] \tag{4.17}$$

Several authors have treated the multilevel deposition process by analytical techniques[26,29–32] and by simulation.[33,34] None of the analytical methods is exact, but one of the more satisfactory approaches is to employ the exact rate of deposition on the first layer

$$r_1(\tau) \equiv \frac{dx_1}{dt} = (I\rho v^2)\exp\left(-\frac{\pi}{3}\tau^3\right) \tag{4.18}$$

to derive an approximation for the asymptotic growth rate R as $t\to\infty$. During the interval $(t', t'+\Delta t)$ a fraction of the nth layer $r_n \Delta t$ is deposited. We then assume that the rate of deposition on this segment of layer n at a later time is equal to that on the substrate after the same elapsed time. Then (4.18) can be used to relate $r_{n+1}(t)$ to $r_n(t)$, that is,

$$r_{n+1}(t)=\int_0^t r_n(t') r_1(t-t')\, dt' \tag{4.19}$$

Actually, the deposition on the various segments of layer n is not exactly equivalent to deposition on the substrate. At $t = 0$, the substrate layer is complete, and no clusters are present at the next level. However, at any finite t', both layer n and layer $n+1$ are partially filled. An incomplete layer n should reduce the deposition rate in comparison with that on the substrate at the corresponding time, but the clusters already deposited at level $n+1$ will have the opposite effect. Because of these opposing influences, the total deposition rate on all levels predicted by (4.19) may be rather accurate. Then we assume that the instantaneous growth rate R_i is

$$R_i(t) = \sum_{n=1}^{\infty} r_n(t) \tag{4.20}$$

where the $r_n(t)$ are obtained from (4.19).

The $r_n(t)$ are most easily evaluated in terms of Laplace transforms, since (4.19) is expressed as a convolution of r_n and r_1. In the limit of $t \to \infty$, the steady-state deposition rate is obtained directly from the Laplace transformations, and the result is (see Ref. 30)

$$R = (I\rho v^2 \pi/3)^{1/3}/\Gamma(4/3) = 1.137(I\rho v^2)^{1/3} \tag{4.21}$$

We emphasize that this is not an exact solution to the multilevel deposition rate, although it has been represented as such. It is perhaps worthwhile to compare this with other estimates. The maximum rate of deposition r_1 in the first layer has been suggested as an approximation for R, and from (4.18) we obtain the value $r_1^{\max} \cong 1.194(I\rho v^2)^{1/3}$. In addition, Nielsen calculated the steady-state deposition rate by assuming a simplified surface structure.[26] His analysis yields the result $R \cong 1.015(I\rho v^2)^{1/3}$. Thus, although (4.21) is only an approximation to the asymptotic multilevel rate, it is apparently quite close to the exact value.

The edge velocity v depends on a number of factors, including the edge orientation, cluster size, and the temperature. Very small clusters have reduced edge velocities resulting from the small number of bonds connecting the atoms at the edge to others in the cluster. Once the cluster is about twice the critical size, however, this effect is small, and the edge velocity may be approximated by that of an infinite step. For our purposes it is sufficient to assume that $v \propto \Delta\mu$. This relation is always valid for small $\Delta\mu$, since the edge of a step is rough at all temperatures and there is a linear response to a small driving force. Large variations in the crystal growth rate due to changes in the temperature and driving force arise primarily from corresponding variations in I, and for this reason we have examined the nucleation process in some detail.

Crystal growth rates calculated by the MC method were discussed previously with reference to the two rate model. At low temperatures, the

MC rates were significantly below the predictions of the two rate model, but it is precisely in this region that the nucleation theory described above is most appropriate. Using I from (4.10) and the relation $v \propto \Delta\mu$, (4.21) becomes

$$R \propto (\beta\Delta\mu)^{5/6} \exp[-\beta\phi^2/3\Delta\mu] \tag{4.22}$$

If a driving force $\Delta\mu$ is applied at $t=0$, the instantaneous growth rate $R_i(t)$ approaches an asymptotic value after an initial transient. Figure 8 is a plot of the average MC growth rates measured on 90 different 60×60 sections of SC (001) faces. Each section, initially in equilibrium with the vapor, was subjected to a sudden application of the driving force. The open triangles represent the growth rates simulated without surface migration; the open squares represent growth rates obtained in the presence of surface migration, where the migration rate to sites of equal coordination is 7.4 times the evaporation rate.

The growth rates exhibit damped oscillations around the asymptotic rate R represented by the dashed lines. The numbers above the curves indicate the total number of monolayers deposited. The minima correspond roughly to the points where a layer is complete, and only small clusters have been nucleated at the next level. The amplitude of the oscillations provides a measure of the correlation in the surface heights at different sites in the array. At the beginning, the majority of the sites are

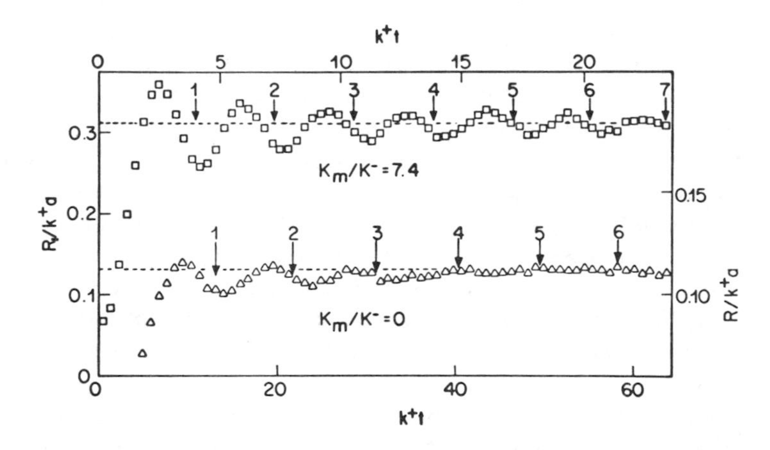

Fig. 8. The transient growth rates normalized by the impingement rate multiplied by the layer spacing. The squares represent data calculated with mobile surface atoms, and the figures at the left and above apply to this case. The triangles represent growth with immobile atoms, and the figures on the right and below apply here. Note that the growth rate scale does not extend to zero in this case. Here $\beta\phi=4$, $\beta\Delta\mu=2.5$. $\square\, k_m/k^-=7.4$ $\triangle\, k_m/k^-=0$.

at the same level, but later they are distributed over a range of levels as a result of statistical variations in the nucleation rates at the various locations. Sites close together remain highly correlated in height, since a cluster that nucleates near one site quickly spreads to the other. However, sites that are widely separated in the lateral dimension may eventually have large differences in height (see the discussion in Section III). In this case there should be little variation in the growth rate, since the different portions of the surface are experiencing the fast and slow growth regimes at different times. When the transient decays and the growth rate approaches R, the mean squared height deviations between such sites should be large and probably exceed one layer spacing.

Note the larger amplitude and greater persistence of the oscillations in the presence of mobile surface atoms. This is a consequence of the increase in the capture region near the cluster edges. Atoms that impinge within a distance of about three atomic diameters of the edge of a cluster have a good chance of migrating to the edge and being captured. This depletes the adatom concentration in this region. The surface heights of different sites remain correlated even after several layers have been deposited, since nucleation is suppressed on top of the smaller clusters. Many of the atoms that impinge on top of a cluster migrate to the edge and are captured at the lower level. The nucleation of clusters in the second layer, for example, generally occurs at a much higher coverage than was observed without surface migration.

The growth rates in Fig. 8 approach the asymptotic values quite rapidly. For example, the average growth rate during the deposition of the third monolayer is $\sim 0.97R$ without migration and $\sim 0.99R$ with migration. The maxima located at a time corresponding to about seven-tenths of a monolayer are greater than the asymptotic rate. This is in agreement with measurements of electrochemical deposition of silver on highly perfected silver crystals.[35] The more gradual increase derived from previous models[31,33,34] is probably a consequence of the idealized cluster shapes that were assumed. The clusters that appear during the MC simulation have a much longer periphery than the squares or circles assumed in the earlier models. As a result, a uniform coverage of the surface with cluster edges may occur with fewer clusters and a shorter transient results.

MC calculations for the asymptotic rates are compared with (4.22) in Fig. 9. The data indicated by triangles were measured in the absence of surface mobility; the dashed curve corresponds to (4.22) with $\beta\phi = 4$, the value employed in the simulation. The theory overestimates the growth rate at the larger values of $\Delta\mu$, because a significant fraction of the sites are adjacent to adatoms and clusters, and hence the assumptions used to

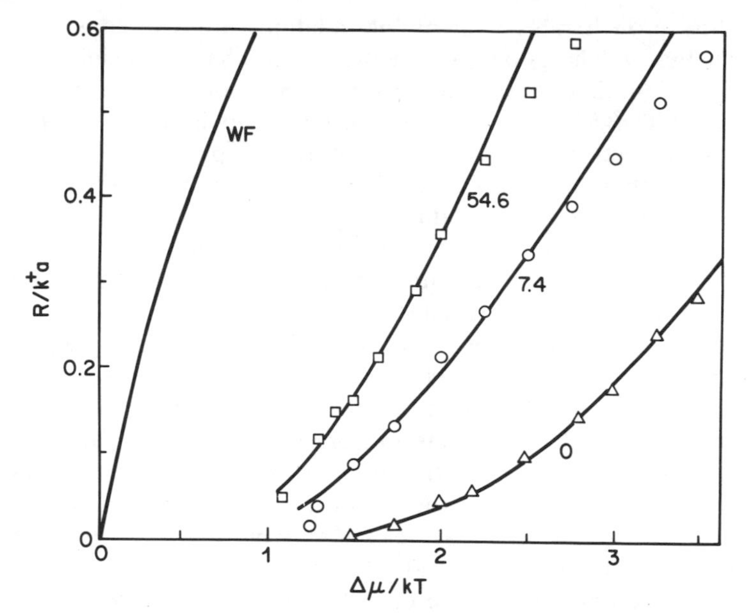

Fig. 9. MC calculations of the growth rates normalized by the impingement flux are indicated by the symbols for three values of the surface migration to evaporation ratio. The curves are derived from (4.22) except for the one labeled *WF*, which is a plot of (3.12). Again $\beta\phi = 4$.

derive the nucleation theory are not accurate. The solid curve through these data points was derived from (4.22), but with $\beta\phi = 4.4$. Thus a naive comparison of measured kinetic data with the 2d nucleation theory would suggest edge energies of clusters that are somewhat higher than the actual values.

Surface mobility causes large increases in the growth rates, and some important changes occur in the form of the growth rate curve. The adatom depletion zone around supercritical clusters can extend over large distances when they are highly mobile. Within this zone the nucleation rate is inhibited, and even a large cluster will expand more slowly if part of its periphery falls within the depletion zone of another cluster. At small values of $\Delta\mu$ the clusters are widely separated, and these effects are not important. In this region surface mobility induces the largest relative increase in the growth rate. At higher driving forces, where a dense array of clusters is nucleated, an appreciable fraction of the crystal surface may fall within the depletion zones. Then an increase in surface mobility causes a much smaller percentage change in R, and in fact, R approaches the maximum Wilson-Frenkel growth rate.

The MC growth rates with mobile adatoms are also compared with (4.22) in Fig. 9. The good agreement is somewhat surprising, since the derivation of (4.22) did not account for surface mobility. The curves were calculated with $\beta\phi = 3.2$ and 3.0 for surface mobility ratios $K_m/K^- = 7.4$ and 54.6, respectively. Thus in these cases the apparent edge energy of steps is *lower* than the actual value. This effect complicates the interpretation of experimental kinetic data.

We can conclude from the discussion of this section that growth rates of perfect crystals at low temperatures are determined primarily by the rate of 2d nucleation of clusters. The classical theory describes this process quite satisfactorily in the more idealized situations, for example when the surface mobility is low and the cluster density is low. In general, however, accurate calculations of edge free energies from kinetic measurements would require detailed information pertaining to the surface mobility. Later we discuss the 2d nucleation process in the presence of impurities, and here again we observe rather large changes in the kinetics. Crystal growth kinetics at low temperatures are extremely sensitive to the conditions in the interfacial region, and reliable interpretation of experimental data is possible only in the case of well-characterized systems.

V. THE SURFACE AT EQUILIBRIUM AND THE ROUGHENING TRANSITION

In the previous sections we discussed several different approaches to the dynamics of crystal growth. However, none of the theories was able to adequately describe the transition region between the nucleated growth regime below T_R and the continuous growth regime above T_R, or indeed even show the existence of the roughening transition. In this and the next section we examine this important transition region in some detail. We first consider the static (equilibrium) properties of a close-packed [SC (001)] face of an impurity-free perfect crystal.

At low temperatures, the interface is microscopically flat, with almost all columns the same height and only a few adatoms and surface vacancies present. As the temperature is increased, more energetic fluctuations allow for an increasing "roughening" of the surface as more and more columns slide up and down. The following argument, due in essence to Burton, Cabrera, and Frank[8], suggests that there should be a qualitative difference in the low and high temperature behavior of the interface because of a roughening phase transition.

We make use of the analogy between the lattice gas and a ferromagnetic Ising model, where an occupied site is represented by an "up" spin and a vacant site by a "down" spin.[36] Then the configuration at $T = 0$ is described in terms of 2d layers of up spins representing occupied sites in the crystal followed by layers of down spins representing the vapor. Consider the final (surface) layer of up spins. The layers directly above and below are magnetized in opposing directions, and their effects on the surface layer cancel; we have effectively an isolated 2d layer. This suggests that the crystal surface might behave like a 2d Ising model with large spin fluctuations (i.e., large regions of surface vacancies and adatoms) and thermodynamic singularities near the 2d critical temperature $kT/\phi = 0.57$.

This argument is merely suggestive, since there can be an exact cancellation only at $T = 0$. Indeed, using this idea, van Beijeren[37] has proven that the 2d critical temperature is a lower bound to T_R. Furthermore, the effects of surface roughening must extend over many layers. Nevertheless, using this one layer model, Burton, Cabrera, and Frank[8] and Jackson[3,38,39] had discussed many of the qualitative implications of surface roughening on crystal growth kinetics and morphology. The recent work reviewed below confirms their general picture and gives new insight into the true nature of the roughening transition.

The most graphic evidence for the effects of surface roughening comes from MC simulations of the SOS model.[9] Figure 10 gives typical equilibrium surface configurations generated by the MC method at various values of kT/ϕ. There seems to be a qualitative change in the surface somewhere between $kT/\phi = 0.57$ and $kT/\phi = 0.67$. At the lower temperature, distinct adatom and vacancy clusters are visible, but at the higher temperature the clusters have grown and merged together to such a degree that the original reference level of the surface is not apparent. It is already clear that crystal growth should be sensitive to the equilibrium structure. On a low temperature surface, the growth of a layer is a difficult process, requiring the formation of a large critical nucleus cluster. On a high temperature surface clusters of arbitrarily large size are already present at equilibrium, so the nucleation barrier disappears and continuous growth is possible. Further discussion of this point is given in Section VI.

Examination of these MC pictures suggests that there are several equivalent ways of characterizing the roughening transition. The size of an "average cluster" should diverge at the roughening temperature T_R. The formation of these arbitrarily large ridges also implies that the edge free energy and edge energy (per unit length) required to form a step on the crystal surface should vanish at T_R.[13] Since large clusters of adatoms and vacancies are equally probable at T_R, the average density of the surface layer should be $\frac{1}{2}$ at and above the roughening temperature. The formation of arbitrarily large clusters in one layer implies a high probability of finding similar large clusters in adjacent layers and the loss of the original reference level. Thus the interface width should diverge at T_R in

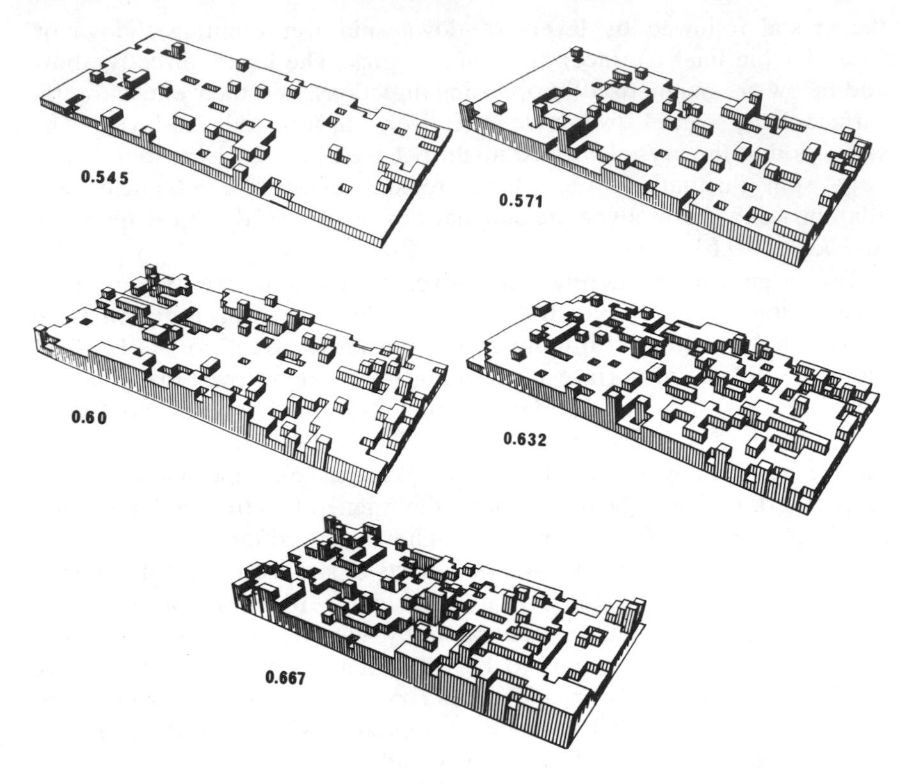

Fig. 10. Computer drawings of typical surfaces generated by the MC method at the indicated values of kT/ϕ.

an infinite system.[40] The disappearance of the nucleation barrier implies that the susceptibility (the partial derivative of the average height with respect to an infinitesimal driving force) should diverge at and above T_R.[41] The motion of the interface at and above T_R can be thought of as similar to that of a drumhead, whose normal modes of vibration correspond to the formation of large clusters of adatoms or vacancies on the surface.

The first theoretical evidence for this picture of the roughening transition using a multilayer model came from the series expansions of Weeks, Gilmer, and Leamy.[40,41] Low temperature series expansions for various measurements of the interface width and the susceptibility of the SOS model and the unrestricted lattice gas were calculated. The first nine terms of the series were evaluated, and over 3000 configurations contributed to the last term. The temperature at which each might diverge was estimated using standard series extrapolation methods which had been successfully applied to study the critical point.[42] The method indeed indicated divergences at a roughening temperature slightly greater than the 2d Ising critical temperature, in agreement with the physical picture discussed above. Of course, these results are not conclusive because of the inherent uncertainties in series extrapolation methods. The different quantities appeared to diverge at slightly different temperature values (differing by less than 10%), so more terms in the series are clearly needed to obtain a completely consistent picture. However, the qualitative agreement obtained from the relatively short series calculated thus far seems most encouraging.

The MC simulations can also be used to estimate T_R.[43] In Fig. 11, we plot the difference in concentration between the surface layer and the layer directly above. As indicated above, this should vanish at T_R, and the data appear to confirm this expectation. Also plotted is the inverse of the fluctuation in particle number (related to the inverse susceptibility) which also appears to vanish at a temperature kT_R/ϕ slightly greater than 0.6. A recent estimate, 0.62 ± 0.01, was obtained using extensive MC data for the surface height correlations.[9B]

Swendsen has also used the MC method to estimate T_R for the SOS model.[14] His estimates for its value were based primarily on an apparent divergence in the specific heat and the step specific heat with system size at a particular T_R. However, the largest system size he used was 40×40, and our experience has been that this is the minimum size needed to obtain

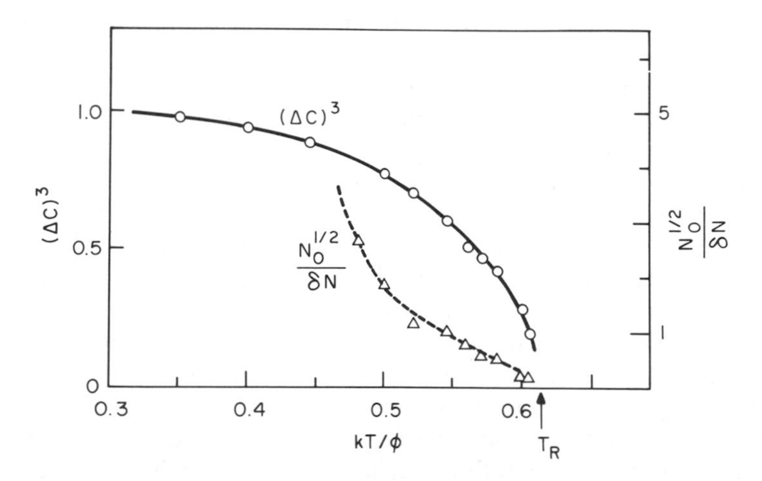

Fig. 11. MC data related to the difference in concentration ΔC in the layers bounding the $z = 0$ plane, and the reciprocal of the standard deviation δN of the number of atoms in the crystal.

accurate simulations near T_R. Thus his extrapolations using smaller system sizes are of questionable value. Furthermore, the best theoretical evidence (see below) indicates that there is no divergence in either of these quantities, but rather a maximum about 10% *below* T_R. Thus we believe that his estimate of $kT_R/\phi = 0.575 \pm 0.025$ is too low and is based on an incorrect assumption about the behavior of the specific heats.

The most definitive theoretical evidence for roughening comes from equivalence relations between models for the roughening transition and other models with a phase transition whose properties are already known. Chui and Weeks[10] considered the discrete Gaussian (DG) model of the interface with an interaction energy

$$E_{DG}(\{h_i\}) = -\frac{J}{2}\sum_{j,\delta} (h_j - h_{j+\delta})^2 \tag{5.1}$$

This differs from the ordinary SOS energy, (2.6), only in the energy assigned to neighboring columns differing in height by two or more units. The MC pictures show that such multiple height jumps between nearest neighbors are rare in the SOS model at the relatively low temperatures around the roughening point. Thus we expect (5.1) to give the same roughening behavior as does the ordinary SOS model, (2.6). (More generally, the roughening transition involves long wavelength fluctuations in the position of different parts of the interface. Changes in the interaction energy between columns that affect only the short wavelength properties should then be irrelevant at the roughening point.) Further, since the DG model assigns a greater energy to the multiple excitations, it is clear that roughening is more difficult than in the ordinary SOS model. If we can establish that the DG model has a roughening transition, then certainly the SOS model does.

Chui and Weeks showed that there is an exact relationship between the partition function for the DG model and the partition function for a 2d lattice coulomb gas. Using renormalization group methods (see Section VI for more details), Kosterlitz[44] had previously established that the coulomb gas has a phase transition from a low temperature dielectric fluid with opposite charges tightly bound together in "diatomic molecules" to a high temperature "metallic" phase with free charges and Debye screening. The properties of this transition can thus be directly related to those of the roughening transition. The DG-coulomb gas equivalence gave the first clear indication that the roughening transition has very different properties from those of the critical point in the 2d Ising model, which the simple argument of Burton, Cabrera, and Frank would associate with roughening.

The most dramatic differences show up in the behavior of the correlation length ξ. Define the height-difference correlation function

$$H(s-s')=\langle[h(s)-h(s')]^2\rangle \tag{5.2}$$

where the brackets indicate an ensemble average in the DG (or SOS) system, and s and s' locate the centers of two columns. $H(s)$ gives a measure of the average fluctuations in height between different regions of the interface separated by a distance s, and the square of the *interface width* is the $s\to\infty$ asymptotic value of $H(s)$. The *correlation length* ξ is proportional to the separation s' at which $H(s)$ is approximately equal to its asymptotic value. Using Kosterlitz's results, Chui and Weeks showed that below T_R the interface width is finite with a finite correlation length ξ. At all temperatures above T_R, however, the interface width diverges as $s\to\infty$, and the correlation length ξ is infinite. For all $T\geqslant T_R$, $H(s)$ is proportional to ln s at large separations s, so the interface width diverges logarithmically.

Kosterlitz[44] further showed that the correlation length ξ diverges very rapidly as $T\to T_R$ from below:

$$\xi\propto\exp\left[c/(T_R-T)^{1/2}\right] \tag{5.3}$$

and of course ξ remains infinite for $T>T_R$. This behavior is very different from that of the 2d Ising model where ξ diverges by a power law only at T_c. Further, the free energy has a similar form near T_R

$$F\propto\exp\left[-\frac{c'}{(|T-T_R|)^{1/2}}\right] \tag{5.4}$$

The free energy is nonanalytic at T_R, but the singularity is a very weak one, with all temperature derivatives of the singular part vanishing at T_R. Thus the roughening transition is an infinite order transition. In particular, there is no anomaly in the specific heat at T_R, again differing from the Ising model result.

As discussed before, we expect this kind of behavior to apply to a wide class of interfacial models with different interaction energies between columns. Indeed, the work of Chui and Weeks strongly suggests that the restriction of the heights of the columns to be discrete integers is not important. A model that allows continuous height variables, but has a term in the Hamiltonian energetically favoring the integer positions should have the same behavior as the discrete models. See Section VI for an analysis of the dynamics of such a model. In the static limit it indeed reproduces the static behavior of the DG model.

Phase transitions in other models can also be related to the roughening transition. Using physically plausible arguments, Kosterlitz and Thouless[44] had previously established a connection between the coulomb gas transition and a phase transition in the XY model (a model of two component spins which can rotate in the plane of the 2d lattice with an interaction energy proportional to cos θ, with θ the difference in angle between nearest neighbor spins). Hence the roughening transition is also related to the transition in the XY model. Recent work by José et al.[45] and Knops[12] have made this connection quite explicit mathematically. They have shown that a generalized SOS model with arbitrary interaction energy between nearest-neighbor columns can be related by an exact "duality transformation" to a generalized XY model in which the spin-spin interaction is some function $f(\theta)$ of the difference of angles. Presumably all these generalized XY models (for which $f(\theta)$ is analytic near $\theta = 0$) have the same behavior at their transition points.

Finally, van Beijeren[11] has established a direct connection between the *exactly solvable* six-vertex model[46] and a model for the roughening transition of the (001) face of a face-centered cubic crystal. In van Beijeren's model, the columns are perpendicular to the (001) plane, and the height of a column can differ by at most one unit from the height of the columns representing the nearest neighbor sites. Thus multiple height excitations of nearest neighbors are rigorously excluded. On physical grounds, we expect this exclusion to have no effect on the behavior at the roughening point. (van Beijeren actually considered a body-centered cubic model with second neighbor forces, but his results also apply to the face-centered cubic model with nearest-neighbor forces. We feel this interpretation is more physically appealing.)

The equivalence of this roughening model and the six-vertex model is important since the existence of the transition and many of its properties can be rigorously established.[46] The properties of the transitions in the XY model and the coulomb gas system have been analyzed by a renormalization group method which gives a very plausible but not completely rigorous description. Thus it is significant that the results of the renormalization group analysis for the XY and coulomb systems appear to be in complete agreement with the exact results of the six-vertex model. In particular, the free energy singularity in the six-vertex model is the same as that of (5.4). The correlation function $H(s)$ in (5.2) goes as ln s above T_R. Further, van Beijeren[11] showed by an explicit calculation for his model that the edge energy of a step goes rigorously to zero as $T \to T_R$, vanishing as $\exp[-c/(T_R - T)^{1/2}]$. There is no divergence in the step specific heat. Swendsen[47] has further shown that the free energy to form a step can be rigorously related to the inverse of the correlation length. Thus these results also confirm the form (5.3) for the divergence of the correlation length.

In summary, generalized SOS models which assign different energies to multiple height excitations can be exactly related to the 2d coulomb gas, the generalized XY model, and the six-vertex model. All of these systems seem to have the same behavior near the phase transition point, in agreement with our physical picture that multiple height excitations are unimportant at the roughening point. The theories confirm the divergence of the interface width, the vanishing of the edge free energy of a step, and other properties suggested by the MC and series expansion results. They provide the precise form of the behavior of these quantities as the roughening temperature is approached and show that the roughening transition has very different properties than does the one-layer model of the 2d Ising model at its critical point.

VI. DYNAMICS OF THE ROUGHENING TRANSITION

In this section we review a theory of crystal growth dynamics near the roughening point introduced by Chui and Weeks.[48] We are thus dealing with the interesting transition between sub-linear (nucleated) growth below T_R and continuous growth above. We assume that the reader has some familiarity with recent developments in the theory of dynamic critical phenomena.[49–51] The sections that follow can be read independently of this one.

We use ideas pioneered by Halperin, Hohenberg, and Ma[50,51] in their study of dynamics at the critical point. The situations are very similar; in both systems below T_R (T_c) there is a finite correlation length ξ which diverges as the roughening (critical) point is approached. However, the correlation length for all $T \geq T_R$ in the roughened phase remains infinite (as at a critical point), so the roughening point can be thought of as the low temperature end point of a line of critical points. As we saw in the previous section, the roughening point is very different from the critical point of an Ising model. Nonetheless, because the correlation length is very large compared to atomic spacings, we expect that many details of the microscopic Hamiltonian used to model the roughening transition are unimportant. Static calculations suggest, for example, that the precise form of the intercolumn interaction is unimportant, as is the restriction to discrete heights and lattice sites.[10] All such systems appear to lie in the same (static) universality class.

The fundamental idea in developing a tractable theory for dynamics at the roughening point is that of *dynamic universality*.[50,51] It is postulated that, in addition to all the properties that affect the static roughening behavior, one need consider in addition only the (hydrodynamic) conservation laws and couplings between the conserved variables. Details of the dynamics which do not affect conservation laws are irrelevant for a description of the long-wavelength low-frequency behavior of the system at the roughening point. For example, systems with and without surface diffusion should exhibit similar behavior at their respective roughening transitions.

Our model for crystal growth is particularly simple, since there are no conserved quantities such as the energy or momentum density to consider. We have postulated from the first a stochastic and purely relaxational model of crystal growth. Assuming dynamic universality, we can thus study, for example, a simple relaxational Langevin model kinetic equation[51] ("Model A") and obtain information about all members of this universality class.

We consider the following generalized SOS model Hamiltonian for the crystal-vapor system

$$H = \frac{J}{2}\sum_{j,\delta} (h_j - h_{j+\delta})^2 + Jg^2 \sum_j h_j^2 - \sum_j \Delta\mu_j h_j - 2y_0 J \sum_j \cos(2\pi h_j) \quad (6.1)$$

The first term gives the interaction energy between a column at site j (and height h_j) and its nearest neighbors at sites $j+\delta$; the second gives the interaction with a dimensionless "stabilizing field" g^2 which tends to localize the interface near $\langle h \rangle = 0$. Usually we consider the limit $g^2 \to 0^+$. The third term gives the interaction with "applied fields" $\Delta\mu_j$ which for generality can be different for different lattice sites. We later associate $\Delta\mu_j$ with the chemical potential driving force for crystal growth. The last term, parameterized by the dimensionless quantity y_0, is a *weighting function* which energetically favors integer values of the h_j. In the usual discrete lattice models, the weighting function is such as to permit only *integer values* of the h_j. Thus (6.1) can be looked on as a more general model in which continuous positions of the adatoms are permitted. Previous work indicates that any periodic weighting function will give the same static behavior at the roughening point.[10] This shows that the roughening phenomenon is not an artifact of the discrete lattice models assumed, but should also occur in more realistic models where atoms can move off their average lattice positions.

We introduce dynamics through the Langevin equation[51]

$$\begin{aligned} \frac{\partial h_j}{\partial t} &= -\frac{\Gamma \delta H}{T \delta h_j} + \eta_j \\ &= -\Gamma K^{-1} \sum_{\delta} (h_j - h_{j+\delta}) - \Gamma K^{-1} g^2 h_j + \Gamma(\Delta\mu_j / T) \\ &\qquad - 2\pi K^{-1} \Gamma y_0 \sin 2\pi h_j + \eta_j \quad (6.2) \end{aligned}$$

Here $K^{-1} = 2J/T$. (We set Boltzmann's constant equal to unity in this section.) The η_j are Gaussain fluctuating white noises which satisfy

$$\langle \eta_j(t) \rangle = 0$$
$$\langle \eta_j(t)\eta_{j'}(t') \rangle = 2\Gamma \delta_{jj'} \delta(t-t') \tag{6.3}$$

where the angular brackets indicate an ensemble average. The parameter Γ is identified later with the equilibrium (kink-site) evaporation rate. We assume that the system starts from equilibrium at $t = -\infty$ and allow the applied fields $\Delta\mu_j$ to be time dependent.

If $y_0 = 0$, then (6.2) is a linear equation and can be solved exactly by Fourier transform methods in terms of a Green's function which, in the long wavelength limit, has the form[52]

$$G(q, \omega) = [K^{-1}(q^2 + g^2) - i(\omega/\Gamma)]^{-1} \tag{6.4}$$

In the limit $g^2 \to 0^+$, which we consider hereafter, G is the Green's function for 2d diffusion. This is not surprising, since when $y_0 = 0$, (6.2) is a finite difference analog of the diffusion equation.

For nonzero y_0, (6.2) can be written

$$h(s, t) = \int_{-\infty}^{\infty} ds' \int_{-\infty}^{\infty} dt' G(s-s', t-t')[\Delta\mu(s't')/T + \eta(s't')/\Gamma - 2\pi K^{-1} y_0 \sin 2\pi h(s't')] \tag{6.5}$$

Here s is a dimensionless 2d lattice vector (the unit of length being the lattice spacing) locating the center of a column. We have taken the limit of an infinite system and replaced sums by integrals.

We analyze (6.5) using linear response theory, assuming that the driving force $\Delta\mu$ is infinitesimally small. Hence we try to predict the limiting slope of the growth rate curve as the driving force tends to zero. In addition, the linear response analysis gives valuable information about spatial and temporal correlations of the interface at equilibrium when $\Delta\mu = 0$.[51,53]

Expanding the solution of (6.5) in powers of $\Delta\mu/T$,

$$h(s, t) = h_0(s, t) + \int ds' \int dt' h_1(st, s't') \frac{\Delta\mu(s't')}{T} + O\left(\frac{\Delta\mu}{T}\right)^2 \tag{6.6}$$

the linear response function $\chi(q, \omega)$ is given by the ensemble average over the noise

$$\chi(q, \omega) = \langle h_1(q, \omega) \rangle \tag{6.7}$$

and using (6.4) to (6.6), the unperturbed ($y_0 = 0$) response function explicitly is

$$\chi_0(q, \omega) = G(q, \omega) = [K^{-1}(q^2 + g^2) - i(\omega/\Gamma)]^{-1} \tag{6.8}$$

The effect of a non-zero y_0 is conveniently expressed in terms of a self-energy $\Sigma(q, \omega)$, defined as

$$\chi^{-1}(q, \omega) = \chi_0^{-1}(q, \omega) + \Sigma(q, \omega) \tag{6.9}$$

Substituting (6.6) into (6.5), we find after some manipulation a formally exact expression for Σ given by

$$\Sigma(q, \omega) = \frac{4\pi^2 y_0 K^{-1} F\{\langle \cos [2\pi h_0(st)] h_1(st, s't')\rangle\}}{\langle h_1(q, \omega)\rangle} \tag{6.10}$$

where $F\{\ \}$ denotes a Fourier transform in space and time. Note that the term transformed is a function only of the differences $s - s'$ and $t - t'$ since the noise ensemble is stationary.

The behavior of Σ in the limit of very low temperatures is easy to analyze. The equilibrium fluctuations of h_0 are very small at low temperatures, and the weighting function localizes the interface very near $h_0 = 0$. Linearizing the sine term in (6.2) then gives a constant value for Σ of

$$\Sigma(q, \omega) \cong 4\pi^2 y_0 K^{-1} \tag{6.11}$$

Thus from Eq. (6.9) there is a finite response even in the $q, \omega \to 0$ limits at low temperatures.

At high temperatures ($T > T_R$) the situation is very different. Here the weighting function has little effect on the system. Thermal fluctuations are large enough that the interface wanders arbitrarily far from its $T = 0$ location (this delocalization characterizes the roughened phase). When $y_0 = 0$, the weighting function vanishes altogether and the response function can be calculated exactly. This divergent response function [(6.8)] presumably gives the limiting high temperature behaviour of a system with a finite y_0.

These qualitative arguments can be put on a much firmer basis by using the renormalization group method of Kosterlitz[44] and José et al.[45] We consider an expansion of the inverse linear response function $\chi^{-1}(q, \omega)$ in powers of y_0. Similar expansions have proved very useful in the static limit.[45,54] The zeroth order term $[\chi_0^{-1}(q, \omega)]$ gives the limiting ($T \to \infty$) behavior, and the higher-order terms give corrections arising from a nonzero weighting function. We use this expansion to derive differential *recursion relations* which relate the response in the original system with parameters K, Γ, and y_0 to that in a system with renormalized parameters K', Γ', and y_0'. Integration of the recursion relations in fact provides a connection for all $T \geq T_R$ between the original system and the exactly solvable system with $y_0 = 0$.

Expanding h_0, h_1, and Σ in powers of y_0, we find, using (6.5) to (6.10), after some straightforward but tedious algebra (much of which can be found in an article by de Gennes[52]), that (6.9) can be written to lowest order in q and ω as

$$\chi^{-1}(q,\omega)=\left[K^{-1}+\pi^3K^{-2}y^2\int_1^\infty ds\, s^{3-2\pi K}\right]q^2 - i\omega\left[\Gamma^{-1}+\Gamma^{-1}\frac{\pi^4y^2}{(\pi K-1)}\int_1^\infty ds\, s^{3-2\pi K}\right]+O(y^4) \quad (6.12)$$

where $y\equiv y_0\exp[-Kc]$ and c is a constant approximately equal to $\frac{1}{2}\pi^2$. Now divide the range of integration of each integral in (6.12) into two parts: 1 to b and b to ∞, with $0<\ln b\ll 1$ (i.e., b is very close to unity). The small s parts of the integration can be combined with the original constant term (either K^{-1} or Γ^{-1}) to yield a new parameter value and the large s part of the integration rescaled so that the integrals again run from 1 to ∞. The scale factor is absorbed in a redefined y variable. Equation 6.12 can thus be rewritten in exactly the same functional form with K, y, and Γ replaced by $K(l)$, $y(l)$, and $\Gamma(l)$, with $l\equiv\ln b$. This equivalence implies the differential recursion relations

$$\frac{dK(l)}{dl}=-\pi^3y^2(l) \quad (6.13)$$

$$\frac{1}{2}\frac{dy^2(l)}{dl}=-[\pi K(l)-2]y^2(l) \quad (6.14)$$

$$\frac{d\ln\Gamma(l)}{dl}=-\frac{\pi^4y^2(l)}{\pi K(l)-1} \quad (6.15)$$

subject to the boundary conditions $K(l=0)=K$, and so on.

The first two equations are essentially identical with the static recursion relations found by José et al.[45] and Nelson and Kosterlitz[54] in their analyses of the planar XY model and the 2d coulomb gas. This shows that the static behavior at the roughening transition is the same as that of the XY and coulomb gas systems at their transition points. It is easy to show by integrating these equations that $y(l)$ is driven to zero as $l\to\infty$ for all temperatures greater than the roughening temperature. This provides a justification for the original expansion in powers of y_0. The roughening temperature can thus be thought of as the low temperature end point of a fixed line of "critical" points where $y(\infty)=0$. For all $T\geqslant T_R$, the correlations in the original system with nonzero y_0 can be described in terms of the divergent response function $\chi_0(q,\omega)$, with renormalized values of the parameters K and (as shown below) Γ.

The third equation describes the behavior of the dynamical parameter Γ. Eliminating $y^2(l)$ between (6.13) and (6.15) and integrating from $l=0$

to $l=\infty$, we have

$$\frac{\Gamma(\infty)}{\Gamma}=\frac{\pi K(\infty)-1}{\pi K-1} \tag{6.16}$$

Here $\Gamma(\infty)$ and $K(\infty)$ are the renormalized values of the bare parameters Γ and K. Thus Γ effectively scales with K, whose behavior has already been discussed by Kosterlitz[44] and José et al.[45] For example, it is easy to show from (6.14) that $K(\infty)=2/\pi$ at the roughening temperature. Equation 6.16 shows that the renormalized Γ is reduced from its bare value, but does not vanish along the entire fixed line of critical points which characterizes the roughened phase including the end point at T_R. Using the language of Hohenberg and Halperin,[51] the dynamics is thus *conventional.* However, the mutual scaling of K and Γ represents an interesting and somewhat unconventional feature of the model. The static calculations have shown that $K(\infty)$ has a square root cusp[44,54] as $T\to T_R$; thus it should be possible to observe a similar anomaly in $\Gamma(\infty)$.

These results have several immediate consequences for the static and dynamic behavior of the crystal-vapor interface. For example, the average growth rate R of a crystal is related to the response to a spatially and temporally uniform driving force when the stabilizing field $g^2=0$. To first order in $\Delta\mu$ it is given by

$$R=\lim_{\omega\to 0} -i\omega\chi(q=0,\omega)\frac{\Delta\mu}{T} \tag{6.17}$$

$$=\Gamma(\infty)\frac{\Delta\mu}{T} \qquad (T\geqslant T_R) \tag{6.18}$$

Thus the theory predicts linear growth at and above T_R in agreement with conventional theories of crystal growth. In particular, note from (3.12) that the Wilson-Frenkel high temperature limiting growth law can be written to first order in $\Delta\mu$ as

$$R_{WF}=k_{eq}\frac{\Delta\mu}{T} \tag{6.19}$$

At very high temperatures, the bare and renormalized parameters become equal. Thus we can identify Γ with the equilibrium evaporation rate k_{eq}. Equation 6.18 then provides an expression for the limiting slope of the growth rate curve at lower temperatures in the continuous growth regime ($T>T_R$).

Below T_R, the situation is very different. Approaching the roughening temperature from below, the response function has the limiting form

$$\chi(q,\omega)=[K'(q^2+\xi^{-2})-i(\omega/\Gamma')]^{-1} \tag{6.20}$$

with a finite correlation length ξ and renormalized coefficients K' and Γ'. Equation 6.17 then predicts a zero growth rate for $T < T_R$ to first order in $\Delta\mu/T$. This result is consistent with the fact that growth at low temperatures occurs by a nucleation mechanism. Nucleation theory gives the result $R \propto \exp(-c/\Delta\mu)$ [see (4.22)], so in fact below T_R all terms in a power series about $\Delta\mu = 0$ should vanish.

This change in growth mechanism is directly related to the change in the *equilibrium* spatial and temporal correlations between different parts of the interface. The height-height correlation function can be immediately calculated from the fluctuation-dissipation theorem[53]

$$\langle |h_0(q, \omega)|^2 \rangle = \frac{2}{\omega} \operatorname{Im}[\chi(q, \omega)] \tag{6.21}$$

where Im[ ] denotes the imaginary part. In particular, for $T \geq T_R$ and large s or large t,

$$\langle [h_0(s, t) - h_0(0, 0)]^2 \rangle \cong \frac{K(\infty)}{2\pi} \ln \left\{ \max \left[s^2, \frac{4\Gamma(\infty)}{K(\infty)} t \right] \right\} \tag{6.22}$$

where we have used some results of de Gennes.[52] Thus there are logarithmically diverging correlations in space and time above T_R. The large distance limiting value of the equal time correlation function gives a measure of the interface width. Equation 6.22 shows that the interface width diverges logarithmically for all $T \geq T_R$. Similar remarks apply to the temporal correlations. Equation 6.22 also implies that the correlation length ξ is infinite for all $T \geq T_R$.

Below T_R, (6.20) holds and the correlation functions reaches *finite* asymptotic values exponentially fast. In particular, the interface width is finite below T_R, and there is a finite correlation length. There are many other interesting features of the roughening point that follow from a more careful analysis of the renormalization group equations. Most of these are mentioned in the preceding section.

Many of these predictions can be checked by currently available computer simulation methods. For example, we can use (6.22) to provide an important quantitative test of the theory. As mentioned before, at the roughening point $K(\infty) = 2/\pi$. Equation 6.22 then predicts for the equal time height-height correlation function at the roughening temperature

$$\langle [h_0(s) - h_0(0)]^2 \rangle \cong \frac{2}{\pi^2} \ln s \tag{6.23}$$

Shugard et al. have provided MC simulations of this correlation function for the solid-on-solid model[9] and indeed find a logarithmic correlation function dependence with a coefficient very nearly $2/\pi^2$ at the roughening point. This agreement strengthens our confidence in the general approach taken here, and in particular the ideas of static and dynamic universality that underly the method.

VII. SPIRAL GROWTH AND THE INFLUENCE OF LINE DEFECTS

The previous sections have been concerned with the properties of low index faces on perfect crystals. This is essential for an understanding of the crystal growth process. Because of the slower kinetics of the low index faces, the morphology of the growing crystal usually consists of a polyhedron bounded with these faces. However, real crystals normally contain impurities and lattice defects such as screw dislocations, and these can have a tremendous effect on the growth kinetics. In 1931 Volmer and Schultz[55] observed iodine crystals growing from the vapor and measured growth rates that were a factor of exp (1000) larger than the prediction of 2d nucleation theory! In the following sections we discuss the relation between defects or impurities and crystal growth kinetics. Much of the quantitative data is obtained by the MC technique, especially in the more complex situations in which several mechanisms contribute to the growth process.

A. Screw Dislocations

The importance of dislocations in crystal growth was first pointed out by Frank,[56] who realized that they could enhance the growth rate of low index faces by many orders of magnitude. A dislocation that terminates with a component of its Burger's vector perpendicular to the surface circumvents the difficult process of nucleating new layers. The step that is connected with this dislocation can wind up into a spiral and thereby provide a continuous source of edge positions. The nucleation of 2d clusters is unnecessary. Dislocations are especially important at low temperatures and driving forces, where the growth rate of a perfect crystal face is essentially zero. Many aspects of the spiral growth process have not been investigated. Vapor deposition experiments are usually performed in the presence of a large driving force, and 2d nucleation can proceed at an appreciable rate. The importance of dislocations is open to question in this case. Also, the relation between dislocations and hillocks on vapor-grown crystals requires consideration.

The formation of a spiral step pattern around a screw dislocation is illustrated in Fig. 12. The temperature is indicated by the ratio L/kT where L is the binding energy per atom in the lattice. This ratio has proved to be useful for comparisons with real crystals.[39] In the SC lattice

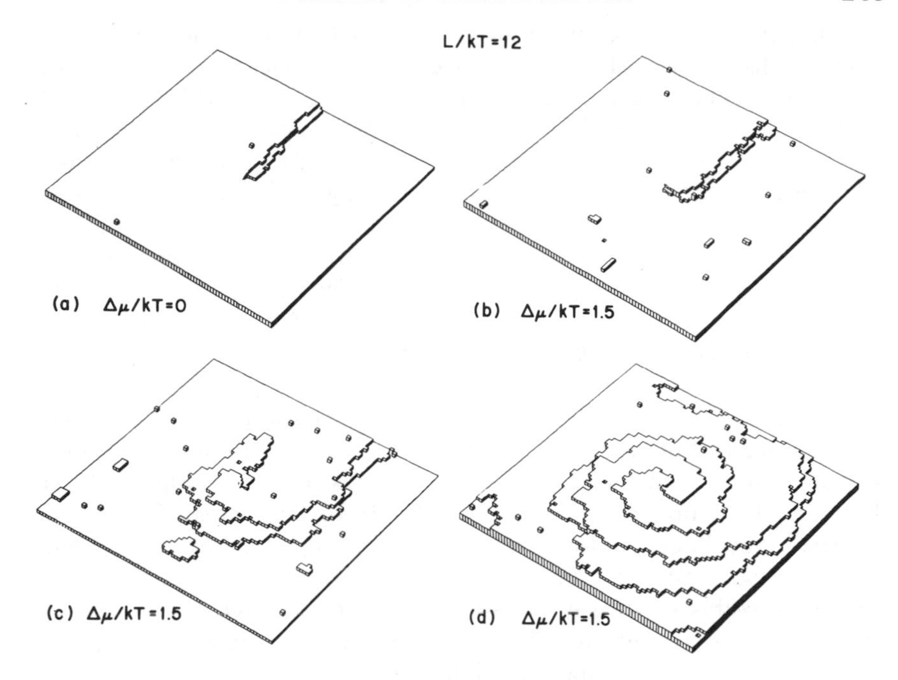

Fig. 12. An illustration of the formation of a double spiral. (*a*) is an equilibrium configuration, whereas (*b*) to (*d*) illustrate the step motion resulting from a driving force of $\Delta\mu = 1.5kT$.

$L = 3\phi$, and hence $L/kT = 12$ corresponds to $kT/\phi = 0.25$, that is, a temperature less than half T_R. These drawings correspond to configurations generated during a MC simulation. The screw dislocation intersects the surface at the center of the section illustrated. The magnitude of the Burger's vector is two, as indicated by the two associated steps. The steps are pinned at the dislocation, but a finite driving force causes them to advance along most of their length. The steps tend to separate from one another both in equilibrium and during growth. This is a consequence of the fact that it is energetically very unfavorable for steps to cross or overlap one another at any point. (The SOS model "no overhang" restriction rules out crossings entirely.)

A segment of the step at the upper level does occasionally move up so that it coincides with the lower step, but atoms impinging at the edge of this segment cause only the lower step to advance, and it then moves ahead of the upper step. This "kinetic repulsion"[57] between neighboring steps tends to produce a stable pattern of separated steps. Since one of the ends is pinned, the steps associated with the screw dislocation eventually assume an irregular spiral pattern that extends over the entire crystal face. The spiral emanating from a single dislocation can provide a uniform distribution of edge sites over faces of macroscopic dimensions, although a very long transient period would be required before steady-state conditions obtain. The spiral step pattern simply rotates about the position of the dislocation as growth proceeds and provides an inexhaustible source of edge positions where impinging atoms may attach to the crystal.

Polygonized spirals may occur at low temperatures when the free energy of a step is sensitive to its orientation.[22] Theories for the kinetics of polygonized spirals are in the literature,[58,59] and for simplicity we limit our consideration to steps with a high density of kink sites and rounded spirals.

The presence of screw dislocations permits measurable growth to occur on close-packed faces under conditions in which the 2d nucleation rate is infinitesimal. The actual growth rate depends on the average distance l between the arms of the spiral steps far from the source. Cabrera and Levine[60] calculated l in the case in which the step could be approximated by a smooth spiral curve. Surface migration and 2d nucleation were excluded in this treatment. Anisotropy of the step free energy was also neglected. The average velocity of a step segment is related to its radius of curvature ρ through the equation

$$v_\perp = v_\infty(1 - \rho_c/\rho) \tag{7.1}$$

where $v_\perp$ is the velocity measured in a direction perpendicular to the segment, ρ_c is the radius of the critical nucleus, and v_∞ is the velocity a straight step. Equation 7.1 includes the first-order correction (in ρ^{-1}) for the curvature of the step. This accounts for the fact that atoms at the edge of the curved step have fewer neighbors and are more likely to evaporate than those along the edge of a straight step. Atoms at the edge of a step with the curvature of the critical nucleus are equally likely to grow or evaporate. Hence, (7.1) is correct when $\rho = \rho_c$, and it is probably accurate for larger values of ρ provided $\rho_c \gg a$.

Apparently there is a unique spiral form that rotates around the screw dislocation without changing shape. The curvature of the step at the point where it is pinned to the dislocation is assumed to be ρ_c, since this point does not move. Dimensional arguments indicate that the asymptotic spacing l between adjacent arms is proportional to ρ_c, and $l = 19\rho_c$ according to the series solution of Ref. 60. The crystal growth rate normal to the close-packed surface is inversely proportional to l, that is,

$$R = \frac{v_\infty a}{l} = K \frac{v_\infty \Delta\mu}{\phi} \tag{7.2}$$

The critical nucleus size $i^* = (\phi/\Delta\mu)^2$ (see Section IV) is employed in the second equality to evaluate ρ_c, and $K \cong 0.1$ is a dimensionless constant determined by the ratio ρ_c/l and the geometry of the cluster.[60] At high temperatures, ϕ must be replaced by 2γ, where γ is the edge free energy per interatomic spacing along the step.

Equation 7.2 predicts a parabolic dependence of R on $\Delta\mu$ in the regime where $v_\infty \propto \Delta\mu$. However, the growth rate on a perfect crystal face in (4.22) contains $\Delta\mu$ in the denominator of a negative exponent, and hence this expression is much smaller than (7.2) in the limit as $\Delta\mu \to 0$. Neither (4.22) nor (7.2) is accurate at large driving force, and the analytical models can not determine which mechanism is the dominant mode of crystal growth in this regime.

MC calculations of growth rates on crystal faces with a screw dislocation are shown in Fig. 13 (closed symbols). 2d nucleation is also possible, and both mechanisms may contribute to the total growth rates measured. Surface migration is excluded. For comparison, the growth rates of perfect crystal faces are also plotted (open symbols). At $L/kT = 6$ ($kT/\phi = 0.5$) dislocations have little effect, even though a Burger's vector of four was used to enhance the difference between the two rates. Very precise data near the origin does reveal significant differences, since the temperature is slightly below the roughening point and a small nucleation barrier is present.

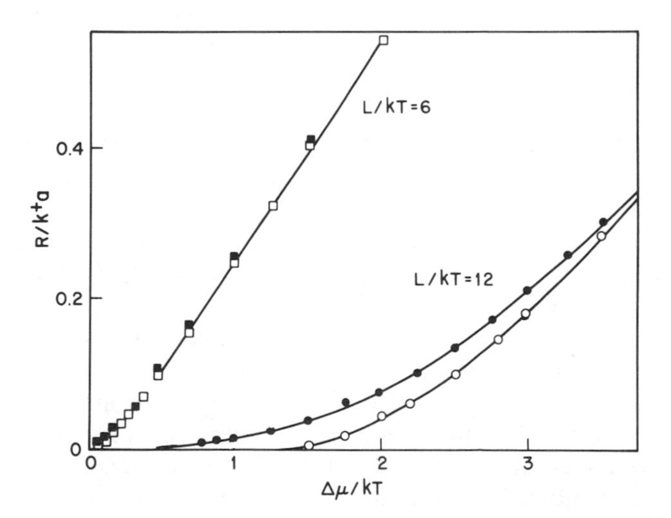

Fig. 13. MC calculations of growth on a perfect crystal and of spiral growth are compared (open and closed symbols, respectively).

The screw dislocations have a large effect on the growth kinetics at $L/kT = 12$. Finite growth rates are observed with $\Delta\mu < kT$, a region in which the perfect surface is essentially static. The relative increase in the growth rate diminishes as $\Delta\mu$ increases, and for $\Delta\mu > 3kT$ the 2d nucleation rate is so large that the additional steps afforded by the screw dislocation have little influence. (In this case a Burger's vector of two was used.)

These results suggest that the surface of a crystal containing a screw dislocation always grows faster than that of a perfect crystal. This is not surprising, since the presence of the extra steps associated with a screw dislocation provide more growth sites than 2d nucleation alone on a perfect crystal surface. Nucleation can occur in parallel with spiral growth; and this is apparent in Fig. 12 where a large cluster is present some distance away from the steps. The increase in the growth rate caused by screw dislocations varies from a fraction of a percent at high temperatures and large $\Delta\mu$ to many orders of magnitude at the opposite extreme.

The spiral growth rate without 2d nucleation can also be measured by the MC method. A special simulation of spiral growth is required in which the 2d nucleation of clusters is suppressed. This is accomplished by inhibiting the deposition of atoms on sites where they would have only one bond to the crystal, that is, $k_1^+ = 0$. (In this and the following sections it is convenient to change the notation to indicate the total coordination number of an atom; e.g., an adatom evaporation rate is k_1^-. Previously k_1^- referred to the evaporation rate of an atom with one *lateral* bond and a coordination number of two.) This does not directly affect the motion of steps, since the sites at the step edge would afford at least two bonds to impinging atoms, but it does prevent the nucleation of clusters that start as single adatoms. (It is also necessary to suppress the annihilation of any one-bonded atoms that are formed by other processes, since the transition probabilities must satisfy microscopic reversibility in equilibrium.)

The growth rate of the crystal with a screw dislocation and $k_1^+ = 0$ is plotted in Fig. 14 (open triangles), together with corresponding data from Fig. 4. The spiral growth rate at values of $\Delta\mu < 2kT$ is not noticeably perturbed by the nucleation of clusters, even though the perfect crystal growth rate is about half the spiral growth rate at $\Delta\mu = 2kT$. At larger values of $\Delta\mu$, nucleation does produce an appreciable increase in the growth rate. Note that the nucleation and spiral growth processes contribute to the combined growth rate in a highly nonlinear fashion, the sum of the two rates being larger than the combined rate. This is to be expected, since 2d clusters and steps interact in a complex way and some clusters may be incorporated into an advancing step at an early stage and contribute little to the process of crystal growth.

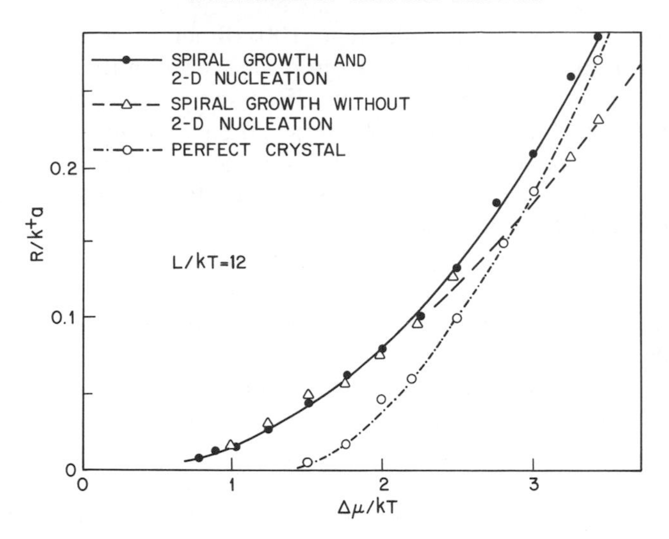

Fig. 14. Normalized growth rates of a surface intersected by a screw dislocation, with and without 2*d* nucleation. The growth rates of a perfect crystal are included for comparison. Here $\beta\phi = 4$.

These data illustrate an important advantage of the MC method over the more idealized analytical models: complex situations can be treated on a quantitative basis. Much experimental data has been expressed in the form

$$R = A(\Delta\mu)^n \tag{7.3}$$

where n is usually an irrational number greater than unity. A least-squares fit to a plot of ln (R) versus ln $(\Delta\mu)$ yields a value $n = 2.00 \pm 0.05$ in the case where $k_1^+ = 0$ (see also Ref. 59). The data of Fig. 14 correspond to a value of $n = 2.17 \pm 0.05$; the contribution of nucleation at large values of $\Delta\mu$ induces a steeper average slope in the plot. This indicates that irrational values of n may occur when several mechanisms operate in the range of driving force that is investigated.

The presence of surface diffusion does not change the fundamentals of spiral growth. The one complication that arises is the competition between adjacent arms of the spiral for the adatom flux. At small values of $\Delta\mu$ adjacent arms may be farther apart than the mean diffusion distance of an adatom along the surface. In this case, the mobility of the adatoms simply increases the value of v_∞, and (7.2) is still valid. At larger values of the driving force the growth rate is reduced by competition, and (7.2) is not applicable in this region.[61]

The value of v_∞ for an isolated step with surface migration can be calculated analytically. We assume for simplicity that the step is perfectly straight, infinitely long, and parallel to a $\langle 100 \rangle$ direction. We also assume that the average evaporation rate is k_3^- per edge atom in the step. This is equivalent to the Wilson-Frenkel law discussed in Section III. It is accurate over a much wider range of temperatures and surface mobilities when applied to steps because the edge of a step is more disordered than the close-packed face. The flux of atoms per site to the edge of the step resulting from direct impingement and evaporation is

$$Q_0 = k^+ - k_3^- \tag{7.4}$$

In addition, we must include the flux resulting from the migration of atoms to and from the step.

The probabilities c_j of finding adatoms on sites j units from the edge of the step are related by the set of conservation equations

$$dc_j/dt = k^+ + \tfrac{3}{8}k_{11}(c_{j+1} + c_{j-1}) - \tfrac{3}{4}k_{11}c_j - k_1^- c_j \tag{7.5}$$

The first term on the right corresponds to addition by direct impingement on the site, and the second accounts for a migration to the site of an atom in an adjacent row. Here k_{11} is the adatom migration rate. The factor $\frac{3}{8}$ appears, since only three of the eight possible migration jumps would move an atom from row $j+1$ to row j, for example. (The simulation permits direct jumps to all eight sites that surround the migrating atom.) The third term accounts for migration from the site, and the factor $\frac{3}{4}$ is required, since only six of the eight jumps remove an atom from row j. Finally, $k_1^- c_j$ is the evaporation rate from the site. Note that we neglect any clustering of adatoms in this formulation.

In the steady state, $dc_j/dt = 0$, and (7.5) is satisfied by an expression of the form

$$c_j = (k^+/k_1^-)(1 - Ae^{-\lambda j}) \tag{7.6}$$

where A and λ are constants. Substitution of (7.6) into (7.5) yields

$$\lambda = 2\ln[(\tfrac{2}{3}k_1^-/k_{11})^{1/2} + (\tfrac{2}{3}k_1^-/k_{11} + 1)^{1/2}] \tag{7.7}$$

The concentration c_1 at a site one unit removed from the edge of the step obeys the relation

$$dc_1/dt = k^+ + \tfrac{3}{8}k_{11}(c_2 + k_3^-/k_1^-) - \tfrac{3}{4}k_{11}c_1 - k_1^- c_1 \tag{7.8}$$

This equation is identical to (7.5) with $j = 1$, except for the replacement of c_0 in (7.5) by k_3^-/k_1^-, the equilibrium density of adatoms. This is consistent with the previous assumption that the average evaporation rate from the sites at the edge of the step is the equilibrium rate. Substitution of (7.6) into the steady-state form of (7.8) yields

$$A = (1 - k_3^-/k^+)/[2e^{-\lambda} + \tfrac{8}{3}(k_1^-/k_{11})e^{-\lambda} - e^{-2\lambda}] \tag{7.9}$$

The surface flux to the step results from the exchange between the atoms in the edge of the step and the adatoms in the adjacent row of sites; that is,

$$Q_s = \tfrac{3}{8}k_{11}(c_1 - c_0)$$
$$= \tfrac{3}{8}(k_{11}/k_1^-)[(k^+ - k_3^-) - k^+Ae^{-\lambda}] \qquad (7.10)$$

where c_1 is eliminated in the expression on the right by the use of (7.6). Equations 7.7 and 7.9 provide the needed expressions for A and λ, and after some manipulation we achieve the simple result

$$Q_s = (k^+ - k_3^-)[(\tfrac{1}{4} + \tfrac{3}{8}k_{11}/k_1^-)^{1/2} - \tfrac{1}{2}] \qquad (7.11)$$

The velocity of the step is a result of direct impingement and migration of adatoms to *both* sides, since in this theory, as in the MC model, adatoms are allowed to jump to sites at different levels. Thus

$$v_\infty = a(Q_0 + 2Q_s)$$
$$= ak^+[1 - \exp(-\Delta\mu/kT)](1 + \tfrac{3}{2}k_{11}/k_1^-)^{1/2} \qquad (7.12)$$

Figure 15 is a plot of spiral and perfect crystal growth rates, both calculated in the presence of surface migration. The dashed curves indicate the growth rates calculated without surface migration. The ratio

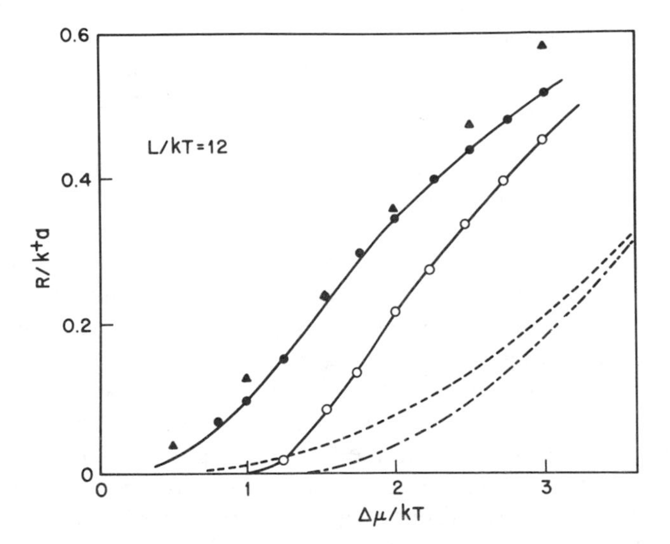

Fig. 15. Spiral and perfect crystal growth rates (closed and open circles) calculated with a surface migration to evaporation ratio of 7.4. The triangles represent the spiral growth rate calculated with (7.2) and (7.12). The dashed curves indicate the growth rates calculated without surface migration.

of the migration to evaporation rate is $k_{11}/k_1 = 7.4$ at the indicated temperature. The spiral growth rate derived from (7.12) and (7.2) is indicated by the triangles. The competition between adjacent arms of the spiral for the surface migration flux was neglected in the derivation of (7.12), and this is an important factor at larger values of $\Delta\mu$ where the arms are closely spaced.[61]

The agreement between theory and the MC data at small $\Delta\mu$ values contrasts with the observation, in the absence of surface diffusion, that the theoretical growth rates were about a factor of two larger than those calculated by the MC method.[6] The reason for the better agreement is related to the evaporation rate for atoms at the edge of the step. The actual evaporation rate during growth is generally higher than the equilibrium value, since the larger impingement rate causes some extra roughening of the step edges. Migration along the edge of the step tends to reduce this effect in a similar fashion to that described in Section III for crystal surfaces. In this case there is sufficient mobility of the atoms in the edge to maintain an evaporation rate close to the equilibrium value. According to the MC data, the addition of surface diffusion causes about a sixfold increase in the growth rate at low values of $\Delta\mu$. The increase that results from adatom diffusion to the step is ~3.5 according to the analysis above. The reduction in the evaporation rate from the step edges must account for the remaining increase. Gilmer and Bennema[62] simulated the growth of stepped surfaces at $L/kT = 6$ with and without surface migration, and in this case adatom diffusion to the steps fully accounts for the increase in the growth rate. At this temperature the edge evaporation rate is approximately k_3^-, even in the absence of surface migration.

The spiral step(s) around a screw dislocation appear as a hillock on the crystal surface when viewed on a macroscopic scale. Hillocks with steep sides may affect the performance of semiconductor devices, for example, and an important practical problem is to find conditions for growth that minimize this effect.[63] The slope of the hillock surrounding the screw dislocation of the MC model is plotted in Fig. 16. The triangles indicate the average slope during steady-state growth. The slope predicted by Cabrera and Levine[60] is indicated by the dashed line. The theory is in good agreement with the data, even at the larger values of $\Delta\mu$ where the screw dislocation has little effect on the kinetics of growth. Large fluctuations in slope are observed in this region however, and these data are less reliable. The open circles represent data on a model that includes surface migration. Although these points can not be distinguished from the other data at small $\Delta\mu$, there is some evidence that surface mobility reduces the slope at large $\Delta\mu$, as expected. Again, this is a result of competition between adjacent arms of the spiral for the adatom flux.[61,64] Thus it

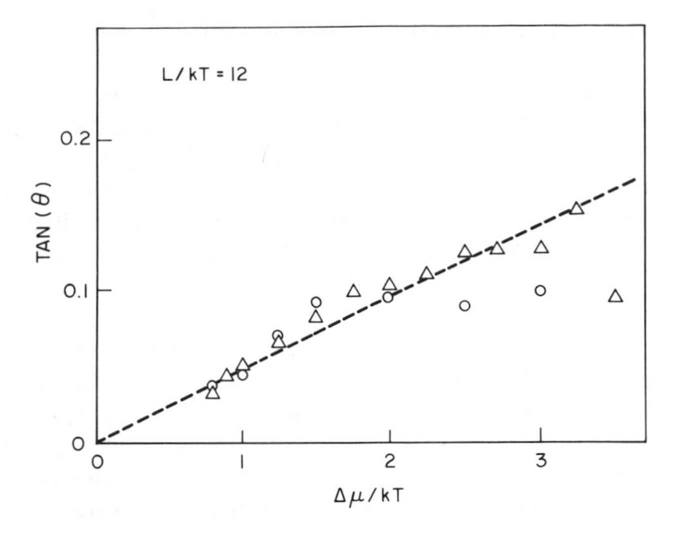

Fig. 16. The average slope of the hillock around a screw dislocation with $b=2$. The triangles and dashed line indicate the MC data in the absence of surface migration, whereas the circles indicate data taken when the adatom jump rate is ~7.4 times the evaporation rate.

appears that the smallest slopes occur on crystals that are grown at small values of $\Delta\mu$ or that have a large surface mobility. There is little or no reduction in the slope when $\Delta\mu$ is increased to the point at which 2d nucleation becomes operative.

B. Asymmetry Between Crystal Growth and Evaporation

The crystal growth mechanisms described in this and the previous sections can also operate during evaporation. Evaporation (dissolution or melting) of crystal lattice planes occurs when the chemical potential is less than the equilibrium value, or $\Delta\mu < 0$. The nucleation of 2d holes in a close-packed surface layer proceeds by the evaporation of single atoms to form surface vacancies, then divacancies, and so forth until a stable cluster of vacant sites is formed. This process is exactly analogous to the nucleation of 2d clusters during growth. Evaporation by a spiral mechanism may also occur. Atoms at the edges of steps that are associated with screw dislocations tend to evaporate preferentially. Here too the steps wind into a spiral pattern, but in the opposite sense to the growth spiral. Thus one might expect the evaporation rate of a crystal at a negative chemical potential $\Delta\mu$ to be equal in magnitude to the growth rate at the positive value $|\Delta\mu|$. However, most experimental evaporation rates are much greater than the corresponding growth rates. The explanation of this asymmetry between growth and evaporation is related to another kind of defect in the crystal lattice.

Edge dislocations are not accompanied by step segments on the surface, and they have only a small influence on the crystal growth rate. The nucleation of new lattice planes is not facilitated by the presence of these defects. In fact, a high density of dislocations may cause a small reduction in the crystal growth rate. This reduction occurs when the strain energy of the dislocations and the lower density of the imperfect crystal lattice produce a significant decrease in the binding energy of the atoms to the crystal.

On the other hand, the rate of evaporation is increased when edge dislocations are present. Part of this increase is simply the direct result of the reduced binding energy per atom in the crystal. However, the core of the dislocation with its high concentration of strain energy may have a more important effect. It can serve as a very efficient site for the 2d nucleation of holes. The reduced nucleation barrier at this point often causes a dramatic change in the evaporation rate of the crystal. Small voids and inclusions are commonly observed in crystals, and these should also reduce the growth rate and increase the evaporation rate.

Figure 17 is a plot of MC calculations of the net rate of gain (or loss) of material by a crystal with a line defect.[65] As in the previous case, $L/kT = 12$. This defect is a small (5×5) columnar hole that lies perpendicular to the (001) face on which the impingement and evaporation events are simulated. The growth rates are indistinguishable from those of the perfect crystal (see Fig. 4). The evaporation rate is appreciable even at $\Delta\mu = -kT$, whereas the growth rate is too small to measure at $\Delta\mu = kT$.

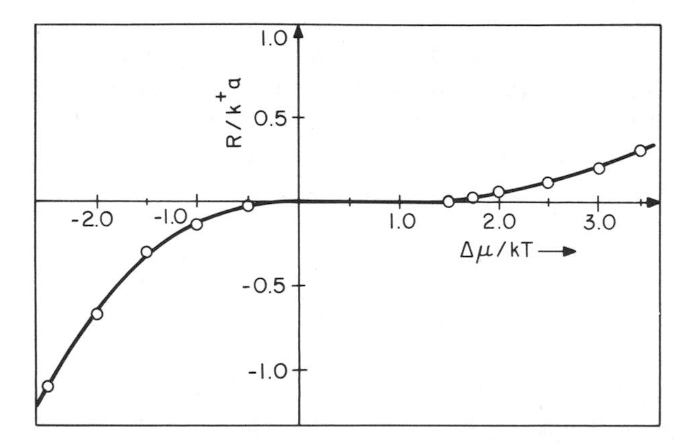

Fig. 17. Growth and evaporation rates of a crystal surface intersected by a columnar 5×5 hole. Here $L/kT = 12$.

It should be noted that the impingement and evaporation rate constants are not symmetric, and this has some effect on the symmetry of the kinetics illustrated here. For example, the formation of a surface vacancy requires the evaporation of a five-bonded atom. This is much slower than the rate of condensation of adatoms by direct impingement, which at equilibrium is equal to the evaporation rate of a three-bonded atom. This asymmetry favors the crystal growth process, but the growth rate in Fig. 17 is normalized by k^+, and the effect of this term is to increase the magnitude of the plotted values of the evaporation rate. These two effects tend to cancel one another, and normalized growth and evaporation curves for perfect crystal faces are approximately symmetrical over a wide range of $\Delta\mu$ values.[62]

The evaporation rate as represented in Fig. 17 is highly nonlinear, and there is a small region ($-0.5 < \beta\Delta\mu < 0$) where the rate is essentially zero. The columnar hole in the crystal represents a cluster of 25 vacant sites in each layer, but in this region of $\Delta\mu$ they do not constitute a critical cluster. The classical expression for the number of atoms in a critical cluster, $i^* = (\phi/\Delta\mu)^2$, applies also to vacancy clusters when $\Delta\mu$ is negative. According to this expression, the critical hole contains 25 vacancies when $\beta\Delta\mu = -0.8$. We would therefore expect a rapid increase in the evaporation rate in the vicinity of this value, as is observed in Fig. 17.

Exposed atoms at the edge of the 5×5 hole readily evaporate at large negative values of $\Delta\mu$, that is, $\beta\Delta\mu \ll -0.8$. As the edge atoms of one layer evaporate, those of the layer below are exposed and soon evaporate also. In this way, an etch pit with steep sides is formed around the hole. This is illustrated in Fig. 18, where $\beta\Delta\mu = -2$. At values of $\Delta\mu$ closer to zero, the enlargement of the hole involves a nucleation event, and only a very shallow pit is formed.

The slope of the etch pit can be calculated from the evaporation rate R_e of the crystal. The growth rate of a vicinal surface, (7.2), must be equal to the rate R_e, and hence the slope at large distances from the center of the etch pit is

$$a/l = R_e/v_\infty \tag{7.13}$$

Since v_∞ is at most linear in $\Delta\mu$, the slope increases rapidly with $\Delta\mu$ in the vicinity of $\beta\Delta\mu = -0.8$. Another phenomenon observed in Fig. 18 is the presence of a steep slope close to the etch pit center. The short step segments in this region recede more slowly than the velocity v_∞. The motion of such a step segment requires a one-dimensional nucleation of a row of vacancies. The rate of nucleation and hence the average step velocity is proportional to the number of nucleation sites or the length of the step. However, when the step is long enough that several such

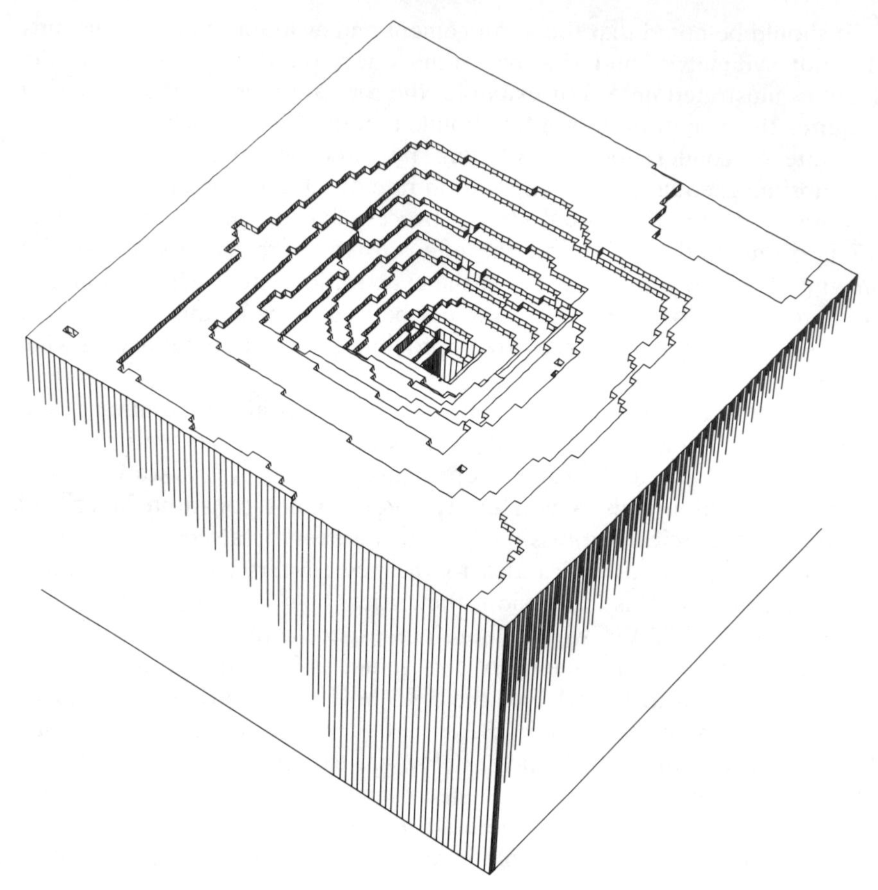

Fig. 18. Computer drawing of the etch pit formed during evaporation of a crystal containing a hole. Here $\beta\phi = 4$ and $\beta\Delta\mu = -2$.

vacancy "nuclei" are present simultaneously, the step velocity approaches a limiting value v_∞ (See Section IV for a discussion of the limiting crystal growth rate as a result of 2d nucleation.) If the dependence of the step velocity on its length were known, a generalization of (7.12) could be employed to calculate the average profile of the etch pit.

Screw dislocations also have a core region with a high concentration of strain energy. This is not included in the SOS model or the MC calculations discussed above, but the strain energy also enhances the evaporation rate. Cabrera and Levine estimated the assymmetry between growth and evaporation in their theory of the spiral mechanism. Because of the strain energy, they found that the shape of the spiral is indeterminate when the chemical potential is below a critical (negative) value. Then the atoms can evaporate from the core region more rapidly than could be accomplished by the rotation of the spiral. Again, an etch pit with steep sides develops under these conditions.

Finally, we point out that crystal faces that are investigated in the laboratory are normally bounded by edges, that is, the intersection of the faces with other low-index faces. Atoms at these edge positions may also evaporate preferentially, and the edges can provide a ready source of steps for the evaporation process. A transient period is required before steps from the edges are distributed over an appreciable fraction of the surface. As a result, the relative importance of line defects, voids, and edges depends on the number of defects, the size of the crystal, and the length of the evaporation period. Edges are not present in the computer simulation models, since periodic boundary conditions provide lateral neighbors for these atoms.

In summary, we have seen that line defects can have different effects on the kinetics of growth and evaporation. Edge dislocations and holes affect primarily the evaporation rate. Screw dislocations enhance both processes, and they are crucial to the growth of crystals at low temperatures and driving forces. Above the roughening temperature these defects have little influence on the kinetics.

VIII. IMPURITIES

The presence of minute quantities of certain impurities can have a dramatic effect on the growth rates. Some impurities are known to act as inhibitors, but others facilitate the growth of high-quality crystals. In this section, we first consider the effect on the kinetics of small quantities of a component that bonds strongly to the crystal surface. Then we discuss the trapping of volatile impurities by the growing crystal. Impurity poisoning of the surface and step bunching have also been studied,[66] but are not included in this review.

A. Impurity-Enhanced Nucleation

The early stages of cluster formation are enhanced by the presence of impurities with strong interactions. Atoms that impinge on a site next to such an impurity have a smaller probability of evaporation than similarly coordinated atoms elsewhere. The crystal growth rates on the SC (001) face with $\phi_{AB}=2\phi_{AA}$, and $\phi_{BB}=\phi_{AA}=4kT$ are shown in Fig. 19. (Detailed condensation and evaporation rates with impurities are provided in Refs. 43 and 65; also see below.) Here a perfect crystal lattice is employed, and both species are immobile. Growth rates with a small quantity of impurity atoms are indicated by the triangular symbols, here

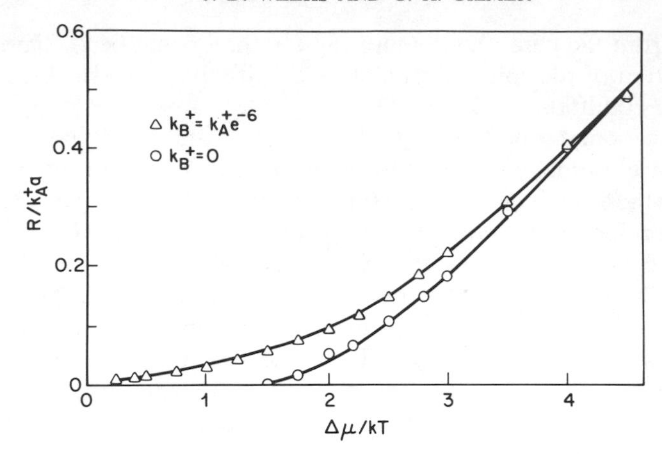

Fig. 19. Growth rates by MC calculations in a system containing impurities. Here $\beta\phi_{AB}=8$, $\beta\phi_{AA}=4$, $\beta\phi_{BB}=4$, and the relative impurity impingement rates are indicated on the figure.

$\mu_B=\mu_A-6kT$. Growth rates without impurities are plotted for comparison. Although the impingement rate of the impurity is small, a high percentage is trapped by the growing crystal because of the large AB bond energy. At small positive values of $\Delta\mu$, the nucleation of clusters occurs only around impurity atoms, and this process permits measurable growth to occur in a region where the (001) face is essentially immobile without the B atoms. The kinetics appear to be similar to those of spiral growth, except that in this case the dependence of R on $\Delta\mu$ is approximately linear at small values of $\Delta\mu$. Consequently, a break in the curve occurs near $\Delta\mu=1.8kT$, where the nucleation rate without impurities becomes appreciable.

This type of impurity reduces the anisotropy of the growth rate with orientation. Vicinal faces and those containing weak bond networks between the atoms in the surface layer normally grow faster than close-packed faces. The anisotropy is especially large at small values of $\Delta\mu$, where the 2d nucleation rate is very slow. Since the impurities facilitate the nucleation of clusters on all types of faces, the growth rate with impurities is less sensitive to orientation. This mechanism may explain the effectiveness of certain impurities as smoothing agents for electroplated deposits on polycrystalline substrates.[67]

A crystal grown under the conditions described above and at a small positive value of $\Delta\mu$ is in a metastable condition. A large reduction in the total free energy would result from a transformation to a sodium chloride structure where each A atom is surrounded by six B nearest neighbors. The formation of the strong AB bonds would more than compensate for the lower chemical potential of the B atoms. In this case the kinetics of growth, that is, the small impingement rate of the B atoms, produces a highly nonequilibrium composition. As $\Delta\mu$ is decreased, the percentage of B atoms in the crystal increases, indicating a tendency to approach the equilibrium composition. If such a crystal remains in contact with a vapor containing B atoms at this chemical potential without growing, it will eventually transform to approximately the 1:1 stoichiometry. In practice, this process may take a very long time, since bulk diffusion rates for these impurities are usually quite low. Furthermore, in most cases the chemical potential of the impurities is even smaller than the value employed here. The value of $\Delta\mu_B$ used in the simulation was selected to induce large growth rates and short computation times, but very large increases in the growth rate also occur at much smaller concentrations.

B. Segregation of Impurities During Crystal Growth

The previous discussion concerned the effect of impurities on the growth rate. Another important aspect of this problem is the effect of the growth rate (or driving force) on the capture of impurities by the crystal. Many industrial applications involve the growth of crystals containing several different atomic species. Composition control is an important objective. For example, the concentration of dopants and the uniformity of their distribution in semiconductor crystals determine the electrical properties of the product. Here we attempt to relate the composition of the crystal to that of the fluid adjacent to the growing crystal surface. Again, we do not discuss mass transport effects in the bulk phases.

The distribution coefficient K is defined as the ratio of the atomic concentration of the impurity in the crystal to that in the fluid. This coefficient is determined by the phase diagram in the limit of an infinitesimal growth rate, since sufficiently slow growth allows time for equilibration between the phases. In the limit of a very large driving force, almost all of the atoms impinging on the surface are incorporated into the growing crystal. Then $K \cong 1$, and the composition has no relation to the phase diagram. We discuss the distribution coefficient at finite driving forces using results derived from an analytical model and from computer simulations of the simple cubic Ising model.

The capture of impurities is intimately related to the details of the crystal growth process. It has been shown that atoms impinging on all of the surface sites have an appreciable chance of being captured.[62] Even the adatoms may be trapped by an advancing step or cluster. In principle, all the surface sites also contribute to the capture of impurities, and an exact model of the process must include all the sites.

For simplicity, however, we first treat a model that employs a single "typical" site. This model can be visualized as a row of atoms at the edge of a straight step terminated by a kink site, where condensation and evaporation occur.[43,68] There is some justification for this drastic oversimplification. First, MC data indicate that a plurality of the atoms captured by the growing crystal actually impinge on kink sites (in the absence of surface migration).[62] Second, an atom in the kink site is intermediate in bonding between the various surface atoms, and we might expect it to approximate the average behavior of the system. The ultimate test of its validity is obtained when we compare this model with a corresponding MC system that is free of such restrictions.

Here we discuss the case in which the atomic concentration of impurities in the crystal is much less than unity, and impurity-impurity interactions may be neglected. Some possible positions of the impurity relative to the kink site are shown in Fig. 20. Accordingly, we define x_n as the fraction of systems in a kinetic ensemble that have an impurity located n atoms down the row from the kink site. The rate of incorporation of impurities (B atoms) at the kink site is

$$Q_B \cong k_B^+ - x_0 k_{AB}^- \tag{8.1}$$

and that for atoms of the host lattice (A atoms) is

$$Q_A \cong k_A^+ - k_{AA}^- \tag{8.2}$$

Here k_B^+ and k_A^+ are the two condensation rates, and k_{AB}^- and k_{AA}^- are the evaporation rates from "pure" kink sites, where the nearest neighbors are

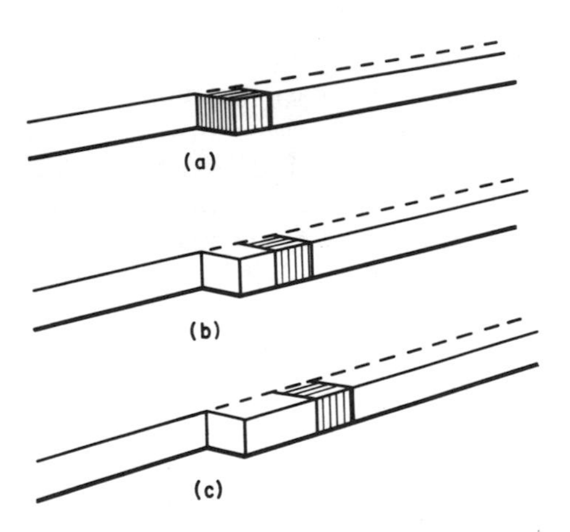

Fig. 20. An impurity atom is illustrated in various positions included in the analytical model: (a) at the kink site, (b) one unit removed, and (c) two units removed.

all A atoms. The distribution coefficient K is

$$K = (Q_B/Q_A)/(k_B^+/k_A^+) \tag{8.3}$$

but the evaluation of Q_B depends on x_0.

The probability x_0 is determined by the rates of transition between the configurations of Fig. 20. Equating the creation and annihilation rates for configuration (a), we obtain

$$k_B^- + x_1 k_{BA}^- = x_0 k^+ + x_0 k_{AB}^- \tag{8.4}$$

The terms on the left represent the rate of creation of configuration (a) by the impingement of an impurity atom at a pure kink site, and by the evaporation of an A atom from configuration (b). The terms on the right account for annihilation of (a) by the impingement of an A atom, and by the evaporation of the B atom at the kink site. The equality must be satisfied in the steady state. Similar equations apply to the other configurations, that is,

$$x_0 k_A^+ + x_2 k_{AA}^- = x_1 k_A^+ + x_1 k_{BA}^- \tag{8.5}$$

applies to (b), where k_{BA}^- is the rate of evaporation of an A atom with one B nearest neighbor. Also, for $n > 1$,

$$x_{n-1} k_A^+ + x_{n+1} k_{AA}^- = x_n k_A^+ + x_n k_{AA}^- \tag{8.6}$$

The solution to these equations is simplified if we realize that the concentration of impurity x_n must approach a constant value x_∞ for large n. Substitution in (8.6) reveals that $x_n = x_\infty$ for all $n > 0$. That is, the concentration of impurities is not a function of position except at the kink site. Substituting $x_1 = x_\infty$ and $x_2 = x_\infty$ in (8.4) and (8.5), we obtain two independent relations between x_0 and x_∞. Solving for x_0 and substituting the expressions for Q_B and Q_A into (8.3), we find

$$K^{-1} = 1 + \frac{k_{AB}^-}{k_A^+} - \frac{k_{AA}^-}{k_A^+} + \frac{k_{AB}^- k_{BA}^-}{(k_A^+)^2} - \frac{k_{AB}^- k_{AA}^-}{(k_A^+)^2} \tag{8.7}$$

Chernov first derived this equation by a different analysis.[68]

We have also calculated the impurity capture rate by the MC technique.[43] Both A and B atoms impinge on the crystal surface, but the AB bond is somewhat weaker than the AA bond. The composition of the deposit is measured, and K is determined by a comparison of this composition with the ratio k_B^+/k_A^+. In every case, k_B^+ is chosen such that the deposit contains less than 5% B atoms.

The parameters in (8.7) are readily identified in terms of the transition probabilities of the MC model. As before, the evaporation rate of an A

atom in a pure kink site is $k^-_{AA} = k^+_A e^{-\beta\Delta\mu}$. Then $k^-_{AB} = k^+_A e^{-3\beta(\phi_{AB}-\phi_{AA})-\beta\Delta\mu}$. That is, we assume that the difference between the two rates is determined entirely by the fact that three *AA* bonds are replaced by *AB* bonds. For simplicity we assume that preexponential factors are identical. Similarly, $k^-_{BA} = k^+_A e^{-\beta(\phi_{AB}-\phi_{AA})-\beta\Delta\mu}$.

A comparison of the two models is presented in Fig. 21.[43] MC data for a stepped plane are represented by the open circles, triangles, and squares for $\phi_{AB} = 3.5kT$, $3kT$, and $2.25kT$, respectively. In all cases $\phi_{AA} = 4kT$. The agreement with the analytical model (dashed lines) is quite good, considering the simplicity of the approach. The discrepancies at both ends of the range of $\Delta\mu$ are expected. When $\Delta\mu \to 0$, the MC data must approach the true bulk equilibrium coefficient $K_{eq} = e^{6\beta(\phi_{AB}-\phi_{AA})}$, and the solid curves were constructed to intersect the axis at this point. The dashed curves of (8.7) intersect the axis at a larger value, $K^s_{eq} = e^{4\beta(\phi_{AB}-\phi_{AA})}$, the equilibrium value for the edge of a step. This is inherent in the model, and the inclusion of the other processes that result in the complete immuring of the impurity is necessary to obtain K_{eq}. At very large $\Delta\mu$, (8.7) also predicts a larger value of K than the MC data. Impurity atoms landing at sites with only one or two bonds to the crystal

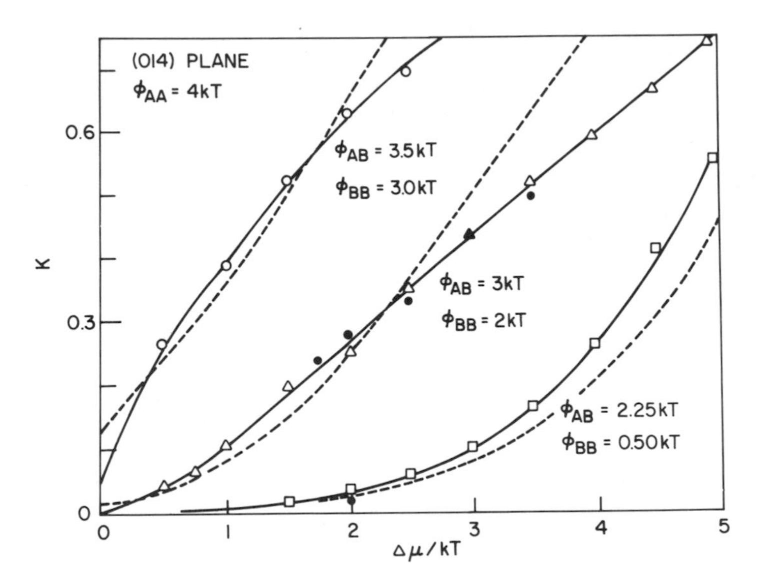

Fig. 21. The distribution coefficient K is plotted as a function of the driving force. Data for three different values of the impurity-host bond energy are illustrated. The solid curves and open symbols are MC data taken on a stepped (041) surface, and the dashed curves correspond to Eq. (8.7). The solid circles are MC data taken on a (001) surface.

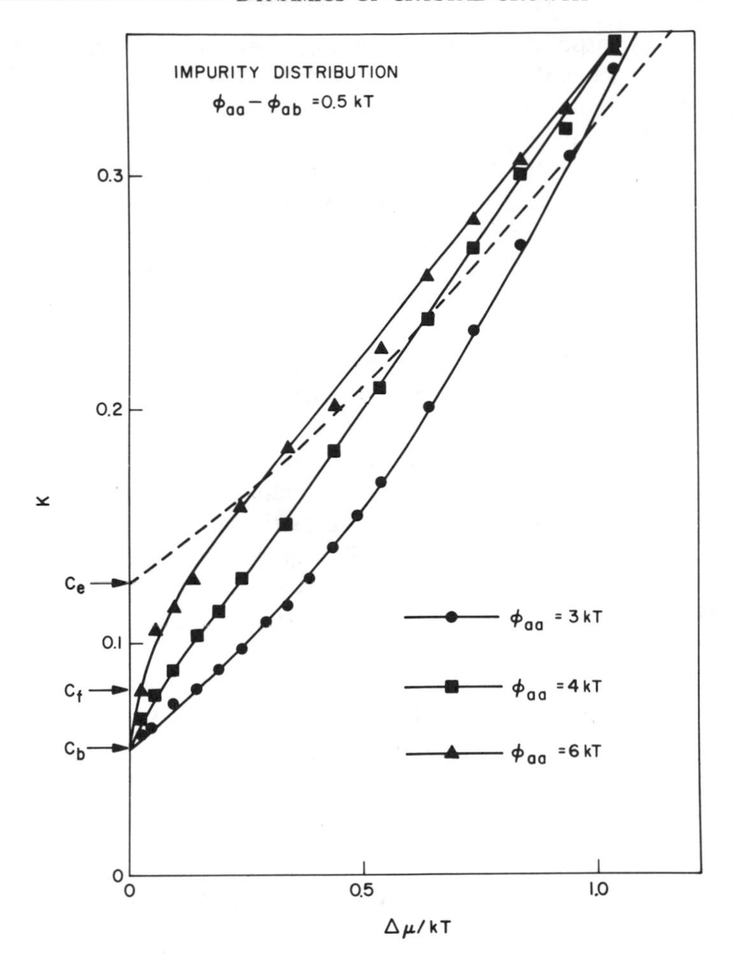

Fig. 22. The distribution coefficient with relatively small values of the driving force.

are rejected with a much higher probability than those landing at kink sites. The nonlinear nature of the exponential function causes the "typical site" rate to deviate from the average rate of the MC model.

The effect of crystallographic orientation on the distribution coefficient was also examined. Values of K at the (001) orientation are indicated by the small black dots. These data are limited to the larger values of $\Delta\mu$ because of the small 2d nucleation rate otherwise. The close agreement between these two sets of data implies that the distribution coefficient is determined primarily by $\Delta\mu$.

Accurate MC data at small values of $\Delta\mu$ are shown in Fig. 22. (The square symbols correspond to the data with $\phi_{AA} = 3.5kT$ in Fig. 21.) The composition approaches the bulk equilibrium value as $\Delta\mu \to 0$. Note that the value of $\beta\phi_{AA}$ affects the composition at small values of $\Delta\mu$, although this parameter does not appear in (8.7). The evaporation of impurities from sites with four or five bonds to the crystal is not included in the analytical model, but these processes are important at small driving force and permit the reduction of the impurity concentration below the value captured by the kink sites. At the smaller values of ϕ_{AA} these processes occur more frequently, and the lowest values of the impurity concentrations are obtained.

The weak dependence of K on orientation affords an explanation of the enhanced impurity concentration at facets. This enhancement is observed in crystals grown by the Czochralski method.[69] The steady-state pulling of a crystal from the melt imposes a constant average growth rate over the entire surface. Except for temporal fluctuations, the center of the facet and the vicinal orientations have growth rates that are nearly identical. A much greater driving force is required at the center of the facet, since 2d nucleation is necessary. This larger driving force then induces a larger value of K. This effect is observed for a large number of impurities in germanium and silicon.

It is evident that surface mobility can cause a drastic change in K. For example, an impurity with a relatively high surface mobility may exhibit a value of K greater than unity, even when $\phi_{AB} < \phi_{AA}$. This has been demonstrated by MC simulations.[25] Rapid surface diffusion allows impurity atoms to migrate to sites where they are strongly bonded to the crystal. We noted in the preceding section the large increase in growth rate (or capture probability) that occurs when surface mobility is included. In some cases this can more than compensate for the weaker bonding of the impurity to the crystal.

In summary, the primary factors that determine the distribution of impurities in a growing crystal are (*1*) the driving force $\Delta\mu$ prevailing in the interfacial region, (*2*) the relative surface mobility of the impurity and host atoms, and (*3*) the interaction energies ϕ_{AA}, and ϕ_{AB}. The crystal surface orientation and crystalline perfection may, in some circumstances, determine the magnitude of the driving force. In these cases they affect the composition indirectly through the driving force.

IX. FINAL REMARKS

In this article we have reviewed theories and computer simulation results for crystal growth using the kinetic SOS model. We seem to have a good theoretical understanding of growth on close-packed faces of perfect crystals that are free of impurities. Theories for the low temperature nucleation regime and the high temperature continuous growth regime are in good agreement with the MC data. Recent work has also greatly advanced our knowledge of the nature of the roughening transition between these two regimes.

We have concentrated on the properties of the SC (001) face on the assumption that it is typical of close-packed surfaces in general. Although we expect universal behavior in the vicinity of the roughening temperature, the absolute magnitude of T_R depends on the details of the atomic interactions and the geometrical arrangement of atoms in the close-packed plane. It has been found in the case of the SOS model that the roughening temperature is slightly greater than the critical temperature of the 2d array of surface atoms. This has been verified in the case of the SC (001) face (Section V) and for the FCC (001) and (111) faces.[43,70] Calculations of T_R for other systems would be very helpful, especially in the case when continuous atomic coordinates are permitted. It is important to determine the degree of localization of atoms in the vicinity of lattice sites at the transition temperature. Molecular dynamics simulations will undoubtedly play an important role in these studies.

Monte Carlo simulations have provided a powerful method for assessing the contribution of defects and impurities to the motion of the interface. The simulations show that often several mechanisms operate in parallel, and the resulting kinetics may not agree with existing theories that assume idealized and simplified conditions. The theories are usually in agreement with MC data on systems that are constrained to permit only a single growth mode (e.g., spiral growth), but there may be large discrepancies between the same theories and MC data on more realistic systems without such constraints.

The development of theories that apply to very general conditions is a goal for the future. It is important to have analytic theories to help clarify the physics of the various processes, to facilitate comparison with experimental data, and to suggest new effects. Existing MC data provide definitive results against which the theories can be tested and improved.

Another important aspect of crystal growth requiring further study is the stability of an array of steps on the crystal surface. The kinetic repulsion mentioned briefly in Section VII has a stabilizing effect on the array, but impurities[66] or the anisotropic capture of migrating atoms[71–73] may have the opposite effect. That is, an instability that leads to step bunching may occur because of impurities or because of the kinetics of surface migration. Relatively steep macrosteps (step bunches) separated by low-index terraces are commonly observed on crystals that have been grown from the vapor and from solution.[74] The theory of the stability of arrays of steps is in a very primitive state.

Improvements in the basic kinetic SOS model are also needed for a more realistic description of the experimental situation. Models that allow continuous coordinates and can consider effects such as strain energy and impurity size effects represent an important area for future work. These improvements are particularly needed for an accurate description of melt growth. Also, the connection between the microscopic approach taken here and macroscopic effects of bulk transport and such questions as morphological stability need clarification.

The theory of crystal growth offers a host of challenging problems of great practical importance. As this article makes clear, we are now beginning to deal with some of the complicated effects that arise in the laboratory, but much work remains to be done. We believe it represents a fruitful area of research for some time to come.

Acknowledgments

Part of the work reviewed herein was done in collaboration with S. T. Chui, K. A. Jackson, and H. J. Leamy. Without their contributions, it would not have been possible to write this review. We are also grateful to P. C. Hohenberg for many helpful and patient discussions.

References

1. R. F. Sekerka, in *Crystal Growth: An Introduction*, North-Holland, Amsterdam, 1973, p. 403.
2. See, for example, K. Huang, *Statistical Mechanics*, Wiley, New York, 1963, p. 329.
3. K. A. Jackson, in *Liquid Metals and Solidification*, American Society for Metals, Metals Park, Ohio, 1958, p. 174.
4. D. E. Temkin, in *Crystallization Processes*, Consultants Bureau, New York, 1966, p. 15.
5. F. F. Abraham and G. H. White, *J. Appl. Phys.*, **41,** 1841 (1970).
6. G. H. Gilmer, *J. Crystal Growth*, **35,** 15 (1976).
7. R. Becker and W. Doering, *Ann. Physick*, **24,** 719 (1935).
8. W. K. Burton, N. Cabrera, and F. C. Frank, *Phil. Trans. R. Soc* (London), **243A,** 299 (1951).
9. H. J. Leamy, G. H. Gilmer, and K. A. Jackson, in J. B. Blakeley, ed., *Surface Physics of Materials I*, Academic Press, New York, 1975. Also see: W. J. Shugard, J. D. Weeks, and G. H. Gilmer, *Phys. Rev. Lett.* **41,** 1399 (1978).
10. S. T. Chui and J. D. Weeks, *Phys. Rev.*, **B14,** 4978 (1976).
11. H. Van Beijeren, *Phys. Rev. Lett.*, **38,** 993 (1977).
12. H. J. F. Knops, *Phys. Rev. Lett.*, **39,** 776 (1977).
13. H. J. Leamy and G. H. Gilmer, *J. Crystal Growth*, **24/25,** 499 (1974).
14. R. H. Swendsen, *Phys. Rev.*, **B15,** 5421 (1977).
15. H. P. Bonzel, in J. B. Blakeley, ed., *Surface Physics of Materials*, Academic Press, New York, 1975.
16. J. D. Weeks, G. H. Gilmer, and K. A. Jackson, *J. Chem. Phys.*, **65,** 712 (1976).

17. The limiting growth law for melt growth was derived by H. A. Wilson, *Philos. Mag.*, **50,** 238 (1900) and by J. Frenkel, *Phys. Z. Sowjetunion*, **1,** 498 (1932). The corresponding limiting law for growth from the vapor phase is due to H. Hertz, *Ann. Phys. (Leipzig)*, **17,** 177 (1882) and M. Knudsen, *Ann. Phys. (Leipzig)*, **29,** 179 (1909). J. W. Gibbs, *Collected Works*, Yale University, New Haven, 1957, p. 325 footnote, realized the limitations imposed on growth by the formation of new layers of a crystal. W. Kossel, *Nachr. Ges. Wiss. Göttingen*, **135,** (1927); I. N. Stranski, *Z. Phys. Chem Leipzig*, **136,** 259 (1928); and J. Frenkel, *J. Phys. USSR*, **9,** 302 (1945) discussed the importance of steps and kink sites in crystal growth.
18. D. E. Temkin, *Soviet Phys. Crystallogr.*, **14,** 344 (1969).
19. D. Walton, *J. Chem. Phys.*, **37,** 2182 (1962).
20. G. H. Gilmer, in *Computer Simulation for Materials Applications*, National Bureau of Standards, Gaithersburg, Md. 1976, p. 964.
21. J. J. Burton, in B. J. Berne, ed., *Statistical Mechanics, Part A*, Plenum, New York, 1977.
22. G. H. Gilmer and J. D. Weeks, *J. Chem. Phys.*, **68,** 950 (1978).
23. C. van Leeuwen and P. Bennema, *Surf. Sci.*, **51,** 109 (1975).
24. A. E. Michaels, G. M. Pound, and F. F. Abraham, *J. Appl. Phys.*, **45,** 9 (1974).
25. G. H. Gilmer, unpublished.
26. A. E. Nielsen, *Kinetics of Precipitation*, Pergamon, Oxford, 1964.
27. E. Budevski, V. Bostanov, T. Vitanov, Z. Stoynov, A. Kotzeva, and R. Kaischev, *Phys. Status Solidi*, **13,** 577 (1966); *Electrochim. Acta*, **11,** 1697 (1966).
28. A. N. Kolmogorov, *Izv. Akad. Nauk Ser. Math.*, No. 3, 355 (1937).
29. W. B. Hillig, *Acta Met.*, **14,** 1968 (1966).
30. S. K. Rangarajan, *J. Electroanalyt. Chem.*, **46,** 125 (1973).
31. L. A. Borovinskii and A. N. Tsindergozen, *Sov. Phys. Crystallography*, **13,** 1191 (1969).
32. M. Hayashi, *J. Phys. Soc. Jap.*, **35,** 614 (1974).
33. U. Bertocci, *Surf. Sci.*, **15,** 286 (1969).
34. J. W. Oldfield, *Electrodeposition and Surf Treatment*, **2,** 395 (1973/74).
35. V. Bostanov, R. Roussinova, and E. Budevski, *J. Electrochem. Soc.*, **119,** 1346 (1972).
36. See, for example, Ref. 2, Chap. 16.
37. H. van Beijeren, *Commun. Math. Phys.*, **40,** 1 (1975).
38. K. A. Jackson, *J. Crystal Growth*, **3/4,** 507 (1968).
39. K. A. Jackson, in *Crystal Growth*, Pergamon, New York, 1967, p. 17.
40. J. D. Weeks, G. H. Gilmer, and H. J. Leamy, *Phys. Rev. Lett.*, **31,** 549 (1973).
41. G. H. Gilmer, K. A. Jackson, H. J. Leamy, and J. D. Weeks, *J. Phys.*, **C7,** L123 (1974).
42. See, for example, J. W. Essam, and M. E. Fisher, *J. Chem. Phys.*, **38,** 8021 (1963) for a good discussion of low temperature series analysis.
43. G. H. Gilmer and K. A. Jackson, in *Crystal Growth and Materials*, North-Holland, Amsterdam, 1977, p. 79.
44. J. M. Kosterlitz, *J. Phys.*, **C7,** 1046 (1974). See also J. M. Kosterlitz and D. J. Thouless, *J. Phys.*, **C6,** 1181 (1973). The method was used earlier for the Kondo problem by P. W. Anderson and G. Yuval, *J. Phys.*, **C4,** 607 (1971).
45. J. V. José, L. P. Kadanoff, S. Kirkpatrick, and D. R. Nelson, *Phys. Rev.*, **B16,** 1217 (1977).
46. E. H. Lieb, *Phys Rev. Lett.*, **18,** 692, 1046 (1967); E. H. Lieb and F. Y. Wu, in C. Domb and M. S. Green, eds., *Phase Transitions and Critical Phenomena*, vol. 1, Academic Press, London, 1972.
47. R. H. Swendsen, *Phys. Rev.*, **B17,** 3710 (1978).
48. S. T. Chui and J. D. Weeks, *Phys. Rev. Lett.*, **40,** 733 (1978).

49. For a general introduction to critical phenomena (including critical dynamics) see S. K. Ma, *Modern Theory of Critical Phenomena*, W. A. Benjamin, Reading, Mass., 1976.
50. B. I. Halperin, P. C. Hohenberg, and S. Ma, *Phys. Rev. Lett.*, **29,** 1548 (1972).
51. P. C. Hohenberg and B. I. Halperin, *Rev. Mod. Phys.*, **49,** 435 (1977).
52. P. G. de Gennes, *Faraday Symposium #5 on Liquid Crystals*, London, 1971, p. 16.
53. See for example, Ref. 49, Chaps. XI–XIV.
54. D. R. Nelson and J. M. Kosterlitz, *Phys. Rev. Lett.*, **39,** 1201 (1977).
55. M. Volmer and W. Schultz, *Z. Phys. Chem.*, **A156,** 1 (1931).
56. F. C. Frank, *Disc. Faraday Soc.*, **5,** 48 (1949).
57. E. E. Gruber and W. W. Mullins, *J. Phys. Chem. Solids*, **28,** 875 (1967).
58. E. Budevski, G. Staikov, and U. Bostanov, *J. Crystal Growth*, **29,** 316 (1975).
59. R. H. Swendsen, P. S. Kortman, D. P. Landau, and H. Müller-Krumbhaar, *J. Crystal Growth*, **35,** 73 (1976).
60. N. Cabrera and M. M. Levine, *Phil. Mag.*, **1,** 450 (1956).
61. N. Cabrera and R. V. Coleman, in J. J. Gilman, ed., *The Art and Science of Growing Crystals*, Wiley, New York, 1963, p. 3.
62. G. H. Gilmer and P. Bennema, *J. Appl. Phys.*, **43,** 1347 (1972).
63. H. T. Minden, *J. Crystal Growth*, **8,** 37 (1971).
64. T. Surek, J. P. Hirth, and G. M. Pound, *J. Crystal Growth*, **18,** 20 (1973).
65. G. H. Gilmer, *J. Crystal Growth*, **42,** 3 (1977).
66. F. C. Frank, in *Growth and Perfection of Crystals*, Wiley, New York, 1958, p. 411.
67. J. D. E. McIntyre and W. F. Peck, Jr., *J. Electrochem. Soc.*, **123,** 1800 (1976).
68. A. A. Chernov, *Sov. Phys.-Uspekhi*, **13,** 101 (1970).
69. J. A. M. Dickhoff, *Solid State Electron*, **1,** 202 (1960).
70. U. Bertocci, *J. Crystal Growth*, **26,** 219 (1974).
71. R. L. Schwoebel, *J. Appl. Phys.*, **40,** 614 (1969).
72. R. Ghez and G. H. Gilmer, *J. Crystal Growth*, **21,** 93 (1974).
73. C. van Leeuwen, R. van Rosmalen, and P. Bennema, *Surf. Sci.*, **42,** 32 (1974).
74. D. L. Rode, *J. Crystal Growth*, **27,** 313 (1974).

NONLINEAR PROBLEMS IN THE THEORY OF PHASE TRANSITIONS

JOHN J. KOZAK

*Department of Chemistry and Radiation Laboratory**
University of Notre Dame, Notre Dame, Indiana 46556

CONTENTS

* The research described herein was supported in part by the Office of Basic Energy Science of the Department of Energy. This is Document No. SR-43 from the Notre Dame Radiation Laboratory.

I. INTRODUCTION

It is traditional nowadays to adopt the point of view that a phase transition manifests itself as a singularity in a thermodynamic function that is otherwise analytic.[1] This view is so commonplace we sometimes tend to forget that in the classical studies of the late 19th and early 20th century, much greater stress was placed on the role of nonlinearity *per se* as the distinctive feature, mathematically, in a theory of phase transitions. For example, the theory proposed by Van der Waals[2] in 1873 to describe the state properties of a gas resulted in an equation

$$\left(p+\frac{a}{V^2}\right)(V-b)=RT$$

which, as is evident, displays a cubic nonlinearity in the variable V (Where V is the volume per mole, p is the pressure, T is the temperature, R is the gas constant, and a and b are constants that reflect the role of attractions and repulsions amongst the molecules comprising the system). Given this algebraic nonlinearity in the Van der Waals equation, it is possible to find three values V that satisfy the equation for a given choice of $\{p, T; a, b\}$. Although at first sight this nonuniqueness would appear to be a major weakness in a theory designed to describe the physical properties of a gas, it is just this feature which accounts for the great importance of the Van der Waals equation in the history of phase transition theory. Van der Waals found, upon plotting pressure versus volume for fixed $\{T; a, b\}$, that isotherms were generated which displayed a remarkable similarity to those reported a few years earlier (1869) by Andrews[2] in his classic study of the critical behavior of carbon dioxide. At the lowest temperatures the Van der Waals equation has three real roots V, and the consequent isotherms have an s-shaped structure (see curve I in Fig. 1). As the temperature is increased, this curve is raised along the p-axis (curve II in Fig. 1), and the three possible values of V for a given p are much closer together. Then, at a certain value of $T=T_c$ (where T_c is denoted the critical temperature), the three real values of V coalesce (curve III), and above T_c there is only one real solution of the Van der Waals equation; for $T>T_c$ the consequent isotherm is a hyperbola corresponding to Boyle's law. The dashed lines in Fig. 1 are drawn so as to construct equal areas and the region enclosed by the set of outermost intercepts can be placed in correspondence with the gas-liquid coexistence region (see Fig. 2); in addition, the portions of the curve I in Fig. 1 denoted by AB and CD can be associated with certain abnormal states, the metastable "supersaturated-vapor" and "superheated-liquid" states, respectively. Indeed, as Uhlenbeck[3] observed a few years ago, when matched against the experimental evidence the qualitative and quantitative successes of the Van der Waals equation "were so remarkable that

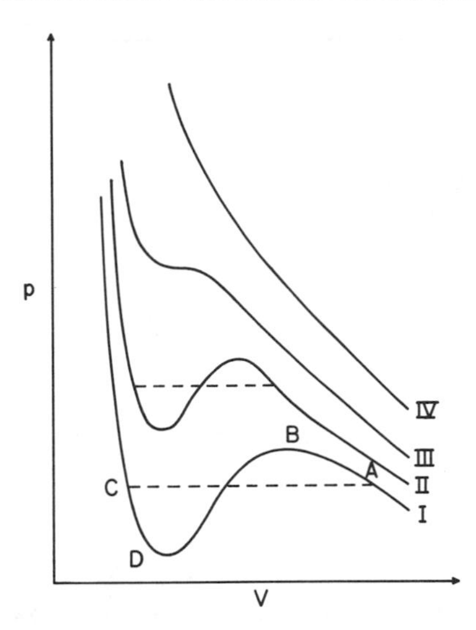

Fig. 1. Isothermal curves according to the Van der Waals equation

$$(p + a/V^2)(V - b) = RT$$

they practically killed the subject for more than fifty years." Somewhat the same situation prevailed in the theory of magnetic systems. Following earlier experimental work by Curie, Hopkinson, and others, Weiss[2] in 1907 proposed a theory of ferromagnetism wherein the constituent "particles," the magnetic moments, interacted with each other via an artificial "molecular field" which is proportional to the average magnetization. Mathematically, the Weiss theory resulted in an equation with a transcendental nonlinearity, and a simple analysis of this equation gave an interpretation of the onset of a net magnetization at a certain critical temperature. In the years following the publication of these studies, disciples of Van der Waals and Weiss developed more refined versions of these theories, and again, from a mathematical point of view, the feature common to these descriptions was the presence of nonlinearity; the general strategy was to associate the occurrence of new solutions of some underlying nonlinear equation, as one or more "strength parameters" of the system were changed, with the onset of new phases.

It is against this background of apparently successful theories that one confronts the work of Onsager[5] on the properties of a two-dimensional Ising lattice, a model for ferromagnetism. In this study, which appeared in 1944, the statistical-mechanical consequences of adopting the Ising Hamiltonian were explored, and exact results were obtained for the case of zero magnetic field; Onsager found that the ferromagnetic transition in two dimensions was characterized by a logarithmic singularity in the

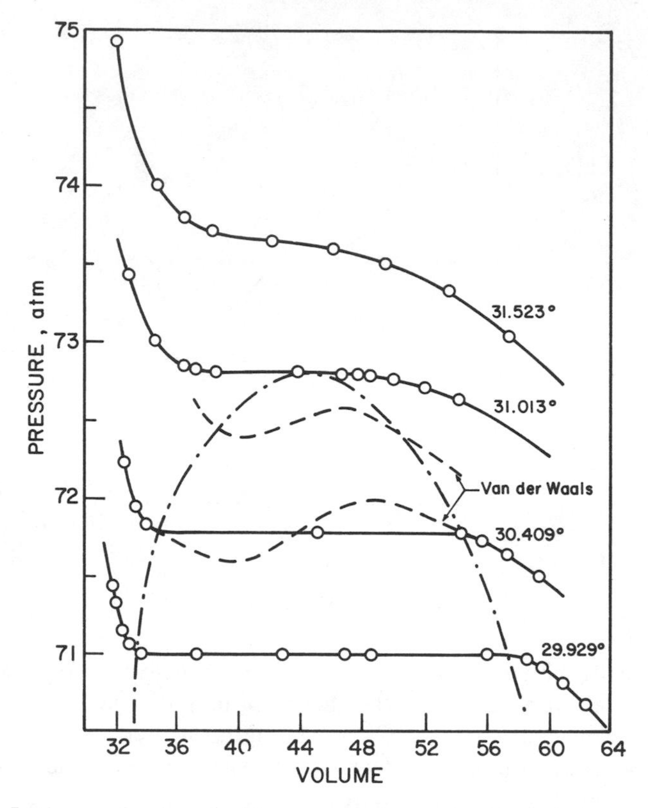

Fig. 2. Isotherms of carbon dioxide near the critical point (data taken from Ref. 4).

specific heat and not, as predicted by classical mean-field theory, a simple "jump" discontinuity in this thermodynamic function. The implications of this result were profound; Onsager's work demonstrated unequivocally that many of the qualitative features inherent in the Van der Waals-Weiss type theories could not be correct, and the simple, intuitive understanding of a phase transition as a crossover from one branch to another branch of solutions to some underlying nonlinear equation at a certain threshold value was apparently lost. Subsequent studies by Van Hove,[6] Yang and Lee,[7] and others[1] clarified the sense in which singularities can arise in thermodynamic functions that are otherwise analytic, and a variety of models[8] have been investigated which enhance our understanding of this important mathematical feature of phase transition theory. We now believe that a theory of phase transitions must consist in proving that the thermodynamic functions are piecewise analytic, and furthermore that a successful theory must identify the nature of the singularities that occur at a point of transition.

The mathematical difficulties encountered in approaching the problem of phase transitions from the point of view expressed in the preceding paragraph are enormous, and it is fair to say that no comprehensive theory of phase transitions in this sense exists at the present time. One of the factors responsible for this dilemma is that in modern, statistical-mechanical theories of phase transitions, one is still challenged by the presence of nonlinearities in the underlying structure of the problem. Although sometimes the nonlinearity is exposed from the very outset, it is often the case that a nonlinear operator equation emerges at the end of a subtle and complicated sequence of arguments. As an example of the latter possibility, we recall that in the renormalization group (RG) approach to the theory of critical phenomena,[9] one studies eventually a nonlinear differential equation for the RG transformation. (Nonlinear problems associated with the RG approach are not discussed in this review, since several excellent discussions are now available.[10]) On the other hand, in studies based on the BBGKY hierarchy, nonlinear equations appear almost at once; one often has to deal with nonlinear integral or integrodifferential equations for the n-particle distribution functions. In fact, the realization that, under a certain closure, the defining equations of the BBGKY hierarchy for the singlet distribution function are nonlinear [and therefore may admit more than one solution for a given set of constraints (e.g., density, temperature)], led Kirkwood and Monroe[11] as early as 1941 to examine whether changes in the analytic structure of the singlet distribution function might be associated with the onset of a phase transition; in particular, they argued that the freezing transition might be signaled by a change from the uniform density solution characteristic of a fluid to a periodic density solution characteristic of a solid. Somewhat later Vlasov[12] and Tyablikov[13] reexamined this possibility, and in these studies the techniques and theorems of modern nonlinear analysis were used for the first time in a discussion of the phase transition problem. In recent years there has been a renaissance of interest in examining the underlying nonlinear structure of statistical-mechanical theories of phase transitions. The motivation behind these studies has classical overtones: Given that modern formulations of the phase transition problem often lead to nonlinear equations, and given that such equations can exhibit multiple solutions in certain situations, what thermodynamic significance (if any) can be attributed to the failure of uniqueness in a given representation? And second, in those situations where relevance to the problem of phase transitions is claimed, can one characterize the thermodynamic functions and their (perhaps singular) behavior in the neighborhood of a transition point? It is to these questions that this treatise is addressed.

Before launching into a discussion of nonlinear problems in the theory of phase transitions, we address a number of informal remarks to the reader regarding the objectives and strategy of this short monograph. First of all, we deal here only with problems in the *equilibrium* theory of phase transitions; to consider problems in the dynamic theory would demand a much broader-based discussion than is attempted here. In any case, the reader now has available the important, recent reviews of Haken,[14] in which many features of the dynamical problem are discussed. Second, although this review has been written for the chemical physicist, the author has tried to keep the mathematician in mind. The reasons for adopting this strategy are entirely selfish: Insights gained from nonlinear functional analysis, bifurcation theory, and dynamical system theory are essential to the continued development of the theory of phase transitions, and it seems advantageous to make mathematicians aware of the mathematical problems that must be solved. To seek out a common ground on which the chemical physicist and methematician can meet, we present in Section II a discussion wherein certain key mathematical ideas are introduced to the chemical physicist and a number of statistical-mechanical concepts are illustrated for the mathematician, all within the context of the Landau theory of phase transitions. The experts can bypass this section and focus attention on Sections III and IV where, sequentially, second- and then first-order phase transitions are discussed. We apologize to the expert for resurrecting this old-fashioned classification of phase transitions, but, as will be seen, the scheme provides a convenient framework within which to develop the basic ideas. Our objective has been to formulate each problem considered with some care, so that the underlying statistical-mechanical issues can be clearly displayed. Then, rather than exhaust the reader with an extensive survey of all the work that has been done on a given problem, we have selected *representative* proofs of existence, uniqueness, and/or bifurcation to illustrate the mathematical techniques and theorems and to demonstrate the kind of conclusions that follow from such analyses. (This remark is offered apologetically to those authors whose contributions we have not cited explicitly, and to authors cited in this review who may feel that I have not sampled the most significant part of their work.) Finally, although an attempt has been made to normalize the notation in this review, a certain degeneracy in notation is unavoidable; the author hopes that this will not prove to be a great inconvenience to the reader. The concluding section is given over to a summary of some of the principal results and some comments on their possible significance.

II. LANDAU THEORY

In this section we use the classical theory of Landau[15] to introduce and illustrate a number of thermodynamic concepts and mathematical problems important to this review. We consider sequentially second- and then first-order phase transitions. We identify the mathematical features which characterize and distinguish these two types of transitions and, in pointing out the deficiencies of Landau theory, we isolate a number of statistical-mechanical issues that play an important role in our later discussion. The following development is a synthesis of ideas drawn from the recent work of H. Haken,[14] R. Thom,[16,17] and K. Wilson.[9]

We regard the free energy F of a system in thermal equilibrium to be dependent on the temperature T and possibly on other parameters such as the volume Λ. We seek to determine the minimum of F subject to a constraint. In the case of a ferromagnet, for example, if we have $\mathcal{N}_1$ elementary magnets pointing upward and $\mathcal{N}_2$ elementary magnets pointing downward, the "magnetization" is

$$M=(\mathcal{N}_1-\mathcal{N}_2)m$$

where m is the magnetic moment of a single elementary magnet. Then the additional constraint imposed is that the *average* magnetization $\bar{M}$ assumes a given value

$$\bar{M}=q$$

In the Landau theory, q is called an "order parameter." The behavior of many physical, chemical, and biological systems can be analyzed using the concept of an order parameter.[14] Suppose we proceed by assuming that q is the order parameter for some transformation and then argue that the functional dependence of the free energy F on this order parameter q is sufficiently well behaved that a Taylor series representation of F in terms of q makes sense; that is, we write

$$F(q, T)=F(0, T)+F'(0, T)q+F''(0, T)\frac{q^2}{2!} + F'''(0, T)\frac{q^3}{3!}+F''''(0, T)\frac{q^4}{4!}+\dots \tag{1}$$

The possibility of such a representation is by no means obvious *a priori*; the series expansion of $F(q, T)$ may have singular, higher-order coefficients, as emphasized by Landau and Lifshitz[15] and discussed by Stanley.[2] Since our objectives in this section are pedagogical, however, we opt for the representation (1), that is, we *assume* that the presence of a singular coefficient(s) does not affect the terms of the expansion that are used.

From statistical mechanics, if F is considered to be a function of the order parameter q, the *most probable* order parameter is determined by the requirement that the free energy be a *minimum* and the associated probability distribution (distribution function)

$$f=N\exp(-F/kT) \tag{2}$$

be a *maximum* (here N is a normalization factor and k is the Boltzmann constant). Accordingly, given the representation (1), we now investigate the minima of $F(q, T)$. In particular, we investigate the positions of the possible minima of $F(q, T)$ as a function of the coefficients $F'(0, T)$, $F''(0, T), \ldots$ appearing in (1). Using this approach we consider sequentially "second-order" and then "first-order" phase transitions.

To treat the case of a second-order phase transition, we proceed by assuming that the function F is invariant under the transformation $q \to -q$. Under inversion symmetry, we have

$$F' = F''' = 0$$

and the expansion (1) assumes the simpler form (to fourth order):

$$F(q, T) = F(0, T) + F''(0, T)\frac{q^2}{2!} + F''''(0, T)\frac{q^4}{4!}$$

If we define $F''(0, T) = \alpha$ and $F''''(0, T) = 3!\beta$, then this expression becomes

$$F(q, T) = F(0, T) + \frac{\alpha}{2}q^2 + \frac{\beta}{4}q^4 \tag{3}$$

In Landau theory, the coefficient α is assumed to have the form

$$\alpha = a(T - T_c) \qquad a > 0 \tag{4}$$

so that α changes sign at the critical temperature, $T = T_c$; the coefficient β is always taken positive in the following discussion. Now suppose a disordered phase exists at a temperature $T > T_c$; here $\alpha > 0$, and the function (3) has the form displayed in Fig. 3a. Notice that F has a single minimum located at $q = q_0 = 0$. At this minimum, the entropy S and the heat capacity C_p may be calculated from the following expressions.

$$S_0 = -\left(\frac{\partial F(q_0, T)}{\partial T}\right)_p \tag{5}$$

$$C_p = T\left(\frac{\partial S_0}{\partial T}\right)_p \tag{6}$$

Let us assume that below the temperature T_c an ordered phase exists; we then repeat the above analysis for $T < T_c$, that is, $\alpha < 0$, starting from

$$F(q, T) = F(0, T) - \frac{|\alpha|}{2}q^2 + \frac{\beta}{4}q^4$$

where we have written $\alpha = -|\alpha|$. A sketch of $F(q, T)$ versus q is displayed

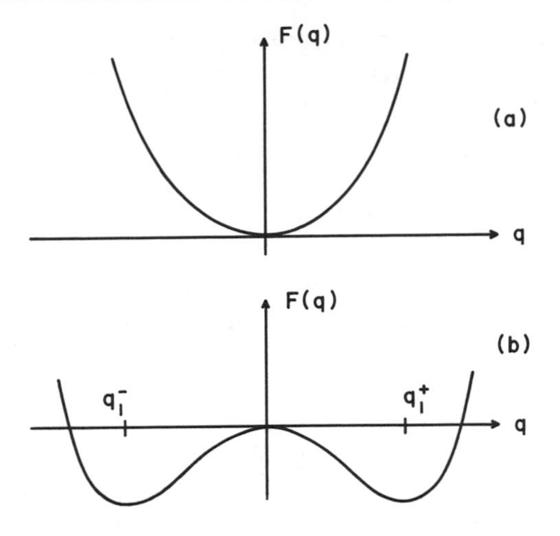

Fig. 3. A sketch of $F(q)$ versus q as determined from

$$F(q, T) = F(0, T) + (\alpha/2)a^2 + (\beta/4)q^4$$

Here we fix $\beta > 0$ and then display $\alpha > 0$ (graph a) and $\alpha < 0$ (graph b).

in Fig. 3b; here we find *two* minima,

$$q_1 = \pm\left(\frac{|\alpha|}{\beta}\right)^{1/2}$$

and one maximum, $q_0 = 0$, in the free energy function. A calculation of the entropy at $F(q, T)$ yields the result

$$S_1 = -\left(\frac{\partial F(q_1, T)}{\partial T}\right)_p = S_0 + \frac{a^2}{2\beta}(T - T_c) \tag{7}$$

and the heat capacity is

$$C_p = T\left(\frac{\partial S_0}{\partial T}\right)_p + \left(\frac{a^2}{2\beta}\right)T \tag{8}$$

Comparing the two expressions obtained for the entropy above and below the transition temperature, (5) and (7), respectively, we notice that the entropy is continuous at the transition temperature $T = T_c$; transitions for which the first derivative of the free energy is continuous at the point of transition are referred to as continuous phase transitions. On the other hand, the second derivative of the free energy (i.e., the heat capacity) is discontinuous at $T = T_c$; such a phase transition is called a phase transition of second order.

The case of first-order transition may also be discussed within the context of Landau theory. To consider this case, we return to (1) and set $F(0, T) = 0$, $F'(0, T) = 0$, $F''(0, T) = \alpha$, $F'''(0, T) = 2!\gamma$ and $F''''(0, T) = 3!\beta$. Then

$$F(q, T) = \frac{\alpha}{2}q^2 + \frac{\gamma}{3}q^3 + \frac{\beta}{4}q^4 \tag{9}$$

Here we maintain constant $\gamma, \beta > 0$, but consider the consequences of sign changes in the coefficient α; again the disordered phase at $T > T_c$ is characterized by $\alpha > 0$ and the ordered phase at $T < T_c$ is characterized by $\alpha < 0$. When we change the temperature, we generate a series of potential curves, representative examples of which are illustrated in Fig. 4. At sufficiently high temperature (Fig. 4*a*), there is only a single minimum in the free energy, at $q = q_0 = 0$. Upon lowering the temperature, some structure begins to appear in the profile of $F(q, T)$ versus q, Figs. 4*b* and 4*c*, corresponding to the emergence of the new minimax points at

$$q = \frac{-\gamma \pm [\gamma^2 - 4\alpha\beta]^{1/2}}{2\alpha}$$

although the system still persists in the state characterized by the local minimum at $q = q_0 = 0$. A further drop in temperature, Fig. 4*d*, allows the system to transit to the global minimum at $q = q_1^+$. Notice that in this case, a calculation of the entropy for the states $q = q_0 = 0$ and

$$q = q_1^+ = \frac{1}{2|\alpha|}[\gamma + (\gamma^2 + 4\,|\alpha|\,\beta)^{1/2}]$$

produces expressions that differ at $T = T_c$; that is, the entropies of the ordered and disordered states differ at the transition temperature. The consequent transition is referred to as a discontinuous phase transition. Further, since the first derivative of the free energy F is discontinuous at $T = T_c$, the phenomenon is termed a "first-order" phase transition. We remark that an important effect predicted by this theory of first-order transition is *hysteresis*. Notice that, upon following the behavior of the system with increase in temperature, Figs. 4*d* to 4*a*, the system remains at q_1^+, through Fig. 4*b*, before transiting to the global minimum at $q = q_0 = 0$, Fig. 4*a*. Thus a plot of the entropy versus temperature would have the profile displayed in Fig. 5.

Having sketched the main ideas of Landau theory, we now use this theory to illustrate a number of mathematical and physical concepts that play an important role in our later discussion. To proceed, let us examine

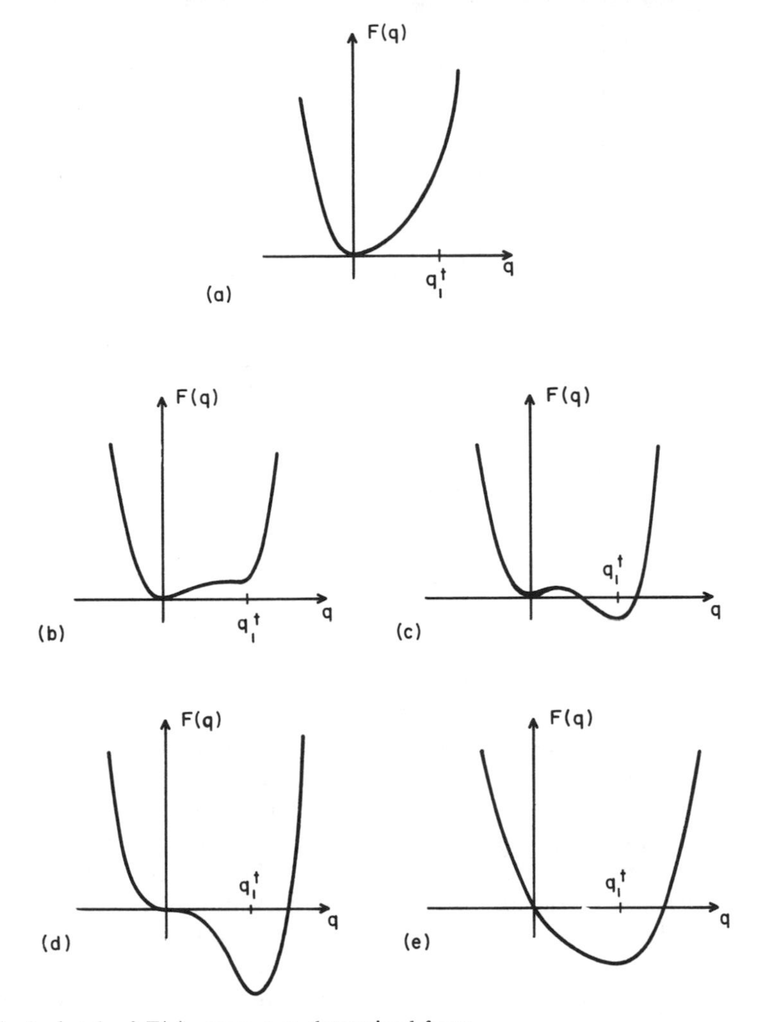

Fig. 4. A sketch of $F(q)$ versus q as determined from

$$F(q, T) = (\alpha/2)q^2 + (\gamma/3)q^3 + (\beta/4)q^4$$

Here we fix γ, $\beta > 0$ and generate the sequence of curves ($a \rightarrow e$) by decreasing α from $\alpha > 0$ to $\alpha < 0$.

the behavior of the equilibrium q_e [the extrema of $F(q, T)$] when the parameter determining the shape of the free energy profile changes. This parameter, as noted above, is α or, given the definition (4), the temperature T. Consider first the case of a second-order transition. We construct a plot of q_e versus $-T$, so that a trajectory left to right in the consequent figure corresponds to a decrease in temperature and transition from the

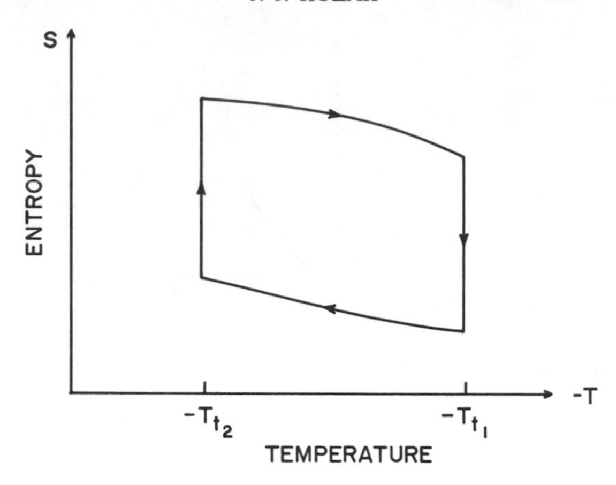

Fig. 5. The variation of entropy versus temperature as predicted by analysis of

$$S_0 = -\left(\frac{\partial F(q_0, T)}{\partial T}\right)_p$$

with

$$F(q, T) = (\alpha/2)q^2 + (\gamma/3)q^3 + (\beta/4)q^4$$

disordered to the ordered state of the system. As portrayed in Fig. 6 for the case $\alpha > 0$, $\beta > 0$, the only stable solution is the one $q_e = 0$. However, for $\alpha < 0$ and $\beta > 0$, $q_e = 0$ becomes unstable (dashed line) and is replaced by two possible stable states (the solid fork), characterized by $q_e = \pm\sqrt{|\alpha|/\beta}$. In the language of bifurcation theory, the graph displayed in Fig. 6 is a *branching diagram* and the point $q = q_0 = 0$ is a *bifurcation point.* Using Fig. 6 we can also introduce the concepts of *symmetry-breaking instability* and *exchange of stability.* With respect to the first of these concepts, notice that when we change α from positive to negative values, the particular value $\alpha = 0$ $(T = T_c)$ gives the point at which the stable equilibrium $q_e = 0$ for $\alpha > 0$ becomes unstable. The corresponding change in the structure of the curve, $F(q, T)$ versus q, from the symmetry displayed in Fig. 3*a* (descriptive of the disordered state) to that displayed in Fig. 3*b* (descriptive of an ordered state) as a certain critical value of the temperature is realized $(T = T_c)$, leads to the description of the whole phenomenon as a *symmetry-breaking instability.* Second, since a decrease of temperature through the critical value $T = T_c$ leads to transition of the system from one stable equilibrium position $(q_e = 0)$ to new possibilities $(q_e = \pm\sqrt{|\alpha|/\beta})$, the process may be referred to as an *exchange of stability.*

We consider now the case of first-order phase transitions and examine the behavior of the equilibrium coordinate q_e with respect to changes in the parameter α. Again the parameter α determines the shape of the

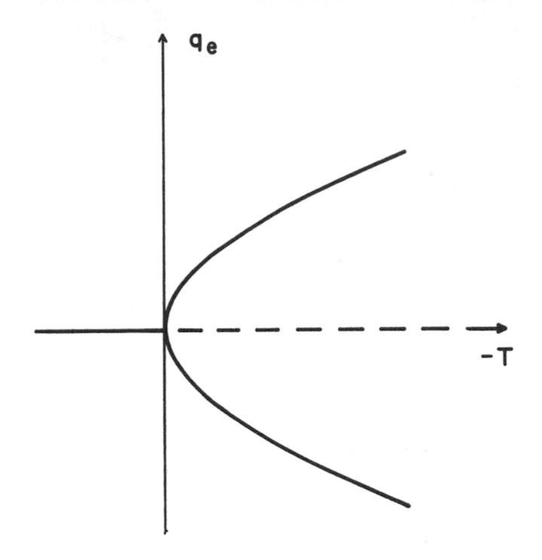

Fig. 6. The branching diagram for a "soft" bifurcation. We start from the equation

$$F(q, T) = F(0, T) + (\alpha/2)q^2 + (\beta/4)q^4 \qquad (\beta > 0)$$

and construct a plot of the equilibrium q_e versus $-T$.

curve generated when the free energy $F(q, T)$ is plotted versus q, and the q_e specify the extrema of $F(q, T)$ for a particular choice of α. The changes in q_e with respect to $-T$ displayed in Fig. 6 may be correlated with the changes in $F(q, T)$ versus q displayed in Fig. 4. Upon lowering the temperature, the system passes from the state portrayed in Fig. 4*a*, which we associate with the high-temperature, disordered phase characterized by a single extremum q_0, to a state characterized by two (Fig. 4*b*), then three (Fig. 4c), then two (Fig. 4*d*) extrema, to the state portrayed in Fig. 4*e*, which we associate with the low-temperature, ordered phase characterized by a single extremum q_1^+. Notice that if one follows the reverse trajectory from the state portrayed in Fig. 4*e* to that in Fig. 4*a*, the system persists in the state q_1^+ until the situation displayed in Fig. 4*b* is realized, at which point the system transits to the disordered phase characterized by the single extremum, $q_e = q_0$; this hysteresis in the response of the equilibrium coordinate q_e to changes in temperature leads directly to the hysteresis in the entropy noted earlier (Fig. 5). If we regard Fig. 7 as a *branching diagram*, we see that there are three possible coordinate values (i.e., solutions) in the range $-T_{t_1} > -T > -T_{t_2}$, whereas for $-T > -T_{t_1}$ and $-T < -T_{t_2}$ there is only one solution. As temperature decreases (i.e., $-T$ increases), one of these solutions (the lower branch) increases, whereas the other (upper branch) decreases, until the latter meshes with the lower branch at $T = T_{t_1}$. The lower and middle branches then vanish as T decreases below T_{t_1}, and only one solution (extremum)

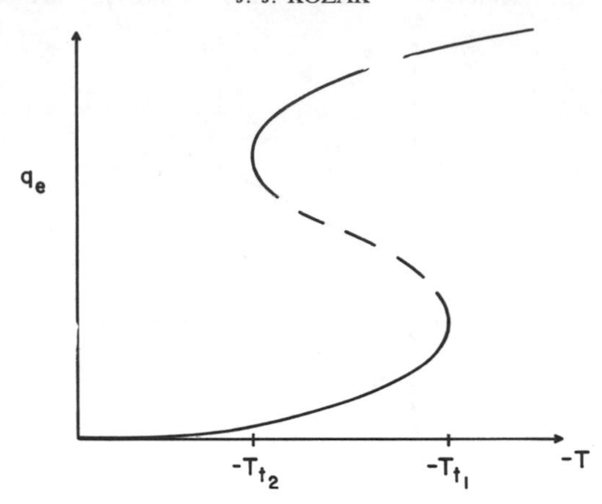

Fig. 7. The branching diagram for a "hard" bifurcation. We start from the equation

$$F(q, T) = (\alpha/2)q^2 + (\gamma/3)q^3 + (\beta/4)q^4 \qquad (\alpha, \beta > 0)$$

and construct a plot of the equilibrium q_e versus $-T$.

persists beyond T_{t_1}. In some applications (e.g., laser theory), the consequent discontinuous transition at $T = T_{t_1}$ (or, $T = T_{t_2}$ for the reverse transition) is referred to as a *hard bifurcation*, whereas the continuous transition at the bifurcation point $q_e = 0$ in the branching diagram Fig. 6 (the case of a second-order transition) is referred to as a *soft bifurcation* (associated with hard and soft modes, respectively).

In the discussions based on (3) for the case of second-order transitions or (9) for first-order transitions, we considered only the response of the free energy $F(q, T)$ (plotted versus q) to changes in the parameter α. It is of interest to study the possible changes in the structure of the function $F(q, T)$ that arise when *arbitrary* variations in the parameters α, β, and γ are permitted, since this allows us to make contact with the theory of stable mappings and their singularities, a theory closely associated with the name of René Thom and usually referred to as *catastrophe theory*. To bring out this relationship, we summarize briefly some of the basic ideas of catastrophe theory. (We emphasize that our intention here is *not* to present an in-depth exposition of catastrophe theory, but rather to give a development sufficient only to demonstrate its relationship to the ideas introduced previously and those discussed later.)

The mathematical theory of catastrophes treats the *local* behavior of smooth functions $f(x)$ near a point x_0 of the domain space. Specifically, two functions f and g are said to be equivalent at x_0 if there is a neighborhood of x_0 throughout which $f(x_0) = g(x_0)$. Two functions f and g are said to be of the same type *locally* at x_0 if there is some neighborhood of x_0 such that a smooth deformation of this neighborhood and the range space takes f into g. Consider now the behavior of a parametrized family of smooth functions, say (3), in the neighborhood of a point q. It may happen that, as the parameters are changed smoothly, the local type of the function changes smoothly. Referring to Figs. 3*a* and 3*b*, consider the behavior near $q_0=0$; when $\alpha>0$, the function $F(q, T)$ locally has a minimum at $q_0=0$, whereas for $\alpha<0$ the function locally has a maximum at $q_0=0$ and two minima to either side. As one decreases the value of the parameter α from positive to negative values, one finds two minima and a maximum arising from a single minimum at $\alpha=0$. That is, the function $\tilde{F}(q, T)\equiv q^4/4$ can be perturbed to yield either one minimum or two by an infinitesimal change in the parameter α. Stated differently, arbitrarily close to the function $\tilde{F}(q, T)=q^4/4$ (in this family) there are functions of several different types locally at $q=q_0=0$.

The fundamental question addressed by catastrophe theory can be stated as follows: For a given smooth function [say, $\tilde{F}(q, T)$], what different types (locally at $q=q_0$) can arise by infinitesimal changes of parameters? Or, how many parameters must there be before a family of functions can be constructed containing $\tilde{F}(q, T)$ as well as all (local) types of functions near to $\tilde{F}(q, T)$? Let us see what this means. Suppose that $\tilde{F}(q, T)$ is increasing near the point q_0. Then, in any smoothly parametrized family of functions of which $\tilde{F}(q, T)$ is a member, all those functions generated from $\tilde{F}(q, T)$ via an infinitesimal changes in parameters will also be increasing near q_0. Only at the points where $\tilde{F}(q, T)$ has a vanishing derivative $\tilde{F}'(q, T)=0$ (or $D\tilde{F}(q, T')=0$ for functions of several variables), can the local type of the functions near $\tilde{F}(q, T)$ change. Such points of functions are called *singularities.* For example, the function $\tilde{F}(q, T)=q^4/4$ is singular at $q=0$. It can be placed in a parametrized family of functions, for example, (3), where several different *local* types of functions occur, each arbitrarily close to $\tilde{F}(q, T)$.

Suppose we introduce some simplifications in notation; since we may shift the origin of the q-axis by the coordinate transformation $q\rightarrow\tilde{q}+\xi$, and since we may shift the zero point of the function $F(q, T)$ so that the constant term in $F(q, T)$ vanishes, the family of functions, (3), may be written

$$\tilde{F}(\tilde{q}, T)=\frac{\tilde{q}^4}{4}+u\frac{\tilde{q}^2}{2}+v\tilde{q} \tag{10}$$

where u and v are now renormalized parameters. Then (3), the free energy function descriptive of second-order transitions in the Landau picture, is just (10) with $v=0$, and (9), the free energy function descriptive of first-order transitions, is (10) with $u, v\neq 0$. It can be shown that the couple (u, v) is the minimum number of parameters that must be specified if all possible nearby types are to be represented by the family, (10). A parametrized family containing all possible nearby types is called an *unfolding* of the singularity. The minimal number of parameters needed to unfold a singularity is the *codimension*, and in the example considered here the codimension is two. A remarkable fact is that the low codimension singularities of many *smooth* functions resemble (under a coordinate transformation) those of simple polynomials. An even more remarkable feature is that almost all singularities of smooth functions of one or two variables, whose codimension is between one and four inclusive, have an unfolding that can be of only seven types (the Thom classification theorem). Certain higher codimension singularities have also been classified.[17]

We are now ready to link our qualitative discussion of catastrophe theory with our earlier description of first-and second-order transitions based on Landau theory. What we say in Landau theory is that the governing potential function [i.e., the free energy $\tilde{F}(\tilde{q}, T)$] is a smooth function $\tilde{F}:\mathcal{R}^k \times \mathcal{R}^n \rightarrow \mathcal{R}$, where k is the dimension of the parameter space (in our example, $k=2$) and n is the dimension of the state space (in our example, $n=1$). In effect, we assert that the states of the system may be monitored by focusing on the order parameter q. Suppose we assume that the order parameter q obeys an evolution equation, perhaps one as simple as the gradient differential equation

$$\dot{q} = -\frac{\partial F(q, T)}{\partial q} \tag{11}$$

(where the dot denotes the derivative with respect to time). The possible equilibrium (stable, metastable, unstable) states of the system governed by the function $\tilde{F}(\tilde{q}, T)$ are then identified by setting the temporal derivative to zero, $\dot{\tilde{q}}=0$; notice that catastrophe theory considers not only minima but also maxima and other stationary values of $\tilde{F}(\tilde{q}, T)$. Incorporating the (renormalized) free energy function $\tilde{F}(\tilde{q}, T)$ defined by (10) into the above evolution equation and setting $\dot{\tilde{q}}=0$, we obtain

$$\tilde{q}^3 + u\tilde{q} + v = 0 \tag{12a}$$

Now consider the manifold $\mathcal{M}_{\tilde{F}} \subset \mathcal{R}^3$ generated by considering all possible variations in the parameters (u, v). Then let $\chi_{\tilde{F}}:\mathcal{M}_{\tilde{F}} \rightarrow \mathcal{R}^2$ be the map induced by the projection of the surface $\mathcal{M}_{\tilde{F}}$ in $\mathcal{R}^3$ to the two-dimensional control space defined by the coordinates (u, v); $\chi_{\tilde{F}}$ is called the *catastrophe map* of $\tilde{F}$. A singularity of the mapping $\chi:\mathcal{M} \rightarrow \mathcal{R}^2$ occurs when two stationary values of $\tilde{F}$ coalesce; in the example, this occurs when

$$\frac{\partial^2 \tilde{F}}{\partial \tilde{q}^2} = 3\tilde{q}^2 + u = 0 \tag{12b}$$

The *singularity set* $\mathscr{S}$ of χ in this case is given by (12); $\mathscr{S}$ consists of two *fold curves*, given parametrically by a nonzero parameter λ such that

$$(u, v, q) = (-3\lambda^2, 2\lambda^3, \lambda), \qquad \lambda > 0 \quad \text{and} \quad \lambda < 0$$

and one *cusp* singularity at the origin. The *bifurcation set* $\mathscr{B}(=\chi\mathscr{S})$ is then given parametrically by

$$(u, v) = (-3\lambda^2, 2\lambda^3)$$

which defines the cusp

$$\frac{u^3}{27} + \frac{v^2}{4} = 0 \tag{13}$$

Both $\mathscr{M}$ and $\mathscr{S}$ are smooth at the origin, so the cusp actually occurs in $\mathscr{B}$ rather than $\mathscr{S}$. The power of Thom's classification theorem as deployed within the context of the present example is this: When $k=2$, the two *elementary catastrophes* are the fold curve and the cusp. Figure 8 thus represents the most complicated *local* behavior possible in the mapping $\chi_{\tilde{F}}$.

Let us consider in more detail the significance of the cusp defined by (13). If we choose a coordinate specification $c_1 = (u_1, v_1)$ outside the cusp, that is,

$$\frac{u^3}{27} + \frac{v^2}{4} > 0$$

then $\tilde{F}_{c_1}$ has a unique minimum; $\mathscr{M}$ is single sheeted over the outside of the cusp. On the other hand, if we choose $c_2 = (u_2, v_2)$ such that we are inside the cusp, that is,

$$\frac{u^3}{27} + \frac{v^2}{4} < 0$$

then $\tilde{F}_{c_2}$ has two minima separated by one maximum (the two minima may differ in depth depending on the size of v); therefore $\mathscr{M}$ is triple sheeted over the inside of the cusp. However, of these three sheets, only the upper and lower sheets matter in Landau theory—they represent minima in $\tilde{F}$; the middle sheet, although relevant for developing the mathematics of catastrophe theory, is assigned no physical relevance in the phase transition theory. If the parameters (u, v) are changed in such a way that one enters the cusp region, the system may transit from one minimum to the other. For fixed $u<0$, this occurs with increasing v, as is evident from Fig. 8; a slice through $\mathscr{M}$ for some fixed $u<0$, under a diffeomorphic distortion, is just the profile sketched in Fig. 7, the figure which in our discussion of first-order transitions gave the ("jump") response in the equilibrium coordinate q_e when the parameter α was

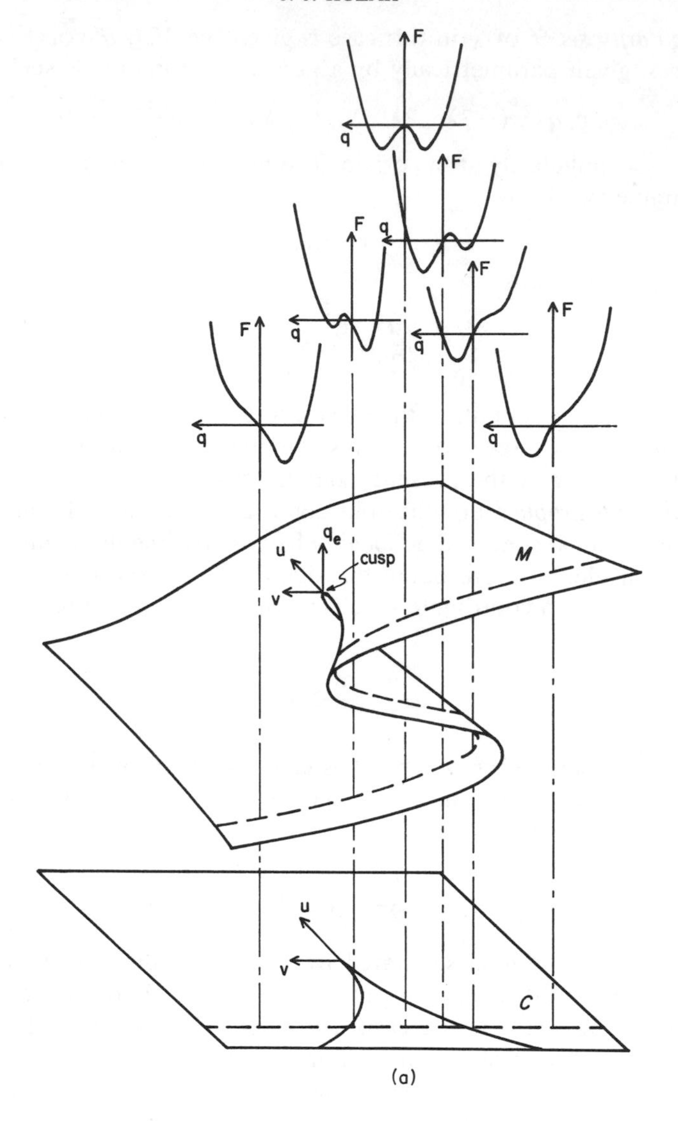

Fig. 8. Two views of the generic form of the Riemann–Hugoniot (cusp) catastrophe, whose potential is

$$\tilde{F}(\tilde{q}, T) = \tilde{q}^4/4 + u\tilde{q}^2/2 + v\tilde{q}$$

The uppermost level of graph (a) displays the potentials corresponding to different settings of (u, v). The middle level represents the surface (manifold) $\mathcal{M}$ of extrema of $\tilde{F}(\tilde{q}, T)$, whereas the lower plane shows the constraint space $\mathcal{C}$, together with the cusp. The bifurcation set $\mathcal{B}$ must lie somewhere within this cusp. The dotted line in this graph (a) shows a "hard" bifurcation, and this curve may be related to the branching diagram in Fig. 7. The changed. On the other hand, for the case of a second-order transition, there is no cubic term in the defining free energy function $F(q, T)$, (3); consequently, $v=0$ in (10) and (12). In Fig. 8, consider the plane defined by $v=0$. A trajectory down the u-axis from $u>0$, through the value $u=0$, to negative u engineers (under a suitable diffeomorphic distortion) the profile displayed in Fig. 4. For $u>0$ only a single real q_e is possible, namely, $q_e=q_0=0$, whereas for $u<0$, two stable branches arise and the continuation of the branch $q_e=0$ (which lies in the middle sheet) is

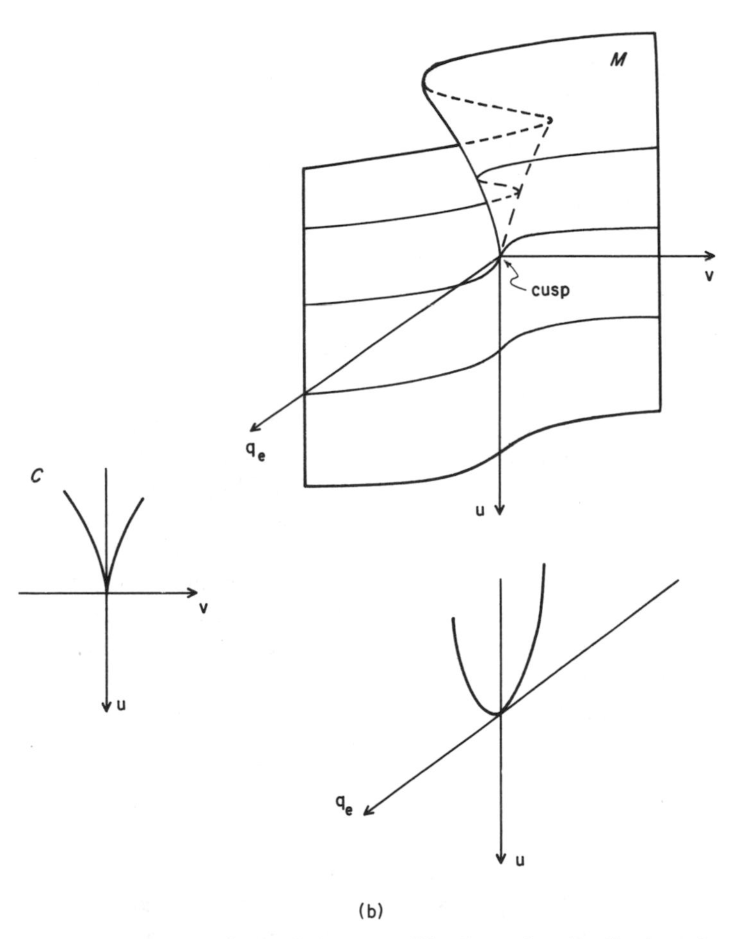

upside-down view of the manifold $\mathcal{M}$ in graph (b) allows the visualization of a "soft" bifurcation; the branching diagram at the bottom of this graph may be put in correspondence with the one displayed in Fig. 6. (Note: for the sake of simplicity, we have dropped the "wiggles" over F and q in these diagrams.)

unstable. In contrast to the discontinuous behavior found in first-order transitions, we find here that a trajectory down the u-axis bifurcates smoothly into two diverging paths, one above and the other below the cusp point.

There are several other aspects of the theory of phase transitions that can be illustrated using the model described in this section. Suppose, for example, we consider the consequence of deforming the free energy function defined by (3) by changing the parameter α from $\alpha > 0$ (Fig. 3*a*), through $\alpha = 0$, to α negative (Fig. 3*b*). Notice that the profile of $F(q, T)$ versus q becomes flatter and flatter as one approaches $\alpha = 0$, from $\alpha > 0$. Hence, a system governed by the evolution equation (11) would, upon being displaced from the stationary state return to the equilibrium state more and more slowly as $\alpha \to 0^+$. This phenomenon is called *critical slowing down.*

Suppose next, instead of considering the behavior in the vicinity of the single minimum in Fig. 3*a* as $\alpha \to 0^+$ we focus on the two minima and one maximum in Fig. 3*b*. As drawn, the point $q = q_0 = 0$ locates the maximum of the function $F(q, T)$ and q_1^+ and q_1^- locate the minima to the right and left, respectively. Again we assume that the system obeys an evolution equation of the form, (11). Now consider the response of the system if displaced to some point $q_1^- < q_i < 0$; given the structure of the governing equation, (11), the system would transit to the point $q = q_1^-$. Alternatively, if the system is displaced to a point $0 < q_i < q_1^+$, the system transits to the point q_1^+. This means that the final position of the system, as modeled by the free energy function (3) and governed by the evolution equation, (11), is a *discontinuous* function of its initial location, q_i. Suppose we compute the time $t_{1/2}$ required for the system to transit halfway from the maximum at $q = 0$ to either minimum, q_1^+ or q_1^-. Notice that as the initial location q_i of the system is set closer and closer to $q = 0$, the time required for descent increases, becoming infinite as $q_i \to 0$. For infinite time, a small change in the initial location from $q_i = 0^+$ to $q_i = 0^-$ can be amplified, with a resulting difference in position of the system given by

$$q_1^+ - q_1^- = 2\left(\frac{|\alpha|}{\beta}\right)^{1/2}$$

Therefore, with an infinite length of time available, one can realize an infinite amount of amplification, leading to a discontinuity in $q(t = +\infty, q_i)$; that is, although $q(t, q_i)$ is continuous for finite t, $q(\infty, q_i)$ is a discontinuous function of q_i. It is important to realize that the function $F(q, T)$ defined by (3) is *analytic* in q, so the discontinuity in $q(\infty, q_i)$ at $q_i = 0$ cannot be blamed on any singularity in $F(q, T)$ itself!

The remarks in the preceding paragraph are essentially a paraphrase of Wilson's qualitative introduction to the renormalization group (RG) approach to critical phenomena.[9] His basic proposal is that the singularities at the critical point of a ferromagnet can be understood as arising from the analogue of the $t \to +\infty$ limit of the solution of a certain differential equation. In fact, in the simplest version of the renormalization group picture (the one based on the Kadanoff block picture[18]), the maximum point in Fig. 3*b*, $q = q_0 = 0$, is analogous to the point $K_c = -J/kT_c$, where J is the "coupling constant" between neighboring spins and T_c is the critical temperature; the time $t_{1/2}$ is analogous to the correlation length $\xi = \xi(K, h = 0)$, where h is the magnetic field variable. Although bringing out these analogies is instructive, it must be stressed that the Kadanoff-Wilson approach goes far beyond Landau theory in providing a correct description of critical phenomena. Therefore we must try to understand why theories such as the RG approach are successful and Landau theory is inadequate for treating phase transitions at thermal equilibrium; the argument developed in the following paragraph is taken from Haken.[14]

We start again with the intuitive assumption that the behavior of a macroscopic system can be monitored by an appropriately identified order parameter q. In many examples it is the average value of q^2 which is the physically interesting quantity. If we adopt the distribution function defined by (2), the average value of q^2 may be computed in the usual way:

$$\langle q^2 \rangle = \frac{\int q^2 \exp[-F/kT]\, dq}{\int \exp[-F/kT]\, dq}$$

Introducing the definition

$$\hat{F} = F/kT$$

and then adopting the representation [based on (3)],

$$\hat{F} = (\alpha - \alpha_c)q^2 + \beta q^4$$

where α_c denotes the point of transition, we notice that

$$\frac{\partial \hat{F}}{\partial \alpha} = q^2$$

so that the above expression for $\langle q^2 \rangle$ may be written

$$\langle q^2 \rangle = -\frac{\partial}{\partial \alpha} \ln\left\{\int \exp(-\hat{F})\, dq\right\} \tag{14}$$

To evaluate the integral in (14), we assume that the function $\exp(-\hat{F})$ is sharply peaked at $q = q_0$. This is a delicate point. Given the polynomial representation of $\hat{F}$, it may happen that there is more than one minimum of $\hat{F}(q)$. In effect, the assumption that there is a single sharp peak in $\exp(-\hat{F})$ implies that only one state is occupied. In a first-order phase transition this means that only one of the two local minima in Fig. 4*c* is occupied. In second-order transitions, this assumes that the symmetry of the problem has been broken and one state is preferred. For both transitions, the assumption is tantamount to the assertion that the system is "away" from the transition point $\alpha = \alpha_c$. Given these qualifications, the program now is to expand the exponential around the minimum value of $\hat{F}$, retaining terms to second order only:

$$\hat{F}(q) = \hat{F}(q_0) + \frac{1}{1!}\left[\frac{\partial \hat{F}(q)}{\partial q}\right]_{q=q_0}(q-q_0) + \frac{1}{2!}\left[\frac{\partial^2 \hat{F}(q)}{\partial q^2}\right]_{q=q_0}(q-q_0)^2 \quad (15)$$

Since the linear term in this expansion vanishes at the minimum q_0 of $\hat{F}$, substitution of (15) into (14) followed by an integration yields the result:

$$\langle q^2 \rangle = \frac{\partial \hat{F}(q)}{\partial \alpha} - \frac{1}{2}\frac{\partial}{\partial \alpha}\ln\left(\frac{\pi}{\hat{F}''(q_0)}\right) \quad (16)$$

In this development, we take $\alpha = -T$, so from (5) we see that the first term on the right-hand side of (16) is proportional to the entropy (which is continuous at $\alpha = \alpha_c$ for a second-order transition and discontinuous for a first-order transition, as noted previously). The second term in (16) derives from the quadratic term in the expansion (15) of the free energy; since qualitatively, $\hat{F}''(q_0)$ scales the difficulty of perturbing a system away from its equilibrium position [characterized by $\hat{F}(q_0)$], it may be anticipated that the second term on the right-hand side of (16) is related to phenomenon of *fluctuations.* As a matter of fact, it has been stressed repeatedly in recent discussions of the RG approach[10] that a main reason for the failure of Landau theory is the inadequate treatment of fluctuations. In mean field theory, *all* fluctuations are neglected, that is, all $k \neq 0$ Fourier components of the local variable $S(\mathbf{r})$ (where $\langle S(\mathbf{r})\rangle = q$), namely,

$$S_{\mathbf{k}} = \sum S(\mathbf{r}) e^{-i\mathbf{k}\cdot\mathbf{r}}$$

are suppressed. [Do not be misled by the apparent divergence in $\langle q^2 \rangle$ at $\alpha = \alpha_c$ in the second term of (16); it is a consequence of the "rough" method used in evaluating (14)]. Therefore Landau theory may be interpreted as a theory in which mean values have been replaced by most probable values, and the failure of the theory to describe correctly equilibrium phase transitions may be attributed in great measure to the neglect of fluctuations. Later on in discussing theories of first- and second-order phase transitions which proceed from a mean field representation of the problem considered, we must be sensitive to the possibility that the neglect of fluctuations in such theories may lead to an incorrect description of the transition being modeled.

There is a final point, related to the problem of fluctuations, that can be brought out within the context of the simple model considered here. Consider, for definiteness, the theory of first-order phase transitions based on (9). Let us label the lowest minimum in Fig. 4*b* as q_1^- and the other minimum as q_1^+. If the system being described were a classical-mechanical system [e.g., suppose (11) described the motion of a ball rolling on a hill with height given by $F(q)$], then it would be customary to label q_1^- as the state of *absolute* stability, given that it would be the state of lowest energy. However, classical mechanics cannot describe the *relative* stability of various competing states of local stability; a particle trapped in the left-hand well of Fig. 4*b* will simply stay there. Once we consider a thermodynamic system, however, and introduce the notion of temperature and fluctuations via the Boltzmann distribution, the state q_1^- becomes a more probable state than q_1^+. Thus, as Landauer has emphasized,[19] a strictly deterministic theory of phase transitions, say one based on catastrophe theory,[16] which classifies points of local stability and their appearance and disappearance as the parameters constraining the system are changed, can tell us nothing about the relative stability problem. Even raising the barrier height in Fig. 4*b* does not change the relative populations in the two wells; the frequency of jumps and hence the relaxation time of the system may change, but not the steady-state distribution in the state(s) near the minima. In effect, catastrophe theory assumes that metastable states remain unaffected by fluctuations until driven to the very limit of existence of a metastable state. Now, although we have introduced the problem of *metastability* within the context of Landauer's critique of catastrophe theory, we alert the reader to a further difficulty of interpretation that arises when one implements a bifurcation approach to analyze nonlinear problems in the theory of phase transitions. As we shall see later, a bifurcation approach can give one more information than one really wants, for example, bifurcation points and emanating branches of new solutions in models for which it is known that no phase transition exists. There has been a temptation to interpret these "unwanted" branches in a more positive way, as defining regimes of metastability or metastable states. The author supports the view[20] that metastability is inherently a dynamic concept, one that can only be described within the context of a time-dependent theory. As emphasized by Landau and Lifshitz,[15] the metastable state is one of *partial* equilibrium, having a certain relaxation time; "the thermodynamic functions in such a state can therefore be defined only without taking account of these processes, and they can not be regarded as the analytic continuation of the functions from the region of stability corresponding to the complete equilibrium states of the substance." Within the framework of *equilibrium* statistical mechanics, fluctuations always drive a system from a state of possible metastability to a state of absolute stability. Theories of *equilibrium* phase transitions based on a bifurcation approach that have invoked the concept of metastability as a working hypothesis are not advanced in this review.

III. SECOND-ORDER TRANSITIONS

A. Prologue

The qualitative discussion of first- and second-order transitions described in the preceding section isolated several physical concepts important in the theory of phase transitions. The most obvious mathematical feature of the problem, as posed, was that the governing equations were *nonlinear* and admitted several different types of solutions depending on the choice of parameters. It was seen that two rather different types of branching diagrams can arise, depending on the order of the transition. In Landau theory, the branching diagram for a second-order transition, Fig. 6, was characterized by a smooth changeover from one branch of solutions to new branches at the bifurcation point. The branching diagram for a first-order transition in Landau theory, Fig. 7, described the possibility of a discontinuous transition from one branch of solutions to another. In fact, two uses of the word "bifurcation" were introduced in the discussion: the so-called "soft" bifurcation at the *bifurcation point* in the case of second-order transitions and the "hard" bifurcation at a point (or points) in the *bifurcation set* for first-order transitions. At the risk of oversimplification, most of the bifurcation theorems or techniques available in the contemporary mathematical literature deal with the phenomenon of "soft" bifurcation, with mathematical procedures for handling discontinuous changes just now becoming available, thanks to the work of René Thom[16] and others.[17] Let us sharpen up this remark by considering the equation[22]

$$F(\lambda, u) = 0 \tag{17}$$

where λ is on the real line $\mathcal{R}$, u is an element of some real Banach space $\mathcal{B}$ with norm $\|\ldots\|$, and F is a *nonlinear* transformation from $\mathcal{R} \times \mathcal{B}$ into $\mathcal{B}$. A solution of (17) is an ordered pair (λ, u), but usually one interprets u as the solution of the problem for fixed λ or as a solution depending parametrically on λ. To study the possible branching of (17), one assumes that a known solution $u(\lambda)$ of (17) is available; suppose, for simplicity, that the known (basic) solution of (17) is $u = 0$ for all λ. As noted above, the most extensive results on the bifurcation of solutions to a given nonlinear equation such as (17) deal with the branching from the basic solution $u = 0$ of new solutions which are of *small norm.* Then one says that $\lambda = \lambda_0$ is a branch point or bifurcation point if in every *small neighborhood* of $(\lambda_0, 0)$ in $\mathscr{R} \times \mathscr{B}$ there exists a solution (λ, u) of (17) with $\|u\| \neq 0$. The nonlinear perturbation theories referred to as the Lyapunov-Schmidt method[22,23] and the Poincaré-Keller method[22,24] all deal with bifurcation defined in the above sense.

From the remarks presented in the preceding section, and given the current state of development of bifurcation theory, the more natural candidate for examining the consequences of nonlinearity in a statistical-mechanical theory of phase transitions would seem to be second-order phase transitions, characterized by "soft" mode bifurcation. In this section we describe several studies that have implemented a bifurcation approach to examine second-order phase transitions.

B. The Ising Model of Ferromagnetism

Considerable theoretical interest attaches to the examination of the second-order phase transition associated with the onset of ferromagnetism. This is because a theoretical model for this phenomenon exists that has been solved exactly in one dimension (by Ising[25]) and in two dimensions for the case of zero field (by Onsager[5]). Moreover, extensive numerical studies have been performed on the three-dimensional Ising model, and the results obtained[2,26] are found to be in excellent accord with experimental evidence on the transition. For these reasons, a study of the D-dimensional Ising model using the methods of bifurcation theory provides an acid test against which the viability of the approach can be assessed, and for this reason we initiate our discussion of second-order phase transitions with a description of the work of Goldstein and Kozak.[27] The general approach described below was suggested to these authors by R. Zwanzig.

We assume that there is a spin σ_i at each site of a regular lattice of N sites. We assume further that the interaction energy between neighboring spins may be written as $-J\sigma_i\sigma_j$, where J is the interaction energy, and each spin can take on the discrete values ± 1. Then we construct the partition function

$$Q_N = \sum_{\{\sigma_j = \pm 1\}} \exp\left(-\Sigma'\{-J\sigma_i\sigma_j/kT\}\right)$$

$$= \sum_{\{\sigma_j = \pm 1\}} \exp\left(+J\beta\Sigma'\sigma_i\sigma_j\right)$$

Here k is Boltzmann's constant, T is the temperature, $\beta = 1/kT$, the set $\{\sigma_j\}$ denotes a particular configuration of spins, and Σ' is written to denote the sum over nearest neighbors. If we introduce the further definition $H \equiv \beta J$ and then normalize the partition function to unity we obtain

$$Q_N = 2^{-N} \sum_{\{\sigma_j = \pm 1\}} \exp\,(H\Sigma'\sigma_i\sigma_j)$$

Berlin and Kac[28] have shown that the Ising partition function can be written in the form

$$Q_N = 2^{-N} \sum_{\{\sigma_j = \pm 1\}} \exp\,(\tfrac{1}{2}H\boldsymbol{\sigma}'\cdot\mathbf{M}\cdot\boldsymbol{\sigma})$$

Here $\boldsymbol{\sigma}$ is a column matrix with components $\sigma_1, \ldots, \sigma_N$. The $\mathbf{M}$ denotes a cyclic matrix with real, orthogonal eigenvectors V_{ks} normalized to unity; for a cubic lattice of side L in D dimensions the eigenvalues m_p of $\mathbf{M}$ are given by

$$m_p = 2\sum_{\nu=1}^{D} \cos\left(\frac{2\pi p_\nu}{L^\nu}\right)$$

where $p = (p_1, p_2, \ldots, p_D)$ and $0 \leqslant p_\nu \leqslant L^\nu - 1$. Siegert[29] has shown how the Kac-Berlin expression for the Ising partition function be transformed further using the following generalization of the Gaussian integral

$$\exp\left(-\tfrac{1}{2}\sum_{k,\ell} \xi_k \zeta_{k\ell} \xi_\ell\right) = (2\pi)^{-\frac{N}{2}}(\det \boldsymbol{\zeta})^{-1/2}$$

$$\times \int_{-\infty}^{+\infty} \cdots \int_{-\infty}^{+\infty} \exp\left(-\tfrac{1}{2}\sum_{k,\ell} y_k(\zeta^{-1})_{k\ell} y_\ell + i\sum_k \xi_k y_k\right)\Pi_j\, dy_j$$

where the identity holds for any set of n (real or complex) variables ξ_k and for any real, symmetric positive-definite matrix $\boldsymbol{\zeta}$. Before this identity can be applied to the problem at hand, we must account for the fact that, since the eigenvalues of $\mathbf{M}$ are cosines, the matrix $\mathbf{M}$ need not be positive definite, although it is real and symmetric. This is done by adding and subtracting a quantity $\alpha(>2D)$ to the diagonal elements of $\mathbf{M}$ and then defining a new matrix $\mathbf{W}$, where

$$\mathbf{W} = \mathbf{M} + \alpha\mathcal{I}$$

with $\mathcal{I}$ the identity matrix, and eigenvalues

$$w_p = m_p + \alpha.$$

By construction, $\mathbf{W}$ is now a real, symmetric, positive-definite matrix, and hence the partition function Q_N may be written in the form

$$Q_N = 2^{-N} \exp\left(-\tfrac{1}{2}NH\alpha\right) \sum_{\{\sigma_j=\pm 1\}} \exp\left(\tfrac{1}{2}H\boldsymbol{\sigma}'\cdot\mathbf{W}\cdot\boldsymbol{\sigma}\right)$$

Applying the above identity directly, we find that

$$Q_N = 2^{-N} \exp\left(-\tfrac{1}{2}NH\alpha\right) \sum_{\{\sigma_j=\pm 1\}} (2\pi)^{-\frac{N}{2}} (\det H\mathbf{W})^{-1/2}$$

$$\times \int_{-\infty}^{+\infty} \cdots \int_{-\infty}^{+\infty} \exp\left(-\tfrac{1}{2}\mathbf{y}'\cdot(H\mathbf{W})^{-1}\cdot\mathbf{y}\right) \exp\left(-\sum_{k=1}^{N} \sigma_k y_k\right) d^N y$$

Since the matrix $\mathbf{W}$ appears everywhere as $\mathbf{W}^{-1}$, it is convenient to identify a new matrix

$$\mathbf{W}^{-1} \equiv \mathbf{A}$$

where the operator $\mathbf{A}$ has the eigenvalues

$$\lambda_p = \frac{1}{\alpha + 2\sum_{\nu=1}^{D} \cos\left(\dfrac{2\pi p_\nu}{L^\nu}\right)}$$

The use of these definitions in the expression for Q_N leads to the standard result

$$Q_N = \exp\left(-\tfrac{1}{2}N\alpha H\right)(2\pi H)^{-\frac{N}{2}} (\det \mathbf{A})^{1/2} \int_{-\infty}^{+\infty} \cdots \int_{-\infty}^{+\infty} \exp[I(y)]d^N y \qquad (18)$$

where

$$I(y) = -\frac{1}{2H}\langle y, \mathbf{A}y\rangle + \sum_{i=1}^{N} \log\cosh y^i \qquad (19)$$

In (19) and in the subsequent discussion y is understood to be the N-component vector

$$y \equiv \begin{bmatrix} y^1 \\ \cdot \\ \cdot \\ \cdot \\ y^i \\ \cdot \\ \cdot \\ \cdot \\ y^N \end{bmatrix}.$$

We refer to (18) and (19) as thc continuum representation of the partition function for the D-dimensional Ising model.

We now prove the following theorem on the asymptotic behavior of the free energy.

Theorem 1. Given the continuum representation of the D-dimensional Ising model, in the limit $H \to 0$, $N \to \infty$, the free energy behaves like

$$-F(H) \cong \lim_{N\to\infty} \frac{1}{N}\Big[\log\Big(\exp(-\tfrac{1}{2}N\alpha H)(H)^{-\frac{N}{2}}(\det \mathbf{A})^{1/2} \times \sum_{y_c}\{\exp[I(y_c) - \tfrac{1}{2}\log(\det[-\nabla^2 I(y_c)])]\}\Big)\Big] + \mathcal{O}(e^{-a/H}) \qquad (H \to 0)$$

In the above expression, $\sum_{y_c}$ is taken over all critical points $\{y_c\}$ of $I(y)$, that is, all $y_c : \nabla I(y_c) = 0$. In deriving this theorem, the importance of considering the limit $H \to 0$ in conjunction with the limit $N \to \infty$ was stressed by Gallavotti.[30] The proof of Theorem 1 proceeds by defining a quantity

$$S = S(N, H) \equiv \int_{-\infty}^{+\infty} d^N y \exp[I(y)] \equiv \sum_{\{y_c\}} S_{y_c,\delta} + S'$$

where

$$S_{y_c,\delta} \equiv \int_{y_c{}^N-\delta}^{y_c{}^N+\delta} \cdots \int_{y_c{}^1-\delta}^{y_c{}^1+\delta} d^N y \exp[I(y)]$$

$$S' \equiv S - \sum_{\{y_c\}} S_{y_c,\delta}$$

We now seek estimates of $S_{y_c,\delta}$ and S'. Consider first the range

$$H < \lambda_1 \equiv (\alpha + 2D)^{-1}$$

It is proven later that the only critical point of $I(y)$ in this range is $y_c = 0$, and furthermore this point is also a maximum point. [Notice that since for any $H > 0$, $I(y) \to -\infty$ as $|y| \to \infty$, and since $I(0) = 0$, all maxima occur at the interior points of the line; therefore when y_c is a maximum point, the estimates for $S_{y_c,\delta}$ and errors are exactly the same as for $S_{0,\delta}$. For y_c not a maximum point, one proceeds exactly as in the one-dimensional steepest-descent technique:[31] we extend y to the complex N plane, $y = (y^1, \ldots, y^N) \varepsilon \mathscr{C}^N (e^{I(y)}$ is then a holomorphic function of y), and in the neighborhood of each critical point y_c, we deform the contour of integration to one of steepest descent; in particular, in a neighborhood of y_c along the contour we write $y^j = y_c^j + r_j \exp(i\alpha_j)(j = 1, \ldots, N)$ and choose α_j so that along this path y_c is a maximum point, that is, $I(y) - I(y_c)$ is real and negative.] Since zero is to be a maximum point of $I(\mathrm{y})$ in the range $H<\lambda_1$, we must have $\nabla^2 I(0)<0$ and therefore

$$\exists \varepsilon_0>0: \quad \text{for} \quad \forall \varepsilon<\varepsilon_0, \qquad \nabla^2 I(0)\pm\varepsilon\mathscr{I}<0$$

where $\mathscr{I}$ is the identity matrix. Choose

$$\delta>0: \qquad |I(\mathrm{y})-I(0)-\tfrac{1}{2}\langle\nabla^2 I(0)\mathrm{y}, \mathrm{y}\rangle|<\tfrac{1}{2}\varepsilon|\mathrm{y}|^2$$

for $|\mathrm{y}^i|<\delta$, $(i=1,\ldots,N)$. If, for example, we set $\delta=(\frac{1}{2}\varepsilon)^{1/2}$, then

$$\int_{|\mathrm{y}^i|<\delta} d^N\mathrm{y}\,\exp\{I(0)+\tfrac{1}{2}\langle[\nabla^2 I(0)-\varepsilon]\mathrm{y}, \mathrm{y}\rangle\}$$
$$<S_{0,\delta}<\int_{|\mathrm{y}^i|<\delta} d^N\mathrm{y}\,\exp\{I(0)+\tfrac{1}{2}\langle[\nabla^2 I(0)+\varepsilon]\mathrm{y}, \mathrm{y}\rangle\} \quad (20)$$

and

$$S'=\int_{|\mathrm{y}^i|\geq\delta} d^N\mathrm{y}\,\exp[I(\mathrm{y})]$$
$$\leq 2^N e^{-N\lambda_1\delta^2/4H}\int_0^\infty d^N\mathrm{y}\,\exp[-(1/4H)\langle\mathbf{A}\mathrm{y}, \mathrm{y}\rangle+\langle\mathbb{1}, \mathrm{y}\rangle]$$

where $\mathbb{1}$ is the N-component vector $\mathbb{1}=(1,\ldots,1)$, and we have used that

$$\langle\mathbf{A}\mathrm{y}, \mathrm{y}\rangle\geq\lambda_1|\mathrm{y}|^2\geq\lambda_1\delta^2 N$$

and

$$\log\cosh \mathrm{y}^i\leq\mathrm{y}^i$$

The last integral may be evaluated by diagonalizing $\mathbf{A}$ and completing the square to obtain

$$S'\leq e^{-N\lambda_1\delta^2/4H}(4\pi)^{\frac{N}{2}}(\det\mathbf{A})^{-1/2}e^{HN/\lambda_1}$$

The integrals on the left and right of the inequalities in (20) may be extended to all $\mathscr{R}^N$, with an error of the form S'. Then, estimating $S_{0,\delta}$ from the right, we have (setting: $\mathscr{E}_0\equiv-H\nabla^2 I(0)\equiv\mathbf{A}-H\mathscr{I}$)

$$S=S_{0,\delta}+S'\leq(2\pi)^{+\frac{N}{2}}H^{+\frac{N}{2}}(\det\mathscr{E}_0)^{-1/2}$$
$$\times[1+2^{\frac{N}{2}}(\det\mathbf{A}^{-1}\mathscr{E}_0)^{1/2}e^{NH/\lambda_1}e^{-N\lambda_1\delta^2/4H}]$$

Hence

$$\log[Q_N^{\frac{1}{N}}]=\left(\frac{1}{N}\right)\log Q_N\leq(1/N)\log[e^{-\frac{N\alpha H}{2}}(\det\mathbf{A}^{-1}\mathscr{E}_0)^{-1/2}]$$
$$+(1/N)\log(1+u^N)$$

where

$$u \equiv 2\,|1-(\alpha-2D)^{-1}H|\, e^{H/\lambda_1} e^{-(\lambda_1\delta^2/4)/H}$$

Exactly the same result is obtained analyzing the integral on the left of (20); therefore, using

$$\log(1+u^N) \leq N \log(1+u) \qquad (u>0)$$

we obtain the following asymptotic estimate for the free energy F:

$$\begin{aligned} -F &= \lim_{N\to\infty} \frac{\log Q_N}{N} \\ &\cong \lim_{N\to\infty} (1/N)\left\{\log\left[\exp\left(-\frac{N\alpha H}{2}\right)(\det \mathbf{A}^{-1}\mathscr{E}_0)^{-1/2}\right]\right\} \\ &\quad + \mathcal{O}(\log(1+e^{-a/H})) \qquad (H\to 0) \end{aligned}$$

which is the result stated in Theorem 1. Notice that the free energy behaves like

$$(1/N)\{\log[\exp(-\tfrac{1}{2}N\alpha H)(\det \mathbf{A}^{-1}\mathscr{E}_0)^{-1/2}]\}$$

either for $H\to 0$ with N fixed, or $N\to\infty$ with H fixed (but $H\to 0$).

Having proved a theorem on the asymptotic behavior of the free energy, we are now ready to proceed with our evaluation of the partition function Q_N. We have the task of finding the maximum points of the integrand in (18) or, what is the same thing, the maximum points of the function $I(\mathrm{y})$. Let us denote the set of all critical points as $\{\mathrm{y}_c\}$; then we require

$$\left.\frac{\partial I(\mathrm{y})}{\partial \mathrm{y}^i}\right|_{\mathrm{y}^i=\mathrm{y}_c} = 0 \qquad (i=1,\ldots,N)$$

or

$$\left[-\frac{1}{H}\sum_{j=1}^{N} A^{ij}\mathrm{y}^j + \tanh \mathrm{y}^i\right]_{\mathrm{y}^i=\mathrm{y}_c} = 0$$

Carrying through this procedure for each of the y^i, we obtain the result

$$\mathbf{A}\mathrm{y}_c = H \tanh \mathrm{y}_c \equiv Hf(\mathrm{y}_c) \tag{21}$$

where

$$\mathrm{y}_c \equiv \begin{bmatrix} \mathrm{y}_c^1 \\ \cdot \\ \cdot \\ \cdot \\ \mathrm{y}_c^N \end{bmatrix} \quad \text{and} \quad f(\mathrm{y}_c) \equiv \begin{bmatrix} \tanh \mathrm{y}_c^1 \\ \cdot \\ \cdot \\ \cdot \\ \tanh \mathrm{y}_c^N \end{bmatrix}$$

Inasmuch as $\tanh(0) = 0$, a particular solution of (21) for all H is

$$\mathbf{y}_0 = \begin{bmatrix} 0 \\ \cdot \\ \cdot \\ \cdot \\ 0 \end{bmatrix}$$

where $\mathbf{y}_0$ is an N-component vector with one zero corresponding to each $y^i (i = 1, \ldots, N)$. This observation leads at once to the following result:

Theorem 2. In the range $H < \lambda_1$, where $\lambda_1 = 1/(\alpha + 2D)$ is the minimum eigenvalue of $\mathbf{A}$, the only critical point of $I(\mathbf{y})$ is zero.

For, suppose there exists a $\mathbf{y}_c \neq 0$ for $H < \lambda_1$. Then, with

$$\mathbf{A}\mathbf{y}_c = Hf(\mathbf{y}_c)$$

we form the inner product of both sides with $\mathbf{y}_c/|\mathbf{y}_c|^2$ to obtain

$$\lambda_1 \leqslant \frac{\langle \mathbf{A}\mathbf{y}_c, \mathbf{y}_c \rangle}{|\mathbf{y}_c|^2} = H \frac{\langle f(\mathbf{y}_c), \mathbf{y}_c \rangle}{|\mathbf{y}_c|^2} \leqslant \frac{H \sum_{i=1}^{N} \tanh (y_c^i)|^2}{\sum_{i=1}^{N} |y_c^i|^2} \leqslant H$$

The left-hand inequality follows at once from the "minimax principle" for the eigenvalues of a positive definite matrix $\mathbf{A}$. But we assumed that $H < \lambda_1$; hence we have a contradiction, and we conclude that the only possible critical point in the range $H < \lambda_1$ is $\mathbf{y}_c = 0$ (Theorem 2).

Having determined the single critical point relative to the range $H < \lambda_1$, we now study the possibility of determining the set of critical points for the range $H > \lambda_1$. This amounts to finding the other solutions of (21); that is, we determine whether new solutions might arise from the known solution $\mathbf{y}_0$ for particular values of the parameter H. Phrased differently, we study the possibility that new solutions can *bifurcate* from the known solution $\mathbf{y}_0$. To approach this problem, we need two fundamental theorems from bifurcation theory.[22] First, we need a precise notion of linearization. Let $\mathbf{B}$ be a (possibly nonlinear) operator from the Banach space $\mathscr{B}$ into $\mathscr{B}$. We say that $\mathbf{B}$ is linearizable at u if there exists a bounded linear operator $\mathbf{L}_u$ such that

$$\mathbf{B}(u + h) - \mathbf{B}u = \mathbf{L}_u h + \mathbf{R}_u h$$

with

$$\lim_{h \to 0} \frac{\|\mathbf{R}_u h\|}{\|h\|} = 0$$

The operator $\mathbf{L}_u$, linear in h, is referred to as the *Fréchet derivative* of $\mathbf{B}$ at u and is also denoted by $\mathbf{B}'(u)$. Then, to investigate branching from the basic solution for the equation

$$\mathbf{B}u = \lambda u$$

we use the following two celebrated theorems:[22]

Theorem I. The number λ_0 can be a branch point of $\mathbf{B}$ only if it is in the spectrum of $\mathbf{L}$.

and

Theorem II (Leray-Schauder). If $\mathbf{B}$ is completely continuous and $\lambda_0 \neq 0$ is an eigenvalue of odd multiplicity of $\mathbf{L}$, then λ_0 is a branch point of the basic solution of the nonlinear problem.

To continue, if bifurcation is to take place in the Ising model problem considered here, it must happen at one of the eigenvalues of the operator $\mathbf{A}$; in this case, bifurcation is guaranteed, since the eigenvalues of $\mathbf{A}$ are of odd multiplicity. Accordingly, we write the general critical point y_{c_v} as a power series in a parameter ε about the known critical point y_0; similarly, we expand the strength parameter H about the value H_0 (the latter to be determined). This perturbation scheme, closely associated with the names of Euler, Poincaré, and Rellich,[32–36] then takes the following form (to third order):

$$y_{c_v} = y_0 + \varepsilon y_1 + \varepsilon^2 y_2 + \varepsilon^3 y_3 + \dots$$
$$H = H_0 + \varepsilon H_1 + \varepsilon^2 H_2 + \varepsilon^3 H_3 + \dots$$

We proceed by substituting these expressions into the operator equation

$$\mathbf{A} y_{c_v} - H \tanh y_{c_v} \equiv \mathbf{F}(\varepsilon)$$

and then, differentiating $\mathbf{F}(\varepsilon)$ successively with respect to ε, we set the result at each order equal to zero. For example, from the condition

$$\left.\frac{\partial \mathbf{F}(\varepsilon)}{\partial \varepsilon}\right|_{\varepsilon=0} = 0$$

we determine the relation

$$\mathbf{A} y_1 = [H \operatorname{sech}^2 y_c \partial_\varepsilon y_c + H_\varepsilon \tanh y_c]_{\varepsilon=0}$$

This leads to the further result

$$\mathbf{A} y_1 = H_0 y_1$$

This equation, which is just the *variational equation* corresponding to

(21), defines a linear eigenvalue problem. We identify

$$H_0 = \lambda_\nu \qquad (\lambda_1 \leqslant \ldots \leqslant \lambda_\nu \leqslant \ldots \leqslant \lambda_N)$$
$$y_1 = a_\upsilon \phi_\upsilon$$

where the ϕ_ν are the eigenvectors corresponding to the eigenvalues λ_ν, and a_υ is some constant, to be determined. Continuing this procedure, we construct

$$\left.\frac{\partial^2 \mathbf{F}(\varepsilon)}{\partial \varepsilon^2}\right|_{\varepsilon=0} = 0$$

which leads to the relation

$$\mathbf{A}y_2 = [-2H \operatorname{sech}^2 y_c \tanh y_c (\partial_\varepsilon y_c)^2 + 2H_\varepsilon \operatorname{sech}^2 y_c\, \partial_\varepsilon y_c + H_{\varepsilon\varepsilon} \tanh y_c + H \operatorname{sech}^2 y_c (\partial_\varepsilon^2 y_c)]_{\varepsilon=0}$$

It is found that

$$\mathbf{A}y_2 = H_0 y_2 + 2H_1 y_1$$

But since $H_0 = \lambda_\nu$, we must have that

$$(\mathbf{A} - \lambda_\upsilon) y_2 = 2H_1 a_\nu \phi_\nu \tag{22}$$

As is well known from the theory of equations,[37] this equation will have a solution iff the right-hand side is orthogonal to the null space of $(\mathbf{A} - \lambda_\nu)$, that is, to the eigenfunctions of $\mathbf{A}$ corresponding to λ_ν. But the eigenfunctions of the homogeneous equation are (to within a constant) just ϕ_ν, that is,

$$\mathbf{A}\phi_\nu = \lambda_\nu \phi_\nu$$

Hence, the right-hand side of (22) can never be orthogonal to the ϕ_ν unless $H_1 = 0$. Accordingly, we write

$$H_1 = 0$$
$$y_2 = c_\nu a_\nu \phi_\nu$$

where c_ν is a constant to be determined. Finally, we construct

$$\left.\frac{\partial^3 \mathbf{F}(\varepsilon)}{\partial \varepsilon^3}\right|_{\varepsilon=0} = 0$$

which leads to the relation

$$\mathbf{A}y_3 = [H_{\varepsilon\varepsilon\varepsilon} \tanh y_c + 3H_{\varepsilon\varepsilon} \operatorname{sech}^2 y_c\, \partial_\varepsilon y_c + 2H_\varepsilon \operatorname{sech}^2 y_c (\partial_\varepsilon^2 y_c) - 2H \operatorname{sech}^4 y_c (\partial_\varepsilon y_c)^3 + H_\varepsilon \operatorname{sech}^2 y_c (\partial_\varepsilon^2 y_c) + H \operatorname{sech}^2 y_c (\partial_\varepsilon^3 y_c)]_{\varepsilon=0}$$

Then we find that

$$\mathbf{A}y_3 = 3H_2y_1 - 2H_0y_1^3 + H_0y_3$$

where y_1^3 is actually a matrix: in particular, if we write

$$\Phi_3 \equiv \begin{bmatrix} (\phi_\nu^1)^3 & 0 & \dots & 0 \\ 0 & (\phi_\nu^2)^3 & \dots & 0 \\ \cdot & \cdot & & \cdot \\ \cdot & \cdot & & \cdot \\ \cdot & \cdot & & \cdot \\ 0 & 0 & \dots & (\phi_\nu^N)^3 \end{bmatrix}$$

where

$$\phi_\nu^j = \cos\left(\frac{2\pi j\nu}{N}\right) + \sin\left(\frac{2\pi j\nu}{N}\right)$$

refers to the jth component of the νth eigenvector of $\mathbf{A}$, then $y_1^3 = a_\nu^3\Phi_3$. Returning to (23) we find

$$(\mathbf{A} - \lambda_\nu)y_3 = 3H_2a_\nu\phi_\nu - 2\lambda_\nu a_\nu^3\Phi_3$$

Given the structure of the corresponding homogeneous equation, this last equation has a solution iff the right-hand side is orthogonal to ϕ_ν; that is, we must have

$$3H_2a_\nu\sum_{j=1}^{N}|\phi_\nu^i|^2 - 2\lambda_\nu a_\nu^3\sum_{j=1}^{N}|\phi_\upsilon^i|^4 = 0 \tag{24}$$

Suppose we define

$$\frac{\frac{2}{3}\sum_{j=1}^{N}|\phi_\nu^i|^4}{\sum_{j=1}^{N}|\phi_\nu^i|^2} = d_\nu$$

Then, given the structure of the ϕ_ν^i, if we set $\chi = 2\pi j/N$ and form

$$d_\nu = \frac{\frac{2}{3}\int_0^{2\pi}|\phi_\nu|^4\,d\chi}{\int_0^{2\pi}|\phi_\nu|^2\,d\chi}$$

we find $d_\nu = 1$. Therefore, solving (24) for H_2, we obtain

$$H_2 = \lambda_\nu a_\nu^2 d_\nu = \lambda_\nu a_\nu^2$$

Collecting all results, the previously noted perturbation expansions for y_{c_ν}

and H can be written

$$y_{c_\nu} = y_0 + (a_\nu \phi_\nu)\varepsilon + (c_\nu a_\nu \phi_\nu)\varepsilon^2 + \mathcal{O}(\varepsilon^3)$$
$$H = H_0 + (\lambda_\nu a_\nu^2)\varepsilon^2 + \mathcal{O}(\varepsilon^3)$$

But $y_0 = 0$ and $H_0 = \lambda_\nu$. Furthermore, we are at liberty to choose $c_\nu = a_\nu$ and then, to simplify the structure of the above expressions, to replace $a_\nu \varepsilon$ by ε. Hence in the neighborhood of λ_ν we can write

$$y_{c_\nu} = \phi_\nu(\varepsilon + \varepsilon^2) + \mathcal{O}(\varepsilon^3) \tag{25}$$

$$H = \lambda_\nu(1 + \varepsilon^2) + \mathcal{O}(\varepsilon^3) \tag{26}$$

We are now ready to determine the asymptotic behavior of the free energy in the ranges $H < \lambda_1$ and $H > \lambda_1$. We consider first the range $H < \lambda_1$. Earlier we proved that in the range $H < \lambda_1$ there exists only one critical point of $I(y)$, namely, the one corresponding to $y_0 = 0$. Before evaluating the free energy, we must prove that the critical point y_0 is also a *maximum point.* For this to be so, the matrix

$$\left.\frac{\partial^2 I(y)}{\partial y_i \partial y_j}\right|_{y=y_0}$$

must be negative definite, that is, $-\nabla^2 I(y_0) > 0$. Now, from before, we have

$$I(y) = \frac{1}{2H}\langle y, \mathbf{A}y\rangle + \sum_{i=1}^{N} \log\cosh y^i$$

so that

$$-\nabla^2 I(y) = \frac{1}{H}(\mathbf{A} - H \operatorname{sech}^2 y)$$

where $\operatorname{sech}^2 y$ is the matrix

$$\operatorname{sech}^2 y = \begin{bmatrix} \operatorname{sech}^2 y^1 & 0 & \cdots & 0 \\ 0 & \operatorname{sech}^2 y^2 & \cdots & 0 \\ \cdot & \cdot & \cdot & \cdot \\ \cdot & \cdot & \cdot & \cdot \\ \cdot & \cdot & \cdot & \cdot \\ 0 & 0 & \cdots & \operatorname{sech}^2 y^N \end{bmatrix}$$

Writing

$$\mathscr{E}_0 \equiv -H\nabla^2 I(y_0) = (\mathbf{A} - H\mathscr{I})$$

we form

$$\langle u, \mathscr{E}_0 u\rangle = \langle u, \mathbf{A}u\rangle - H|u|^2 \geqslant (\lambda_1 - H)|u|^2 \geqslant 0$$

where we are assured that the right-hand inequality is satisfied since we are considering the range $H<\lambda_1$. Therefore it follows that $-\nabla^2 I(y_0)>0$, and hence in the range $H<\lambda_1$ the critical point $y_0=0$ is also a maximum point. This clears the way to determine the asymptotic behavior of the free energy in the range $H<\lambda_1$. In effect we must study

$$\frac{1}{N}\log\left[e^{-\frac{N\alpha H}{2}}(\det \mathbf{A}^{-1}\mathscr{E}_0)^{-1/2}\right]=-\frac{\alpha H}{2}-\frac{1}{2N}\log(\det \mathbf{A}^{-1}\mathscr{E}_0)$$

in the limit $N\to\infty$. Since

$$\det \mathbf{A}^{-1}\mathscr{E}_0=\det(\mathscr{I}-\mathbf{A}^{-1}\mathscr{E}_0)=\prod_{i=1}^{N}(1-\lambda_i^{-1}H)$$

we have

$$\left(\frac{1}{2N}\right)\log(\det \mathbf{A}^{-1}\mathscr{E}_0)=\frac{1}{2N}\sum_{i=1}^{N}\log(1-\lambda_i^{-1}H)$$

which in the limit $N\to\infty$ can be written in the form

$$\frac{1}{2(2\pi)^D}\int_0^{2\pi}d^D\omega\,\log\left[1-H\Sigma_D^{\alpha}(\omega)\right]$$

where

$$\Sigma_D^{\alpha}(\omega)\equiv\alpha+2\sum_{j=1}^{D}\cos\omega_j\qquad(0\leqslant\omega_j\leqslant 2\pi)$$

From Theorem 1, we have

$$\begin{aligned}-F&\equiv\lim_{N\to\infty}\frac{\log Q_N}{N}\\&\cong-\frac{1}{2(2\pi)^D}\int_0^{2\pi}d^D\omega\{\alpha H+\log\left[1-H\Sigma_D^{\alpha}(\omega)\right]\}\\&\equiv G_\alpha(H)\qquad(H\to 0)\end{aligned}$$

Expanding $G_\alpha(H)$ around $\alpha=2D$, we find that in the limits $H\to 0, N\to\infty$

$$-F\cong G_0(H)$$

where

$$G_0(H)\equiv-\frac{1}{2(2\pi)^D}\int_0^{2\pi}d^D\omega\,\log\left[1-H\Sigma_D(\omega)\right]$$

and

$$\Sigma_D(\omega)\equiv 2\sum_{j=1}^{D}\cos\omega_j$$

We now observe that $G_0(H)$ is just the free energy of the corresponding D-dimensional Gaussian model; we state this result as a theorem

Theorem 3. Given the continuum representation of the D-dimensional Ising model, in the limit $H \to 0$, $N \to \infty$, we have $-F(H) \cong G_0(H)$, where $G_0(H)$ is the free energy of the corresponding D-dimensional Gaussian model.

We are now ready to consider the asymptotic behavior of the free energy in the range $H > \lambda_1$. We begin with the observation that in the range $H > \lambda_N$, $-\nabla^2 I(y_0) < 0$; this implies that $y_0 = 0$, although a maximum point for $I(y)$ in the range $H < \lambda_1$ is in fact a *minimum point* for $I(y)$ in the range $H > \lambda_1$. Now, via the method of steepest descent,[31] we know that the dominant contribution to the integral comes from the neighborhood of the maximum points, whereas the minimum points contribute negligibly. We mention this explicitly to point out why the Gaussian model yields the behavior of the Ising model for $H < \lambda_1$, but fails to represent the Ising model in the range $H > \lambda_1$. It will be seen, however, that it is the model *bifurcating from* the Gaussian model which yields the behavior in the range $H > \lambda_1$. It is here that we can point to the usefulness of a bifurcation approach in the theory of phase transitions, since we can determine not only whether bifurcation will occur from the known solution $y_0 = 0$, but we can also obtain an estimate of the structure of the new solutions that spring from $y_0 = 0$ at $H = \lambda_1$ and from thence to determine their thermodynamic relevance. Once again, the main problem in the determination of the partition function (and hence the free energy) is the evaluation of the determinant

$$-\nabla^2 I(y_{c_\nu}) = \frac{\mathbf{A}}{H} - \operatorname{sech}^2(y_{c_\nu}) \equiv \mathscr{E}(y_{c_\nu})$$

This task can be accomplished by determining the eigenvectors and eigenvalues of the matrix $\mathscr{E}$, or, what is more convenient here, the eigenvectors U_j and eigenvalues η_j of the associated matrix

$$H\mathscr{E}(y_{c_\nu}) = \mathbf{A} - H \operatorname{sech}^2(y_{c_\nu})$$

This is achieved via another use of bifurcation theory. We write

$$U_j(\varepsilon) = \sum_{k=0}^{\infty} U_{j,k} \frac{\varepsilon^k}{k!}$$

$$\eta_j(\varepsilon) = \sum_{k=0}^{\infty} \eta_{j,k} \frac{\varepsilon^k}{k!}$$

Then, substituting these expansions into the operator equation

$$H\mathscr{E}U_j = (\mathbf{A} - H\,\mathrm{sech}^2\, y_{c_\nu})U_j = \eta_j U_j \tag{27}$$

we obtain

$$[(\mathbf{A}-H\,\mathrm{sech}^2\,y_{c_\nu})U_{j,0}]+\frac{\varepsilon}{1!}[(\mathbf{A}-H\,\mathrm{sech}^2\,y_{c_\nu})U_{j,1}]$$

$$+\frac{\varepsilon^2}{2!}[(\mathbf{A}-H\,\mathrm{sech}^2\,y_{c_\nu})U_{j,2}]+\frac{\varepsilon^3}{3!}[(\mathbf{A}-H\,\mathrm{sech}^2\,y_{c_\nu})U_{j,3}]+\dots$$

$$=(\eta_{j,0}U_{j,0})+\frac{\varepsilon}{1!}(\eta_{j,0}U_{j,1}+\eta_{j,1}U_{j,0})$$

$$+\frac{\varepsilon^2}{2!}(\eta_{j,0}U_{j,2}+2\eta_{j,1}U_{j,1}+\eta_{j,2}U_{j,0})$$

$$+\frac{\varepsilon^3}{3!}(\eta_{j,0}U_{j,3}+3\eta_{j,1}U_{j,2}+3\eta_{j,2}U_{j,1}+\eta_{j,3}U_{j,0})+\dots$$

Making use of the limits [see (25) and (26)]

$$\lim_{\varepsilon\to 0} H = \lambda_\nu \qquad \lim_{\varepsilon\to 0} y_{c_\nu} = 0$$

$$\lim_{\varepsilon\to 0} \frac{\partial H}{\partial \varepsilon} = 0 \qquad \lim_{\varepsilon\to 0} \frac{\partial y_{c_\nu}}{\partial \varepsilon} = \phi_\nu$$

$$\lim_{\varepsilon\to 0} \frac{\partial^2 H}{\partial \varepsilon^2} = 2\lambda_\nu \qquad \lim_{\varepsilon\to 0} \frac{\partial^2 y_{c_\nu}}{\partial \varepsilon^2} = 2\phi_\nu$$

and using the same formal solution procedure to analyze (27) as described earlier, we find that the eigenvalues η_j of the determinant $H[-\nabla^2 I(y_{c_j})]$ are given by

$$\eta_j(\varepsilon) = (\lambda_j - \lambda_\nu) + \lambda_\nu \varepsilon^2 [L_{\nu,j} - 1]$$

where

$$L_{\nu,j} \equiv \frac{1}{N}\langle \phi_j, \phi_\nu^2 \phi_j \rangle$$

and the ϕ_j are the eigenfunctions and the λ_j are the eigenvalues of the operator $\mathbf{A}$, that is,

$$\mathbf{A}\phi_j = \lambda_j \phi_j$$

Upon evaluating the determinant

$$\{\det H[-\nabla^2 I(y_{c_j})]^{1/2}\} = \left(\prod_{j=1}^{N} \eta_j\right)^{1/2}$$

One finds to order ε^3 that

$$\sum_{j=1}^{N} \eta_j = \varepsilon^2(L_{\nu,\nu}-1)\prod_{j=1}^{N} \lambda_j \prod_{j\neq\nu}^{N}\left(1-\frac{\lambda_\nu}{\lambda_j}\right)+\mathcal{O}(\varepsilon^3)$$

Using this result and Theorem 1, we obtain the following expression for the partition function Q_N in the range $H>\lambda_1$.

$$Q_N \cong \sum_{y_{c_\nu}} \exp\left(-\tfrac{1}{2}N\alpha H\right)\varepsilon^{-1}(L_{\nu,\nu}-1)^{-1/2}$$
$$\times \exp\left\{I(y_{c_\nu})-(1/2)\sum_{j\neq\nu}\left[1-\lambda_\nu\left(\alpha+2\sum_{\tilde{\nu}=1}^{D}\cos\left(\frac{2\pi j_{\tilde{\nu}}}{L^{\tilde{\nu}}}\right)\right)\right]\right\} \quad (28)$$

where we have incorporated as well the earlier expression for the eigenvalues λ_j. The free energy may be obtained from this result by constructing the limit $N\to\infty$ of $\dfrac{\log Q_N}{N}$. Given this objective, it is convenient to cast the expression (28) for Q_N into the alternative form

$$Q_N \cong \mathcal{O}(N)\exp\left(-\tfrac{1}{2}N\alpha H\right)\int_{(\alpha+2D)^{-1}}^{(\alpha-2D)^{-1}} d\lambda\left(\frac{\lambda}{H-\lambda}\right)^{1/2}\exp\left(NE(\lambda)\right)C(\lambda) \quad (29)$$

where

$$E(\lambda) = -\frac{1}{2}+\frac{\lambda}{2H}+\frac{1}{2}\log\lambda-\frac{1}{2}\log(2\lambda-H)$$
$$-\frac{1}{2(2\pi)^D}\int_0^{2\pi} d^D\omega \log\left[1-\lambda\left(\alpha+2\sum_{j=1}^{D}\cos\omega_j\right)\right] \quad (30)$$

with $0\leq\omega_j\leq 2\pi$, and $C(\lambda)^{-1}$ is an analytic function whose zeros are the set $\{\lambda_\nu\}$, constructed such that the residue of $C(\lambda)$ at any λ_ν is unity. This representation is achieved by replacing the sum $\sum_{j\neq\nu}$ by an integral over ω, and replacing the sum over y_{c_ν}, or in fact λ_ν, by an integral over λ; the form (29) follows after application of Cauchy's integral formula and the residue theorem[37] and a deformation of the contour of integration such that the range of λ_1 is (λ,∞). To evaluate the integral in (29) via the method of steepest descent, we look for the maximum points of $E(\lambda)$; that is, we seek the $\lambda_s=\lambda_s(H)$ for which

$$E'(\lambda_s)=0 \quad (31)$$

$$E''(\lambda_s)<0 \quad (32)$$

Now our bifurcation analysis yields only critical points, but the coordinate transformation described earlier allows the identification of the maximum points on the path of steepest descent. Once the maximum points have been found, the free energy is determined as

$$-F = \lim_{N\to\infty} \frac{\ln Q_N}{N} \cong E(\lambda_s) - \tfrac{1}{2}\alpha H$$

By expanding $E(\lambda_s)$ around $\alpha = 2D$ (the same procedure as was followed for the range $H < \lambda_1$), we may then examine the asymptotic behavior of the free energy in the range $H > \lambda_1$. Or, since we are interested here in the problem of phase transitions, we seek to determine whether the heat capacity

$$C_v = kH^2\left(\frac{\partial^2\lambda}{\partial H^2}\right)\frac{\partial E(\lambda)}{\partial\lambda} + kH^2\left(\frac{\partial\lambda}{\partial H}\right)^2\frac{\partial^2 E(\lambda)}{\partial\lambda^2}$$

on the path of steepest descent, $\lambda_s = \lambda_s(H)$, exhibits a singularity and, if so, to isolate the temperature at which transition takes place, and then to characterize the behavior in the neighborhood of the singularity.

To proceed with the examination of the thermodynamic behavior of C_v in the range $H > \lambda_1$, we construct $E'(\lambda_s)$ and $E''(\lambda_s)$; the condition (31) then takes the form

$$\frac{1}{2H} = \frac{1}{2\lambda - H} - \frac{1}{2\lambda}K(\lambda) \tag{33}$$

where

$$K(\lambda) = \frac{1}{(2\pi)^D}\int_0^{2\pi} d^D\omega\left[1 - \lambda\left(\alpha + 2\sum_{j=1}^{D}\cos\omega_j\right)\right]^{-1} \tag{34}$$

and

$$E''(\lambda) = \frac{2}{(2\lambda - H)^2} - \frac{1}{2\lambda^2}K(\lambda) + \frac{1}{2\lambda}K'(\lambda) \tag{35}$$

We remark that the integral $K(\lambda)$ is reminiscent of the one analyzed by Berlin and Kac (Ref. 28, Appendix C), and we take advantage of this drawing on the same analytical techniques for characterizing the behavior of the heat capacity within the framework of a steepest-descent calculation as used by these authors in characterizing the thermodynamic behavior of the spherical model.

Let us consider first the one-dimensional case. Examining the behavior of $E(\lambda)$ around $\alpha = 2D$, one finds that $E(\lambda)$ has a "potential" singularity when $\lambda \to \frac{1}{4}$ and $H \to \frac{1}{2}$. In order that this point be a true singularity, it must lie on the saddle curve $(H, \lambda_s(H))$; this, in turn, requires that $E''(\lambda)$ must be negative at $\lambda = \frac{1}{4}$ and $H = \frac{1}{2}$. Notice that the integral $K(\lambda)$ is well behaved except possibly when $\lambda = \lambda_c = \frac{1}{4}$; specifically, around $\lambda = \lambda_c$, $K(\lambda)$ behaves like

$$K(\lambda) \cong -\frac{1}{\lambda^{1/2}}(4\lambda - 1)^{-1/2}$$

Moreover, from (33), we see that $(2\lambda - H)^{-1}$ behaves like $K(\lambda)$. Hence, considering the behavior of $K(\lambda)$ and $K'(\lambda)$, we find

$$\lim_{\substack{\lambda \to 1/4 \\ H \to 1/2}} E''(\lambda) = +\infty$$

That is, the point $(\frac{1}{2}, \frac{1}{4})$ does *not* lie on the saddle curve $(H, \lambda_s(H))$. Hence we conclude that the free energy and all derivative properties are well behaved throughout the range $H > \lambda_1$, because the "potential" singularity at $H = \frac{1}{2}$, $\lambda = \frac{1}{4}$ corresponds to a *minimum* of the integrand and consequently contributes negligibly to the thermodynamic behavior of the system.

In two dimensions, we consider again the behavior of $E(\lambda)$ around $\alpha = 2D$ and notice that $E(\lambda)$ has a "potential" singularity when $\lambda \to \frac{1}{8}$ and $H \to \frac{1}{4}$. Examination of the structure of $K(\lambda)$ around $\lambda = \lambda_c = \frac{1}{8}$ reveals that $K(\lambda)$ behaves like

$$K(\lambda) \cong -\frac{1}{\lambda}\log\left(\frac{8\lambda - 1}{\lambda}\right)$$

To determine whether the point $(\frac{1}{4}, \frac{1}{8})$ lies on the saddle curve $(H, \lambda_s(H))$, we consider the behavior of $E''(\lambda)$ in the neighborhood of this point. Using (33) to infer the behavior of $(2\lambda - H)^{-1}$ and considering the behavior of $K(\lambda)$ and $K'(\lambda)$, we find that

$$\lim_{\substack{\lambda \to 1/8 \\ H \to 1/4}} E''(\lambda) = -\infty$$

Since $E''(\lambda) < 0$ corresponds to a *maximum* of the integrand, a singularity in the thermodynamic heat capacity is achieved in two dimensions. Moreover, the behavior of C_v in the neighborhood of the singularity is *logarithmic*, in agreement with the behavior found by Onsager in his exact solution of the two-dimensional Ising model. Notice that, whereas the singularity in λ occurs at the bifurcation point, $\lambda = \frac{1}{8}$, the singularity in H is shifted; the heat capacity is well behaved until the strength parameter H achieves a value 0.25. As it happens, this estimate of H_c (valid to $\mathcal{O}(\varepsilon^2)$ in our analysis) is exactly the same as the prediction of classical mean-field theory; by way of comparison, the result obtained by Onsager in his exact analysis of the two-dimensional Ising model was $H_c = 0.44$. Therefore, we see that although the *qualitative* description of the singularity is correct, to this order in perturbation theory the *quantitative* estimate of H_c is not in good agreement with the Onsager result.

Finally, we treat the three-dimensional case. Again we examine the behavior of $E(\lambda)$ around $\alpha = 2D$ along the saddle curve, $\lambda_s = \lambda_s(H)$. Notice that one possible singularity of $E(\lambda)$ occurs at the intersection of $\lambda_s = \lambda_s(H)$ with the line $\lambda = \frac{1}{2}H$; this intersection occurs at $\lambda = H = 0$ and hence at infinite temperature, a point which is never reached. A second possible singularity occurs at $\lambda = \frac{1}{12}$, given the structure of $K(\lambda)$. To determine the H_c corresponding to this λ_c, we use the result of Watson,[38] namely,

$$\frac{1}{(2\pi)^3}\int_0^{2\pi} d\omega_1\, d\omega_2\, d\omega_3 \frac{1}{(3-\cos\omega_1-\cos\omega_2-\cos\omega_3)} = 0.50546$$

in conjunction with (33) to find that $H_c = 0.12$. Again the question is whether or not the point $(0.12, \frac{1}{12})$ lies on the saddle curve $(H, \lambda_s(H))$. Around $\lambda = \lambda_c$, $K(\lambda)$ behaves like

$$K(\lambda) \cong -\frac{1}{\lambda^{3/2}}(12\lambda - 1)^{1/2}$$

and using the behavior of $K(\lambda)$ to infer the behavior of $(2\lambda - H)^{-1}$ we find that

$$\lim_{\substack{\lambda\to 1/12\\ H\to 0.12}} E''(\lambda) = -\infty$$

Since $E''(\lambda) < 0$, a singularity in the heat capacity is realized when $\lambda \to \lambda_c = \frac{1}{12}$ and $H \to H_c = 0.12$. This estimate for the critical strength parameter is somewhat displaced from the estimate obtained from numerical studies on the three-dimensional Ising model,[26] namely, $H_c = 0.22$. Note further that the behavior of C_v in the neighborhood of the singularity is *algebraic*, behaving like $(\lambda - \lambda_c)^{-1/2}$ [a different exponent may characterize the behavior with respect to $(H - H_c)$]. Finally, by considering λ_{HH}, one can show (see the procedure in Ref. 21) that λ_{HH} has a jump discontinuity; this implies that there exists a discontinuity in the slope of the heat capacity at $H = H_c$.

The use of a bifurcation approach within the framework of an asymptotic analysis of the partition function for the D-dimensional Ising model described above leads to results that are in *qualitative* accord with the known results in one, two, and three dimensions, namely no phase transition in one dimension, a logarithmic singularity in two dimensions, and an algebraic singularity in three dimensions. However, the *quantitative* estimates of the critical strength H_c in two and three dimensions are displaced from known results; perhaps this is to be expected, since the bifurcation analysis was carried through only to order ε^2. The qualitative picture that emerges in the above analysis may be summarized. The asymptotic behavior of the Ising model is described correctly by the Gaussian model in the range

$$H<\lambda_1=\frac{1}{\alpha+2D}$$

In the range $H>\lambda_1$ a certain similarity exists between this asymptotic representation of the Ising model and the spherical model of Berlin and Kac, and in the above analysis one finds a spherical-like model (but *not* the spherical model per se) bifurcating from the Gaussian model at

$$H=\lambda_1=\frac{1}{\alpha+2D}$$

with the singularity in the thermodynamic heat capacity for $D\geqslant 2$ occurring in the range $H>\lambda_1$. It may be recalled[39] that in the renormalization group approach, for a system with a scalar-order parameter and short-range forces, there are three fixed points: the (trivial) Gaussian fixed point (in which the order parameter *fluctuations* are completely uncorrelated), the nontrivial Gaussian fixed point which is related to the spherical model, and the Ising fixed point, which is presumed to describe the ferromagnetic (and gas-liquid) transition. Given the underlying "fixed point" basis of our analysis (specifically with respect to the use of the Leray-Schauder Theorem), it is difficult to avoid the impression that a connection may exist between the approach of the renormalization group and the approach described above: in both approaches stress is placed on examining the behavior of certain equations in the limit as a variable of the theory tends to infinity (for example, referring again to Wilson's formulation[9] of the Kadanoff block picture, one imagines an infinite cubic lattice, L lattice sites on a side and considers (eventually) the behavior of a certain differential equation of motion in the limit $L\to\infty$; in the above study, it is the behavior of solutions to the operator equation (27) that are examined (eventually) in the limit $N\to\infty$), and in both approaches fixed point theorems play an essential role.

Finally, our analysis has isolated the importance of attributing thermodynamic significance to bifurcation iff the associated "potential" singularity lies on the path of steepest descent in the asymptotic evaluation of the partition function. Recall that, although a bifurcation point was found in the one-dimensional problem, the "potential" singularity did not lie on the saddle curve $(H, \lambda_s(H))$; in fact, the potential singularity at $H=\frac{1}{2}$, $\lambda=\frac{1}{4}$ corresponded to a minimum of the integrand in the continuum representation of the Ising partition function and consequently contributed negligibly to the thermodynamic behavior of the system. From this it would appear that bifurcation is a necessary but not sufficient condition for a phase transition; we return to this point later in our discussion of the model of a one-dimensional line of hard rods.

C. Orientational Phase Transitions

It is found experimentally that crystal-crystal orientational phase transitions are of second order (or very nearly so), with no discontinuities in the thermodynamic potentials,[40] although Landau and Lifshitz[15] have pointed out that this classification, strictly speaking, is incorrect, since "at a transition point of the second kind the state of the body must change continuously, and so there can be no sharp change in the nature of the motion." In those cases in which the orientational coupling between molecules is relatively weak, as in crystalline heavy methane CD_4, however, the "ordered" and "disordered" phases approach the same physical state at the transition point and are very similar in the immediate neighborhood of the transition temperature. In 1959 James and Keenan[41] proposed a self-consistent (mean-field) theory of the successive phase transitions that occur in solid CH_4 and CD_4. Their justification for using a mean-field theory hinged on the observation that X-ray studies showed that in all phases of solid CD_4, for example, the carbon atoms were positioned on a face-centered cubic lattice and each molecule in this structure had a large number (12) of nearest neighbors—hence fluctuations in the intermolecular coupling were believed to be relatively small compared to the overall or total coupling. In 1973 Lemberg and Rice[42] showed how the James and Keenan model could be cast into a form in which the orientational transitions in solid heavy methane could be interpreted as a problem in bifurcation theory. The interplay between mathematical concepts and physical insights that has come to characterize many studies using a bifurcation approach is particularly well displayed in the study of Lemberg and Rice, and for this reason we summarize their approach here.

As in our discussion of the ferromagnetic transition, we devote some attention to the formulation of the problem. (For a detailed specification of the model see Ref. 41.) The James and Keenan theory was predicated on four assumptions: (1) the dominant contribution to the intermolecular coupling in crystalline CD_4 was assumed to be electrostatic octupole-octupole interactions; (*2*) only interactions between nearest neighbors needed to be considered; (*3*) dynamical effects on the intermolecular coupling could be neglected. Finally, it was assumed that each CD_4 molecule is subjected only to the mean field of its neighbors; hence the distribution functions that describe the probability of finding particular molecules in particular orientations are independent of one another. This means that the total orientational distribution function $\Phi(\boldsymbol{\omega}_i)$ for all the molecules in the crystal can be expressed as a simple product of distribution functions for each of the molecules; that is, we write

$$\Phi(\boldsymbol{\omega}_i) = \prod_i f_i(\boldsymbol{\omega}_i)$$

where the functional form of the single-molecule distribution function $f_i(\boldsymbol{\omega}_i)$ is indexed to allow for phases in which not all molecules are subjected to the same mean field (and hence may have different distributions in $\boldsymbol{\omega}$ space). Within the framework of mean-field theory, James and Keenan derived a set of self-consistency conditions that must be satisfied by the distribution functions $f_i(\boldsymbol{\omega}_i)$ to guarantee that the free energy is a extremum. These conditions, which are necessary conditions for the existence of a stable phase, are

$$f_i(\boldsymbol{\omega}) = G_i \exp\left[-\sum_\tau \beta^i_\tau u_\tau(\boldsymbol{\omega})\right]$$

where the $\{u_\tau(\boldsymbol{\omega})\}_{\tau=1,\ldots,7}$ represent the set of tetrahedral rotator functions which depend on the Euler angles $\boldsymbol{\omega}$ that relate the orientation of a CD_4 molecule to a set of standard crystal axes. Here

$$\beta^i_\tau = \frac{I_3^2}{kTR^7} \sum_j \sum_\nu C^{ij}_{\tau\nu} \int d\boldsymbol{\omega}' u_\nu(\boldsymbol{\omega}') f_j(\boldsymbol{\omega}')$$

where k is Boltzmann's constant, T is the temperature, R is the distance between molecular centers, I_3 is the effective octupole moment, and the $C^{ij}_{\tau\nu}$ are coefficients (evaluated by James and Keenan) that are nonzero only if molecules i and j are nearest neighbors. The G_i is a factor that normalizes $f_i(\boldsymbol{\omega})$ to unity, namely,

$$G_i^{-1} = \int d\boldsymbol{\omega} \exp\left[-\sum_\tau \beta^i_\tau u_\tau(\boldsymbol{\omega})\right]$$

Lemberg and Rice recast the above formulation of the problem by defining

$$g_i(\boldsymbol{\omega}) = \ln\left[\frac{f_i(\boldsymbol{\omega})}{G_i}\right]$$

and then noticing that $g_i(\boldsymbol{\omega})$ obeys the following equation

$$\begin{aligned} g_i(\boldsymbol{\omega}) &= -\sum_\tau \beta^i_\tau u_\tau(\boldsymbol{\omega}) \\ &= -\lambda \sum_\tau \sum_\nu \sum_j C^{ij}_{\tau\nu} \int d\boldsymbol{\omega}' u_\nu(\boldsymbol{\omega}') G_j e^{g_j(\boldsymbol{\omega}')} u_\tau(\boldsymbol{\omega}) \end{aligned}$$

with

$$\lambda = \frac{I_3^2}{kTR^7}$$

If we now define a kernel

$$K^{ij}(\boldsymbol{\omega}, \boldsymbol{\omega}') \equiv \sum_{\tau} \sum_{\nu} C^{ij}_{\tau\nu} u_{\tau}(\boldsymbol{\omega}) u_{\nu}(\boldsymbol{\omega}')$$

the integral equation for the defined function $g_i(\boldsymbol{\omega})$ can be written

$$g_i(\boldsymbol{\omega}) + \lambda \sum_j G_j \int d\boldsymbol{\omega}' K^{ij}(\boldsymbol{\omega}, \boldsymbol{\omega}') \exp [g_i(\boldsymbol{\omega}')] = 0 \tag{36}$$

where, again, the sum j is over the 12 nearest neighbors of a given molecule. Except for the sum over j and the fact that the normalizing factors G_j are *functionals* of $g_j(\boldsymbol{\omega})$, (36) is of the form of a standard Hammerstein equation[43]

$$\psi(x) + \int_{\Lambda} K(x, y) F[y, \psi(y)]\, dy = 0 \tag{37}$$

where K is the kernel, F is a nonlinear functional, and the integration is over a finite domain (Λ is a bounded closed set in a finite dimensional space).

We pause in our development to comment on (37). It will be seen in this and the following chapter that the statistical-mechanical formulation of a number of different problems in the theory of phase transitions can be recast, under certain simplifying approximations, into an equation of Hammerstein form. Mean-field theories, such as the one described above, often find a natural expression in terms of a Hammerstein-like equation, but a number of theories based on a distribution-function approach lead as well to equations of the form of (37). For this reason, theorems on the existence, uniqueness, or possible bifurcation of solutions to the Hammerstein equation have played a central role in the development of theories of second- *and* first-order phase transitions based on a bifurcation approach. The 1930 paper by Hammerstein[43] (the principal results of which are summarized in the book by Tricomi[44]), the theorems of Dolph,[45] the review articles by Krasnosel'skii[46] and Vainberg and Trenogin,[23] and the treatises by Krasnosel'skii,[47] Krasnosel'skii and Rutickii,[48] and Vainberg[49] provide the mathematical background on equations of the Hammerstein type against which many physical problems are discussed. We certainly cannot include here a summary of all the theorems that have been proved regarding this equation, although we cite theorems of Dolph and Krasnosel'skii later on. With respect to the problem of bifurcation, however, we stress that the theorems that have been proved deal with the possibility of finding new solutions of *small norm* that arise continuously from the basic solution at a certain point, that is, they deal with "soft" bifurcation.

Let us return now to the "generalized" Hammerstein equation (36) and consider the situation in which all molecules in the crystal are characterized by the same orientational distribution function, that is, we assume each molecule is subjected to the same mean field. This is referred to as the case of uniform order; a second case in which the structure is characterized by nonuniform order, that is, where there is periodicity of orientation within the basic face-centered cubic structure, but different molecules may have different distribution functions, is also treated by Lemberg and Rice but is not considered here.

The existence of uniformly ordered phases implies that in such phases the distribution function $f_i(\boldsymbol{\omega})$ for the molecule on the ith site of the lattice is independent of the site index i. This in turn implies that the normalizing constants G_j must be the same for all lattice sites, and under this simplification (36) reduces to the form

$$g(\boldsymbol{\omega})+\lambda G\int d\boldsymbol{\omega}' K(\boldsymbol{\omega},\boldsymbol{\omega}')\exp[g(\boldsymbol{\omega}')]=0 \tag{38}$$

where

$$K(\boldsymbol{\omega},\boldsymbol{\omega}')=\sum_j K^{ij}(\boldsymbol{\omega},\boldsymbol{\omega}')$$

The new kernel $K(\boldsymbol{\omega},\boldsymbol{\omega}')$ can be evaluated using the interaction coefficients $C^{ij}_{\tau\nu}(\boldsymbol{\omega}_R)$ calculated by James and Keenan (where $\boldsymbol{\omega}_R$ is shorthand for the set of Euler angles relating the orientations of the molecules), with the result

$$K(\boldsymbol{\omega},\boldsymbol{\omega}')=\sum_{\tau=1}^{7}\Gamma_\tau u_\tau(\boldsymbol{\omega})u_\tau(\boldsymbol{\omega}')$$

where

$$\Gamma_1=117$$

$$\Gamma_2=\Gamma_3=\Gamma_4=\frac{195}{4}$$

$$\Gamma_5=\Gamma_6=\Gamma_7=-\frac{351}{4}$$

The physical arguments introduced earlier suggest that the ordered and disordered phases are likely to be very similar in the immediate neighborhood of the transition temperature and that both phases approach the same physical state as one approaches the transition temperature. Therefore we expect the normalization constant

$$G^{-1} = \int d\boldsymbol{\omega} \exp\left[-\lambda G \int d\boldsymbol{\omega}' K(\boldsymbol{\omega}, \boldsymbol{\omega}') e^{g(\boldsymbol{\omega}')}\right]$$

to change continuously, and very close to the transition point we expect the values of G for the two phases to be almost equal. Hence we expect the orientational distribution function for one phase to change over into that of the other phase in a "continuous" manner.

The remarks in the preceding paragraph have led us to consider the possibility of "soft" bifurcation of a known (basic) solution of the nonlinear problem to one or more new solutions (branches). To identify a basic solution, notice that, since the integral of any odd number of u_τ is identically zero, the trivial solution $g(\boldsymbol{\omega}) = 0$ will always be a solution of (38). Physically, this trivial solution corresponds to a hypothetical phase in which the $f(\boldsymbol{\omega})$ is a constant, independent of $\boldsymbol{\omega}$, that is, a phase with complete orientational disorder. We have used the word "hypothetical," since the trivial solution $g(\boldsymbol{\omega}) = 0$, although representing the thermodynamically stable phase in a certain range of temperature, may not represent a stable phase outside this range. With the identification of the basic solution as $g(\boldsymbol{\omega}) = 0$, the normalization constant G for the disordered phase is $(8\pi^2)^{-1}$, and we assume that close to the transition temperature this value characterizes the (possible) new phase as well. This completes the specification of the problem, and we now study the possibility of finding new solutions of

$$g(\boldsymbol{\omega}) + \frac{\lambda}{8\pi^2} \int d\boldsymbol{\omega}' K(\boldsymbol{\omega}, \boldsymbol{\omega}') \exp[g(\boldsymbol{\omega}')] = 0 \qquad (39)$$

which arise continuously from the known, disordered-phase solution at a (possible) bifurcation point.

Given the expository approach taken in this review, we now proceed to analyze (39) in three stages. First, we discuss the existence and uniqueness properties of (39) in light of Dolph's theorems.[45] Then we consider the problem of bifurcation from the more general point of view developed by Krasnosel'skii and co-workers[46,47] in their topological approach to the study of nonlinear integral equations. Finally, we study the solutions of small norm of (39) by a linearization procedure. It should be emphasized at the very outset that the Hammerstein equation, (39), is characterized by a nonlinear function, $F[\boldsymbol{\omega}, g(\boldsymbol{\omega})] = \exp[g(\boldsymbol{\omega})]$, which grows more rapidly than any power of $g(\boldsymbol{\omega})$. In principle, one must use the complicated machinery of Orlicz space to handle problems with such a "strong" nonlinearity. We return to this point in more detail later, but, as regards the present problem, progress can be made by recognizing that only *bounded* $g(\boldsymbol{\omega})$ can represent real phases of the crystal. Accordingly, we define a bounded functional

$$F^*[\boldsymbol{\omega}, g(\boldsymbol{\omega})] = \begin{cases} e^M & g(\boldsymbol{\omega}) > M \\ e^{g(\boldsymbol{\omega})} & |g(\boldsymbol{\omega})| \leq M \\ e^{-M} & g(\boldsymbol{\omega}) < -M \end{cases}$$

where M is a positive constant (notice that $F^*[\boldsymbol{\omega}, g^*(\boldsymbol{\omega})] = F[\boldsymbol{\omega}, g^*(\boldsymbol{\omega})]$ for $|g^*(\boldsymbol{\omega})| \leq M$) and then study the solutions of the nonlinear integral equation

$$g^*(\boldsymbol{\omega}) + \frac{\lambda}{8\pi^2} \int d\boldsymbol{\omega}' K(\boldsymbol{\omega}, \boldsymbol{\omega}') F^*[\boldsymbol{\omega}', g^*(\boldsymbol{\omega}')] = 0 \tag{40}$$

We now carry out an analysis of (40) based on the application of the following two theorems proved by Dolph:

Theorem III (Dolph). Let L_2 be a Hilbert space whose elements are real measurable functions $h(x)$ defined over $a \leq x \leq b$ for which the norm

$$\|h\| = \left(\int_a^b |h(x)|^2 \, dx \right)^{1/2}$$

is finite. Let

$$\mathscr{K}^* = \int_a^b K^*(x, y)[\ldots] \, dy$$

be a completely continuous Hermitian operator in L_2 with (real) characteristic values λ_N with $N = 0, 1, 2, \ldots$, and the ordering

$$\ldots \lambda_{-2} \leq \lambda_{-1} < 0 < \lambda_0 < \lambda_1 \leq \lambda_2 \ldots$$

Let $F(x, y)$ be a continuous function of x and y over the intervals $a \leq x \leq b$, $-\infty \leq y \leq +\infty$, respectively, for which there exist positive constants A, y_0 and numbers μ_N, μ_{N+1} in the relation

$$\lambda_N < \mu_N < \mu_{N+1} < \lambda_{N+1}$$

so that the inequalities

$$\mu_N y - A \leq F(x, y) \leq \mu_{N+1} y + A \qquad y \geq y_0$$

$$\mu_{N+1} y - A \leq F(x, y) \leq \mu_N y + A \qquad y < y_0$$

hold for any fixed N. Then there exists at least one solution to the integral equation

$$\psi(x) = \int_a^b K^*(x, y)F[y, \psi(y)]dy$$

(Note that in this representation of the Hammerstein equation, the kernel K^* is the negative of the kernel K in (37); thus the positive eigenvalues of K^* referred to in the theorem correspond to negative eigenvalues of the kernel K in (37).)

Theorem IV (Dolph). If $F(x, y)$ satisfies the stronger inequalities

$$\lambda_N < \mu_N \leqslant \frac{F(x, y_2) - F(x, y_1)}{y_2 - y_1} \leqslant \mu_{N+1} < \lambda_{N+1}$$

for all y_1, y_2 instead of the conditions cited above and if the rest of the hypotheses of Theorem III are satisfied, then
(i) There exists a unique solution to the integral equation.
(ii) This solution can be obtained by successive approximations.

These theorems, stated here in one-dimensional form, are proved in Ref. 45 using general topological fixed-point methods; hence they apply with obvious modifications to multidimensional (but finite) domains Λ as well.

To mobilize Dolph's existence theorem, we must find positive constants A, y_0 and numbers μ_N, μ_{N+1} in the relation

$$\lambda_N < \mu_N < \mu_{N+1} < \lambda_{N+1}$$

such that the following inequalities hold for any fixed N:

$$\mu_N y - A \leqslant F^*(x, y) \leqslant \mu_{N+1} y + A \qquad y \geqslant y_0$$

$$\mu_{N+1} y - A \leqslant F^*(x, y) \leqslant \mu_N y + A \qquad y < y_0$$

Choosing $N = -1$, λ_{-1} and λ_0 are identified as the first negative and positive eigenvalues of the integral kernel

$$-\left[\frac{\lambda}{8\pi^2}\right] K(\boldsymbol{\omega}, \boldsymbol{\omega}')$$

respectively; these are

$$\lambda_{-1} = -\left[\frac{195}{28}\right]\lambda$$

$$\lambda_0 = \left[\frac{351}{28}\right]\lambda$$

Notice that the conditions of the theorem are satisfied if we select

$$y_0 = \varepsilon > 0 \quad \text{(with } \varepsilon \text{ small)}$$

$$A = e^M$$

and prescribe constants μ_{-1}, μ_0 such that

$$\lambda_{-1} < \mu_{-1} = \alpha < 0 \qquad \text{(with } |\alpha| \text{ small)}$$

$$\lambda_0 > \mu_0 = \delta > 0 \qquad \text{(with } \delta \text{ small)}$$

This specification of constants guarantees the existence of a solution $g^*(\boldsymbol{\omega})$ of (40) for all temperatures. Notice further that this solution is always bounded, since for

$$|g^*(\boldsymbol{\omega})| \leq \frac{\lambda}{8\pi^2} \int d\boldsymbol{\omega}' |K(\boldsymbol{\omega}, \boldsymbol{\omega}')| F^*[\boldsymbol{\omega}', g^*(\boldsymbol{\omega}')] \leq \frac{\lambda}{8\pi^2} e^M K_0$$

where

$$K_0 \geq \int d\boldsymbol{\omega}' |K(\boldsymbol{\omega}, \boldsymbol{\omega}')|$$

Hence whenever

$$\frac{\lambda}{8\pi^2} K_0 \leq Me^{-M} \tag{41}$$

so that

$$F^*[\boldsymbol{\omega}, g^*(\boldsymbol{\omega})] = F[\boldsymbol{\omega}, g^*(\boldsymbol{\omega})]$$

the existence of at least one *bounded* solution to the earlier equation, (39), is guaranteed as well. Finally, since the maximum value of Me^{-M} is e^{-1}, whenever the condition

$$\frac{\lambda}{8\pi^2} K_0 = \frac{I_3^2 K_0}{(8\pi^2)kTR^7} \leq e^{-1} \tag{42}$$

is satisfied, some $M \geq 0$ can always be found such that the satisfaction of (41) is guaranteed; in effect, this demonstrates the existence of physical (bounded) solutions $g(\boldsymbol{\omega})$ for (39) for sufficiently *high* temperatures. The failure of (42) to be satisfied at lower temperatures says nothing about the existence or nonexistence of solutions to (39).

To apply the Dolph uniqueness theorem (IV), we must find constants μ_N and μ_{N+1} such that

$$\lambda_N < \mu_N < \mu_{N+1} < \lambda_{N+1}$$

and the following inequalities are satisfied:

$$\mu_N \leq \frac{F^*(x, y_2) - F^*(x, y_1)}{y_2 - y_1} \leq \mu_{N+1} \tag{43}$$

Again we choose $N=-1$ and prescribe

$$\mu_{-1}=\alpha<0 \qquad (\text{small } |\alpha|)$$
$$\mu_0=e^M$$

to satisfy the condition (43). To satisfy the inequality $\lambda_{-1}<\mu_{-1}$, we choose $|\alpha|$ to be "small enough." To satisfy $\lambda_0>\mu_0$, we require

$$e^M<\frac{351\lambda}{28}$$

for $M\geq 0$; this follows if $351\lambda/28>1$. Thus, provided $351\lambda/28>1$, we can prove the uniqueness of solutions to (40) via a proper choice of M; if, on the other hand, this requirement is not satisfied, solutions of (40) are not necessarily unique. This latter remark indicates (but does not prove) that we may have a bifurcation point of the nonlinear equation, (40), when the value $351\lambda/28=1$ is realized. Moreover, if the condition (41) is satisfied as well, then solutions of (40) satisfy (39), and the point $kT=(351/28)I_3^2/R^7$ would be a bifurcation point of the nonlinear equation, (39), and from this point new solutions of small norm would branch continuously from the basic solution $g(\boldsymbol{\omega})=0$.

Let us now go on to a topological analysis of (39). We cite the following composite result, given in Krasnosel'skii:[47]

Theorem V (Krasnosel'skii). Let $\mathbf{B}$ be a completely continuous operator having a Fréchet derivative $\mathbf{L}$ at the point θ and satisfying the condition $\mathbf{B}\theta=\theta$. Then each characteristic value of odd multiplicity of the linear operator $\mathbf{L}$ is a bifurcation point of the operator $\mathbf{B}$, and to this bifurcation point there corresponds a continuous branch of eigenvectors of the nonlinear operator $\mathbf{B}$.

Since we restrict ourselves to consider only physically meaningful (bounded) solutions of (39), the natural setting within which to pose the problem is the Banach space $\mathscr{C}_\omega$ of all bounded, continuous functions defined over the set of Euler angles $\boldsymbol{\omega}=(\xi,\eta,\zeta)$, where $\xi\in[0,\pi]$, $\eta\in[0,2\pi]$, and $\zeta\in[0,2\pi]$. If the domain of the Hammerstein operator in (39), say

$$\mathbf{B}\equiv-\int d\boldsymbol{\omega}' K(\boldsymbol{\omega},\boldsymbol{\omega}')\exp(\ldots)$$

is taken to be $\mathscr{C}_\omega$, the range of the operator is also $\mathscr{C}_\omega$; no information is sought on the possible existence of new solutions of (39) that are infinite or discontinuous (with respect to norm) relative to the basic solution. A procedure for determining whether the conditions of the Theorem V are satisfied in the problem at hand consists in decoupling the full nonlinear operator **B** into an operator product of the nonlinear functional

$$F[g(\boldsymbol{\omega})]=\exp[g(\boldsymbol{\omega})]$$

and the linear integral operator

$$K=-\int d\boldsymbol{\omega}' K(\boldsymbol{\omega},\boldsymbol{\omega}')(\ldots)$$

Then the properties of **B**, where

$$\mathbf{B}=\mathbf{KF}$$

are analyzed by focusing on the properties of the component operators **K** and **F**. A sufficient condition for the complete continuity (completeness and compactness) of **B** is that

$$\mathbf{F}:\mathscr{C}_\omega\to\mathscr{C}_\omega$$

be continuous and bounded (i.e., that **F** maps any bounded set of $\mathscr{C}_\omega$ into another bounded set) and that

$$\mathbf{K}:\mathscr{C}_\omega\to\mathscr{C}_\omega$$

be completely continuous. The conditions on **F** are no problem, and the complete continuity of **K** follows from the continuity of the integral kernel $K(\boldsymbol{\omega},\boldsymbol{\omega}')$; more sophisticated cases involving the Hammerstein operator defined in other function spaces are indicated in Section IV. Finally, again following Krasnosel'skii, the existence of a Fréchet derivative **L** at the null point of the function space, $\theta(\boldsymbol{\omega})=0$, is guaranteed provided $\mathbf{F}(g)$ and $\mathbf{F}'(g)$ are continuous and **K** is completely continuous; specifically **L** is given by

$$\mathbf{L}=-\int d\boldsymbol{\omega}' K(\boldsymbol{\omega},\boldsymbol{\omega}')\left[\frac{\partial F(g)}{\partial g}\right]_{g=\theta}(\ldots)=-\int d\boldsymbol{\omega}' K(\boldsymbol{\omega},\boldsymbol{\omega}')(\ldots)$$
$$\equiv\mathbf{K}$$

The symmetry of the kernel $K(\boldsymbol{\omega},\boldsymbol{\omega}')$ guarantees that $\mathbf{B}\theta=0$. Having satisfied the conditions of Theorem V, bifurcation of solutions of (39) occur at the eigenvalues of odd multiplicity of the linear integral operator **L**. The results of this analysis may be correlated with one obtained in the analysis based on Dolph's uniqueness theorem (Theorem IV), and the two are found to be consistent.

The above correlation can be made more explicit by studying the "small" solutions of (39). We make the approximation

$$\exp[g(\boldsymbol{\omega})]\simeq 1+g(\boldsymbol{\omega})$$

and, using the symmetry properties of $K(\boldsymbol{\omega}, \boldsymbol{\omega}')$, we determine the linearized integral equation for the problem to be

$$g(\boldsymbol{\omega})+\frac{\lambda}{8\pi^2}\int d\omega' K(\boldsymbol{\omega}, \boldsymbol{\omega}')g(\boldsymbol{\omega}')=0 \tag{44}$$

This eigenvalue equation can be solved directly. Recall that the kernel $K(\boldsymbol{\omega}, \boldsymbol{\omega}')$ is a symmetric bilinear form in the seven $u_\tau(\boldsymbol{\omega})$ functions and hence must have eigenfunctions that lie entirely in the subspace of Hilbert space spanned by these seven functions. Given the earlier specification of the $\{\Gamma_i\}$, one finds nontrivial solutions of (40) only when

$$kT=-\frac{117}{7}\frac{I_3^2}{R^7}$$

$$kT=-\frac{195}{28}\frac{I_3^2}{R^7}$$

or

$$kT=\frac{351}{28}\frac{I_3^2}{R^7}$$

The first two possibilities represent branching from $g(\boldsymbol{\omega})=0$ at negative temperatures, a situation which, for this problem, is clearly unphysical; the remaining possibility represents the only positive temperature at which continuous branching can take place from the disordered phase to a uniformly ordered phase. Now the physically interesting eigenvalue is a triply degenerate eigenvalue of the linear integral equation, (44), and hence the corresponding eigenfunction of (44) may be taken to be any three linear independent combinations of the form

$$g(\boldsymbol{\omega})=\sum_{k=5,6,7} A_k U_k(\boldsymbol{\omega}) \tag{45}$$

Here the coefficients A_k are functions of the temperature and are determined by the requirement that $g(\boldsymbol{\omega})$ must satisfy the seminal equation, (38), (rather than (39), since the use of the approximation $G \simeq G_0 = 1/8\pi^2$ in (39) forces an inconsistency when $A_k \neq 0$). Given Theorem V, with the attendant emphasis on continuous branches, we know that $A_k \to 0$ as the transition temperature

$$T_b=\frac{351}{28}\frac{I_3^2}{kR^7}$$

is approached; accordingly, we calculate the $A_k(T)$ explicitly to first-order in $(T-T_b)$ by inserting (45) into (39) and expanding in powers of A_k. The results of the calculations may be summarized:

(a) Only one A_k different from zero:

$$A_k = \pm\left[3003\left(\frac{T_b - T}{T_b}\right)\right]^{1/2}$$

(b) Two A_k different from zero:

$$A_k = A_{k'} = \pm\left[\frac{924}{41}\left(\frac{T_b - T}{T}\right)\right]^{1/2}$$

(c) Three A_k different from zero:

$$A_k = A_{k'} = A_{k''} = \pm\left[\frac{2002}{117}\left(\frac{T_b - T}{T}\right)\right]^{1/2}$$

These results of Lemberg and Rice are striking. The three classes of limiting solutions obtained in the bifurcation analysis correspond exactly to the class (a), (b), and (c) solutions reported by James and Keenan, and the transition temperature T_b is exactly that found by James and Keenan! Within the framework of mean-field theory, Lemberg and Rice conclude that the uniform order-disorder transition in solid CD_4 can be interpreted as a problem in bifurcation theory, with the bifurcation point placed in one-to-one correspondence with the transition temperature kT_b and the behavior of the system in the neighborhood of the transition point characterized in terms of bifurcating branches of solutions to the underlying nonlinear problem. Against these successful correlations, one must recognize that a principal deficiency of mean-field theory is the neglect of (short-range) fluctuations, and at the present time it is not known to what extent the predictions of the above analysis would change were the role of fluctuations taken explicitly into account. Nonetheless, the above calculation is of great interest in that Lemberg and Rice display in a clear and physically intuitive way the factors that must be taken into account in formulating a statistical-mechanical theory of second-order phase transitions based on a mean-field approach, and they show that the bifurcation theory of nonlinear integral equations is particularly well adapted to an analysis of the consequent mathematical problem.

D. Polymorphic Phase Transitions in Crystals

Transitions from one crystalline modification to another may be of first or second order. Here we focus on the latter possibility and point out that arbitrarily small, translational displacements of atoms from their original positions in a crystal is sufficient to change the symmetry of the lattice. One may find situations (e.g., the crystal $BaTiO_3$) where the configuration of atoms in the crystal changes continuously, although such changes are accompanied by no discontinuous change in state of the system; rather, it is the *symmetry* of the crystal that changes discontinuously at the transition point. Alternatively, one can find situations (e.g., the alloy CuZn) in which the number of lattice points that can be occupied by atoms of a given kind exceeds the number of such atoms; hence, upon increasing the temperature, a transition from an ordered state to a disordered state may occur in which the latter state is characterized by a nonzero probability of finding atoms at every lattice site, with a concomitant increase in symmetry of the lattice. The above types of transitions may be modeled using a generalized mean-field approach, and this procedure leads to equations of the Hammerstein form, say,

$$\Phi(\mathbf{r}_1)+\mu\int K(\mathbf{r}_1,\mathbf{r}_2)F[\mathbf{r}_2,\Phi(\mathbf{r}_2)]\,d\mathbf{r}_2=0 \tag{46}$$

Here

$$\Phi(\mathbf{r})=\ln\{z\rho(\mathbf{r})\}$$

where $\rho(\mathbf{r})$ is a density (probability), z a normalization constant, $\mu = 1/zkT$, and $K(\mathbf{r}_1, \mathbf{r}_2)$ is some kind of effective potential which specializes the mean-field approach to the particular problem being considered.

In 1963 Khachaturyan[50] presented a discussion of the order-disorder transition in a binary solid solution, starting from a certain nonlinear integral equation (essentially a mean-field equation which can easily be cast in Hammerstein form). In his approach, the crystalline state of the system is specified through the kernel of problem. Although he did not use the results of bifurcation theory *per se*, he did develop theorems that allow one to decide whether a given solution or solutions are consistent with a given choice of kernel, that is, a given space group of the superlattice. In this section, we describe a later paper by Tareyeva[51] in which a bifurcation approach is taken to discuss second-order polymorphic phase transitions in crystals. Tareyeva's paper is of particular interest, given the objectives of this review, since it illustrates the use of the Lyapunov-Schmidt method for the study of the nonlinear equation, (46). (For a general discussion of this method, see the review article of Vainberg and Trenogin,[23] and for a simple introduction to the approach see Section 1.4 of that article.)

Tareyeva's objective is to describe, by studying the bifurcation of solutions to (46), the possible structural phase transitions in a system in which the space group G_1 of the final state is a subgroup of the space group G_0 of the initial state. Since she is only interested in qualitative results on the symmetry properties of the initial and final functions $\Phi(\mathbf{r})$

and their temperature dependence, she considers (46) without specifying a particular kernel. She does require that the kernel characterizing the initial phase be an integrable function, invariant under the operations of the group G_0; specifically, she writes

$$K(\mathbf{r}_1, \mathbf{r}_2) = \sum_{jn} K_j(r_1^0, r_2^0)\psi_j^n(\mathbf{r}_1)\psi_j^{*n}(\mathbf{r}_2)$$

where the $\psi_j^n(\mathbf{r})$ are basis functions of the jth irreducible representation of the group G_0 and r^0 is G_0-invariant. As regards the nonlinear function $F[\mathbf{r}_2, \Phi(\mathbf{r}_2)]$, Tareyeva assumes that it is a bounded function; given this constraint, Dolph's theorems then guarantee that (46) has at least one solution. Now suppose that for $\mu = \mu_0$ there exists a continuous G_0-invariant solution $\Phi_0(\mathbf{r})$ of (46); then we seek all continuous solutions

$$\Phi_\mu(\mathbf{r}) = \Phi_0(\mathbf{r}) + u(\mathbf{r})$$

for

$$\mu = \mu_0 + \lambda$$

such that

$$\Phi_\mu(\mathbf{r}) \to \Phi_0(\mathbf{r}) \quad \text{when} \quad \mu \to \mu_0.$$

We proceed by constructing the Taylor expansion for the nonlinear function $F[\mathbf{r}, \Phi_0(\mathbf{r}) + u(\mathbf{r})]$, namely,

$$F[\mathbf{r}, \Phi_0(\mathbf{r}) + u(\mathbf{r})] = \sum_m A_m(\mathbf{r})u^m(\mathbf{r})$$

where the coefficients $A_m(\mathbf{r})$ are G_0-invariant. Using this representation for F, we rewrite (46) as

$$\begin{aligned} u(\mathbf{r}_1) - \mu_0 \int B(\mathbf{r}_1, \mathbf{r}_2)u(\mathbf{r}_2)\, d\mathbf{r}_2 \\ = \lambda f_1(\mathbf{r}_1) + \mu_0 \sum_{m=2}^{\infty} \int K(\mathbf{r}_1, \mathbf{r}_2)A_m(\mathbf{r}_2)u^m(\mathbf{r}_2)\, d\mathbf{r}_2 \\ + \lambda \sum_{m=1}^{\infty} \int K(\mathbf{r}_1, \mathbf{r}_2)A_m(\mathbf{r}_2)u^m(\mathbf{r}_2)\, d\mathbf{r}_2 \end{aligned}$$

where

$$B(\mathbf{r}_1, \mathbf{r}_2) = K(\mathbf{r}_1, \mathbf{r}_2)A_1(\mathbf{r}_2)$$

$$f_1(\mathbf{r}_1) = \int K(\mathbf{r}_1, \mathbf{r}_2)A_0(\mathbf{r}_2)d\mathbf{r}_2$$

In the Lyapunov-Schmidt method one shows that bifurcation is possible when $\mu = \mu_0$ is an eigenvalue of the operator $\mathbf{B}$. Specifically, let the rank of μ_0 be unity and let $\phi_1(\mathbf{r}_1)$ be the eigenfunction corresponding to μ_0, so that

$$\phi(\mathbf{r}_1) - \mu_0 \int B(\mathbf{r}_1, \mathbf{r}_2)\phi_1(\mathbf{r}_2)\, d\mathbf{r}_2 = 0 \tag{47}$$

By constructing the Fredholm resolvent $R(\mathbf{r}_1, \mathbf{r}_2)$ for the kernel

$$\mu_0 B(\mathbf{r}_1, \mathbf{r}_2) - \chi(\mathbf{r}_1)\overline{\phi_1(\mathbf{r}_2)}$$

where $\chi_1(\mathbf{r}_1)$ is the eigenfunction of $\overline{B(\mathbf{r}_2, \mathbf{r}_1)}$ with eigenvalue μ_0, one shows (see Ref. 51) that $u(\mathbf{r})$ can be written in the form

$$u(\mathbf{r}) = \xi\phi_1(\mathbf{r}) + \lambda g(\mathbf{r}) + \sum_{i+k\geqslant 2} a_{ik}(\mathbf{r})\lambda^k \xi^i \tag{48}$$

where

$$\xi = \int u(\mathbf{r})\overline{\phi_1(\mathbf{r})}\, d\mathbf{r}$$

$$g(\mathbf{r}_1) = f(\mathbf{r}_1) + \int R(\mathbf{r}_1, \mathbf{r}_2) f(\mathbf{r}_2)\, d\mathbf{r}_2$$

and the bar in the above equations denotes complex conjugate. One determines the $a_{ik}(\mathbf{r})$ via an iteration procedure. Then substitution of (47) into the normalization condition for the eigenvector $\phi_1(\mathbf{r})$ yields the branching or bifurcation equation for the possible $\xi(\lambda)$, namely,

$$\sum_{m=2}^{\infty} L_{m0}\xi^m + \sum_{m=0}^{\infty} \xi^m \sum_{n=1}^{\infty} L_{mn}\lambda^n = 0 \tag{49}$$

where

$$L_{01} = \int g(\mathbf{r})\overline{\phi_1(\mathbf{r})}\, d\mathbf{r}$$

$$L_{mn} = \int a_{mn}(\mathbf{r})\overline{\phi_1(\mathbf{r})}\, d\mathbf{r}$$

We remark that Hammerstein,[43] Schmidt,[52] and Iglisch[53] studied the branching equation for the Hammerstein problem and showed how the number of small solutions of (48) can be determined, even in cases in which the eigenvalue μ_0 has multiplicity greater than 1; Vainberg and Trenogin[23] describe how this analysis may be extended (via the use of the Newton diagram technique) so that all *small* solutions of (48) can be represented as a series of integral or fractional powers in the parameter λ.

Given the structure of $B(\mathbf{r}_1, \mathbf{r}_2)$ in this problem, Tareyeva notes that the linear integral equation, (47), can be written as a system of linear integral equations with kernels $B_j(\mathbf{r}_1^0, \mathbf{r}_2^0)$, where r^0 is assumed G_0-invariant. Apart from accidental degeneracy, she argues that $\phi_1(\mathbf{r})$ will have nontrivial components only in one irreducible representation, say the j_0th. In analyzing the coefficients L_{mn} in (49), using only the symmetry properties of the functions $f_1(\mathbf{r})$, $\phi_1(\mathbf{r})$ and the resolvent

$$R(\mathbf{r}_1, \mathbf{r}_2) = \sum_{jn} R_j(r_1^0, r_2^0)\psi_j^n(\mathbf{r}_1)\psi_j^n(\mathbf{r}_2)$$

she finds that all the L_{0n} vanish, all the L_{1n} are nonzero, and the properties of the other coefficients L_{mn} depend on the properties of the j_0th irreducible representation. For example, if the representation of the product group $[j_0]^3$ does not contain the irreducible representation j_0 of the initial group G_0, she shows that $L_{2n} = 0$. Then, if the coefficient $L_{3n} \neq 0$, there are three branches of $\Phi(\mathbf{r})$; $\Phi_{1,2}(\mathbf{r})$ are series in the parameter $\sqrt{\lambda}$ with the broken symmetry, and the third is a series in λ with the initial symmetry. In this case, Tareyeva notes that if, in the low symmetry phase, the order parameter is represented by a series in $\sqrt{T_c - T}$, one recovers the phenomenological result of Landau.

Although for pedagogical reasons we have focused on an early paper by Tareyeva in this section, we should emphasize that Tareyeva and her colleagues have considered several applications of the ideas described here. Of particular importance are her studies on orientational ordering in molecular hydrogen: the transition temperature and change of symmetry,[54] the quantum case,[55] and the first-order phase transition.[56] Also of interest is her discussion of the applicability of a bifurcation approach to the theory of first-order, isotropic-nematic phase transitions.[57]

IV. FIRST-ORDER TRANSITIONS

A. Prologue

We turn now to an examination of certain nonlinear problems that arise in the statistical-mechanical theory of first-order phase transitions. From our earlier discussion of Landau theory and from the remarks on the status of bifurcation theory (in particular, the comment that the available theorems deal, for the most part, with "soft" bifurcation), one might expect that the discontinuous changes that characterize first-order transitions may not be accessible to the kind of analysis described in the preceding section. It is found experimentally, however, that there are certain regions of the thermodynamic phase plane where the change in density across a first-order phase boundary is "almost" continuous. For example, in the fluid-solid phase transition, the change in density between the two phases at coexistence is small[58]; the density of solid argon at 80 K is 1.636 g/cm^3, whereas the density of liquid argon at 84.5 K is 1.407 g/cm^3; for metals, where long-range coulombic forces play a crucial role, the change is even less; for example, the density of solid sodium at 371 K is 0.951 g/cm^3, whereas that of liquid sodium at 373 K is 0.927 g/cm^3. With respect to the gas-liquid transition, if one is well away from the critical point, the change in density is significant and the transition is first order. However, in the neighborhood of the critical point the difference in density of the two phases becomes small, vanishing *at* the critical point; the transition in this limit becomes second order. Assuming, then, that one can identify a parameter (say, the density) which changes in an almost continuous manner as one crosses a phase boundary in a first-order transition, the question arises whether the underlying statistical-mechanical theory (for the probability density) can be investigated using the methods of "soft" bifurcation theory. We remark that an alternative approach to the study of first-order phase transitions using the methods of nonlinear functional analysis (and in particular the concept of the turning point) has been developed by Tareyeva.[56,57]

In this chapter we examine the problem of first-order phase transitions, again by considering in detail a few representative contributions, and then try to correlate the insights gained from these elaborations. As in the preceding section, we devote a great deal of attention to the formulation of each problem to expose clearly the underlying statistical-mechanical issues; in fact, because a greater literature exists here, we begin our discussion by formulating the problem in several different representations before proceeding to discuss the applications. As we shall see, the formal theorems of statistical mechanics as well as certain exact results play an important role in interpreting the results obtained in these analyses.

1. *Distribution Function Theories*

Consider a classical system of N molecules in a volume Λ at a temperature T. Let $\boldsymbol{\xi}=(\xi_1,\ldots,\xi_N)$ be the set of parameters $(0\leq\xi_i\leq 1)$ which scales the coupling of the N molecules in the system; $\boldsymbol{\xi}=\mathbf{0}$ denotes a completely uncoupled system and $\boldsymbol{\xi}=\mathbf{1}$ denotes full coupling of the intermolecular forces. Then, as is shown in many places,[59] the probability $\rho_N^{(n)}(\mathbf{r}_1,\ldots,\mathbf{r}_n)d\mathbf{r}_1\ldots d\mathbf{r}_n$ of observing molecules in $d\mathbf{r}_1\ldots d\mathbf{r}_n$ at $\mathbf{r}_1\ldots\mathbf{r}_n$ when the volume Λ contains N molecules is given by the following expression

$$\rho^{(n)}(\mathbf{r}_1,\ldots,\mathbf{r}_n;\xi)=\frac{1}{Z(\xi)}\frac{N!}{(N-n)!}\int_{\Lambda}\cdots\int\exp[-\beta U(\xi)]\,d\mathbf{r}_{n+1}\ldots d\mathbf{r}_N \tag{50}$$

with $\boldsymbol{\xi}=\mathbf{1}$, multiplied by the volume element $d\mathbf{r}_1 \ldots d\mathbf{r}_n$ (as before, $\beta = 1/kT$ where k is the Boltzmann constant). In this definition, U is the sum of potentials due to molecular pairs

$$U(\mathbf{r}_1, \ldots, \mathbf{r}_N; \boldsymbol{\xi}) = \sum_{1 \leq i < j \leq N} \xi_i \xi_j \Phi(r_{ij})$$

where $\Phi(r_{ij})$ is the potential of intermolecular force between molecules i and j as a function of the distance r_{ij} between the molecules, and $Z(\xi)$ is the configurational partition function

$$Z_N(\xi) = \int_{\Lambda} \cdots \int \exp\left[-\beta U_N(\xi)\right] d\mathbf{r}_1 \ldots d\mathbf{r}_N$$

These statements, valid for closed thermodynamic systems, may be generalized to open systems. For $\boldsymbol{\xi}=\mathbf{1}$, the probability that an open system actually does contain exactly N molecules is

$$P_N = \frac{z^N Z_N}{N! \Xi}$$

where Ξ is the grand canonical partition function

$$\Xi = \sum_{N \geq 0} \frac{z^N}{N!} \cdot Z_N$$

In these expressions, z is the thermodynamic activity

$$z = \frac{e^{\mu/kT}}{h^3} (2\pi mkT)^{3/2}$$

where h is Planck's constant, μ is the chemical potential, and m is the mass per molecule of the species involved. In this case, the probability $\rho^{(n)}(\mathbf{r}_1, \ldots, \mathbf{r}_n)\, d\mathbf{r}_1 \ldots d\mathbf{r}_n$ of observing molecules in $d\mathbf{r}_1 \ldots d\mathbf{r}_n$ at $\mathbf{r}_1, \ldots, \mathbf{r}_n$, properly averaged over N, is given by the following expression.

$$\rho^{(n)}(\mathbf{r}_1, \ldots, \mathbf{r}_n) = \frac{1}{\Xi} \sum_{N \geq n} \frac{z^N}{(N-n)!} \int_{\Lambda} \cdots \int \exp\left[-\beta U_N\right] d\mathbf{r}_{n+1} \ldots d\mathbf{r}_N \quad (51)$$

multiplied by the volume element $d\mathbf{r}_1 \ldots d\mathbf{r}_n$. (The probability density $\rho_N^{(n)}$ is zero for $N \leq n$.)

The $\rho^{(n)}$ appearing in (50) and (51) are the n-particle distribution functions, and these functions can be shown to satisfy various systems (hierarchies) of integral or integrodifferential equations. In this review, we consider explicitly three hierarchies:

1. *Kirkwood-Salsburg Hierarchy*[60]

$$\rho^{(n)} = z \exp[-\beta U_n^{(1)}]\rho^{(n-1)}(\mathbf{r}_2, \ldots, \mathbf{r}_n) + \int_\Lambda K^{(n)}(\mathbf{r}_1, \ldots, \mathbf{r}_{n+1})\rho^{(n)}(\mathbf{r}_2, \ldots, \mathbf{r}_{n+1}) d\mathbf{r}_{n+1} \quad (52)$$

where

$$K^{(n)}\rho^{(n)} = zf_{1,n+1} \exp[-\beta U_n^{(1)}] \cdot \left\{\rho^{(n)}(\mathbf{r}_2, \ldots, \mathbf{r}_{n+1}) + \sum_{s\geq 1} \frac{1}{(s+1)!} \int \cdots \int_\Lambda \rho^{(n+s)}(\mathbf{r}_2, \ldots, \mathbf{r}_{n+s+1}) \prod_{\sigma=n+2}^{n+s+1} f_{1\sigma}\, dr_\sigma\right\}$$

$$U_n^{(1)}(1, \ldots, n) = \sum_{i=2}^{n} \Phi(r_{1i})$$

$$f_{ij} = e^{-\beta\Phi(r_{ij})} - 1$$

2. *Kirkwood Coupling-Parameter Hierarchy*[61]

$$-\frac{1}{\beta}\frac{\partial\rho^{(n)}}{\partial\xi} + \rho^{(n)}\left(\frac{\partial A}{\partial\xi}\right)_{N,\Lambda,T} = \rho^{(n)} \sum_{i=2}^{n} \Phi(r_{1i}) + \int_\Lambda \Phi(r_{1,n+1})\rho^{(n+1)}(\mathbf{r}_1, \ldots, \mathbf{r}_{n+1}; \xi)\, d\mathbf{r}_{n+1} \quad (53)$$

where in the grand canonical ensemble the derivative of the Helmholtz free energy A with respect to the coupling parameter ξ is

$$-kT\left(\frac{\partial \ln \Xi}{\partial\xi}\right)_{z,\Lambda,T}$$

3. *Yvon-Born-Green Hierarchy*[62]

$$-\frac{1}{\beta}\nabla_1\rho^{(n)} = \rho^{(n)} \sum_{i=2}^{n} \nabla_1\Phi(r_{1i}) + \int_\Lambda \nabla_1\Phi(r_{1,n+1})\rho^{(n+1)}(\mathbf{r}_1, \ldots, \mathbf{r}_{n+1})\, d\mathbf{r}_{n+1} \quad (54)$$

where ∇_1 denotes gradient with respect to $\mathbf{r}_1$.

Equations 52 to 54 have been derived both for closed and open thermodynamic systems and apply to any pure phase (classical gas, liquid, or solid). The sense in which they apply to phases in *coexistence* is described in the following subsection. From a mathematical point of view, the main thing to note about these three systems of inhomogeneous equations is that they are *linear* in the n-particle distribution functions $\rho^{(n)}$.

2. *Distribution Functions at a Phase Boundary*

In what follows, we focus on first-order phase transitions. Our objective here is to discuss the formal structure of the solution to the hierarchy equations, (52) to (54), for closed and open thermodynamic systems. In addition, to link this discussion with the earlier qualitative remarks on the importance of *fluctuations* in phase transition theory (presented in Section II), we comment on the behavior of the compressibility in the region of coexistence. The following synopsis is taken mainly from Hill,[59] where the reader may find many more details and further insights.

Consider a one-component system confined to a volume Λ and at a temperature T. The volume Λ is understood here to be extremely large (the reason for this requirement is brought out later in drawing attention to Van Hove's work on the importance of the "thermodynamic limit," whereby a system possessing N degrees of freedom and confined to a volume Λ is considered to be properly defined only in the limit $N \to \infty$, $\Lambda \to \infty$ while $N/\Lambda = \rho = \text{constant}$). Moreover, we assume that no gravitational (or other) field is present. (This requirement is discussed later in conjunction with the translational and rotational invariance properties of the singlet and pair distribution functions.)

Suppose it is known that only a single phase a exists when $N \leqslant N_a$ particles are confined to a volume Λ. Similarly, when $N \geqslant N_b$, we presume that only a single phase b is present. In this particle specification, the coexistence region of the two phases a and b can be characterized by the inequality

$$N_a < N < N_b$$

Now the canonical-ensemble (generic) n-particle distribution functions [defined in (50)] associated with the pure phases a and b are

$$\rho_N^{(n)} = \begin{cases} \rho_{(a)N}^{(n)} & N \leqslant N_a \\ \rho_{(b)N}^{(n)} & N \geqslant N_b \end{cases}$$

The pure-phase, n-particle distribution functions $\rho_{(a)N}^{(n)}$ and $\rho_{(b)N}^{(n)}$ are the solutions of the hierarchy equations, (52) to (54), in the respective regimes $N \leqslant N_a$ and $N \geqslant N_b$. Now, from considerations of surface tension (see the discussions in Landau and Lifshitz[15] and Huang[63]), it is anticipated that the most stable two-phase configurations in a coexistence region will be those with the smallest interfacial areas. This favors the formation of large single-phase regions, so we may safely assume that a set of points $\mathbf{r}_1, \ldots, \mathbf{r}_n$ may be identified located within the same single-phase region at any time. If the probability that any *local* region in Λ is occupied by phase a when $N_a < N < N_b$ can be scaled by the volume fraction

$$x_a = \frac{N_b - N}{N_b - N_a}$$

of phase a in the *global* (two-phase) system Λ (with a similar remark pertaining to phase b), then, as Mayer[64] has stated, the n-particle distribution function in the two-phase region may be written

$$\begin{aligned}\rho_N^{(n)} &= x_a(N)\rho_{(a)N_a}^{(n)} + x_b(N)\rho_{(b)N_b}^{(n)} \\ &= x_a(N)\rho_{(a)N_a}^{(n)} + [1 - x_a(N)]\rho_{(b)N_b}^{(n)}\end{aligned} \tag{55}$$

The mathematical content of (55) is that in a two-phase region of a closed system, the overall $\rho_N^{(n)}$ is a *linear* combination of the two distribution functions $\rho_{(a)N_a}^{(n)}$ and $\rho_{(b)N_b}^{(n)}$ characterizing the pure phases in thermal equilibrium with each other. Notice that (55) defines a one-parameter family of functions; the coefficient $x_a(N)$ may assume any value between zero and unity. We now point out that in a closed system, for any given average density $\rho = N/\Lambda$, with $N_a < N < N_b$, the coefficient $x_a(N)$ assumes a fixed value, and the *linear* combination (55) *is* the solution of the hierarchy equations, (52) to (54), all other quantities in (52) to (54) having the same values in the two phases a and b.

The development of the above argument for open systems proceeds along similar lines, but with an important difference. From the conditions of thermal, mechanical and chemical equilibrium, we have at the transition point

$$T = T'$$

$$p = p'$$

$$\mu = \mu' \qquad (z = z')$$

Suppose phase a is the stable phase when $z < z'$, and phase b is the stable phase when $z > z'$. As noted earlier [(51)], in an open system

$$\rho^{(n)} = \sum_{N \geq n} P_N \rho_N^{(n)} \tag{56}$$

so that when $z < z'$ only terms for which $N \leq N_a$ contribute appreciably to (56); moreover, in the limit $\Lambda \to \infty$, only the term $N = \bar{N}$ (bar stands for average) counts. Similar remarks pertain to the situation when $z > z'$, so that we may write for the pure phases a and b:

$$\rho^{(n)} = \begin{cases} \rho_{(a)}^{(n)} = \rho_{(a)\bar{N}}^{(n)} & z < z', \quad \bar{N}(z, \Lambda, T) \leq N_a \\ \rho_{(b)}^{(n)} = \rho_{(b)\bar{N}}^{(n)} & z > z' \quad \bar{N}(z, \Lambda, T) \geq N_b \end{cases}$$

Now, in the coexistence region, neglecting all surface effects, we write

$$\rho^{(n)} = \chi_a \rho^{(n)}_{(a)N} + \chi_b \rho^{(n)}_{(b)N} \qquad z = z' \tag{57}$$

where

$$\chi_a = \frac{C_a}{C_a + C_b} \qquad \chi_b = \frac{C_b}{C_a + C_b}$$

and the C_i are quantities that may be regarded as constants (although in principle they can change slowly with N, Λ if N, Λ are finite[59]). Given that

$$\lim_{z \to z'_-} \rho^{(n)}_{(a)}(z) = \rho^{(n)}_{(a)N_a}$$

$$\lim_{z \to z'_+} \rho^{(n)}_{(b)}(z) = \rho^{(n)}_{(b)N_b}$$

the result (57) may be written

$$\rho^{(n)}(z') = \chi_a \rho^{(n)}_{(a)}(z'_-) + \chi_b \rho^{(n)}_{(b)}(z'_+) \tag{58}$$

Here again we have that at the point of transition, the overall distribution function is a *linear* combination of two separate distribution functions, with the important difference [vis a vis the earlier result (55)] that in an *open* system the coefficients χ_a, χ_b are not variable, since in the grand canonical ensemble an average over the possible values of N has been taken. Because of the linearity expressed in (58) (and the equality in the two phases of all other quantities appearing in the hierarchy equations (52–54)), the linear combination (58) is the solution of (52) to (54) when $z = z'$. Although in the neighborhood of $z = z'$ the n-particle distribution functions are different (i.e., the set $\rho^{(1)}_{(a)}(z'_-)$, $\rho^{(2)}_{(a)}(z'_-), \ldots$ computed as $z \to z'_-$ versus the set $\rho^{(1)}_{(b)}(z'_+)$, $\rho^{(2)}_{(b)}(z'_+), \ldots$ computed as $z \to z'_+$), exactly *at* $z = z'$ the solution of the hierarchy equations *is* the linear combination (58) with $n = 1, 2, \ldots$.

At this point we may draw attention to one of the central difficulties involved in tackling the problem of phase transitions using one of the distribution function theories of statistical mechanics. If we approach the phase transition point $z = z'$ from below, where the distribution function $\rho^{(n)}_{(a)}(z'_-)$ descriptive of phase a pertains, we may enquire whether a new solution $\rho^{(n)}(z')$ of the hierarchy appears as a certain critical activity is reached; if such a solution is identified, we may investigate whether there is an exchange of stability between the old and new solutions at $z = z'$. From the above discussion, we know that the new solution will be of the form (58) since, due to the linearity of (58), *any* linear combination of $\rho^{(n)}_{(a)}(z'_-)$ and $\rho^{(n)}_{(b)}(z'_+)$ will satisfy the equations of the hierarchy at $z = z'$.

However, *rigorously*, it is only the linear combination (58) with the specific coefficients χ_i that has the *physical significance* of a two-phase distribution function in an open system, that is, the mere identification of a new solution (distribution function) at a certain $z = z'$ *may* have no physical significance in *equilibrium* statistical mechanics, and this point must be kept in mind in our later discussions where bifurcating branches of solutions may arise in a given (nonlinear) representation of the problem.

Let us conclude now with a consideration of *fluctuations* at a point of phase transition. From statistical mechanics we have the result[59]

$$\frac{\langle N^2 \rangle - \langle N \rangle^2}{\langle N \rangle} = \rho k T \kappa \tag{59}$$

where κ is the compressibility

$$\kappa = -\frac{1}{\Lambda}\left(\frac{\partial \Lambda}{\partial p}\right)_T$$

In a one-phase system, κ is of the order of $1/\rho kT$, whereas in a two-phase system κ is of the order of $\langle N \rangle/\rho kT$. In the thermodynamic limit, $\Lambda \to \infty$, $\langle N \rangle \to \infty$, while ρ, the mean density of the system, remains finite. This in turn means that in a coexistence region, the two-phase portion of a p-Λ isotherm is *flat*, with the profile of the isotherm entering and leaving the coexistence region characterized by a discontinuous change in slope (see Fig. 2). The fact that the numerator in (59) is of the order of $\langle N \rangle$ for a pure phase means that thermodynamic fluctuations are of negligible importance in influencing property calculations away from a point of phase transition, whereas since $\langle N^2 \rangle - \langle N \rangle^2$ is of the order of $\langle N \rangle^2$ *at* a phase transition, fluctuations are of crucial importance there. This observation makes more explicit the earlier criticism of Landau theory (or certainly any deterministic theory of phase transitions, e.g., the theory presented in Ref. 16). The failure to take into account the importance of fluctuations in the neighborhood of a coexistence region may lead to incorrect property predictions.

B. Formulation of the Nonlinear Problem

The distribution functions satisfy various systems of integrodifferential or integral equations, among which the more familiar are the Kirkwood-Salsburg hierarchy (52), the Kirkwood coupling-parameter hierarchy (53), and the Yvon-Born-Green hierarchy (54). As noted in the preceding section, each hierarchy comprises a linear inhomogeneous system for the (generic) distribution functions, defined via (50) for closed thermodynamic systems and via (51) for open systems. In a very important paper, Ruelle[1,65] has taken advantage of the linear structure of the Kirkwood-Salsburg (KS) equations and has shown how this system of equations may be transformed into a single linear operator equation; within this framework, he was able to obtain certain bounds on the existence of a pure (low-density) phase. It is very instructive, in light of the issues discussed in this review, to recall the principal results of the Ruelle analysis (Section IV.B.1), since the results of his exact analysis of the KS equations serve as a guide in our later discussions, and since similar results of comparable rigor seem not to be available on the Kirkwood (K) and Yvon-Born-Green (YBG) hierarchies. Following this reflection, the nonlinear problem is posed within the framework of the KS hierarchy, followed by corresponding treatments of the Kirkwood (Section IV.B.2) and Yvon-Born-Green (Section IV.B.3) problems. Finally, in Section IV.B.4 we describe several formulations based on the direct correlation function. Our motivation for focusing initially on the formulation of the nonlinear problem in several different representations is that the underlying statistical-mechanical issues can be more conveniently stressed and the common features of the consequent mathematics can be exposed prior to getting involved in the description of detailed results. The latter task is deferred until Section IV.C.

1. *The Kirkwood-Salsburg Hierarchy*

In order that the reader be able to correlate the present discussion with the earlier work of Ruelle, we adopt his more explicit notation in this subsection. Thus, for example, we write

$$U(x)_n = \sum_{1\leq i<j\leq n} \Phi(x_j - x_i)$$

$$\psi(x) = \exp[-\beta U(x)_n] \quad \text{with} \quad (x)_n = (x_1, \ldots, x_n)$$

where molecular positions are denoted by the variable x, the pair potential Φ is assumed to be stable and regular, and $\beta = 1/kT > 0$ as before. Let us introduce as well the characteristic function χ_Λ of the bounded measureable set $\Lambda \varepsilon R^\nu$ (where ν is the dimension) and write

$$\chi_\Lambda(x)_n = \prod_{i=1}^{n} \chi_\Lambda(x_i)$$

Using these notations, the grand partition function

$$\Xi(\Lambda, z, \beta) = \sum_{n=0}^{\infty} \frac{z^n}{n!} \int_{\Lambda^n} dx_1 \ldots dx_n \exp[-\beta U(x)_n]$$

and the m-particle distribution function

$$\rho_\Lambda(x_m) = \Xi(\Lambda, z, \beta)^{-1} \sum_{n=0}^{\infty} \frac{z^{m+n}}{n!} \int_{\Lambda^n} dx_{m+1} \ldots dx_{m+n} \exp\left[-\beta U(x)_{m+n}\right]$$

(where z is the activity) may be rewritten as

$$\Xi(\Lambda, z, \beta) = \sum_{n=0}^{\infty} \frac{z^n}{n!} \int d(x)_n \chi_\Lambda(x)_n \psi(x)_n \tag{60}$$

and

$$\rho_\Lambda(x)_m = \Xi(\Lambda, z, \beta)^{-1} \sum_{n=0}^{\infty} \frac{z^{m+n}}{n!} \int dx_{m+1} \ldots dx_{m+n} \chi_\Lambda(x)_{m+n} \psi(x)_{m+n} \tag{61}$$

Then the KS equations assume the form

$$\rho_\Lambda(x_1) = \chi_\Lambda(x_1) z\left[1 + \sum_{n=1}^{\infty} \frac{1}{n!} \int dy_1 \ldots dy_n K(x_1, (y)_n) \rho_\Lambda(y)_n\right] \tag{62}$$

$$\begin{aligned} \rho_\Lambda(x)_m = {} & \chi_\Lambda(x)_m \, z \exp\left[-\beta W^1(x)_m\right] \\ & \cdot \left[\rho_\Lambda(x)'_{m-1} + \sum_{n=1}^{\infty} \frac{1}{n!} \int dy_1 \ldots dy_n K(x_1, (y)_n) \rho_\Lambda((x)'_{m-1}, (y)_n)\right] \end{aligned} \tag{63}$$

where

$$W^1(x)_m = \sum_{i=2}^{m} \Phi(x_i - x_1)$$

$$(x)'_{m-1} = (x_2, \ldots, x_m)$$

$$K(x_1, (y)_n) = \prod_{j=1}^{n} \{\exp\left[-\beta\Phi(y_j - x_1)\right] - 1\}$$

Ruelle has shown how the above equations may be transformed into a single equation for

$$\boldsymbol{\rho}_\Lambda = (\rho_\Lambda(x)_n)_{n \geq 1}$$

in the Banach space E_ξ (given $\xi > 0$, E_ξ is the space of sequences

$$\phi = (\phi(x)_n)_{n \geq 1}$$

where $\phi(x)_n$ is a complex Lebesque measurable function on $R^{n\nu}$ such that

$$\|\phi\|_\xi = \sup_{n \geq 1} (\xi^{-n} \operatorname{ess} \sup_{(x)_n \varepsilon R^{n\nu}} |\phi(x)_n|) < +\infty).$$

In particular, he shows that $\boldsymbol{\rho}_\Lambda$ satisfies the *linear* equation

$$\boldsymbol{\rho}_\Lambda = z\chi_\Lambda\boldsymbol{\alpha} + z\chi_\Lambda\Pi K\boldsymbol{\rho}_\Lambda \tag{64}$$

where $\boldsymbol{\alpha}$ is such that $\alpha(x)_1 = 1$, $\alpha(x)_n = 0$ for $n > 1$, and Π is a permutation operator, and considers as well the equation

$$\boldsymbol{\rho} = z\boldsymbol{\alpha} + z\Pi K\boldsymbol{\rho} \tag{65}$$

where $\boldsymbol{\rho}$ is the sequence of "infinite volume correlation functions" $\rho(x)_n$. The theorem proved by Ruelle, of importance in our later discussion, can now be stated.

Theorem VI (Ruelle). Let Φ be a stable regular potential, and let z be a complex number satisfying

$$|z| < e^{-2\beta B - 1}C(\beta)^{-1} \tag{66}$$

where

$$C(\beta) = \int dx|e^{-\beta\Phi(x)} - 1| < +\infty$$

$$\Phi(x) = U(0, x) \geqslant -2B$$

Then, the grand partition function, (60), has no zero in (66). If the distribution functions $\rho_\Lambda(x)_n$ are defined by (61) for z in (66), there exist "infinite volume correlation functions" $\rho(x)_n$ and a function $\varepsilon(\cdot)$ positive and decreasing, such that

$$\lim_{\lambda\to\infty} \varepsilon(\lambda) = 0$$

and

$$|\rho_\Lambda(x)_n - \rho(x)_n| \leqslant \xi^n \varepsilon(\lambda)$$

if ξ, z satisfy

$$|z| < e^{-2\beta B}\xi \exp[-\xi C(\beta)] \tag{67}$$

and λ is the minimum distance from $x_1, \ldots, x_m \varepsilon \Lambda$ to the boundary of Λ. The sequence $\boldsymbol{\rho}_n$ of the $\rho_\Lambda(x)_n$ and the sequence $\boldsymbol{\rho}$ of the $\rho(x)_n$ are the only solutions of (64) and (65), respectively, in E_ξ when ξ satisfies (67) (in particular, if $\xi = C(\beta^{-1})$) and $\boldsymbol{\rho}_\Lambda$ and $\boldsymbol{\rho}$ depend analytically on z.

It should be noted that if $|z|$ actually attains the value given by the right-hand side of (67), this does *not* of necessity mean that a phase transition will occur. Rather, the bound must be interpreted as a lower bound on the limit of stability of a pure phase (i.e., in the notation of the preceding subsection, the bound may be displaced from the actual transition activity $z = z'$).

Let us turn now to the problem of phase transitions, formulated as a problem in nonlinear analysis, and set within the framework of the KS hierarchy. (The following discussion is drawn from Cheng and Kozak[66] and Kozak.[67]) Suppose, for definiteness, we consider the first equation in the KS hierarchy (62), and then reexpress this equation in terms of the n-particle correlation function $g_\Lambda(y)_n$ where

$$\rho_\Lambda(y)_n = \rho_\Lambda(y_1)\ldots\rho_\Lambda(y_n)g_\Lambda(y)_n \tag{68}$$

By taking advantage of the symmetry of the functions $K(x_1;(y)_n)$ and $g_\Lambda(y)_n$ with respect to interchange of the $(y)_n$, (62) may be written as a homogeneous integral equation; in particular, if we introduce the transformation

$$\rho_\Lambda(x_1)\to\tilde{\rho}_\Lambda(x_1)+\alpha$$

and define

$$\tilde{K}(x_1,(y)_n)=\sum_{m=n}^{\infty}\frac{1}{n!(m-n)!}\alpha^{m-n}\int_{\Lambda^{m-n}}K(x_1,(y)_m)g_\Lambda(y)_m\,dy_{n+1}\ldots dy_m$$

where here the shifting parameter α is a number which satisfies the equation

$$\frac{\alpha}{z}=1+\sum_{n=1}^{\infty}\frac{1}{n!}\alpha^n\int_{\Lambda^n}K(x_1,(y)_n)g_\Lambda(y)_n\,dy_1\ldots dy_n$$

then the first equation in the KS hierarchy can be recast as

$$\tilde{\rho}_\Lambda(x_1)=z\sum_{n=1}^{\infty}\frac{1}{n!}\int_{\Lambda_n}\tilde{K}(x_1,(y)_n)\tilde{\rho}_\Lambda(y_1)\ldots\tilde{\rho}_\Lambda(y_n)\,dy_1\ldots dy_n \tag{69}$$

It was pointed out in Ref. 66 that the structure of (69) is formally the same as an integral power series. Recall that an integral power term of order n relative to the function f is defined as

$$L(f)=\int K(s,(y)_n)f^{\gamma_0}(s)\,f^{\gamma_1}(y_1)\ldots f^{\gamma_n}(y_n)\,dy_1\ldots dy_n \tag{70}$$

if

$$\gamma_0+\gamma_1+\ldots+\gamma_n=n$$

where the $\gamma_0,\ldots,\gamma_n$ are nonnegative numbers; an integral power series is a summation from $n=1$ to infinity of terms of the type in (70). In operator notation, (69) can be written

$$\tilde{\rho}_\Lambda(x_1)=z\mathbf{A}\tilde{\rho}_\Lambda(x_1) \tag{71a}$$

where

$$\mathbf{A}\tilde{\rho}_\Lambda(x_1) = \sum_{n=1}^{\infty} \frac{1}{n!}\int_{\Lambda^n} \tilde{K}(x_1, (y)_n)\tilde{\rho}_\Lambda(y_1)\dots\tilde{\rho}_\Lambda(y_n)\,dy_1\dots dy_n \quad (71b)$$

Nonlinear operators having the structure of $\mathbf{A}$ were first studied by Lyapunov and later by Lichtenstein, Sobolev, and Vainberg, and in the literature such operators are referred to as Lichtenstein-Lyapunov integral power series operators.[68]

We now draw attention to the fact that although the *full system* of KS equations could be written as a *linear* operator equation, that is, an operator $\mathbf{\Pi K}$ was identified by Ruelle that mapped E_ξ into itself with norm

$$\|\mathbf{\Pi K}\|_\xi \leq e^{-2\beta B}\xi^{-1}\exp[\xi C(\beta)]$$

consideration of a *single* equation in the KS hierarchy (namely, the first) leads to a *nonlinear* operator equation with the operator $\mathbf{A}$ being of the Lichtenstein-Lyapunov type. As we shall see shortly, this same general feature characterizes the other previously mentioned hierarchies; for example, although the system of integral equations referred to as the Kirkwood coupling-parameter hierarchy forms an inhomogeneous system, linear in the m-particle distribution functions, the first equation in this hierarchy can be written *formally* as a nonlinear operator equation of the classic Hammerstein type. In any case, the operative word here is *formal*, since the resultant nonlinear problem can be specified precisely only if certain assumptions are made regarding the m-body correlation problem. For example, within the context of the KS hierarchy, an examination of the kernel $\tilde{K}(x_1, (y)_n)$ for the case $n=1$, namely,

$$\tilde{K}(x_1, (y)_1) = \sum_{m=1}^{\infty} \frac{1}{(m-1)!}\alpha^{m-1}\int_{\Lambda^{m-1}} K(x_1, (y)_m)g_\Lambda(y)_m\,dy_2\dots dy_m$$

reveals the essential role played by the $g_\Lambda(y)_m$, the m-body correlation functions. If the $g_\Lambda(y)_m$ were completely determined, then $\tilde{K}(x_1, (y)_1)$ would be a perfectly well-defined kernel, and existence and uniqueness properties of the nonlinear operator equation, (71), could be studied in a straightforward way. Unfortunately, the $g_\Lambda(y)_m$ are *not* known; to have the full set of $g_\Lambda(y)_m$ available from the outset would be tantamount to having solved the m-body problem for the system under study. In other words, as it stands, (71) is *not* a closed equation for the unknown function $\tilde{\rho}_\Lambda(x_1)$; we have simply swept our ignorance of the m-body problem into the kernel $\tilde{K}$. Therefore to make any progress whatever in analyzing this nonlinear equation (or in fact any *finite* number of such equations) of the KS hierarchy, one must introduce some assumptions regarding the m-body correlation problem.

The problem identified in the preceding paragraph is common to all studies based on the BBGKY hierarchy and is handled in various ways depending on one's objectives, that is, whether one is seeking numerical solutions to one of the equations of the hierarchy, or whether one is seeking analytic results on existence and uniqueness properties. An approximation (either explicit or implicit) common to most studies is the introduction of a *closure* to restrict the vector

$$\boldsymbol{\rho}_\Lambda = (\rho_\Lambda(x)_n)_{n\geq 1} \varepsilon E_\xi$$

to a finite number of elements. The most commonly used closure is the one suggested by Kirkwood,[69]

$$g_\Lambda(x_1, x_2, x_3) = g_\Lambda(x_1, x_2) g_\Lambda(x_2, x_3) g_\Lambda(x_1, x_3) \tag{72}$$

called the superposition approximation. This approximation, when used within the context of one of the hierarchies mentioned previously, effectively decouples the infinite system of integral or integrodifferential equations, with the consequence that one then has the option of studying the properties of a single, closed *nonlinear* equation for the distribution functions $\rho_\Lambda(x)_m$ rather than an infinite hierarchy of equations *linear* in the m-body distribution functions. Before this option can be exploited, however, one must assess whether the contraction of the m-body problem specified by the above (or some other) closure is sensible. It is well known that the closure specified by the superposition approximation is actually *exact* in a few simple cases, but this is by no means a general result; indeed, the use of the superposition approximation in different versions of the BBGKY hierarchy for the same problem is known to lead to different quantitative results in property calculations.[69] Accordingly, one must be sensitive to possible errors introduced into an analysis based on the superposition (or some other) approximation. To illustrate the kind of difficulty that can arise, suppose we examine the consequences of introducing the following closure in each equation of the KS hierarchy:

$$\rho(y)_m = 0 \qquad \forall n > m. \tag{73}$$

One obtains

$$\rho(x)_m = z\chi_\Lambda(x)_m \exp[-\beta W^1(x)_m] \cdot \left[\rho(x)_{m-1} + \int_\Lambda dy_1 K(x_1, y_1) \rho((x)'_{m-1}, y_1) \right]$$

Then, for the resultant hierarchy, the Banach spaces E_ξ and the various norms can be defined in the same manner as Ruelle, and the formal methods he introduced can also be followed to yield the following bound on $|z|$, namely,

$$|z| < e^{-2\beta B} \xi \frac{1}{1+\xi C(\beta)} \tag{74}$$

Incidentally, the approximate generic distribution functions generated via the closure, (73), can be related to corresponding, approximate "infinite volume correlation functions" provided the conditions specified by Ruelle in the previously quoted theorem are satisfied.[66] Now, if we embed ρ_Λ in the same Banach space as Ruelle, that is, the one for which

$$\xi = C(\beta)^{-1}$$

the bound, (74), obtained with the closure, (73), is

$$|z| < e^{-2\beta B} C(\beta)^{-1} 2^{-1}$$

whereas the largest bound that can be constructed for $|z|$ is

$$|z| < e^{-2\beta B} C(\beta)^{-1}$$

Hence it would appear that one consequence of introducing an approximate closure is that the bound that determines a range of analyticity of the low-density (gas-phase) correlation functions tends to move around. To specialize these remarks to a specific case, recall the results obtained for the case of a one-dimensional gas of hard rods, each of diameter d. The bound on $|z|$ determined via the Ruelle theorem[1,65] is

$$|z| \leq \frac{1}{2e} d^{-1}$$

whereas the maximum bound on $|z|$ determined via the approximate closure, (73), is

$$|z| \leq \tfrac{1}{2} d^{-1}$$

The last result is interesting inasmuch as Penrose[70] determined that the corresponding Mayer series

$$\beta p = \sum_{n=1}^{\infty} b_n z^n$$

for the hard-rod problem had a radius $\mathcal{R}$ of convergence

$$\frac{1}{2e} d^{-1} \leq \mathcal{R} \leq \tfrac{1}{2} d^{-1}$$

It can be speculated that the consequence of introducing the closure, (73), has been to shift the Ruelle bound (which corresponds to a lower bound on the radius $\mathcal{R}$ of convergence of the Mayer expansion) to a point that corresponds to the upper bound on $\mathcal{R}$ determined by Penrose for the hard-rod problem. Whether or not we take seriously this speculation, the point to be stressed is that the quantitative estimate of the radius $|z|$ differs from that of Ruelle, despite the fact that the closure investigated here would not appear to be that serious an approximation for the hard-rod problem, since the structure of the resultant kernel $K(x_1, (y)_n)$ is such that only *one* term in each of the equations of the full hierarchy is neglected using this closure. Yet the bound on $|z|$ does shift, indicating that the introduction of an approximate closure may lead to results that are in quantitative disagreement with those obtained in a rigorous analysis.

From the above considerations, two factors have emerged. First, to achieve a contracted representation of the n-body problem, a closure (or some other approximation) is necessary, and the results obtained via a poor choice of closure may be quantitatively in error. Second, although a major simplification would appear to have been realized in contracting the full hierarchy of linear equations to a single nonlinear equation for $\rho_\Lambda(x)_n$, it must be recognized that, from a technical point of view, nonlinear operator equations are much more difficult to analyze than linear ones, and formal results on existence and uniqueness may be much more difficult to obtain. At the same time, since nonlinear problems can exhibit multiple solutions, it is difficult to avoid the impression that their possible occurrence in a nonlinear integral equation obtained from the BBGKY hierarchy, for example (52) to (54), may be linked in some way with the phenomenon of phase transformations. Of course such a remark is only a speculation at this point; the sense in which the conjecture can be formulated precisely and then explored systematically within the framework of the BBGKY hierarchy is being studied in this chapter. Apart from its intrinsic appeal, however, there are two auxiliary reasons why an approach to the theory of phase transitions based on a nonlinear formulation of the hierarchy problem and implemented using the methods of nonlinear functional analysis would seem worth pursuing. The first reason has already been offered in the introduction to this chapter: no theorem approaching the rigor of Ruelle's result is available for the full Kirkwood or YBG hierarchies; indeed, even for the case of the KS hierarchy, the methods developed by Ruelle cannot be pressed into effective service beyond the dilute gas regime. The second reason is even more pragmatic and can best be introduced by recalling the formal expression for the equation of state of a system characterized by an intermolecular pair potential $\Phi(x_{12})$:

$$\beta\bar{p} = \frac{\bar{N}}{\Lambda} - \frac{\beta}{6\Lambda}\int_{\Lambda^2} x_{12}\Phi'(x_{12})\rho_\Lambda(x_1, x_2)\, d(x)_2$$

From this expression it is evident that once the pair distribution function $\rho_\Lambda(x_1, x_2)$ for a system is known, the pressure may be determined explicitly. In effect, then, although the notion of a distribution function becomes precise only within the context of the full n-body problem, detailed information about the spatial distribution of two particles is sufficient to generate the thermodynamics. In the last three decades, a rather extensive bank of data, both experimental and computer generated [the latter encompassing Monte Carlo and molecular dynamics studies and numerical solution of the Kirkwood and YBG equation for the pair correlation function using the closure, (72)], has been compiled on the structure of the function $g_\Lambda(x_1, x_2)$ in various regions of the (T, ρ)-plane.[69] From these data, one can monitor the apparent radial and angular correlations between particles; it is found (see, for example, Ref. 71 and Fig. 9) that the pair correlation function changes in a smooth and regular way in certain regimes of the thermodynamic parameter space, but as certain critical (T, ρ)-loci are crossed, rather more dramatic changes in structure can occur, and these changes have been interpreted as signaling the onset of a new phase. Given that the existence of correlation functions has been established rigorously only in the regime of low density (see Theorem VI), it has been suggested that one way of characterizing the occurrence of new phases in the intermediate and high-density regimes, is to try first to define the notion of a phase; this led to the description of the states of a physical system as positive linear functionals on a B^*-algebra and to the characterization of the group invariance properties of physical states.[1] An alternative approach to the problem, one that would be guided by the available structural information on the pair correlation function, would be to conjecture that a phase transition can be associated with the occurrence of multiple solutions to one of the nonlinear equations obtained from the BBGKY hierarchy. We postulate that if in a certain region of thermodynamic parameter space a unique solution (a basic solution) of the underlying nonlinear problem has been identified, and if, upon varying the thermodynamic parameters, the equation admits other real solutions, the new distribution functions may characterize a different phase of the system. Supposing that bifurcation can be demonstrated in a given representation of the statistical-mechanical problem, two separate questions must be addressed: (*1*) What can one say about the stability of the bifurcating solutions relative to the initial pure phase? and (*2*) Given that one does find an exchange of stability between a pure-phase solution and one of the bifurcating branches, what are the thermodynamic consequences of this behavior? These questions are important, since, as we have seen in our discussion of second-order transitions, bifurcation appears to be a necessary but not

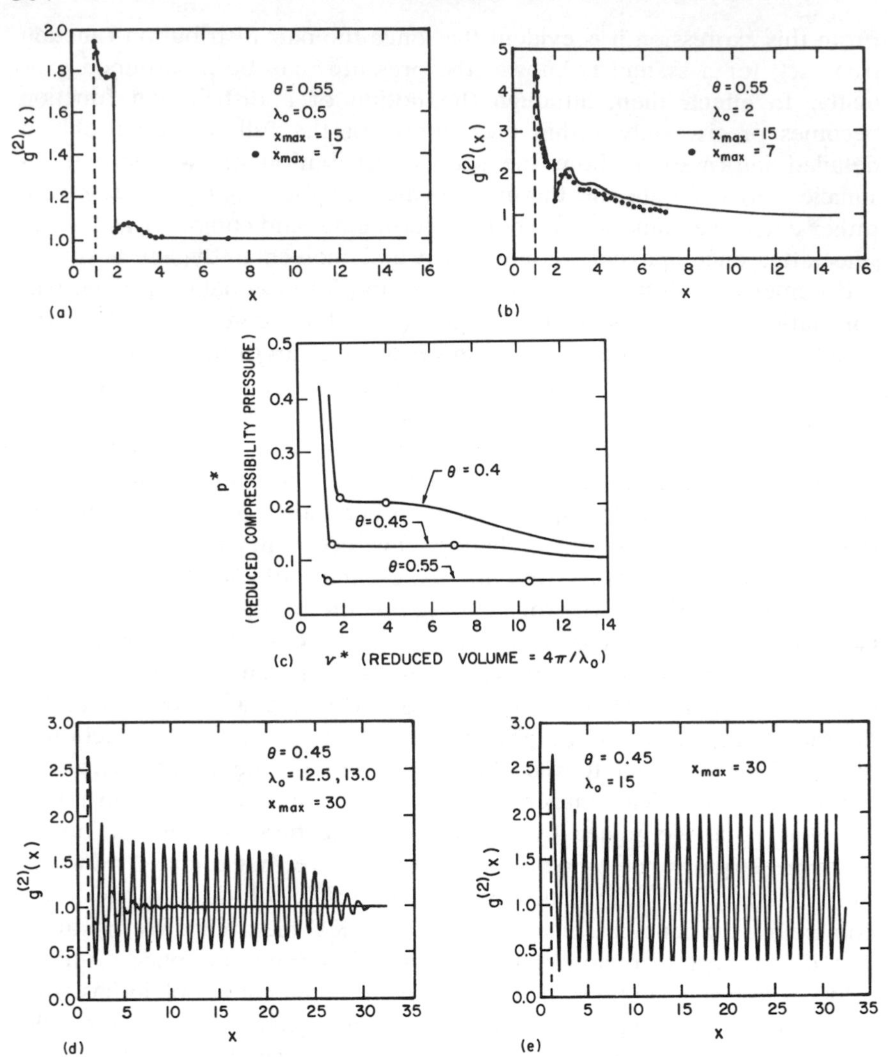

Fig. 9. The behavior of solutions to the Yvon-Born-Green equation 71 for the pair correlation function $g^{(2)}(x)$ for a system of molecules interacting via the reduced square-well potential $\gamma(x)$ (see Fig. 10), where

$$\gamma(x)=\begin{cases} +\infty & x \leqslant 1 \\ -1 & 1 < x \leqslant R \\ 0 & x > R \end{cases}$$

($x = r/\sigma_1$, $R = \sigma_2/\sigma_1$, $\sigma_2 > \sigma_1$). Here, (a) and (b) display the behavior found in the gas and coexistence regions, respectively. The plot (c) shows isotherms generated in the coexistence region. The *damped* solution in (d) is a typical liquid-like solution. The *undamped* sufficient condition for a phase transition (as evidenced by the occurrence of a singularity in an associated thermodynamic function).

Before implementing the grand strategy of using the apparatus of nonlinear functional analysis to attack one of the nonlinear integral or integrodifferential equations obtained from the BBGKY hierarchy, a number of qualifications (over and above those stated previously) must be faced squarely. First, as emphasized several times in this review, most of the available theorems on bifurcation deal with "soft" rather than "hard" bifurcation. Moreover, theorems such as Theorems I to V cited earlier are *local* theorems, that is, they describe only what happens *locally* in the immediate neighborhood of a bifurcation point. Global results on the bifurcation of nonlinear problems are few and far between (see the recent lecture notes by Nirenberg[72]).

Second, the available bifurcation theorems require that certain continuity conditions be satisfied. As seen later, this demands in turn that the intermolecular potential function be a "smooth" function (or be reasonably approximated by a smooth function).

A third qualification is much more serious. In the traditional applications of bifurcation theory, one seeks new solutions of a nonlinear operator equation of the form

$$\mathbf{B}u = \lambda u.$$

In the problem, the nonlinear operator $\mathbf{B}$ is assumed to be invariant, that is, one does not usually assume that $\mathbf{B}$ has one analytic structure below the (possible) bifurcation point and a different structure immediately above the bifurcation point. This simple remark is of great consequence as regards the underlying statistical-mechanical problem. Notice first that even after a closure has been specified in (71), one still must specify some *particular* correlation function to fix the kernel, else the problem is not well defined. Now suppose, for the sake of argument, that one need specify only the pair correlation function, say, $g_\Lambda^{(2)}(x_1, x_2)$ to fix the kernel appearing in (71). Suppose we choose $g_\Lambda^{(2)}(x_1, x_2)$ to be the correlation function descriptive of a single pure phase in a given regime of thermodynamic parameter space, a procedure that goes back to Kirkwood and Monroe[11] and used with great effectiveness by Brout[73] and Jancovici.[74] [In principle, a particular pure phase $g_\Lambda^{(2)}(x_1, x_2)$ can be determined independently from experimental data (electron, neutron, or x-ray diffraction studies[58]), from numerical simulations (e.g., Monte

---

solution in (d) is representative of solutions generated in the high-density, low-temperature region if the boundary condition at $x_{\max} = 30$ is set at $g^{(2)}(x) = 1$; if this constraint is replaced by a periodic boundary condition, the behavior displayed in (e) is realized.

Carlo) or from (exact or approximate) expressions obtained in other statistical-mechanical studies.] We know from our earlier discussion of the behavior of distribution functions at a phase transition that solutions of the hierarchy equations are *discontinuous* at a point of phase transition, with the actual distribution function descriptive of coexistence for a phase transition of order n given by (55) or (58), depending on whether the problem is posed in the canonical or grand canonical ensemble. In the latter formulation, for example, below the activity $z = z'$ one should use the correlation function $g_\Lambda^{(2)}(x_1, x_2; z'_-)$, whereas above $z = z'$ one should specify $g_\Lambda^{(2)}(x_1 x_2; z'_+)$. Hence, depending on whether one were fixing the pure phase correlation function above or below the transition activity $z = z'$, the Lichtenstein-Lyapunov nonlinear operator $\mathbf{A}$, specified by fixing the kernel in (71), would have a different structure above and below the transition point. This procedural problem complicates the relationship between bifurcation points and points of actual phase transition. First, if one were to carry out a bifurcation analysis for the transition, phase 1→phase 2, by specifying the correlation function for pure phase 1, and then were to invert the analysis (i.e., proceeding from a specification of the correlation function for pure phase 2 to consider the transition, phase 2→phase 1), there is no guarantee that the results of the two calculations would be the same; simply stated, if you change the analytic structure of the nonlinear operator in the problem, there is no reason to expect *a priori* that the bifurcation points in the spectra of the two nonlinear operators would necessarily coincide. One can speculate that if one were to use the *exact* $g_\Lambda^{(2)}(z'_-)$ for phase 1 in the first calculation and the *exact* $g_\Lambda^{(2)}(z'_+)$ for phase 2 in the second calculation, the bifurcation points for the two problems would coincide and correspond to a point of phase transition, but no rigorous results are available to support this conjecture. We return to this point in the next paragraph. A second difficulty is that if one decides to go ahead and consider the transition, phase 1→phase 2, by fixing the correlation function of phase 1 in the problem, the requirement that the nonlinear operator not change its structure as one crosses a bifurcation point implies that something like metastable correlation functions may have to be assumed (either explicitly or implicitly) in carrying through the analysis. Although metastable correlation functions can be obtained from Monte Carlo calculations, there are serious conceptual questions, noted earlier in this review, associated with the use of nonequilibrium concepts (namely, metastability) within the framework of an equilibrium theory of statistical mechanics, here a theory of correlation functions based on the ideas of the Gibbsian ensemble.

Notice that both of the problems identified in the preceding paragraph become less serious if the bifurcation point is *not* coincident with the actual point of phase transition, *provided* it is understood that a traditional search for bifurcation points must be complemented by an analysis of the stability of the bifurcating branches of solutions. For example, if bifurcation of the nonlinear problem were to take place in a pure-phase region, with the actual thermodynamic singularity occurring on a branch at a point somewhat removed from the bifurcation point (as was found in our study of the Ising model problem), then one would only have to require that the bifurcating branches "from below" and "from above" intersect at the same transition activity $z = z'$, and questions of metastability need not arise. Alternately, one might argue that both problems are finessed, given the short-ranged properties of the intermolecular potential function and the structure of the kernels appearing in the hierarchy equations, (53 to 55). For, in addition to the correlation function (to be specified for a pure phase of interest), there appears as well in the kernel of the nonlinear integral equation obtained from the first equation of each hierarchy either the intramolecular potential function *per se* (in equations derived from the Kirkwood coupling parameter hierarchy) or simple functions of the intermolecular potential function [in the KS hierarchy it is the Mayer f function, and in the YBG hierarchy it is $\Phi'(x_{12})$]. Bear in mind that the intermolecular potential $\Phi(x_{12})$ [also f or $\Phi'(x_{12})$] damps rapidly to zero with increasing interparticle separation. Notice also from the representative correlation functions plotted in Fig. 9 that the most pronounced difference among the gas, liquid, and "periodic" correlation functions is in the down-range behavior of $g_{\Lambda}^{(2)}(x_{12})$ [or, as is plotted, $g^{(2)}(x)$]. Therefore, taking into account the rapid truncation induced by the intermolecular potential function, it may be argued with respect to the down-range behavior that there will be essentially no difference in the *kernels* constructed using g_{gas}, g_{liquid} or g_{solid}, severally, as the pure phase correlation function. Moreover, in that range for which $\Phi(x_{12})$ is significantly different from zero, the fluid versus periodic correlation functions display reasonably similar short-range behavior; hence one anticipates that with respect to their short-range behavior, the various kernels will preserve a certain structural similarity as well. Thus it appears that if one takes into account the short- and long-range behavior of the (known) correlation function *and* the intermolecular potential function, the *composite* kernel has a structure that may be relatively unchanged as one crosses a phase boundary. This observation modulates somewhat the seriousness of the ambiguity noted in the preceding paragraph.

A fourth factor that enters in the implementation of a bifurcation approach to the theory of phase transitions is related to the fact that most studies based on a distribution function theory have focused on nonlinear integral equations for the singlet and/or pair distribution functions. Consider now the first components $\rho_\Lambda(x_1)$, $\rho(x_1)$ of $\boldsymbol{\rho}_\Lambda$, $\boldsymbol{\rho}$ in the exact representations, (64) (65), of the KS hierarchy. Because of the *translational invariance* of these equations, it is evident that $\rho(x_1)$ is a constant that depends analytically on the activity z in the circle

$$|z| < e^{-2\beta B - 1} C(\beta)^{-1}$$

so that fluids are characterized by a uniform density. A seemingly obvious distinction between a fluid (gas or liquid) and a solid is that the latter is characterized by a spatially periodic function. It is important to recognize, however, that the apparent spatial periodicity in $\rho(x_1)$ for a solid will be realized iff the translational symmetry of the problem is broken, that is, if the solid is localized as a whole by some external field (e.g., a gravitational field) and/or boundary conditions. To see this, consider a solid in a "box" with periodic boundary conditions; the statistical-mechanical ensemble then includes systems translated as a whole in *all* possible ways, and the singlet density function is uniform for a solid as well as for a fluid. Several authors have addressed this problem. Kirkwood and Boggs[75] show that a nonuniform (periodic) singlet density function can appear when proper account is taken of external forces (i.e., the walls) keeping the system in the volume Λ. Brout[76] has pointed out that the free energy of a restricted ensemble in which the first unit cell is centered at the origin differs from the unrestricted translationally invariant ensemble by thermodynamically negligible terms of $\mathcal{O}(\log N)$ only. We now emphasize that in most studies based on the examination of a nonlinear integral equation for the singlet distribution function, a nonuniform density approach is taken by assuming implicitly the existence of appropriate external forces (Ryzhov and Tareyeva[77] have taken the external field explicitly into account in a discussion of order-disorder transitions based on a mean-field approach). Given this constraint, Jancovici[74] has pointed out that in an *exact* theory the onset of the freezing transition is signaled in equivalent ways by the occurrence of a periodicity either in the singlet distribution function or in the pair distribution function in a translationally invariant ensemble; however, he stresses that if the underlying theory has *approximations*, the two points of view may predict different results.

A final conceptual problem that arises in a formulation based on the theory of distribution functions, a problem which must be addressed in adopting a bifurcation approach to study phase transitions, has to do with the so-called thermodynamic limit. This concept, by which one means that the number N of degrees of freedom of a system and the size of Λ of the system become infinite while their ratio $N/\Lambda = \rho$ remains finite, is fundamental to an understanding of the theory of phase transitions (and the theory of irreversibility). Van Hove[6] has proved that it is only in this limit that the intensive thermodynamic quantity $-A/NkT$ can be defined rigorously in terms of the canonical partition function. In the theory of distribution functions Ruelle has proved[1,65] (Theorem VI) that

$$|\rho_\Lambda(x)_n - \rho(x)_n| \leq \xi^n \varepsilon(\lambda)$$

where again $\rho(x)_n$ is an "infinite volume" distribution function and $\rho_\Lambda(x)_n$ is the corresponding finite volume distribution function. The absolute magnitude of the difference between these two functions is scaled by a product function of $\varepsilon(\lambda)$ and ξ, both defined in the statement of Theorem VI. One possible choice of ξ is $C(\beta)^{-1}$, where

$$C(\beta) = \int dx |e^{-\beta\Phi(x)} - 1| < +\infty$$

is a condition that must be satisfied if the pair potential Φ is to be regular; notice that if Φ is stable and short ranged this inequality is certainly satisfied. Then, for a stable, regular, short-ranged pair potential Φ, the condition

$$\lim_{\lambda \to \infty} \varepsilon(\lambda) = 0$$

(where λ is the minimum distance from the set of points $x_1, \ldots, x_n \in \Lambda$ to the boundary of the volume Λ) guarantees that the above difference becomes progressively smaller with increase in λ. We pause here to recall that in its present state of development, theorems on the bifurcation of nonlinear integral equations are proved for equations defined on a bounded closed set, that is, a finite domain. Therefore, in applying bifurcation theorems to one of the nonlinear integral equations of the hierarchy, we must argue that the domain of definition, although finite, is large enough that the (possible) occurence at particular points in the spectrum of the governing nonlinear operator of new finite volume distribution functions, reflects a changeover to new "infinite volume" correlation functions. To develop a rigorous treatment of phase transitions using a bifurcation approach within the framework of a distribution function theory, one must estimate the errors introduced in this approximation. To the author's knowledge no exact results are available here, but it is certainly clear that two characteristic lengths will play a crucial role in constructing precise estimates. The first length is the range of intermolecular forces, and, as we have described, this length dictates that the kernels appearing in the various nonlinear equations are short ranged; in fact, as we shall see later, great practical advantage can be taken of this feature: the limits of integration at a certain stage in the analysis of the associated linear operator equation can be extended to infinity, thereby allowing the mobilization of the Fourier transform technique. The second important length is the correlation length; as noted in our earlier discussion of the compressibility, fluctuations become infinite in a region of coexistence, and correlations between distant particles become important. Hence we are led back once again to the importance of considering the *stability* of the basic solution to perturbations in opting for a bifurcation approach to the theory of phase transitions.

With the above qualifications in mind, we now go on to consider the formulation of the corresponding nonlinear problem for the Kirkwood and Yvon-Born-Green hierarchies and for approaches based on the direct correlation function.

2. *The Kirkwood Coupling-Parameter Hierarchy*

The lowest member of the Kirkwood coupling-parameter hierarchy, (53), in the canonical ensemble is

$$\ln \rho^{(1)}(\mathbf{r}_1, \xi) = \ln z(\xi) - \beta \int_0^{\xi} d\eta \int_{\Lambda} d\mathbf{r}_2 \Phi(r_{12}) g^{(2)}(\mathbf{r}_1, \mathbf{r}_2, \eta) \rho^{(1)}(\mathbf{r}_2, \eta) \quad (75)$$

where

$$\ln z(\xi) = \ln \rho + \frac{\beta}{N} \int_0^{\xi} d\eta \int_{\Lambda} d\mathbf{r}_1\, d\mathbf{r}_2 \Phi(r_{12}) g^{(2)}(\mathbf{r}_1, \mathbf{r}_2, \eta) \rho^{(1)}(\mathbf{r}_1, \eta) \rho^{(2)}(\mathbf{r}_2, \eta) \quad (76)$$

Here $\rho^{(1)}(\mathbf{r}, \xi)$ is the singlet density at $\mathbf{r}$ when the molecule at $\mathbf{r}_1$ is partially coupled to a degree ξ, and $g^{(2)}$ is the pair correlation function; when the coupling parameter ξ is unity, (76) gives the standard definition of the activity z. We follow the development of Weeks et al.[78] and proceed by defining a new function

$$\Psi(\mathbf{r}_1, \xi) = \ln \left[\frac{\rho^{(1)}(\mathbf{r}_1, \xi)}{z(\xi)} \right]$$

so that (75) may be rewritten as

$$\Psi(\mathbf{r}_1, \xi) = -\beta \int_0^{\xi} d\eta\, z(\eta) \int_{\Lambda} d\mathbf{r}_2 \Phi(r_{12}) g^{(2)}(\mathbf{r}_1, \mathbf{r}_2, \eta) \exp [\Psi(\mathbf{r}_2, \eta)] \quad (77)$$

To obtain a solution $\Psi(\mathbf{r}_1, \xi)$ of (77) we must know the pair correlation function $g^{(2)}(\mathbf{r}_1, \mathbf{r}_2, \eta)$ which in turn requires the solution of the next equation of the hierarchy; in other words, as it stands, (77) is *not* a closed integral equation. Thus the closure problem encountered earlier in dealing with the first equation in the KS hierarchy also presents itself here in a formulation based on the first equation in the Kirkwood coupling-parameter hierarchy. In fact, not only are the various other difficulties described in the preceding subsection encountered, but in addition, (77) as it stands is difficult to treat in any practical way because the pair correlation function is not known for partial couplings. To make some headway, an approximation first suggested by Kirkwood and Monroe[11] and used later by Jancovici[74] may be introduced: the effects of partial coupling of the molecule $\mathbf{r}_1$ on the singlet distribution function $\rho^{(1)}(\mathbf{r}_1, \eta)$ at $\mathbf{r}_2$ in the integrand of (77) are neglected. The justification for this approximation is that, if the system size is large, the effects of partial coupling should be small, and in the thermodynamic limit these effects should be strictly negligible. Under this rubric, one replaces $\rho^{(1)}(\mathbf{r}_2, \eta)$ in (77) by $\rho^{(1)}(\mathbf{r}_2)$; then (77) can be written in the simpler form:

$$\psi(\mathbf{r}_1) + \beta z \int_\Lambda \Phi(r_{12}) \int_0^1 d\xi g^{(2)}(\mathbf{r}_1, \mathbf{r}_2, \xi) \exp[\psi(\mathbf{r}_2)]\, d\mathbf{r}_2 = 0 \tag{78a}$$

with

$$\psi(\mathbf{r}_1) = \ln\left[\frac{\rho^{(1)}(\mathbf{r}_1)}{z}\right] \tag{78b}$$

If we choose a fluid pair correlation function

$$g^{(2)}(\mathbf{r}_1, \mathbf{r}_2, \xi) = g^{(2)}(r_{12}, \xi)$$

to specify the kernel in (78) and adopt the approximation of Kirkwood and Monroe, and Jancovici (i.e., neglect of partial coupling) the kernel in (78) symmetrizes and simplifies to

$$K_{KJ}(r_{12}) = \beta z \Phi(r_{12}) \int_0^1 d\xi g^{(2)}(r_{12}, \xi) \tag{79}$$

An equation of the general structure of (78) has already been encountered in our discussion of mean-field theory and, in particular, the work of Lemberg and Rice.[42] Equation 78 is of the classic Hammerstein form of nonlinear integral equation, (37), with the nonlinearity again an exponential one. In fact, a common feature of several different, approximate theories of the singlet and/or pair distribution function is that the underlying mathematical problem can be collapsed to a Hammerstein equation of the form

$$\psi(\mathbf{r}_1) + \int_\Lambda K(\mathbf{r}_1, \mathbf{r}_2) \exp[\psi(\mathbf{r}_2)]\, d\mathbf{r}_2 = 0, \tag{80}$$

that is, an integral equation with an exponential nonlinearity. Tyablikov[13] seems to have been the first to state explicitly that the singlet density equation in the Bogolubov version of the BBGKY hierarchy is of the Hammerstein form, (80). As noted above, the singlet equation, (78), obtained from the Kirkwood coupling-parameter hierarchy, is of Hammerstein form. It is shown in our discussion of the YBG hierarchy that the nonlinear integral equation for the singlet (and pair) distribution function can be cast into the Hammerstein form, (80), or, following Brout,[73] one can start with an approximate mean-field representation of the singlet density function and derive an equation of the form of (80), but with the kernel given by

$$K_B(r_{12}) = \beta z \Phi^{(1)}(r_{12}) g_0^{(2)}(r_{12}) \tag{81}$$

where $g_0^{(2)}(r_{12})$ is the correlation function of a reference fluid (say a hard-sphere fluid) and $\Phi^{(1)}(r_{12})$ is a perturbation potential that carries the reference system into the system of interest. Or consider the representation of the grand partition function Ξ developed by Morita and Hiroike[79]:

$$\begin{aligned}
\ln \Xi = {} & \int d\mathbf{r}\rho(\mathbf{r}) \ln z^*(\mathbf{r}) - \int d\mathbf{r}\rho(\mathbf{r}) [\ln \rho(\mathbf{r}) - 1] \\
& + \tfrac{1}{2}\int\int d\mathbf{r}_1\, d\mathbf{r}_2 \rho(\mathbf{r}_1)\rho(\mathbf{r}_2)[1 + v(\mathbf{r}_1, \mathbf{r}_2)] \ln [1 + b(\mathbf{r}_1, \mathbf{r}_2)] \\
& - \tfrac{1}{2}\int\int d\mathbf{r}_1\, d\mathbf{r}_2 \rho(\mathbf{r}_1)\rho(\mathbf{r}_2)\{[1 + v(\mathbf{r}_1, \mathbf{r}_2)] \\
& \times \ln [1 + v(\mathbf{r}_1, \mathbf{r}_2)] - v(\mathbf{r}_1, \mathbf{r}_2)\} \\
& \pm \ldots
\end{aligned} \tag{82}$$

where

$$\begin{aligned}
z^*(\mathbf{r}) &= z \exp [-\beta \phi(\mathbf{r})] \\
b(\mathbf{r}_1, \mathbf{r}_2) &= \exp [-\beta \Phi(\mathbf{r}_1, \mathbf{r}_2)] - 1 \\
v(\mathbf{r}_1, \mathbf{r}_2) &= g(\mathbf{r}_1, \mathbf{r}_2) - 1
\end{aligned}$$

Here $\phi(\mathbf{r})$ is the external field (set to zero in this discussion), and all other quantities have their usual interpretation. Provided all terms in the expression for $\ln \Xi$ are taken into account, this expression is an exact representation of $\ln \Xi$. If one uses the variational condition

$$\left[\frac{\delta \ln \Xi}{\delta \rho(\mathbf{r})}\right]_{z*,b,v} = 0$$

in conjunction with (82), one obtains the following expression:

$$\begin{aligned}
\ln z^*(\mathbf{r}_1) = \ln \rho(\mathbf{r}_1) - \int d\mathbf{r}_2 \rho(\mathbf{r}_2)\{&[1 + v(\mathbf{r}_1, \mathbf{r}_2)]\ln [1 + b(\mathbf{r}_1, \mathbf{r}_2)] \\
&- [1 + v(\mathbf{r}_1, \mathbf{r}_2)]\ln [1 + v(\mathbf{r}_1, \mathbf{r}_2)] + v(\mathbf{r}_1, \mathbf{r}_1)\} \pm \ldots
\end{aligned} \tag{83}$$

As it stands, (83) represents an exact integral equation for $\rho(\mathbf{r})$ if $v(\mathbf{r}_1, \mathbf{r}_2)$ is known. In practice, one neglects all terms beyond the second on the right-hand side of (83), thus forming the hypernetted chain approximation. Cheng and Kozak[80] have pointed out that within the framework of the hypernetted chain approximation, the integral equation, (83), can be cast into the Hammerstein form, (80), with the kernel

$$K_{HNC}(\mathbf{r}_1, \mathbf{r}_2) = \beta z \Phi(r_{12}) g^{(2)}(\mathbf{r}_1, \mathbf{r}_2) + z\{[1+v(\mathbf{r}_1, \mathbf{r}_2)] \ln [1+v(\mathbf{r}_1, \mathbf{r}_2)] - v(\mathbf{r}_1, \mathbf{r}_2)\} \quad (84)$$

with $\psi(\mathbf{r}_1)$ given by (78b). Finally, in the "indirect coupling" theory developed by Weeks et al.[81] one can obtain an integral equation of the form (80) with a kernel

$$K_{IC}(r_{12}) = -zf(r_{12}) Y(r_{12}) \quad (85)$$

where $f(r_{12})$ is the Mayer f function, and

$$Y(r_{12}) = g^{(2)}(r_{12}) \exp [\beta \Phi(r_{12})]$$

Although in this review we are concerned only with classical theories, it should be noted that Girardeau[82] has developed a quantum mechanical theory of the one-particle distribution function based on the Peierls variational principle, employing an operator formalism that transforms annihilation and creation operators into linear combinations of one another. It turns out that the nonlinear integral equation for the singlet distribution function that results from the Girardeau development can be cast into the form of a Hammerstein equation. This nonlinear equation was analyzed by Gartenhaus and Stranahan,[83] and Weeks et al.[78] state that it was this paper which stimulated their study of the classical Kirkwood singlet equation, (75).

Given the variety of theories for the singlet distribution function that can be cast into Hammerstein form and given the availability of existence, uniqueness, and bifurcation theorems for the Hammerstein equation, it is not surprising that some of the earliest and most detailed results obtained using a bifurcation approach to the theory of phase transitions have emerged from studies based on (80). Let us examine the structure of (80) in more detail. As illustrated above, the feature common to all the approximations cited is that the structure of the nonlinear function appearing in the associated nonlinear integral equation remains unchanged. That is, although the kernel changes from one approximate theory to the next [recall the representations (79), (81), (84), and (85)], the nonlinearity remains exponential in every case. Now it is known that very different requirements must be satisfied to prove (existence, uniqueness, bifurcation) theorems for nonlinear equations with exponential, as opposed to power, nonlinearities. As noted by Krasnosel'skii,[46] two approaches may be taken:

1. Given a class of nonlinear equations, associated nonlinear operators **B**, and a definite function space E, one can try to find the conditions under which the operator of the given class will act in E and possess a required property P.
2. Given a class of nonlinear operators **B**, one can try to construct such function spaces in which the operators **B** act and possess required properties P.

With respect to the first approach, theorems can be proved on existence, uniqueness, and bifurcation properties of (80) when that equation is set in a "restricted" Banach space E such as L^p space, *provided* one imposes sufficiently strong conditions on the kernel and/or nonlinear function appearing in (80). This is the approach taken by Lemberg and Rice[42] in their study of (39); the methods used by these authors were introduced in the paper by Weeks et al.[78] cited earlier in this subsection. With respect to the second approach, the natural framework for discussing equations with exponential nonlinearities is that of Orlicz space; the theory of nonlinear operators acting in this Banach space has been developed by Krasnosel'skii and Rutickii.[48] Theorems on existence, uniqueness, and bifurcation are much more difficult to establish in this "broader" Banach space, and fewer formal results are available. Cheng and Kozak[80] have examined the complete continuity of the various nonlinear operators that are formed when (80) is used in conjunction with one of the kernels (79), (81), (84), and (85), using theorems developed by Krasnosel'skii and Rutickii,[48] and have obtained for a certain class of potentials a general theorem on the existence of at least one solution to (80) for the singlet distribution function. They also display explicitly the conditions that must be satisfied in order that a given solution of (80) for the singlet distribution function be unique. Although these results on the existence and uniqueness properties of the singlet density function are among the most general that have been obtained for the problem formulated as (80), the proofs are not reproduced here, since a review of the theory of convex (N) functions would be a prerequisite for such a discussion; the reader interested in the technical discussion may consult Ref. 80.

3. *The Yvon-Born-Green Hierarchy*

The first equation in the YBG hierarchy in the canonical ensemble has the structure

$$-\frac{\partial}{\partial \mathbf{r}_1} \ln \rho^{(1)}(\mathbf{r}_1) = \beta \int_\Lambda \frac{\partial \Phi(\mathbf{r}_{12})}{\partial \mathbf{r}_1} \rho^{(1)}(\mathbf{r}_2) g^{(2)}(r_{12})\, d\mathbf{r}_2 \tag{86}$$

that is, the equation for the singlet distribution function is an integrodifferential equation. The theory of nonlinear equations of this type is much less developed than for nonlinear integral equations *per se*,[84] and, although Kirkwood earlier had derived an instability criterion for the melting transition based on equations of the YBG hierarchy, it was thought[78] that analytic studies based on (86) or higher-order equations in the YBG hierarchy would encounter discouraging technical difficulties from the very outset. However, in relatively rapid succession, three reports appeared which showed how YBG equations for the singlet or pair distribution function could be handled within the framework of a bifurcation study. For convenience, we review these contributions in chronological order. First, Hiroike[85] (1970) pointed out that the lowest member of the hierarchy could be transformed into an equation derived by H. S. Green, which was of Hammerstein form.[86] Notice that the right-hand side of (86) can be written as

$$-\beta\int_\Lambda d\mathbf{r}_2\rho^{(1)}(\mathbf{r}_2)\frac{\partial}{\partial\mathbf{r}_1}\int_{r_{12}}^{\infty} dr\Phi'(r)g^{(2)}(r)$$

The substitution of this form then permits integration of the resulting equation for $\rho^{(1)}(\mathbf{r}_1)$; the result is

$$\ln\{C\rho^{(1)}(\mathbf{r}_1)\}=\beta\int_\Lambda d\mathbf{r}_2\rho^{(1)}(\mathbf{r}_2)\int_{r_{12}}^{\infty} dr\Phi'(r)g^{(2)}(r)$$

where C is a constant. This result had been obtained earlier by H. S. Green,[87] and if we identify

$$\psi(\mathbf{r}_1)=\ln\{C\rho^{(1)}(\mathbf{r}_1)\}$$

$$K_{\text{YBG}}(r_{12})=-\frac{\beta}{C}\int_{r_{12}}^{\infty} dr\Phi'(r)g^{(2)}(r) \tag{87}$$

we recover the Hammerstein form, (80), for the case of a fluid kernel. Hence the above kernel may be added to the list itemized in the preceding subsection, and a discussion of the complete continuity of the kernel (87) and of the properties of the corresponding nonlinear operator acting in Orlicz space has been presented in Ref. 80.

The second contribution of interest is a study based on the YBG equation for the pair correlation function for a fluid

$$-kT\nabla_1\ln g^{(2)}(r_{12},\xi)=\xi\nabla_1\Phi(r_{12})$$
$$+\xi\rho\int_\Lambda \nabla_1\Phi(r_{13})g^{(2)}(r_{13},\xi)g^{(2)}(r_{23})\,d\mathbf{r}_3 \tag{88}$$

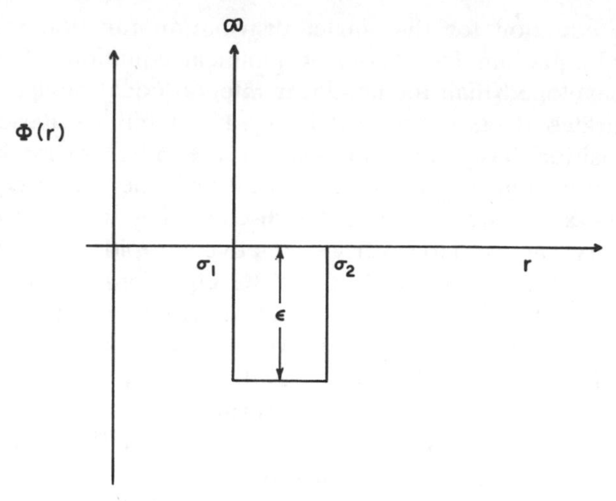

Fig. 10. The square-well potential, characterized by a strength parameter ε (the depth of the well) and range parameters σ_1 and σ_2 (the boundaries of the well).

by Lincoln et al.[71,88] For the case of the square-well potential (sketched in Fig. 10), characterized by range parameters σ_1 and σ_2 ($\sigma_1 < \sigma_2$) and the strength parameter ε, (88) can be written in the form

$$\ln g^{(2)}(x) = -\theta\gamma(x) + \frac{\lambda_0}{4x}\int_{-\infty}^{+\infty} K(x-s)s[g^{(2)}(s)-1]\,ds \tag{89}$$

where

$$K(t) = \theta\int_{|t|}^{\infty} d\omega(\omega^2 - t^2)g^{(2)}(\omega)\left[\frac{d\gamma(\omega)}{d\omega}\right]$$

$$\lambda_0 = 4\pi\rho\sigma_1^3$$

$$\theta = \frac{\varepsilon}{kT}$$

and where $\gamma(x)$ is the square-well potential in reduced units; specifically with $x = r/\sigma_1$ and $R = \sigma_2/\sigma_1$

$$\gamma(x) = \begin{cases} +\infty & x \leqslant 1 \\ -1 & 1 < x \leqslant R \\ 0 & x > R \end{cases}$$

Taking advantage of results obtained in an extensive numerical study of (89) over a wide region of thermodynamic parameter space (θ, λ_0), Lincoln et al.[88] show that simple operations allow (89) to be cast into a Hammerstein equation which then can be used to study the fluid-solid as well as the liquid-gas transition from the standpoint of bifurcation theory. This formulation is not reproduced here, since an extensive review of this and other aspects of the statistical mechanics of square-well fluids has been presented by Luks and Kozak[71] in a previous volume of this series. We do, however, refer to some of their conclusions in a later discussion.

Finally, we emphasize the studies of Raveché and Stuart[89–92] on the first equation of the YBG hierarchy. These authors proceed by taking advantage of an earlier result of Raveché and M. S. Green,[93] namely, if (86) is to have a solution, then the curl of the right-hand side of (86) must vanish. If one fixes the kernel by specifying a fluid correlation function $g^{(2)}(r_{12})$, then since

$$\nabla_1 \times [g^{(2)}(r_{12})\nabla_1\Phi(r_{12})]=0$$

there must exist a scalar χ such that

$$g^{(2)}(r_{12})\nabla_1\Phi(r_{12})=\nabla_1\chi(r_{12}) \tag{90}$$

(the dependence of χ on the temperature and density is suppressed). The use of the result (90) in the equation for the singlet distribution function leads to the result

$$\nabla_1 \ln g^{(1)}(\mathbf{r}_1)=-\rho\beta\nabla_1\int d\mathbf{r}_2 g^{(1)}(\mathbf{r}_2)\chi(r_{12}) \tag{91}$$

Notice that $\int d\mathbf{r}_2 g^{(1)}(\mathbf{r}_2)\chi(r_{12})$ is independent of $\mathbf{r}_1$, so clearly $g^{(1)}(\mathbf{r}_1)=1$ is a solution of (91). Raveché and Stuart associate this constant solution with the pure liquid phase and then define a new function

$$h^{(1)}(\mathbf{r}_1)=g^{(1)}(\mathbf{r}_1)-1$$

and show that this new function $h^{(1)}(\mathbf{r}_1)$ must satisfy the nonlinear equation

$$\nabla_1 \ln [h^{(1)}(\mathbf{r}_1)+1]=-\rho\beta\nabla_1\int d\mathbf{r}_2 h^{(1)}(\mathbf{r}_2)\chi(r_{12}) \tag{92a}$$

with the attendant normalization and condition

$$\lim_{\Lambda\to\infty}\frac{1}{|\Lambda|}\int_\Lambda d\mathbf{r}\, h^{(1)}(\mathbf{r})=0 \tag{92b}$$

the latter result following from the definition of the number density ρ in the grand canonical ensemble

$$\lim_{\Lambda\to\infty}\frac{1}{|\Lambda|}\int_\Lambda d\mathbf{r}\,\rho^{(1)}(\mathbf{r})=\rho$$

They then cast the vector equation (92a) into a form in which the transcendental nonlinearity can be handled more easily. In particular, if $h^{(1)}(\mathbf{r})$ is a solution of the pair of equations, (92), then it must satisfy

$$h^{(1)}(\mathbf{r}_1)+1=C(h)\exp\left[-\beta\rho\int d\mathbf{r}_2 h^{(1)}(\mathbf{r}_2)\chi(r_{12})\right]$$

where $C(h)$ is obtained by integrating this result over all space subject to the condition (92b), with the consequence that

$$1=C(h)\lim_{\Lambda\to\infty}\frac{1}{|\Lambda|}\int_\Lambda d\mathbf{r}_1\exp\left[-\beta\rho\int d\mathbf{r}_2 h^{(1)}(\mathbf{r}_2)\chi(r_{12})\right]$$

Raveché and Stuart note that the preceding pair of equations can be simplified further if one focuses on the hard-sphere problem. They show that the problem then reduces to the following nonlinear integral equation.

$$h(r)=-1+\frac{|\omega|\exp[-\mu(q)K(q)h(r)]}{\int_\omega \exp[-\mu(q)K(q)h(r)]\,dr} \tag{93a}$$

for $r\in\mathcal{R}^n$, where

$$K(q)h(r)=\left(\frac{q}{d}\right)^n\int_{|r'|\le\frac{d}{q}} h(r-r')\,dr' \tag{93b}$$

and where ω denotes the basic cell of some space lattice in $\mathcal{R}^n$. Here d is the diameter, q is the ratio of the characteristic length in the lattice to the diameter of the spheres ($q=1$ corresponds to close packing), and μ is regarded as a known function of q, related to the pressure in the system. Raveché and Stuart then seek solutions of (93a) that satisfy the condition

$$\int_\omega h(r)\,dr=0 \tag{93c}$$

and have all the symmetries of the lattice under consideration. A formal bifurcation analysis of the nonlinear equation (93a) is carried through after having established the compactness and self-adjointness of the linear operator **K** [defined by (93b)] acting in Hilbert space. The results of their study are reviewed in a later discussion in light of a recent exact result of Raveché and Stuart[94] on the uniqueness of solutions of the YBG hierarchy for the problem of a one-dimensional line of hard rods.

4. *The Direct Correlation Function*

In each of the formulations described thus far, we have seen that the kernel of the consequent nonlinear integral equation must be specified completely before analysis can proceed. In particular, for studies based on an equation for the singlet distribution function, knowledge of the pair distribution function was required; in effect, this specification closed the relationship between the nonlinear integral equation for $\rho(\mathbf{r}_1)$ and the underlying system of integral equations for the *same* hierarchy. However, one could imagine completing this specification in a given nonlinear equation for $\rho(\mathbf{r}_1)$ by using a pair correlation function determined from some *other* integral-equation theory of distribution functions. Jancovici[74] was perhaps the first to exploit this possibility; he suggested that a theory of freezing could be realized if an exact equation for $\rho(\mathbf{r}_1)$,

$$\rho(\mathbf{r}_1) = z \exp\left\{-\beta \int \Phi(\mathbf{r}_1 - \mathbf{r}_2)\left[\int_0^1 g(\mathbf{r}_1, \mathbf{r}_2; \xi)\, d\xi\right]\rho(\mathbf{r}_2)\, d\mathbf{r}_2 \right. \tag{94}$$

were coupled with an approximate equation for the pair correlation function and the two equations solved simultaneously. Given the formal relationship between the pair correlation function and the direct correlation function $c(\mathbf{r}_1, \mathbf{r}_2)$, namely,

$$g(\mathbf{r}_1, \mathbf{r}_2) - 1 = c(\mathbf{r}_1, \mathbf{r}_2) + \int [g(\mathbf{r}_1, \mathbf{r}_2) - 1]\rho(\mathbf{r}_3)c(\mathbf{r}_3, \mathbf{r}_2)\, d\mathbf{r}_3 \tag{95}$$

he explored the possibility of using expressions derived for the direct correlation function in Percus-Yevick theory[95]

$$c(\mathbf{r}_1, \mathbf{r}_2) = g(\mathbf{r}_1, \mathbf{r}_2)[1 - e^{\beta\Phi(\mathbf{r}_1 - \mathbf{r}_2)}]$$

and in the hypernetted chain theory

$$c(\mathbf{r}_1, \mathbf{r}_2) = g(\mathbf{r}_1, \mathbf{r}_2) - 1 - \log g(\mathbf{r}_1, \mathbf{r}_2) - \beta\Phi(\mathbf{r}_1 - \mathbf{r}_2)$$

(after the above expressions were generalized to the case of partial coupling) to specify an approximate pair distribution function through (95) and thence to close the nonlinear equation (94).

Within the context of an analytic approach to the theory of phase transitions, the above idea was mobilized by Weeks et al.[78] and Kozak et al.[86] These authors took advantage of the fact that in the indirect coupling (IC) approximation[81] [see (85)], the kernel of the resulting Hammerstein equation for the singlet distribution function is simply related to the Percus-Yevick approximation to the Ornstein-Zernicke direct correlation function. As described in detail later, this relationship could then be exploited to yield information on the possible bifurcation of solutions to (80) with the kernel (85) in the neighborhood of the liquid-solid transition and in the neighborhood of the critical point for the liquid-gas transition. These authors also suggested[86] that the results of Percus-Yevick theory could be used in a more schematic way to specify the kernel (79); they noticed that when the composite function $r_{12}^2\Phi(r_{12})g^{(2)}(r_{12})$ was plotted as a function of the interparticle separation using the PY pair correlation function for a hard-sphere fluid, the resulting profile was sensibly described by a monotonically decaying function; essentially the same result was found when a PY Lennard-Jones pair correlation function was considered. Upon replacing the full kernel (79) by a matching exponential function, they studied the branching *and* stability of solutions of the resulting nonlinear model equation using rules set down by Dean and Chambré[96] that systemize the results of the Vainberg-Trenogin generalization[23] of Lyapunov-Schmidt branching theory. Although the results of this study are interesting in that they dramatize in a simple model that different bifurcating branches may respond differently to fluctuations we turn now to a recent study of Lovett where this point is made in a much more convincing way.

Lovett's analysis[97] proceeds from a general relation between the singlet density and direct correlation function

$$\nabla_1[\ln \rho(\mathbf{r}_1) + \beta\phi(\mathbf{r}_1)] = \int d\mathbf{r}_2 c(\mathbf{r}_1, \mathbf{r}_2)\nabla_2\rho(\mathbf{r}_2) \tag{96}$$

derived by Moo et al.[98]; in this equation, $\phi(\mathbf{r}_1)$ is a weak external field. and $c(\mathbf{r}_1, \mathbf{r}_2)$ is the direct correlation function. In the absence of an external field, a constant $\rho^{(1)}(\mathbf{r})$ is always a solution to this equation, and the question is whether there are ever any additional (say, solid-like) solutions. In a fluid, $c(\mathbf{r}_1, \mathbf{r}_2) = c(r_{12})$, and the above equation may be simplified to read

$$\nabla_1 \ln \rho(\mathbf{r}_1) = \int d\mathbf{r}_2 c(r_{12})\nabla_2\rho(\mathbf{r}_2) = -\int d\mathbf{r}_2\rho(\mathbf{r}_2)\nabla_2 c(r_{12})$$

$$= \nabla_1 \int d\mathbf{r}_2 c(r_{12})\rho(\mathbf{r}_2)$$

which, upon integration, gives a nonlinear integral equation for $\rho(\mathbf{r})$ in terms of a (presumed known) direct correlation function,

$$\ln \rho(\mathbf{r}_1) = \int d\mathbf{r}_2 c(r_{12})\rho(\mathbf{r}_2) + C \tag{97}$$

where C is a constant of integration. Lovett then develops a bifurcation theory for the nonlinear integral equation (97), described later on in this review. However, Lovett goes much further: he complements the bifurcation analysis with a separate stability analysis, the latter similar in spirit to one introduced earlier by Brout[73] in his discussion of phase transitions in a mean-field model. In particular, Lovett examines the response of a fluid to a weak external field that has solid-like periodicity. The resulting mechanical stability criterion is then correlated with the results of the bifurcation analysis, and he finds that a new solution of the nonlinear equation (97) (which fixes the singlet density for a given direct correlation function) appears (bifurcates off the fluid solution) at the same state at which mechanical instability first appears. This striking result is important, not only because it suggests a simple physical interpretation of the bifurcation point, but also because understanding the response of a given state to a perturbation casts light on the role of fluctuations in determining the thermodynamically stable branch in a study of phase transitions based on a bifurcation approach.

C. Representative Results

In this section we review a number of representative results obtained in analyzing nonlinear problems in the statistical-mechanical theory of first-order phase transitions. In particular, we consider sequentially each of the nonlinear integral equations for the singlet (or pair) distribution function formulated in the preceding section. The main features of each analysis are identified and the principal conclusions noted. We hope to illustrate some of the technical problems involved in treating first-order phase transitions using a bifurcation approach and to provide a basis for comparison with the results obtained in the preceding chapter on second-order phase transitions.

A principal concern of this section is to examine whether the results obtained in an analysis carried out using bifurcation theory are consistent with known theorems and/or exact results in statistical mechanics. Since it is often the case that the essential features of a problem can be exposed via a carefully chosen counterexample, we pay particular attention in each of the following analyses to the model of a one-dimensional line of hard rods. Although physically uninteresting, this model is of great importance in statistical mechanics, since its properties are well understood. The equation of state of a one-dimensional line of hard rods (each of length d) in a box of size L, was determined by Tonks[99] to be

$$\frac{pdl}{NkT}=\frac{l}{l-1} \tag{98a}$$

where

$$l=\frac{L}{Nd} \tag{98b}$$

and Hauge and Hemmer[100] have shown that the zeros of the grand partition function $\Xi(\Lambda, z, \beta)$ of the problem fill the *negative* real axis from

$-1/e$ to $-\infty$. Therefore, either from the formal result of Tonks or via application of the Yang-Lee theorems,[7,63] there should exist *no* transition in a gas of hard rods. Accordingly, this case provides an acid text against which the results generated in a bifurcation analysis of an equation obtained from one of the hierarchy equations, (52) to (54), can be compared.

1. *Kirkwood-Salsburg Hierarchy*

We have seen that the first equation of the KS hierarchy under a particular closure is of the form of a Lichenstein-Lyapunov nonlinear operator equation. In accordance with the philosophy of a bifurcation approach, we seek to determine the (possible) bifurcation point(s) of this nonlinear operator equation for a particular class of potential functions, and then, should bifurcation occur, to determine whether this mathematical behavior has anything to do with the problem of phase transitions. In this subsection we deal specifically with the hard-rod gas, since many of the features of more complicated problems are already encountered in this simple case.

Before one can carry through a traditional bifurcation analysis of the nonlinear operator equation (71), and in particular before one can use Theorem V, it is important to establish first the conditions under which the operator $\mathbf{A}$ is completely continuous. For the case that $e^{-\beta\Phi}$ is assumed continuous on R^{ν}, a proof has been presented by Cheng and Kozak using Banach spaces of bounded continuous functions (see Ref. 66 for the technical details). For all intents and purposes, we may regard the hard-rod problem to be sensibly covered by this analysis, say, by considering the hard-rod potential to be a limiting case of the potential

$$\Phi(r)=\begin{cases} m^m & 0<r<d+1/m-1 \\ \dfrac{1}{(r-d+1)^m} & d+1/m-1\leq r \end{cases}$$

when $m\to\infty$; officially, of course, one should use the Banach space of bounded measurable functions for the hard-rod problem.[1]

An examination of the homogeneous integral equation (71) for $\tilde{\rho}_\Lambda(x_1)$ reveals that $\tilde{\rho}_\Lambda(x_1)=0$ is a solution of that equation; this trivial solution is referred to as the basic solution, and it corresponds to the choice $\rho_\Lambda(x_1)=\alpha$. Following the notation of Krasnosel'skii,[47] we identify the trivial solution with the null vector θ, so that in operator form (71) is written

$$z\mathbf{A}\theta=\theta \tag{99}$$

where z is understood to be a variable parameter (here it is just the activity). For small values of the parameter z, it is straightforward to show that the null solution is unique. However, it is of interest to see if one can go beyond this minimal result and ask whether for increasing values of the parameter z, starting at some z_0, a nonzero solution makes its appearance in the neighborhood of θ. Recall that μ_0 will be a bifurcation point of the nonlinear equation

$$\mu\mathbf{B}\theta = \theta$$

if, for every $\varepsilon > 0$, $\delta > 0$, there exists a characteristic value μ of the nonlinear operator $\mathbf{B}$ such that $|\mu - \mu_0| < \varepsilon$, and such that this characteristic value has at least one eigenfunction ϕ,

$$\mu\mathbf{B}\phi = \phi$$

with norm

$$\|\phi\| < \delta$$

Stated explicitly in terms of the (99), our hypothesis is that such z_0, if they exist, have something to do with the onset of a phase transition. To determine the possible existence of a bifurcation point(s) for the problem under study, we may use the results of Krasnosel'skii.[46] In particular, Krasnosel'skii shows that for the Lichenstein-Lyapunov operator $\mathbf{A}$, the Fréchet derivative $\mathbf{L}$ at the origin of the space $\mathscr{C}$ is the linear operator

$$\mathbf{L}\tilde{\rho}_\Lambda(x_1) = \int_\Lambda \tilde{K}(x_1; y_1)\tilde{\rho}_\Lambda(y_1)\, dy_1 \tag{100}$$

Therefore, according to Theorem V, the bifurcation points of the full nonlinear operator $\mathbf{A}$ are determined by the characteristic values of odd multiplicity of the linear equation

$$z_0 \int_\Lambda \tilde{K}(x_1; y_1)\phi(y_1)\, dy_1 = \phi(x_1) \tag{101}$$

In this last statement, we have a concrete equation from which the bifurcation points of the first equation in the KS hierarchy can be determined, at least in principle.

Our earlier discussion of the kernel $\tilde{K}$ stressed the crucial role played by $g_\Lambda(y)_n$. There we pointed out that before the bifurcation problem could be posed correctly, a closure had to be introduced and the pure phase correlation function(s) had to be specified. However, even after these problems have been addressed, there is a further difficulty that must be faced. In the distribution function theories of statistical mechanics, the ρ_Λ are functionals of the density; however, the activity z, which plays the role of a strength parameter in the nonlinear problem (99), *also* depends on the density. Therefore, in seeking possible characteristic values z_0 of the linear equation (101), we may consider only those activities consistent with a given specification of the density. Keeping this additional self-consistency condition in mind, we now investigate the possible existence of a bifurcation point for the hard-rod problem in two limiting cases. In the first case, we introduce a very naive approximation, since, in exploring the consequences, we are able to isolate some features of the overall problem. In the second calculation we do the problem exactly.

In the first calculation, our naive approximation consists in neglecting all many-body correlation effects, and in addition, suppressing the previously noted self-consistency condition. Mathematically, this amounts to setting $\forall g_\Lambda(x)_n = 1$. Since we impose no device to break the translational invariance of the problem here, a nontrivial solution of (99), should one exist, must be a constant (different from α). Simple manipulations then allow calculation of the bifurcation point z_0 as the nonzero characteristic value of (99); the result is

$$z_0 = e^{-1}\tilde{C}(\beta)$$

where

$$\tilde{C}(\beta) = \int (e^{-\beta\Phi(r)} - 1)\, dr$$

Thus, for the particular case of a one-dimensional gas of hard-rods, the bifurcation point calculated using the extreme closure introduced above is

$$z_0 = -\frac{1}{2e}\, d^{-1}$$

We encountered this number earlier; the absolute value of the right-hand side is the same bound on $|z|$ determined by Ruelle[1,65] in his exact analysis of the full KS hierarchy. Is this correspondence with Ruelle's result coincidental? Or, more to the point, what does the existence of a bifurcation point mean in a problem for which we know that *no* phase transition exists? First, Pastur[101] has noted that the spectrum of the operator **K** of (64) (looked on as a bounded linear operator with domain and range in the Banach space) is an eigenvalue spectrum consisting of the inverses of the complex values of z for which $\Xi(z) = 0$ (his proof has been generalized recently by Moraal[102]). Now, as noted previously, Hauge and Hemmer have shown that the zeros of $\Xi(z)$ are distributed along that part of the negative real axis stretching from $-1/e$ to $-\infty$. Hence the bifurcation point of the operator equation (99) calculated using the closure $\forall g_\Lambda(x)_n = 1$ would appear to be a point in the spectrum of the operator **K** and accordingly might be interpreted in light of the Ruelle theorem as representing a lower bound on the limit of stability of a pure phase. Given this interpretation, the existence of a bifurcation point in the present problem might then be regarded as a necessary but not sufficient condition for the onset of a phase transition. This interpretation of a bifurcation point is obviously in contrast to the interpretation given in such *deterministic* theories as hydrodynamics, chemical network theory, and elasticity; for example, if one achieves a critical load in the buckling problem, the rod bends. One wonders whether we are led to this relaxed interpretation of the bifurcation point in the present problem because of the underlying *probabilistic* nature of statistical mechanics, or if the interpretation has arisen because of the introduction of approximations in the calculation of the bifurcation point. Stated explicitly, what would be the result in a bifurcation analysis of the hard-rod problem *if* the exact closure were introduced, *if* the exact pure-phase correlation function were specified, and *if* self-consistency were preserved?

To answer this question, we recall the work of Salsburg et al.[103] These authors showed that if one labels the particles along the one dimension of the problem so that $y_1 < y_2 < \ldots < y_i < \ldots$, the superposition law

$$\frac{g^{(n+1)}(y_1, y_2, \ldots, y_{n+1})}{g^{(n)}(y_1, y_2, \ldots, y_n)} = \frac{g^{(2)}(y_n, y_{n+1})}{g^{(1)}(y_n)} \tag{102}$$

is *exact* for a one-dimensional system in which the interactions are restricted to nearest neighbors. Using this *exact* closure and the fact that the Mayer f functions satisfy the relations

$$f_{1\sigma} = \begin{cases} 0 & r_{1\sigma} > d \\ -1 & r_{1\sigma} < d \end{cases}$$

Salsburg et al. computed an *exact* expression for the pair correlation function

$$g^{(2)}(x) = l \sum_{k=1}^{\infty} A(x-k) \frac{1}{(l-1)^k} \frac{(x-k)^{k-1}}{(k-1)!} \exp\left[-\frac{(x-k)}{(l-1)}\right] \tag{103}$$

(x is a reduced interparticle separation distance and $A(x-k)$ is the step function) from which the thermodynamics of the system can be calculated exactly, the result being the Tonks equation of state. We now specify the $g^{(2)}(x)$ given by (103) as the pure-phase correlation function in our formulation. Notice that the functional dependence of $g^{(2)}(x)$ on the interparticle separation is parametrized by the density $\rho = 1/ld$. Corresponding to this one-parameter family of functions $g^{(2)}(x)$, there will exist a one-parameter family of kernels $\tilde{K}$ and in turn a *possible* one-parameter family of bifurcation points z_0. However, as stressed earlier, there also exists a functional relationship between the activity z and the density ρ; for the hard-rod problem Penrose[70] has specified this relationship as

$$z = \frac{p}{kT} \exp\left(\frac{dp}{kT}\right) = \frac{1}{d(l-1)} \exp\left(\frac{1}{(l-1)}\right) \tag{104}$$

Hence only those activities z_0 which are consistent with the particular choice of density $\rho = 1/ld$ specified in the prior identification of the kernel $\tilde{K}$ are acceptable solutions of the linearized equation. Upon examining the linear equation (101), taking into account the properties of the Mayer f function for the hard-rod gas together with the result (104), and recognizing that the shifting parameter α may be identified explicitly as $1/ld$, we determine that a nontrivial solution of the original nonlinear operator equation can arise when

$$z_0 = \{+2 + 2l[\exp(-\beta\mu^E) - 1]\}^{-1} d^{-1} \tag{105}$$

where the term involving the excess chemical potential μ^E is given (exactly)[103] by

$$\exp(-\beta\mu^E) = \frac{l-1}{l} \exp\left(\frac{1}{1-l}\right) \tag{106}$$

However, when the result (105) is analyzed, it is found that there is in fact *no* density $\rho = 1/ld$ for which self-consistency between the thermodynamic constraint (104) and the bifurcation result (105) can be realized. Hence we may conclude that there is no density at which new solutions of the nonlinear equation appear, and from our hypothesis we argue that there is *no* phase transition in a one-dimensional gas of hard rods predicted by the bifurcation analysis.

Let us try to isolate the factor (or factors) in the above analysis that led to the conclusion of *no* bifurcation in the hard-rod problem. From a qualitative point of view, the closure (102) or the choice (103) of $g^{(2)}(x)$ would not appear to be crucial factors, since their introduction led to the identification of a (possible) one-parameter family of bifurcation points. In retrospect, it was the self-consistency condition (104) that was crucial, and consequently it is important to understand how the particular relationship between the activity and density specified by (104) arises. First, the relationship between the activity z and the pressure p, as defined by (104), is a formal result; the explicit expression of z in terms of the density $\rho = 1/ld$ follows from the incorporation of the exact equation for the pressure of a hard-rod system, namely, (98). Now, as it happens, Salsburg et al.[103] studied the asymptotic behavior of the partition function for the hard-rod problem by using the method of steepest descent. The saddle point in this determination was located at a point c on the positive real axis. For a system of hard rods, one can show that

$$c = \frac{1}{d(l-1)}$$

and in fact Gürsey[104] had noted earlier that the constant c introduced by the method of steepest descent is identically equal to βp. In other words, the Tonks equation of state, used in specifying the self-consistency relationship for the hard-rod problem, gives that pressure which follows from a steepest-descent analysis of the partition function. Therefore, the fact that there is no density ρ for which self-consistency between the thermodynamic constraint (104) and the bifurcation result (105) can be realized, implies that the z_0 given by (105) do not correspond to points on the path of steepest descent in the asymptotic evaluation of the partition function. This conclusion is reminiscent of the one that emerged in our earlier discussion of the asymptotic behavior of the D-dimensional Ising model (Section III.B). There we found that, although bifurcation occurred in all dimensions, there was no phase transition for $D=1$, since in one dimension the point $(H, \lambda(H)) = (\frac{1}{2}, \frac{1}{4})$ did *not* lie on the saddle curve $(H, \lambda_s(H))$, and the "potential" singularity at $(\frac{1}{2}, \frac{1}{4})$ was never reached. In this problem, the one-dimensional line of hard rods, we assert that the points $(\rho, z_0(\rho))$ [the relation specified by (105)] do *not* lie on the saddle curve $(\rho, z_s(\rho))$ [the relation specified by (104)], and hence have no significance in the equilibrium thermodynamic description of the problem.

2. *The Kirkwood Coupling-Parameter Hierarchy*

The nonlinear equation for the singlet distribution function, as formulated within the framework of the Kirkwood coupling-parameter hierarchy and in several approximations, is of the classic Hammerstein form

$$\psi(\mathbf{r}_1) + \int_\Lambda K(\mathbf{r}_1, \mathbf{r}_2) F[\mathbf{r}_2, \psi(\mathbf{r}_2)]\, d\mathbf{r}_2 = 0 \tag{107}$$

with an exponential nonlinearity

$$F[\mathbf{r}, \psi(\mathbf{r})] = \exp[\psi(\mathbf{r})] \tag{108}$$

The discussion we give here, based on the paper by Weeks et al.,[78] applies equally well to any of the kernels (79), (81), and (85), but is not limited to these and can be considered a general discussion of equations for the singlet distribution function which can be cast into Hammerstein form [e.g., one with a kernel given by (84)]. In what follows we need to know only very general properties of the kernel, such as its symmetry in

$\mathbf{r}_1$ and $\mathbf{r}_2$, its dependence for fluids on $|\mathbf{r}_1-\mathbf{r}_2|$ only, and the short-ranged nature of the kernel for short-ranged potentials. Thus we are able to treat a wide class of equations within a single theoretical framework.

We show first how Dolph's theorems (Theorems III and IV) on the existence and uniqueness properties of the Hammerstein equation set in the space L^2 (i.e., Hilbert space) can be applied to obtain results on systems whose molecules interact via purely repulsive forces. We assume that a known fluid pair correlation function is used to specify the kernel; then the kernel will be symmetric and a function only of $|\mathbf{r}_1-\mathbf{r}_2|=r_{12}$. Notice that when $\Phi(r_{12})\geqslant 0$, then $K(r_{12})\geqslant 0$ and $F[\mathbf{r}, \psi(\mathbf{r})]\geqslant 0$, and hence any solution ψ^* to

$$\psi(\mathbf{r}_1)+\int_\Lambda K(r_{12})F[\mathbf{r}_2, \psi(\mathbf{r}_2)]d\mathbf{r}_2=0 \tag{109}$$

must be nonpositive. Therefore, ψ^* satisfies (109) iff it satisfies also

$$\psi^*(\mathbf{r}_1)+\int_\Lambda K(r_{12})F^*[\mathbf{r}_2, \psi(\mathbf{r}_2)]\, d\mathbf{r}_2=0 \tag{110}$$

where

$$F^*(\mathbf{r}, s)=\begin{cases} F(\mathbf{r}, s)=e^s & s\leqslant 0 \\ 1 & s>0 \end{cases} \tag{111}$$

This follows because for any solution ψ^* of (109),

$$F^*(\mathbf{r}, \psi^*)=F(\mathbf{r}, \psi^*).$$

Since for a fluid kernel the integral

$$\int_\Lambda K(r_{12})\, d\mathbf{r}_2$$

is essentially a constant and certainly bounded, and since (111) shows that $F^*\leqslant 1$ everywhere [$F^*(\mathbf{r}, s)$ is a bounded nonlinear function], we conclude that any solution to (110) must be bounded, that is, we identify a constant K_0 such that

$$|\psi^*(\mathbf{r}_1)|\leqslant\int_\Lambda K(r_{12})\, d\mathbf{r}_2<K_0 \tag{112}$$

With the above results at our disposal, we now examine (110) in light of Dolph's existence theorem. Theorem III requires that the integral operator $\mathscr{K}^*$ be completely continuous. A sufficient condition that this requirement be satisfied is that the kernel K^* be continuous. In turn, the kernel will be continuous if the potential function $\Phi(r_{12})$ is continuous. As noted in our earlier discussion of the KS hierarchy, such discontinuous cases as the hard-sphere potential can be represented with negligible error by continuous functions; thus we expect the main features of the following discussion to hold for this important special case as well.

The bounded nonlinear function F^* clearly satisfies the conditions of Theorem III if we choose

$$\lambda_N^* = \lambda_{-1}^*$$

the smallest negative eigenvalue (in absolute magnitude) of K^* and then

$$\lambda_{N+1}^* = \lambda_0^*$$

the first positive eigenvalue. We choose a constant $A > 1$ (a value greater than F^*) and

$$\mu_{-1} < 0, \ \mu_0 > 0$$

with $y_0 = \varepsilon > 0$ (where ε is a small constant). Under these constraints, the conditions of Theorem III are satisfied, and therefore (110) always has at least one solution.

We can go further and show that for small densities and/or high temperatures this solution is unique. Note that the function F^* satisfies the bounds

$$0 \leqslant \frac{[F^*(\mathbf{r}, s_2) - F^*(\mathbf{r}, s_1)]}{(s_2 - s_1)} \leqslant 1 \qquad (113)$$

Then, according to Theorem IV, the solution of the Hammerstein equation (in Dolph's representation) will be unique if $\lambda_0^* > 1$. Since the eigenvalues of K are the negative of the eigenvalues of K^*, we have the following uniqueness criterion: (110) has a unique solution when λ_{-1}, the lowest negative eigenvalue (in absolute magnitude) of K, obeys the condition

$$\lambda_{-1} < -1$$

that is,

$$|\lambda_{-1}| > 1$$

We have proved therefore that the solution of (110) is unique provided there are no eigenvalues of K in the range

$$-1 < \lambda_{-1} < 0 \qquad (114)$$

Now, as noted earlier, a purely repulsive potential implies that the kernel of the nonlinear equation (110) satisfies the relation $K(r_{12}) > 0$

everywhere. If this were sufficient to imply that the kernel has no negative eigenvalues, we could immediately prove the solution unique using the following theorem of Hammerstein:

Theorem VII (Hammerstein). Given that the kernel in the Hammerstein equation (37) is symmetric, positive definite, and sufficiently bounded and continuous so that the theory of linear integral equations is applicable to it, (37) will have at most one solution if for every fixed point y, $F[y, \psi(y)]$ is a monotonic nondecreasing function of ψ.

However, as is well known, kernels obeying the relation $K(r_{12}) \geq 0$ may still have negative eigenvalues, and hence the result (114) obtained via Theorem IV shows that the solution is always unique if the kernel has negative eigenvalues but none in the range of (114). Conversely, to find multiple solutions of (109) with a nonlinearity specified by (108), the kernel must have negative eigenvalues in the range of (114), otherwise the Hammerstein uniqueness theorem would apply.

To consider the case of systems whose molecules interact via repulsive *and* attractive forces, some modifications of the above procedure must be introduced. In this case, the kernel $K(r_{12})$ is not positive everywhere, and the bound (112) on solutions of (109) does not apply. Given the physical meaning of ψ (see (78b)), however, we may focus our attention on *bounded* solutions to (109); the same point of view was taken by Lemberg and Rice[42] (for the same mathematical reason) in their study of orientational phase transitions in crystalline methane (see Section III.C). We now proceed by defining a bounded function F^* as follows:

$$F^*(\mathbf{r}, s) = \begin{cases} e^{-M} & s < -M \\ F(\mathbf{r}, s) = e^s & |s| \leq M \\ e^{+M} & s > +M \end{cases} \tag{115}$$

for some positive constant M. Note that for a function ψ^* satisfying the bound $|\psi^*| \leq M$, we have

$$F^*(\mathbf{r}, \psi^*) = F(\mathbf{r}, \psi^*) \tag{116}$$

Then, F^* has an upper bound e^M, and, as before, Theorem III guarantees the existence of a solution ψ^* of the equation

$$\psi^*(\mathbf{r}_1) + \int_\Lambda K(r_{12}) F^*[\mathbf{r}_2, \psi^*(\mathbf{r}_2)]\, d\mathbf{r}_2 = 0 \tag{117}$$

This solution has the bounds

$$|\psi^*(\mathbf{r}_1)| \leq \int_\Lambda |K(r_{12})| F^*[\mathbf{r}_2, \psi^*(\mathbf{r}_2)]\, d\mathbf{r}_2$$

or

$$|\psi^*(\mathbf{r}_1)| \leq e^M K_0 \tag{118}$$

where

$$\int_\Lambda |K(r_{12})|\, d\mathbf{r}_2 < K_0$$

To show that the solution ψ^* also satisfies (109), we must demonstrate that (116) holds. This requires that $|\psi^*| \leq M$; but, from the bound (118) we see that (116) will be guaranteed if we have

$$e^M K_0 \leq M$$

or

$$K_0 \leq Me^{-M} \leq e^{-1} \tag{119}$$

where the last inequality follows from the specification $M=1$. Thus we conclude that when $K_0 \leq e^{-1}$, we have a solution to (109). Now recall the structures of the representative, approximate kernels, (79), (81), (85). Since $z \to \rho$ as $\rho \to 0$, we have $K(r_{12}) \to 0$ as $\rho \to 0$; further, $K(r_{12}) \to 0$ as $\beta \to 0$ (i.e., $T \to \infty$). Hence, for reasonably small ρ and/or β we are assured that (109) has a solution that is bounded, namely, $|\psi^*| < 1$. It is to be emphasized that this result (for a system characterized by repulsive *and* attractive interactions) is weaker than the one obtained earlier, since in the case of purely repulsive interactions the existence of a bounded solution could be proved for all K_0.

We now consider the conditions specified in Dolph's uniqueness theorem (Theorem IV) for the case of repulsive and attractive interactions. We construct the analogue of (113) for F^* defined as in (115); thus we write

$$0 \leq \frac{[F^*(\mathbf{r}, s_2) - F^*(\mathbf{r}, s_1)]}{s_2 - s_1} \leq e^M \tag{120}$$

For small (ρ, β), the first negative eigenvalue λ_{-1} in absolute magnitude is very large, and (119) and (120) are easily satisfied with $e^M < |\lambda_{-1}|$. For this regime, Theorem IV guarantees a unique solution ψ^* to (117), and thus ψ^* will be the only solution to (109) which is also bounded by (118). For small (β, ρ), the solution ψ^* is the *only* bounded solution that tends to zero as $\rho \to 0$, as expected physically. The main difference in the argument used in this case as compared to that used in the case of purely repulsive forces is that for repulsive potentials ($\Phi(r_{12}) > 0$), we could prove that (109) had a solution with $F = e^s$ iff (109) had a solution with F^* given by (111). In the present case, we lose the "only if" part of the result: (109) with $F = e^s$ may have other solutions, but at least they are not bounded by (118) and thus appear unphysical. [Note that the failure of the inequality (118) to hold does not imply that the equation (109) has no solutions.]

Having established the densities and temperatures for which the solution to the basic equation (109) is unique, we now take up the problem of bifurcation. We consider the problem of determining the (possible) bifurcation points of (109), and try to learn something about the character of the (possible) bifurcating branches. [Of course, the failure of uniqueness theorems (e.g., $\lambda_{-1} > -1$ for repulsive potentials) does not necessarily imply the existence of one or more new solutions.] To proceed, we follow an approach suggested by the work of Hammerstein.[43] Suppose that (109) has a given solution ψ_0 corresponding to a fluid phase. Suppose further that there exists a range of (ρ, β) for which there exists another solution ψ_1 to (109). That is, we assume

$$\psi_0(\mathbf{r}_1) + \int_\Lambda K(r_{12}) \exp[\psi_0(\mathbf{r}_2)]\, d\mathbf{r}_2 = 0 \tag{121a}$$

$$\psi_1(\mathbf{r}_1) + \int_\Lambda K(r_{12}) \exp[\psi_1(\mathbf{r}_2)]\, d\mathbf{r}_2 = 0 \tag{121b}$$

The difference in solutions

$$\psi_1(\mathbf{r}) - \psi_0(\mathbf{r}) \equiv \chi(\mathbf{r}) \tag{122}$$

must satisfy the nonlinear integral equation

$$\chi(\mathbf{r}_1) + \int_\Lambda K(r_{12}) \exp[\psi_0(\mathbf{r}_2)] \cdot \{\exp[\chi(\mathbf{r}_2) - 1]\}\, d\mathbf{r}_2 = 0 \tag{123}$$

Let us define a new kernel

$$H(r_{12}) \equiv K(r_{12}) \exp[\psi_0(\mathbf{r}_2)] \tag{124}$$

and a new nonlinear function

$$G(\mathbf{r}, s) \equiv e^s - 1 \tag{125}$$

The resulting Hammerstein equation

$$\chi(\mathbf{r}_1) + \int_\Lambda H(r_{12}) G[\mathbf{r}_2, \chi(\mathbf{r}_2)]\, d\mathbf{r}_2 = 0 \tag{126}$$

will always have the trivial solution $\chi(\mathbf{r})=0$ [namely, when $\psi_0(\mathbf{r})=\psi_1(\mathbf{r})$]; when the solution to the original nonlinear equation (109) is unique, this will be the only solution to (126). The existence of a *new* solution $\psi_1(\mathbf{r})$ will be signalled by the existence of a nonzero $\chi(\mathbf{r})$ satisfying (126).

We pause in this development to give a physical interpretation to the function $\chi(\mathbf{r})$. Let us assume that the fluid solution ψ_0 is known; from (78b) we have that

$$\psi_0(\mathbf{r}_1)=\ln\left[\frac{\rho_\ell(\mathbf{r}_1)}{z}\right]=\ln(\rho/z)$$

where ρ_ℓ is the density of the fluid. That is, ψ_0 is a constant, and the new kernel $H(r_{12})$ in (126) differs from the original kernel $K(r_{12})$ only by a constant factor; specifically, we have that

$$H(r_{12})=\left(\frac{\rho_\ell}{z}\right)K(r_{12}) \tag{127}$$

Given the explicit dependence of the approximate kernels, (79), (81), and (85), on the activity z, one notices that in constructing the kernel $H(r_{12})$ defined by (127), there is a cancellation of the factor z, and consequently the composite kernel $H(r_{12})$ does *not* depend explicitly on the activity. Moreover, in this case, the function $\chi(\mathbf{r})$ assumes the simpler form

$$\chi(\mathbf{r})=\ln\left[\frac{\rho^{(1)}(\mathbf{r})}{\rho_l}\right] \tag{128}$$

Thus $\chi(\mathbf{r})$ can be interpreted as the logarithm of the ratio of the new singlet distribution function to the (constant) density characteristic of the fluid phase. The new Hammerstein equation (126) therefore has no explicit dependence on the activity z, an interesting result in light of the remarks in the previous subsection on the role of the activity in the singlet equation obtained from the first equation in the KS hierarchy.

Let us motivate the bifurcation analysis with some physical arguments. We introduce the physical fact that in the freezing transition the change in density between liquid and solid is small. Thus, in considering (126), we are interested in solutions $\chi(\mathbf{r})$ which, from (128), are *small* in magnitude. For this case, then, it is natural to consider a Taylor series expansion of the nonlinear function $G(\mathbf{r}, s)$ about $s=0$, namely,

$$G(\mathbf{r}, s)=s+\frac{1}{2!}s^2+\frac{1}{3!}s^3+\ldots$$

For s sufficiently small, we consider only the first nonvanishing term in this representation and, returning to (126), we produce the following linear integral equation.

$$\chi(\mathbf{r}_1)=(-1)\int_{\Lambda} H(r_{12})\chi(\mathbf{r}_2)\,d\mathbf{r}_2 \tag{129}$$

For $\chi(\mathbf{r})$ sufficiently small in magnitude, we expect the solution to the linearized equation (129) to be a good approximation to the solution of the full nonlinear equation (126). Incidentally, since (129) is a homogeneous linear equation, it will have a solution *only* when -1 is an eigenvalue of the kernel $H(r_{12})$; this observation is consistent with the condition established earlier concerning the uniqueness of solutions of (109).

The above physical arguments lead to a criterion for the freezing transition: We consider the eigenvalues of $H(r_{12})$ as functions of (ρ, β). When first we have $\lambda=-1$, there exists the possibility of a phase transition. We anticipate that the new distribution function may be determined to good approximation by the solution to the linear integral equation (129). Of course, beyond this point the method of fixing the kernel described above breaks down, and the theory developed can no longer give reliable information on the solid singlet distribution function away from the liquid-solid coexistence curve; *near* the coexistence curve, however, we expect that solutions to the linear equation (129) should yield information on the structure of the solid $\rho^{(1)}(\mathbf{r})$.

The above conclusions have followed from a single physical premise, namely, that in the freezing transition the change in density between solid and liquid is small. Within the framework of bifurcation theory, this premise provides as well the rationale for regarding the fluid-solid transition as a "soft" bifurcation, and hence accessible to analysis via Theorems I, II, and V. We may now ask whether these formal theorems, when applied to (126), strengthen or challenge the principal conclusions set down in the preceding paragraph. First, it is clear that the existence of solutions to the linear operator equation obtained from the full nonlinear operator equation is a necessary but *not* sufficient condition for the existence of a bifurcation point of the nonlinear equation. In particular, the eigenvalues of the associated linear problem must be of *odd* multiplicity. This qualification is important since, as we now describe, it allows a reinterpretation of an instability criterion derived many years ago by Kirkwood[11,105,106] for the freezing transition. To amplify this remark, let us return to the linear equation (129). Since the kernel is short ranged, it is consistent with the above discussion to extend the limits of integration to infinity. (N.B. The use of the limit ∞ in this way does *not* correspond to constructing the thermodynamic limit.) Taking the Fourier transform of the resulting equation, it follows that

$$\tilde{\chi}(\mathbf{k})+\tilde{H}(\mathbf{k})\tilde{\chi}(\mathbf{k})=0$$

or

$$\tilde{\chi}(\mathbf{k})[1+\tilde{H}(\mathbf{k})]=0 \tag{130}$$

where

$$\tilde{\chi}(\mathbf{k})=\int_{-\infty}^{+\infty} \chi(\mathbf{r}) \exp(i\mathbf{k}\cdot\mathbf{r})\, d\mathbf{r} \tag{131a}$$

$$\tilde{H}(\mathbf{k})=\int_{-\infty}^{+\infty} H(\mathbf{r}) \exp(i\mathbf{k}\cdot\mathbf{r})\, dr \tag{131b}$$

Since we assumed a fluid kernel $H(|\mathbf{r}|)$ in deriving (129), the last result may be written

$$\tilde{H}(k)=\frac{4\pi}{k}\int_0^{\infty} H(r)r\sin(kr)\, dr \tag{132}$$

(Thus $\tilde{H}(k)$ depends only on the magnitude of $\mathbf{k}$.) From (130) we see that $\tilde{\chi}(\mathbf{k})$ is nonzero iff

$$1+\tilde{H}(\mathbf{k})=0 \tag{133a}$$

that is,

$$1+\frac{4\pi}{k}\int_0^{\infty} H(r)r\sin(kr)\, dr=0 \tag{133b}$$

If we specify $H(r)$ using one of the model kernels described previously, we can regard (133) as defining a curve in the (β, ρ) plane where the underlying nonlinear equation permits new solutions of small magnitude which signal crystallization.

The above development, based on the first equation in the Kirkwood coupling-parameter hierarchy, has much in common with Kirkwood's original derivation of an instability criteria for the melting transition, a derivation based on the first equation in the YBG hierarchy (in fact, as we shall see shortly, the criterion (133) collapses to one derived by Kirkwood and Boggs[105] for hard-sphere systems). However, by interpreting the insta bility criterion (133) in light of the fundamental theorems of bifurcation theory, we see that an equation similar to Kirkwood's represents a necessary but *not* sufficient condition for the existence of bifurcation points of the nonlinear equation (126); Theorem V states that it is only for eigenvalues of *odd* multiplicity that bifurcation is guaranteed. Moreover, Kirkwood's original derivation suffers because he considers perturbations in $g^{(2)}(r_{12})$ and $\rho^{(1)}(\mathbf{r}_1)$ as independent, a difficulty pointed out by Rice and Lekner[107] and later by Kunkin and Frisch.[108] In the present interpretation of (133), it is seen that at a bifurcation point one needs no perturbations at all in the fluid $g^{(2)}(r_{12})$ for there to be other solutions of the nonlinear equation (126). On the other hand, if one could characterize the response of a fluid to a periodic perturbation, it is possible that the point at which the fluid becomes unstable with respect to such a perturbation might be related to the bifurcation point. It is exactly this possibility that has been explored by Lovett.[97] As we describe later, he shows that a new solution of a nonlinear integral equation which fixes the singlet density for a given direct correlation function appears (bifurcates off the fluid solution) at the *same* state at which *mechanical* instability first appears.

Let us specialize the result (133) to the particular case of hard spheres. When one is outside the hard-sphere radius a, that is, when $r>a$, we have $H(r)=0$. Inside the interval $0<r<a$, $H(r)$ is nonvanishing, and here Kirkwood makes the *further* assumption that $H(r)$ can be replaced by its average value H_0. Then (133) becomes

$$1+\left(\frac{4\pi}{k}\right)H_0\int_0^a r\sin(kr)\,dr=0$$

or

$$1-\frac{\lambda_K}{\tilde{z}^3}(\tilde{z}\cos\tilde{z}-\sin\tilde{z})=0 \tag{134}$$

where $\tilde{z}=ka$ and $\lambda_K=4\pi a^3/H_0$. For hard spheres, this equation is the same as the stability criterion given by Kirkwood and Boggs.[105] Later Kirkwood,[106] using this equation, predicted a phase transition for a system of hard spheres. Unfortunately, as Kunkin and Frisch[108] have pointed out, the Kirkwood criterion also predicts a phase transition in a one-dimensional line of hard rods, a system which, as we know, does *not* exhibit a phase transition. Hence we have uncovered an apparent contradiction with a known statistical-mechanical result and, as regards the present discussion, one that must be addressed before proceeding further in our discussion of phase transitions, analyzed as a problem in bifurcation theory. Weeks et al.[78] have pointed out that although a bifurcation approach leads to a criterion similar to that derived by Kirkwood, the specific result (134) was obtained by adopting Kirkwood's further assumptions on the structure of the kernel. They point out that a quite different result may be obtained in the hard-rod problem if different assumptions are made on the structure of the kernel. Consider, for example, the "indirect coupling" kernel, (85), for the hard-rod problem. From before, the criterion for a bifurcation point requires that

$$1+\tilde{H}(\mathbf{k})=0$$

where again $\tilde{H}(\mathbf{k})$ is the Fourier transform of $H(\mathbf{r})$ and where the latter is related to the kernel of the original problem via (127). Using the IC expression for $K(r)$, one obtains for one-dimensional hard rods the new requirement

$$1-\rho_\ell \int_{-\infty}^{+\infty} f(x) Y(x) \exp(ikx)\, dx = 0 \tag{135a}$$

or

$$1-2\rho_\ell \int_0^a Y(x) \cos(kx)\, dx = 0 \tag{135b}$$

where the evenness of the functions $Y(x)$ and $f(x)$ and the fact that $f(x)=-1$ when $x<a$ (a, the hard-rod radius) have been used. To examine the consequences of the representation (135), we change first to reduced units $ak \to k^*$, $a\rho_\ell \to \rho_\ell^*$, $x/a \to x^*$, so that for $a=1$ we have

$$\tilde{H}(k^*) = 2\rho_l^* \int_0^1 Y(x^*) \cos(k^* x^*) dx^* \tag{136}$$

Here, ρ_ℓ^* varies between zero and unity (close packing). We now examine whether the equation $1+\tilde{H}(k^*)=0$ has real solutions. Meeron and Siegert[109] have given an *exact* expression for

$$Y(x^*) \equiv g^{(2)}(x^*) \exp[\beta\Phi(x^*)]$$

namely,

$$Y(x^*) = (1+q) \exp[q(1-x^*)] \qquad 0 \leqslant x^* \leqslant 2$$

where

$$q = \frac{\rho_\ell^*}{1-\rho_\ell^*}$$

Note that $\rho_\ell^*(1+q)=q$. It follows from (136) that

$$1+\tilde{H}(k^*) = 1+2qe^q \int_0^1 e^{-qx^*} \cos(k^* x^*)\, dx^*$$

or, on performing the integration,

$$1+\tilde{H}(k^*) = 1+2qe^q \left[\frac{e^{-q}(k^* \sin k^* - q \cos k^*) + q}{q^2+(k^*)^2}\right] \tag{137}$$

As ρ_ℓ^* varies between zero and unity, q varies between zero and infinity. To show that (137) has no real zeros in k^* for any q, we simply note

$$\begin{aligned} 1+\tilde{H}(k^*) &> 1+2qe^q \left[\frac{e^{-q}(-k^*-q)+q}{q^2+(k^*)^2}\right] \\ 1+\tilde{H}(k^*) &> 1+2qe^q \left[\frac{-k^* e^{-q}}{q^2+(k^*)^2}\right] \\ 1+\tilde{H}(k^*) &> 1-\left[\frac{2qk^*}{q^2+(k^*)^2}\right] \end{aligned} \tag{138}$$

But, since $(q-k^*)^2 = q^2 - 2qk^* + (k^*)^2 \geqslant 0$, $1 \geqslant 2qk^*/(q^2+k^{*2})$. Then, from the result (138),

$$1+\tilde{H}(k^*)>0.$$

Therefore, *no* bifurcation points for the hard-rod problem are found if one adopts the "indirect coupling" kernel and then carries through the consequences of this choice *without* further approximation.

The above discussion shows that qualitatively different conclusions on the occurrence of a bifurcation point in the one-dimensional hard-rod problem are reached depending on the structure of the kernel assumed in the formulation of the problem. The same kind of sensitivity can be seen in the three-dimensional problem, a system of hard spheres. As noted previously, Kirkwood predicted a phase transition in such a system based on his analysis of (134). Suppose, however, we consider again the "indirect coupling" approximation for $H(r)$ given by (85) and (127). For the case of hard spheres $H_{\mathrm{IC}}=0$ for $r>a$ and within the interval $0<r<a$, $f(r)=-1$, so that the IC bifurcation equation is

$$1+\frac{4\pi\rho_\ell}{k}\int_0^a Y(r)r\sin(kr)\,dr=0 \tag{139}$$

If we approximate $Y(r)$ by some constant value Y_0 (the assumption of Kirkwood), we recover the Kirkwood criterion (134). If, on the other hand, we take the *exact* solution to the Percus-Yevick equation[110,111] for $Y(r)$ for hard spheres, the numerical computations show that (139) has *no* solutions. Given the philosophy of the approach, this result would be interpreted as meaning that there is *no* phase transition in a hard-sphere system. However, the PY solution for $Y(r)$ inside the core is apparently not as accurate as it is outside the core (as shown by calculating virial coefficients[112] and examing density expansions[113]); thus the prediction of no phase transition in a system of hard spheres based on the use of the IC kernel is not entirely convincing either. Again we see that approximations over and above those inherent in the use of a bifurcation approach can change *qualitative* predictions, namely, the prediction of a phase transition in a problem where it is known that one does not exist. We add that the present status of numerical computations on the hard-sphere system is such that the possible existence of the "Kirkwood transition" is still an open question. Wood,[114] in reviewing the status of the Monte Carlo calculations, states that, as to the existence of a first-order phase transition in a system of hard spheres, "there is yet little concrete evidence." In the two-dimensional case, Alder and Wainwright[115] found a "Van der Waals" loop for 870 hard discs, and Wood believes that if a transition exists in this case, "one would be inclined to expect a phase transition for three-dimensional hard cores." We return to this problem later in the review.

The final point we wish to address in this subsection concerns the structure of the (possible) new solutions that can arise in the neighborhood of a bifurcation point; the structure of these new solutions should be related to the structure of the distribution functions characterizing the new phase. As shown in (130) and (133), the emergence of nonzero solutions of (129) (and hence the existence of bifurcation points) occurs at the zeros of the function $1+\tilde{H}(\mathbf{k})$. For a rigorous proof of this result within the context of the general theory of integral equations see Morse and Feshbach;[116] these authors show that in one dimension the solutions are of the form

$$\phi(x)=\sum_n A_n \exp(ik_n x)$$

where the k_n are the zeros of the function $1+\tilde{H}(\mathbf{k})$ inside some strip of analyticity. It is important to note that only the real zeros give bounded solutions for all x. In three dimensions, a similar result holds:

$$\chi(\mathbf{r})=A\exp(i\mathbf{k}_0\cdot\mathbf{r})$$

where $\mathbf{k}_0$ is a solution to the equation $1+\tilde{H}(\mathbf{k})=0$ as specified in (133). Since (133) has solution, which depend on the absolute magnitude of $\mathbf{k}$ only, it was noted by Weeks et al.[78] that an approach based on (133) can *not* specify the orientation of the resulting lattice, although the periodicity (position of minima or maxima for a *given* orientation) is determined; this is just what one would expect for a volume in which no fixed and prefered axis of orientation has been specified. Complex zeros of (133) do *not* give solutions that are bounded everywhere and can be rejected on physical grounds. The only bounded solutions of (133) are of the form $A\exp(i\mathbf{k}\cdot\mathbf{r})$, where the magnitude of $\mathbf{k}$ is given by a solution to (133). Using the definition of $\chi(\mathbf{r})$, (128), we determine that

$$\rho^{(1)}(\mathbf{r}_1)=\rho_\ell \exp[A\exp(i\mathbf{k}\cdot\mathbf{r}_1)]$$

or, consistent with the linearization of (126) to (129), we have

$$\rho^{(1)}(\mathbf{r}_1)=\rho_\ell[1+A\exp(i\mathbf{k}\cdot\mathbf{r}_1)]$$

One cannot obtain the magnitude of the constant A from the linearized theory, since (129) is satisfied for any A. In order to have a more accurate description of the transition, including the change in volume on freezing, one must return to the full nonlinear equation (126).

3. *The Yvon-Born-Green Hierarchy*

In this section we focus on two different studies based on the Yvon-Born-Green hierarchy (54). First we discuss the recent contributions of

Raveché and Stuart,[89–92] based on the nonlinear problem formulated in (93). The principal results of these studies may be re-examined in light of a recently obtained exact result of these authors[94] on the uniqueness of solutions to the YBG equation for the problem of a one-dimensional line of hard rods. In fact, it is the latter study that is described in detail here, since their use of symmetry arguments in analyzing possible solutions to the hierarchy equations for the hard-rod problem is particularly clear and can serve therefore as a model *par excellence* for further studies. A second reason for focusing on their more recent study is that it provides a continuing focus on the hard-rod problem, which, as emphasized several times in this review, serves as a model against which the predictions of a bifurcation analysis can be assessed. The second study reviewed (briefly) in this section is that of Luks and Kozak[71] on the bifurcation analysis of the second equation in the YBG hierarchy for the square-well fluid; the latter is a simple, model system whose molecules interact via both attractive and repulsive forces (see Fig. 10 for the potential energy diagram). However, the main emphasis here is the work of Raveché and Stuart[94] to which we now turn.

Upon using the exact closure for the hard-rod problem, namely, (102) obtained earlier by Salsburg et al.,[103] Raveché and Stuart show that the full hierarchy of integrodifferential equations (54) can be reduced to the (coupled) pair of equations

$$\nabla_1 \ln g^{(1)}(r_1) = \rho \int dr_2 \frac{\tilde{g}^{(2)}(r_1, r_1)}{g^{(1)}(r_1)} \nabla_1 \exp[-\beta\Phi(r_{12})] \tag{140}$$

and

$$\nabla_1 \ln \tilde{g}^{(2)}(r_1, r_2) = \rho \int dr_3 \exp[-\beta\Phi(r_{23})] \frac{\tilde{g}^{(3)}(r_1, r_2, r_3)}{\tilde{g}^{(2)}(r_1, r_2)} \nabla_1 \exp[-\beta\Phi(r_{12})] \tag{141}$$

In these equations,

$$\Phi(|r_{ij}|) = \begin{cases} +\infty & |r_{ij}| < d \\ 0 & |r_{ij}| \geq d \end{cases}$$

where d is the length of the rod, and

$$\tilde{g}^{(n)}(r_1, \ldots, r_n) \equiv \{\exp[\beta \sum_{i \leq i < j \leq n} \sum \Phi(|r_{ij}|)]\} g^{(n)}(r_1, \ldots, r_n)$$

so that $g^{(1)}$ and $\tilde{g}^{(1)}$ are identical (since the interaction contains no terms that depend on the position of just a single particle) and $\tilde{g}^{(n)} = g^{(n)}$ if all $|r_{ij}| \geq d$. To take into account all orderings of the infinite one-dimensional system, they consider configurations where $r_1 > r_2$ and also where $r_2 < r_1$. Using the properties of the hard-rod potential, they show that (140) reduces to

$$\frac{d \ln g^{(1)}(r_1)}{dr_1} = \frac{\rho \tilde{g}^{(2)}(r_1, r_1 - d)}{g^{(1)}(r_1)} - \frac{\rho \tilde{g}^{(2)}(r_1, r_1 + d)}{g^{(1)}(r_1)} \tag{142}$$

Using (102) they show that (141) can be specified, depending on whether $r_{12} > 0$ or $r_{21} > 0$, as follows:

$$\frac{\partial \ln \tilde{g}^{(2)}(r_1, r_2)}{\partial r_1} = \frac{\rho \tilde{g}^{(2)}(r_1 - d, r_1)\tilde{g}^{(2)}(r_2, r_1 - d)}{\tilde{g}^{(2)}(r_1, r_2) g^{(1)}(r_1 - d)} - \frac{\rho \tilde{g}^{(2)}(r_1, r_1 + d)}{g^{(1)}(r_1)} \qquad r_{12} \geqslant 2d \tag{143a}$$

$$\frac{\partial \ln \tilde{g}^{(2)}(r_1, r_2)}{\partial r_1} = -\frac{\rho \tilde{g}^{(2)}(r_1, r_1 + d)}{g^{(1)}(r_1)} \qquad 0 \leqslant r_{12} < 2d \tag{143b}$$

or

$$\frac{\partial \ln \tilde{g}^{(2)}(r_1, r_2)}{\partial r_1} = \frac{\rho \tilde{g}^{(2)}(r_1 - d, r_1)}{g^{(1)}(r_1)} - \frac{\rho \tilde{g}^{(2)}(r_1 + d, r_2)\tilde{g}^{(2)}(r_1, r_1 + d)}{g^{(1)}(r_1 + d)\tilde{g}^{(2)}(r_1, r_2)} \qquad r_{21} \geqslant 2d \tag{143c}$$

$$\frac{\partial \ln \tilde{g}^{(2)}(r_1, r_2)}{\partial r_1} = \frac{\rho \tilde{g}^{(2)}(r_1 - d, r_2)}{g^{(1)}(r_1)} \qquad 0 \leqslant r_{21} < 2d \tag{143d}$$

Raveché and Stuart then investigate whether the set (142) to (143) has a unique solution for the pair $(g^{(1)}, g^{(2)})$. They proceed by considering first the set (142), (143b), and (143d); that is, they consider the coupled set of equations for those configurations where two particles are separated by distances less than twice their length. They show that the solution is unique in this interval; then, by considering (143a) and (143c), they show that there is a unique extension of this solution to all of one-dimensional space. Since, as noted above, their proof is quite instructive, we summarize here the main features of the approach. From (143b) and (143d),

$$\ln \tilde{g}^{(2)}(r_1, r_2) = e(r_1) + f(r_2) \qquad 0 \leqslant r_{12} < 2d \tag{144a}$$

$$\ln \tilde{g}^{(2)}(r_1, r_2) = k(r_1) + \ell(r_2) \qquad 0 \leqslant r_{21} < 2d \tag{144b}$$

Given that the probability density must be invariant under exchange of the positions of identical particles, that is,

$$\tilde{g}^{(2)}(r_1, r_2) = \tilde{g}^{(2)}(r_2, r_1) \tag{145}$$

(144b) is equivalent to

$$\ln \tilde{g}^{(2)}(r_1, r_2) = \ell(r_1) + k(r_2) \qquad 0 \leqslant r_{12} < 2d \tag{146}$$

For a uniform fluid or a perfect crystalline structure, the probability density determined from $\tilde{g}^{(2)}$ must also remain unchanged if the position vectors are reflected through the origin, namely,

$$\tilde{g}^{(2)}(r_1, r_2) = \tilde{g}^{(2)}(-r_1, -r_2) \tag{147}$$

For a uniform fluid, the origin can be located anywhere, whereas for a perfect crystalline array the origin may be located at a lattice site. Taking (147) in conjunction with (146) shows that

$$\ell(-r) = k(r) + \kappa \qquad (\kappa \text{ is a constant}) \tag{148}$$

so that one finds

$$\ln \tilde{g}^{(2)}(r_1, r_2) = k(r_1) + k(-r_2) + \kappa \qquad 0 \leqslant r_{21} < 2d \tag{149a}$$

$$\ln \tilde{g}^{(2)}(r_1, r_2) = k(-r_1) + k(r_2) + \kappa \qquad 0 \leqslant r_{12} < 2d \tag{149b}$$

Collecting the results (142), (143b), (143d), and (146), one determines that

$$\frac{d \ln g^{(1)}(r_1)}{dr_1} = \frac{dk(r_1)}{dr_1} + \frac{d\ell(r_1)}{dr_1} \tag{150}$$

Upon integration of (150) and application of (148), there results

$$k(-r) = \ln g^{(1)}(r) - k(r) - \lambda - \kappa \tag{151}$$

(with λ a constant of integration). This result, together with (148) yields

$$g^{(2)}(r_1 - d, r_1) = g^{(1)}(r_1) \exp[-\lambda - k(r_1) + k(r_1 - d)] \tag{152a}$$

$$g^{(2)}(r_1 + d, r_1) = g^{(1)}(r_1 + d) \exp[-\lambda + k(r_1) - k(r_1 + d)] \tag{152b}$$

which, together with (150), (143b), and (143d), leads to

$$\frac{dg^{(1)}(r_1)}{dr_1} = \rho\left\{\exp[-\lambda - k(r_1) + k(r_1 - d)] - \frac{g^{(1)}(r_1 + d)}{g^{(1)}(r_1)} \exp[-\lambda + k(r_1) - k(r_1 + d)]\right\} \tag{153}$$

and

$$\frac{dk(r_1)}{dr_1} = \rho \exp[-\lambda - k(r_1) + k(r_1 - d)] \tag{154}$$

In fact, using (146), (148), and (152), the result (153) can be simplified to read

$$\frac{d \ln \tilde{g}^{(1)}(r_1)}{dr_1} = \rho\{\exp[-\lambda - k(r_1) + k(r_1 - d)] - \exp[-\lambda - k(-r_1) + k(-r_1 - d)]\} \qquad (155)$$

Raveché and Stuart then point out that the right-hand side of this equation involves only $k(r)$, and therefore they need consider only a single functional differential equation to realize a solution of the full YBG hierarchy over the interval $|r_{12}| < 2d$. The problem is now clearly exposed, and the question is whether solutions of (155) are unique or whether, starting at some critical density, multiple solutions can arise. To address this question Raveché and Stuart proceed by recasting (154) in the form

$$\frac{d\zeta(r)}{dr} = \rho e^{-\lambda}\zeta(r-d) \qquad (156)$$

where

$$\zeta(r) = \exp[k(r)] \qquad (157)$$

Since $\tilde{g}^{(2)}(r_1, r_2)$ is a probability, and hence always positive, $k(r)$ must be a real-valued function and hence

$$\zeta(r) > 0 \qquad (158)$$

It may be recalled that up to this point in the proof two symmetry conditions have been mobilized, namely, (145), which ensures invariance under exchange of particle positions, and (147), which ensures invariance under reflection through the origin. Raveché and Stuart now posit a third symmetry condition, a condition on translational invariance:

$$\tilde{g}^{(2)}(r_1, r_2) = \tilde{g}^{(2)}(r_1 + a, r_2 + a) \qquad (159)$$

and they note that this condition is satisfied for a uniform fluid and also for a crystal with periodicity a. Then, using (159) in conjunction with (149) and (157), they find

$$\frac{\zeta(r)}{\zeta(r+a)} = e^{\mu} \qquad (160)$$

(μ a real constant); defining the new function $\tau(r)$ by

$$\zeta(r) = \tau(r) \exp(-\mu r/a) \qquad (161)$$

they determine that $\tau(r)$ must be periodic

$$\tau(r) = \tau(r+a) \qquad (162)$$

and

$$\tau(r) > 0 \qquad (163)$$

the latter condition following from the inequality (158). Returning to (156), the use of the defining relation (161) yields

$$\frac{d\tau(r)}{dr}=\left[\frac{\mu}{a}\right]\tau(r)+\rho[\exp(-\lambda+\mu d/a)]\tau(r-d) \tag{164}$$

Given the periodicity in $\tau(r)$, one may adopt the representation

$$\tau(r)=\sum_m A(m)\exp(2\pi imr/a) \tag{165}$$

and this, used in conjunction with (164), requires that

$$A(m)\left[\frac{2\pi im}{a}-\frac{\mu}{a}-\rho\exp\left(-\lambda+\frac{\mu d}{a}\right)\exp\left(-\frac{2\pi imd}{a}\right)\right]=0 \tag{166}$$

for all m. Since $\tau(r)>0$, the integral of $\tau(r)$ over one period cannot vanish, and therefore $A(0)\neq 0$; consequently, from (166) we must have

$$-\frac{\mu}{a}=\rho\exp\left(-\lambda+\frac{\mu d}{a}\right) \tag{167}$$

For $A(m)\neq 0$, (166) will be satisfied if

$$\frac{2\pi m}{a}=\frac{\mu}{a}\sin\left(\frac{2\pi md}{a}\right) \tag{168a}$$

$$1=\cos\left(\frac{2\pi md}{a}\right) \tag{168b}$$

are satisfied. Now, the only solution of (168) is $m=0$ and therefore

$$\tau(r)=e^{\nu} \tag{169}$$

where ν is a real constant. With (169), (157), and (161) there results

$$k(r)=-\left(\frac{\mu r}{a}\right)+\nu \tag{170}$$

and from (151)

$$\ln g^{(1)}(r)=2\nu+\lambda+\kappa \tag{171}$$

Given the normalization of $g^{(1)}(r)$,

$$\lim_{L\to\infty}\frac{1}{L}\int_L drg^{(1)}(r)=1$$

the constants in (171) must be such that

$$2\nu+\lambda+\kappa=0. \tag{172}$$

Therefore, applying the results (170) and (172) to (149), one determines that

$$\tilde{g}^{(2)}(r_1, r_2) = \exp\left[-\lambda + \frac{\mu}{a} r_{21}\right] \qquad 0 \leq r_{21} < 2d \tag{173a}$$

$$\tilde{g}^{(2)}(r_1, r_2) = \exp\left[-\lambda + \frac{\mu}{a} r_{12}\right] \qquad 0 \leq r_{12} < 2d \tag{173b}$$

From (167), we find that

$$\exp(-\lambda) = g^{(2)}(d) \exp\left(-\frac{\mu d}{a}\right) \tag{174a}$$

and

$$\frac{\mu}{a} = -\rho g^{(2)}(d) \tag{174b}$$

where $g^{(2)}(d)$ denotes the contact value of $g^{(2)}$, the latter a physically significant value, since it determines, through the virial theorem, the pressure of the system in the limit of the uniform fluid. Given the results (174), the expression (173) for the pair correlation function may be written

$$\tilde{g}^{(2)}(r_1, r_2) = g^{(2)}(d) \exp[-\rho g^{(2)}(d)(r_{21} - d)] \qquad 0 \leq r_{21} < 2d \tag{175a}$$

$$\tilde{g}^{(2)}(r_1, r_2) = g^{(2)}(d) \exp[-\rho g^{(2)}(d)(r_{12} - d)] \qquad 0 \leq r_{12} < 2d \tag{175b}$$

Raveché and Stuart note that, since the particle positions are ordered, r_{21} (and r_{12}) can be replaced by its modulus. Thus, using the results (171) to (173), they obtain the solution of the YBG hierarchy for *all* densities over the interval $|r_{12}| < 2d$:

$$g^{(1)}(r_1) = 1 \tag{176a}$$

$$\tilde{g}^{(2)}(r_1, r_2) = g^{(2)}(d) \exp[-\rho g^{(2)}(d)(|r_{12} - d|)] \qquad |r_{12}| < 2d \tag{176b}$$

Since this pair of functions is invariant under all transformations, (176) corresponds to the uniform fluid.

There is a unique extension of the above results to all of one-dimensional space. Consider the case in which $r_1 > r_2$ and suppose

$$3d > r_1 - r_2 > 2d$$

Since $g^{(1)}(r) = 1$ for all r, the results (143a) and (175b) yield

$$\frac{\partial}{\partial r_1} \tilde{g}^{(2)}(r_1, r_2) = \rho[g^{(2)}(d)]^2 \exp[-\rho g^{(2)}(d)(r_{12} - 2d)] \\ -\rho g^{(2)}(d) \tilde{g}^{(2)}(r_1, r_2) \tag{177a}$$

which, upon integration, yields

$$\tilde{g}^{(2)}(r_1, r_2) = \rho[g^{(2)}(d)]^2 r_1 \exp[-\rho g^{(2)}(d)(r_{12} - 2d)] + \exp[-\rho g^{(2)}(d) r_1] F(r_2) \tag{177b}$$

The arbitrary function $F(r_2)$ which arises in the integration can be determined by using the fact that $\tilde{g}^{(2)}$ is continuous and then setting

$$r_1 = r_2 + 2d$$

in (177b). The final result is

$$\tilde{g}^{(2)}(r_1, r_2) = \rho[g^{(2)}(d)]^2 (r_{12} - 2d) \exp[-\rho g^{(2)}(d)(r_{12} - 2d)] + g^{(2)}(d) \exp[-\rho g^{(2)}(d)(r_{12} - 2d)] \tag{178}$$

Hence $\tilde{g}^{(2)}(r_1, r_2)$ is determined in the interval

$$r_2 + 3d > r_1 > r_2 + 2d$$

and, as in the earlier result (176), it is dependent only on the modulus $|r_{12}|$ in this region. This procedure may be continued, yielding the value of $\tilde{g}^{(2)}(r_1, r_2)$, for all values of r_1 and r_2. Raveché and Stuart state that this procedure leads to the same result as that determined earlier by Salsburg et al.,[103] namely, (103).

The principal conclusions of the above analysis may be summarized. Raveché and Stuart have shown that the use of the closure (102): (a closure which is *exact* for hard rods) within the framework of the full hierarchy (54) leads to a pair of coupled equations for $g^{(1)}$ and $g^{(2)}$, namely, (140) and (141). They then demonstrate that within the symmetry class

$$\tilde{g}^{(2)}(r_1, r_2) = \tilde{g}^{(2)}(r_1 + a, r_2 + a)$$

(where a is the periodicity), a symmetry class that pertains either to a uniform fluid or a crystalline array, the pair of equations (140) and (141) has a unique solution for all densities. Moreover, they show that the solution is such that $g^{(1)}(r_1) = 1$ and $g^{(2)}(r_1, r_2) = g^{(2)}(|r_1 - r_2|)$ for all positions r_1 and r_2. They conclude that the full system (54) of integrodifferential equations has the uniform fluid as the unique state of the one-dimensional line of hard rods.

We now consider the above result in light of the earlier studies of Raveché and Stuart[89–92] based on (93). In these studies, they investigate whether the nonlinear equation (93) displays bifurcation from a fluid phase to states that have crystalline symmetry. They find that as the density of the fluid increases from zero, there *is* bifurcation in *one*, two, and three dimensions, and they determine that this bifurcation does not occur at the equilibrium coexistence of two phases. With respect to the hard-rod problem, Raveché and Stuart believe that the apparent contradiction between the results of their exact analysis (summarized above) and the analysis based on (93) may be understood if one recognizes that in the study based on (93) an approximate closure was used in specifying the problem. In particular, they chose $g^{(2)}(\mathbf{r}_1, \mathbf{r}_2)$ to be the known pair correlation function of a uniform fluid and then specified the kernel in the first equation of the YBG hierarchy using this known, approximate $g^{(2)}(|\mathbf{r}_1 - \mathbf{r}_2|)$. From our discussion of the first equation in the Kirkwood coupling-parameter hierarchy, and in particular from the derivation of the Kirkwood instability criterion (recall the analysis based on the "indirect coupling" kernel), we have seen that the existence or nonexistence of bifurcation is intimately linked to the structure of the kernel assumed in the formulation of the problem; it may be that the prediction of bifurcation in the studies based on (93) is a consequence of approximations introduced into the analysis. Alternately, it can be conjectured (see the remarks in Refs. 27, 42, 89–92) that the existence of a bifurcation point in a problem for which it is known that there is no phase transition can be associated in some sense with the problem of metastability. However, from the studies of Langer,[20] it is clear that metastability is *not* an equilibrium concept and hence, at first sight, cannot be described within the framework of an equilibrium representation of the many-body problem, for example (52) to (54). However, Raveché and Stuart argue, in effect, that the representation (54) may have a wider range of applicability than is superficially apparent, and in the higher-dimensional problem support their arguments by correlating bifurcating branches of (93) with known, computer-generated, hard-disk and hard-sphere isotherms. It is probably safe to say that this question will receive much attention in ongoing research.

Finally, there is a rather important conceptual point that should be raised regarding the use of symmetry arguments within the framework of a distribution function theory of phase transitions. In the analysis of (93), the authors require that $h(r)$ have maxima at, and only at, the site of a given lattice. Lovett[97] has pointed out that this requirement, in effect, restricts the domain of integration of the nonlinear equation (93) and represents an artificial supplementary criterion not inherent in the original problem. The gist of his argument is the following: Suppose the lattice vector $\mathbf{k}$ can lie only on a prescribed reciprocal lattice. With the lattice vector $\mathbf{k}$ fixed by

$$|\mathbf{k}| = 2\pi |n_1\mathbf{A} + n_2\mathbf{B} + n_3\mathbf{C}|$$

with (n_1, n_2, n_3) integers, there will be a finite number of solutions on the given lattice $\mathbf{A}$, $\mathbf{B}$, $\mathbf{C}$. But the *number* of solutions depends on the choice of **A**, **B**, **C**, and, although experimental observation may suggest a particular lattice, the underlying formulation of the problem contains no suggestion for a particular lattice.

We turn now to a brief discussion of the work of Luks and Kozak[71] on the nonlinear integral equation (89) for the pair distribution function for the square-well fluid. The equilibrium and transport properties of the square-well fluid were reviewed in an earlier volume of this series, along with the formulation of the bifurcation problem. Here we state briefly that if one takes advantage of the structural information on the pair correlation function, as determined in the numerical analysis of (89) over a wide range of thermodynamic parameter space, and then performs an analysis similar in spirit to that described in Section IV.C.2, one is led to the following criterion for determining the (possible) bifurcation points of (89):

$$1-\mu\tilde{L}(u)\equiv F(u;\theta,\lambda_0)=0 \tag{179}$$

where

$$L(x-s)=\int_{|t|}^{\infty} d\omega(\omega^2-t^2)\left[\frac{d\gamma(\omega)}{d\omega}\right]$$

$$\mu=\frac{\theta\lambda_0}{4}$$

where u is the Fourier transform variable, and the remaining symbols in these equations have been introduced in our earlier statement of the square-well problem. As regards the possible existence of a high-density bifurcation point (marking a limit of stability of a pure fluid phase), (179) is formally equivalent to the Kirkwood criterion. One can take advantage of this correlation to guess the possible ranges of the wavelength $\tilde{\lambda}=2\pi/u$ at which the function $F(u;\theta,\lambda_0)$ vanishes. Specifically, in their studies on the YBG equation, Young and Rice[117] found that the criterion (179) is satisfied for the hard-sphere fluid, the square-well fluid, and the Lennard-Jones fluid at wavelengths $\tilde{\lambda}$ corresponding to one sphere diameter, that is, for $u\cong 6$. Lincoln et al.[88] found a high-density limit of stability of $u=5.4$ for a reduced temperature of $\theta=0.45$ (see Figs. 5 and 6 of Ref. 88). Alternately, to identify a possible $\tilde{\lambda}$ (or set of $\tilde{\lambda}$) which signals a low-density limit of stability of a pure-fluid phase (i.e., to bracket the coexistence region), one recalls that in the neighborhood of a critical point, interparticle correlations are known to become extremely long ranged. This suggests that for some (θ,λ_0) one search for the vanishing of $F(u;\theta,\lambda_0)$ in the range $\tilde{\lambda}\rightarrow\infty$, or equivalently $u=0$. It was found by Lincoln et al. that when $u=0$ the function $F(u;\theta,\lambda_0)$ vanishes for two

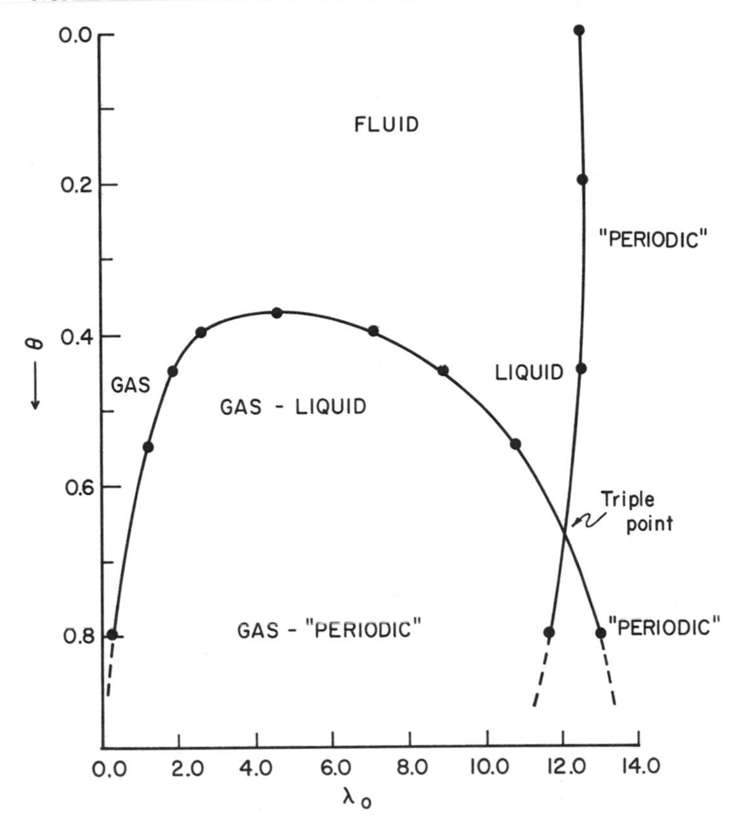

Fig. 11. Phase diagram for the square-well fluid. The locus of points in the reduced temperature ($\theta = \varepsilon/kT$), reduced-density ($\lambda_0 = 4\pi\rho/\sigma_1^3$, where ρ is the number density) plane is generated via a bifurcation analysis of the YBG nonlinear integral equation for $g^{(2)}(r)$.

values of λ_0 over a certain range of reduced temperature θ (see Figs. 5, 6, and 8 of Ref. 88). Figure 11 summarizes the results of the search for bifurcation points (here interpreted as marking the limits of stability of a pure phase) for the square-well fluid. These results are chiefly of significance in demonstrating that the fluid-"periodic" regime as well as the gas-liquid coexistence regime (with the latter characterized by *nonclassical* values of certain of the critical exponents[71]) can be identified via a traditional search for the eigenvalues of the linear operator **L** constructed as the Fréchet derivative of the underlying nonlinear operator of the problem.

4. *The Direct Correlation Function*

In this section, we discuss two different uses of the direct correlation function in bifurcation studies on a nonlinear equation for the singlet distribution function. The first discussion is drawn from the work of Kozak et al.,[86] and the second is based on the recent work of Lovett.[97]

One of the problems dealt with in Ref. 86 concerns the behavior of a fluid in the neighborhood of its critical point; in terms of the broad classification scheme adopted in this review, this study would more properly be placed in Section III, but the emphasis below on the properties of the direct correlation function makes it more convenient to discuss this problem here. At the gas-liquid critical point two fluid phases merge into one phase, that is, the difference in density of the two phases vanishes. Very near the critical point, then, we expect that the approximation that the short-ranged correlations in one phase be similar to those in the other phase should be accurate. This belief is supported by the equality of the pressure and near equality of the internal energies of the two phases, computations of which, using the virial and energy equations, probe the nature of the short-ranged correlations. Of course, only near the critical point is this true, and the theory described below for the liquid-gas transition fails for conditions where the correlations in the two phases are very different.

We consider again the nonlinear equation (126):

$$\chi(\mathbf{r}_1)+\int_\Lambda H(r_{12})[e^{\chi(\mathbf{r}_2)}-1]\,d\mathbf{r}_2=0 \tag{180}$$

with the kernel $H(r_{12})$ given by (127) and the function $\chi(\mathbf{r}_1)$ given by (128). To examine the critical region, where our physical assumptions appear valid, we look for a constant nonzero solution χ_1 of the nonlinear equation (180), since the singlet densities in both fluid phases are then constant. We attribute physical meaning to the new solution only in the limit that $\chi_1 \rightarrow 0$, that is, when the densities of the two solutions approach each other. In the case of constant solutions χ, (180) becomes a transcendental equation which may be written as

$$\chi+[e^{\chi}-1]\tilde{H}(0)=0 \tag{181}$$

where

$$\tilde{H}(0)=\int_\Lambda H(r_{12})\,d\mathbf{r}_2 \tag{182}$$

Here, $H(r_{12})$ is a short-ranged function of $|\mathbf{r}_1-\mathbf{r}_2|$, so the integration over $\mathbf{r}_2$ effectively gives a constant equal to the $k=0$ value of the Fourier transform of $H(r)$ (this statement is exact in the infinite-volume limit). For $\tilde{H}(0)\neq 0$ we rewrite (181) as

$$-\chi/\tilde{H}(0)=e^{\chi}-1 \tag{183}$$

Constant solutions χ are determined by the intersection of the curve, $e^{\chi}-1$ and the line with slope $-1/\tilde{H}(0)$. As shown in Fig. 12, if $\tilde{H}(0)>0$, there is only one solution, $\chi_0=0$, whereas if $\tilde{H}(0)<0$, there is another solution χ_1. Within the framework of this analysis, the critical point occurs in the limit that $\chi_1 \rightarrow \chi_0=0$, and only then do we attribute physical meaning to the new solution. For this limiting value of χ_1 to occur, the line must have a slope $-1/\tilde{H}(0)$ such that for some ρ_c and T_c

$$\tilde{H}(0)=-1 \tag{184}$$

as can be seen from Fig. (12) and (183). This result can also be obtained by looking for constant solutions to the linearized equation (129) that we used earlier to study the liquid-solid transition. There we predicted that the liquid-solid instability occurred when for some $k_0 \neq 0$,

$$\tilde{H}(k_0)=-1 \tag{185}$$

and here we associate the $k=0$ instability with the critical point, an association that goes back to Brout.[76]

If $\Phi(r_{12}) \geqslant 0$ everywhere, all the approximate kernels introduced earlier [namely, (79), (81), (84), and (85)] are nonnegative. Then $\tilde{H}(0)>0$, and, as discussed before, the nonlinear equation (181) has only the trivial solution $\chi=0$. Thus (184) can never be satisfied, and the theory quite sensibly predicts that a system with purely repulsive forces can have no critical point. (Note, however, that a fluid-solid instability for this system is not ruled out. The Fourier transform of the positive kernel $H(r_{12})$ can

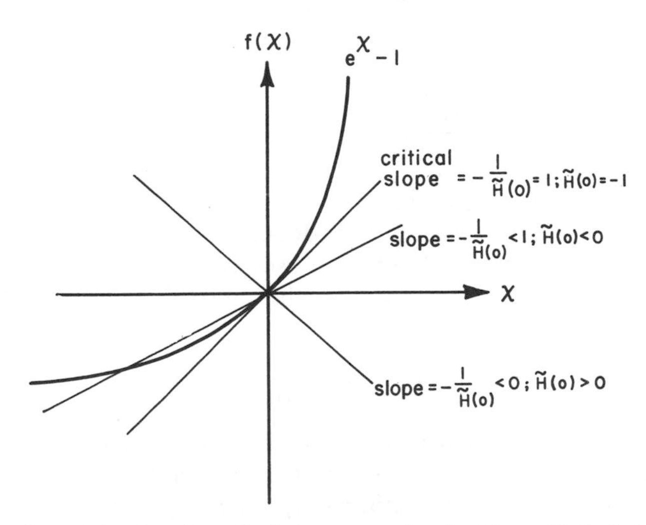

Fig. 12. A plot of the function $e^{\chi}-1$ versus χ, showing the critical slope (given by $\tilde{H}(0)=-1$).

become negative for nonzero k, and (185) can be satisfied for some ρ, T and k_0.)

In the general case where $\Phi(r_{12})$ can have both positive and negative contributions, $\tilde{H}(0)$ can be negative and (184) can be satisfied. To consider this possibility, we consider the "indirect coupling" approximation to the kernel $K(r_{12})$, (85), and notice that then

$$H_{IC}(r_{12}) = -\rho_\ell c_{PY}(r_{12}) \tag{186}$$

where $c_{PY}(r_{12})$ is the Percus-Yevick (PY) approximation[95] to the direct correlation function $c(r_{12})$. We recall that $c(r_{12})$ is defined in terms of the "total correlation" $h(r_{12})$

$$h(r_{12}) \equiv g^{(2)}(r_{12}) - 1 \tag{187}$$

by the integral equation

$$h(r_{12}) = c(r_{12}) + \rho \int_\Lambda h(r_{23})c(r_{13})\, dr_3 \tag{188}$$

Given the definition of the liquid structure factor $S(k)$, viz.

$$S(k) = 1 + \rho \int_\Lambda e^{i\mathbf{k}\cdot\mathbf{r}} [g(r) - 1]\, d\mathbf{r}, \tag{189}$$

Fourier transformation of the Ornstein–Zernike equation (188) yields

$$-\rho\tilde{c}(k) = \frac{1}{S(k)} - 1 \tag{190}$$

Finally, fluctuation theory predicts for the compressibility[58]

$$\left[\frac{\partial\rho}{\partial(\beta p)}\right]_{\Lambda,T} = S(0) = \frac{1}{1 - \rho\tilde{c}(0)} \tag{191}$$

If we assume that $c_{PY}(r)$ provides a fairly good approximation to the exact $c(r)$, which seems true for a Lennard-Jones fluid at moderate density and temperature, (190) and (191) may be used to give a qualitative description of the region in which (184) could be satisfied. Notice that (186) and (190) suggest that

$$\tilde{H}_{IC}(0) \cong \frac{1}{S(0)} - 1 \tag{192}$$

so that when $S(0) < 1$, $\tilde{H}_{IC}(0) > 0$, and there is only one fluid solution χ_0.

At very high temperatures the effect of attractive forces is small, and the positive part of integral defining $\tilde{H}_{IC}(0)$ must dominate the negative part, so that $\tilde{H}_{IC}(0) > 0$. Here the theory correctly predicts that the high-temperature phase is stable. At low temperature and high densities (i.e., in the dense-liquid regime), the compressibility is very small and $S(0)<1$. Then (191) and (192) show that no other solutions are possible, and the existence of a stable liquid phase is predicted. However, there is an intermediate range of density and temperature where $S(0)>1$, and here there is the possibility of a critical point. As noted earlier, at a true critical point the compressibility $[\partial\rho/\partial(\beta p)]_{\Lambda,T}\to+\infty$. Comparison of (191) with (186) and (184) shows that the above criterion for the critical point substitutes $\tilde{c}_{PY}(0)$ for the exact $\tilde{c}(0)$ but is otherwise in agreement with the exact theory.

The equality of $H_{IC}(r)$ and $c_{PY}(r)$ can be exploited still further. If we define

$$\hat{h}(r_{12})=\rho h(r_{12}) \tag{193}$$

$$\hat{c}(r_{12})=-\rho c(r_{12}) \tag{194}$$

then the Ornstein-Zernike equation (188) can be written

$$\hat{c}(r_{12})+\hat{h}(r_{12})=(-1)\int_{\Lambda}\hat{h}(r_{23})\hat{c}(r_{13})\,d\mathbf{r}_3 \tag{195}$$

From the theory of linear integral equations,[44] we know that the resolvent (or reciprocal) kernel $R(r_{12},\lambda)$ of a given kernel $K(r_{12})$ obeys the integral equation

$$K(r_{12})+R(r_{12},\lambda)=\lambda\int_{\Lambda}R(r_{23},\lambda)K(r_{13})\,d\mathbf{r}_3 \tag{196}$$

Comparison of the two equations (195) and (196) shows that $\rho h(r_{12})$ is a resolvent kernel to the kernel $-\rho c(r_{12})$ with $\lambda=-1$. We recall that the resolvent kernel is finite and well defined for all λ except when λ is an eigenvalue of the direct kernel, at which points the resolvent kernel has simple poles. Thus, if we take $-\rho c(r_{12})$ as given and finite, the corresponding resolvent kernel $\rho h(r_{12})$ is well behaved except when -1 is an eigenvalue of $-\rho c(r_{12})$, and $c(r_{12})$ and $h(r_{12})$ then cannot both be bounded.

One derivation of the PY equation defines $c_{PY}(r_{12})$ in terms of $h(r_{12})$ and substitutes this definition into the Ornstein-Zernike equation (188). Then $c_{PY}(r_{12})$ can be written in the form

$$c_{PY}(r_{12})=[1-e^{\beta\Phi(r_{12})}][h(r_{12})+1] \tag{197}$$

and when (197) is substituted into (188), the PY equation results. The PY direct correlation function $c_{PY}(r_{12})$ is bounded and well defined for any physically realizable $h(r_{12})$; hence the PY substitution should lead to a stable nonlinear equation except at those temperatures and densities for which -1 is an eigenvalue of $-\rho c_{PY}(r_{12})$. If this should occur, the basic PY assumption of the simultaneous existence of $h(r_{12})$ and $c(r_{12})$ must fail, and the PY equation has no solution.

The above discussion is based solely on Fredholm theory and refers to a system in a finite volume Λ. In the limit of infinite volume, one can take the Fourier transform of (188) to find

$$\tilde{h}(k)=\frac{\tilde{c}(k)}{1-\rho\tilde{c}(k)}$$

assuming that both $\tilde{c}(k)$ and $\tilde{h}(k)$ exist. Under the PY approximation, $c_{\mathrm{PY}}(r_{12})$ is short ranged, so $\tilde{c}_{\mathrm{PY}}(k)$ certainly exists. However, $\tilde{h}_{\mathrm{PY}}(k_0)$ is undefined if for some k_0

$$1-\rho\tilde{c}_{\mathrm{PY}}(k_0)=0 \tag{198}$$

This condition represents the infinite-volume limit of the finite-volume condition that -1 be an eigenvalue of $-\rho c(r_{12})$. When (198) is met, the PY equation should have no solution, and therefore one can associate divergences in an attempted numerical solution at $k_0=0$ with a PY critical point. In conclusion, the PY critical point should be the same as that predicted above when the IC kernel is used in (185).

Finally, we consider the important recent work of Lovett.[97] Lovett's bifurcation analysis proceeds from (97). He identifies a new function

$$W(\mathbf{r})=\frac{1}{\rho}[\rho(\mathbf{r})-\rho]$$

where ρ is a known constant solution of (97), so that the latter equation may be rewritten as

$$\ln[1+W(\mathbf{r}_1)]=\rho\int d\mathbf{r}_2 c(r_{12})W(\mathbf{r}_2) \tag{199}$$

At a bifurcation point, new nonconstant solutions for $W(\mathbf{r})$ may branch off continuously from the known solution, $W(\mathbf{r})=0$. Then, near the bifurcation point, $W(\mathbf{r})$ will be small and (199) may be linearized to read

$$W(\mathbf{r}_1)=\rho\int d\mathbf{r}_2 c(r_{12})W(\mathbf{r}_2) \tag{200}$$

a linear eigenvalue problem for $W(\mathbf{r})$. The states at which there are nonzero solutions $W(\mathbf{r})$ to (200) represent points of "soft" bifurcation of the full nonlinear equation (199); at these states a new singlet density function, given by the corresponding $W(\mathbf{r})$, branches off the uniform density solution ρ, given by $W(\mathbf{r})=0$. Now suppose (199) has a solution with the crystal symmetry

$$W(\mathbf{r}+n_1\mathbf{a}+n_2\mathbf{b}+n_3\mathbf{c})=W(\mathbf{r}) \tag{201}$$

where (n_1, n_2, n_3) are integers. Then $W(\mathbf{r})$ may be Fourier represented as

$$W(\mathbf{r}) = \sum_{n_1,n_2,n_3} a_{n_1,n_2,n_3} \exp\left[2\pi i(n_1\mathbf{A}+n_2\mathbf{B}+n_3\mathbf{C})\cdot\mathbf{r}\right] \tag{202}$$

where the $\mathbf{A}$, $\mathbf{B}$, $\mathbf{C}$ represent vectors generating the reciprocal lattice and where the Fourier coefficients a_{n_1,n_2,n_3} are given by

$$a_{n_1,n_2,n_3} = \int_0^1 d\xi \int_0^1 d\eta \int_0^1 d\zeta \exp\left[-2\pi i(n_1\mathbf{A}+n_2\mathbf{B}+n_3\mathbf{C}) \cdot (\xi\mathbf{a}+\eta\mathbf{b}+\zeta\mathbf{c})\right] \cdot W(\xi\mathbf{a}+\eta\mathbf{b}+\zeta\mathbf{c}) \tag{203}$$

Given the representation (202) for $W(\mathbf{r})$, the linear equation (200) becomes

$$a_{n_1,n_2,n_3}\left[1-\rho\int dv\, c(r)\exp\left[2\pi i(n_1\mathbf{A}+n_2\mathbf{B}+n_3\mathbf{C})\cdot\mathbf{r}\right]\right]=0 \tag{204}$$

Thus the Fourier coefficient a_{n_1,n_2,n_3} must be zero unless

$$1 = \rho\tilde{c}(k) \tag{205a}$$

with

$$k = 2\pi\,|n_1\mathbf{A}+n_2\mathbf{B}+n_3\mathbf{C}| \tag{205b}$$

At this point, Lovett strikes out in a new direction. He notes that the linear response of the singlet number density $\rho(\mathbf{r})$ to a weak external field $\phi(\mathbf{r})$ is

$$\delta\rho(\mathbf{r}_1) = \int d\mathbf{r}_2 \left[\frac{\delta\rho(\mathbf{r}_1)}{\delta\phi(\mathbf{r}_2)}\right]_{\phi(\mathbf{r})=0} \phi(\mathbf{r}_2) \tag{206}$$

The functional derivative here may be evaluated in terms of the pair number density $\rho^{(2)}(\mathbf{r}_1, \mathbf{r}_2)$ as[118]

$$\frac{\delta\rho(\mathbf{r}_1)}{\delta\phi(\mathbf{r}_2)} = -\beta\left[\rho^{(2)}(\mathbf{r}_1,\mathbf{r}_2)-\rho^2+\rho\delta(\mathbf{r}_2-\mathbf{r}_1)\right] \tag{207}$$

Since this derivative is to be evaluated for the field-free system (or a translationally-invariant fluid state), it is just a function of the relative distance r_{12}, and hence Fourier transformation of (206) yields

$$\delta\tilde{\rho}(\mathbf{k}) = -\rho\beta\mathscr{X}(k)\tilde{\phi}(\mathbf{k}) \tag{208}$$

where

$$\mathscr{X}(k) = 1+\rho\int dv\, e^{i\mathbf{k}-\mathbf{r}}\left[g^{(2)}(r)-1\right] \tag{209}$$

is a **k**-dependent linear susceptibility, and $g^{(2)}(r)$ is the pair correlation function. Stability for the fluid means that a weak external field will induce a weak response. The larger the response of the system to a given perturbation, the less stable the system is. Then for small **k**, if $\mathscr{X}(k)$ becomes infinite, the system becomes unstable. (When the response of the system is very large, the linear analysis above must fail; however, the divergence of the linear susceptibility signals, at the least, that the original state can not be realized: it has become unstable with respect to infinitesimal *fluctuations*.) The strategy of this analysis goes back to Brout,[73] and more recently Schneider et al.[119] have extended Brout's idea by investigating the response of a system to a *dynamical* instability.

The above remarks, phrased in terms of the pair distribution function, can be recast in terms of the direct correlation function. Suppose we denote by $\tilde{c}(r)$ the Fourier transform of the direct correlation function. Then the corresponding Ornstein-Zernike relation yields

$$\mathscr{X}(k)=\frac{1}{1-\rho\tilde{c}(k)} \tag{210}$$

Thus instability is associated with the zeros of $1-\rho\tilde{c}(k)$. Lovett goes on to point out that mechanical instability (large response to an external field) is closely related to thermodynamic instability (the failure of the free energy to be a minimum). The appropriate free energy for a system of specified chemical potential, temperature, and external field $\phi(\mathbf{r})$ is the grand canonical potential Ω. The response of a system to an external field $\phi(\mathbf{r})$ may be monitored by examining the change in Ω; through terms quadratic in $\delta\rho(\mathbf{r})$, we have

$$\Delta\Omega=[1-\rho\tilde{c}(0)]\delta\tilde{\rho}(0)+\frac{1}{16\pi^3\rho}\int d^3k[1-\rho\tilde{c}(k)]\,|\delta\tilde{\rho}(k)|^2+\dots \tag{211}$$

At low density, Lovett notes that $[1-\rho\tilde{c}(k)]$ is always positive, and hence the free energy is a minimum with respect to small perturbations about the uniform state. However, a zero in the quantity $[1-\rho\tilde{c}(k)]$ signals the transition into a region in which the original free energy function is no longer a stable minimum. (Of course, the absolute position of the true minimum cannot be determined without including the changes in $c(\mathbf{r}_1, \mathbf{r}_2)$ that occur when the system becomes nonuniform.) Notice that the two stability analyses have led to the same result: The fluid will not exist beyond the smallest density for which $1=\rho\tilde{c}(k)$ for some k. Lovett emphasizes that only instability with respect to infinitesimal changes has been addressed in the above analysis; the possibility of instability with respect to a large perturbation is still an open question. Moreover, since the true liquid-solid transition represents a finite change (in principle the bifurcation is "hard"), instability with respect to small fields is not necessarily related to true liquid-solid coexistence (see the remarks in the next section).

The results of Lovett's stability analysis become even more meaningful when related to the earlier result (205) of the bifurcation analysis. The common result, $1=\rho\tilde{c}(k)$, obtained in these studies shows that bifurcation is expected at just those densities where an instability first arises and further that the new solution is expected to start with the same symmetry as the perturbing field with respect to which the system becomes unstable. The important conclusion that mechanical instability and bifurcation occur together is critically assessed by Lovett. He points out that the fundamental Theorem V is valid for equations defined on a bounded, closed set. Under this circumstance, if the multiplicity of the eigenvalues of (200) is odd, then "soft" bifurcation to a finite number of new branches is guaranteed by Theorem V. Without the restriction to a finite domain Λ of integration, however, the degeneracy of the eigenvalues of the associated linear equation can become infinite (see the discussion in section 4.1 of reference 44); Theorem V is then inapplicable. Moreover, rotational and translational invariance of a fluid means that $c(\mathbf{r}_1, \mathbf{r}_2)=c(r_{12})$, and hence if $\rho(\mathbf{r})$ is a solution of the equation (199), then $\rho(R\mathbf{r}+T)$ is also a solution for any three-dimensional rotation R and translation T. Thus, when the domain of integration is not restricted, if there is one nonconstant solution to (200) there is an infinity of solutions. These remarks are linked to a second point, raised in the preceding subsection. Lovett argues that, although an effective domain restriction procedure may result if $c(r)$ is sufficiently short ranged, the prescription that $\mathbf{k}$ can only lie on a given reciprocal lattice introduces an artificial supplementary condition *not* inherent in the original problem; within the context of the present discussion the equation (211) for $\Delta\Omega$ contains no prescription for a particular lattice.

Lovett then carries through a Lyapunov-Schmidt analysis to determine the actual amplitude of the bifurcating solutions. He introduces the linear operator $\hat{K}$

$$\hat{K}W(\mathbf{r}_1)=W(\mathbf{r}_1)-\rho_0\int d\mathbf{r}_2 c(r_{12})W(\mathbf{r}_2) \tag{212}$$

so that (199) can be rewritten

$$\frac{\rho}{\rho_0}W-\ln(1+W)=\frac{\rho}{\rho_0}\hat{K}W \tag{213}$$

In the traditional way, one assumes that W and ρ have analytic expansions in an "amplitude" A

$$W = \sum_{n=1}^{\infty} w_n A^n \tag{214}$$

$$\frac{\rho}{\rho_0} = 1 + \sum_{n=1}^{\infty} a_n A^n \tag{215}$$

and then incorporation of (214) and (215) into (213) yields

$$\begin{aligned} \hat{K}w_1 &= 0 \\ \hat{K}(w_2 + a_1 w_1) &= a_1 w_1 + \tfrac{1}{2} w_1^2 \\ \hat{K}(w_3 + a_1 w_2 + a_2 w_2) &= a_1 w_2 + a_2 w_1 + w_1 w_2 - \tfrac{1}{3} w_1^3 \end{aligned} \tag{216}$$

At each stage, $\hat{K}w_n$ is determined using the preceding $w_i (i < n)$ so that

$$\begin{aligned} \hat{K}w_1 &= 0 \\ \hat{K}w_2 &= a_1 w_1 + \tfrac{1}{2} w_1^2 \\ \hat{K}w_3 &= -a_1^2 w_1 - \tfrac{1}{2} a_1 w_1^2 + a_1 w_2 + a_2 w_1 + w_1 w_2 - \tfrac{1}{3} w_1^3 \end{aligned} \tag{217}$$

Notice that w_1 is determined by the first of the equations in this last sequence; this is just the linear eigenvalue problem considered in (205). Now ρ_0 is to be chosen such that there is a solution

$$w_1 = \cos(\mathbf{k} \cdot \mathbf{r}) \tag{218}$$

for some $\mathbf{k}$. Given this, w_2 is determined by

$$\hat{K}w_2 = a_1 w_1 + \tfrac{1}{2} w_1^2 = a_1 \cos(\mathbf{k} \cdot \mathbf{r}) + \tfrac{1}{4} + \tfrac{1}{4} \cos(2\mathbf{k} \cdot \mathbf{r}) \tag{219}$$

In the notation

$$K_n = 1 - \rho_0 \tilde{c}(nk)$$

$K_1 = 0$ is the same as (205). Since $K_1 = 0$, no term in $\hat{K}w$ can give a cos $(\mathbf{k} \cdot \mathbf{r})$ term, hence this term must be absent in (219), or $a_1 = 0$. Accordingly, we must have

$$w_2 = \frac{1}{4K_0} + \frac{\cos(2\mathbf{k} \cdot \mathbf{r})}{4K_2} \tag{220}$$

Continued iteration yields

$$\begin{aligned} \hat{K}w_3 &= a_2 w_1 + w_1 w_2 - \tfrac{1}{3} w_1^3 \\ &= \left(a_2 + \frac{1}{4K_0} + \frac{1}{8K_2} - \frac{1}{4}\right) \cos(\mathbf{k} \cdot \mathbf{r}) + \left(\frac{1}{8K_2} - \frac{1}{12}\right) \cos(3\mathbf{k} \cdot \mathbf{r}) \end{aligned} \tag{221}$$

which gives the results

$$a_2+\frac{1}{4K_0}+\frac{1}{8K_2}-\frac{1}{4}=0 \tag{222}$$

$$w_3=\frac{1}{K_3}\left(\frac{1}{8K_2}-\frac{1}{12}\right)\cos(3\mathbf{k}\cdot\mathbf{r}) \tag{223}$$

Hence we obtain

$$\frac{\rho}{\rho_0}=1+\frac{A^2}{4}\left(1-\frac{1}{K_0}-\frac{1}{2K_2}\right)+\mathcal{O}(A^3)$$

so that

$$A=\left\{\frac{4}{1-\dfrac{1}{K_0}-\dfrac{1}{2K_2}}\cdot\frac{\rho-\rho_0}{\rho_0}\right\}^{1/2}+\mathcal{O}\left(\frac{\rho-\rho_0}{\rho_0}\right)$$

Consequently, we have determined

$$W(\mathbf{r})=A\cos(\mathbf{k}\cdot\mathbf{r})+\mathcal{O}(A^2)$$

Notice that if

$$1-\frac{1}{K_0}-\frac{1}{2K_2}>0$$

the new solution appears as ρ is increased beyond ρ_0, where ρ_0 is the density fixed by the linear eigenvalue problem (200); this new solution bifurcates smoothly out of the old solution with amplitude increasing as $(\rho-\rho_0)^{1/2}$.

Finally, to examine in more detail the condition (205), Lovett calculated explicitly the Fourier transform $\rho\tilde{c}(k)$ for the problem of a one-dimensional line of hard rods, using the known, exact expression for the direct correlation function. He determined that *no* point of instability is ever actually reached. Then he used the Thiele-Wertheim solutions[110,111] to the Percus-Yevick equation to give a simple approximation to $c(r)$ for the three-dimensional hard-sphere system. Numerical studies on the resulting expression for the Fourier transform $\rho\tilde{c}(k)$ showed that the three-dimensional hard-sphere system remained stable at *all* densities, but that an additional attractive pair interaction may induce instability. The possible realization of an instability in the latter case deserves further comment. Strictly speaking, the relation

$$\frac{N}{1-\rho\tilde{c}(k)}=\left\langle\left|\sum_{i=1}^{N}e^{ik\cdot r_i}\right|^2\right\rangle\geq 0$$

dictates that $\rho\tilde{c}(k) \leq 1$. Thus if bifurcation occurs, $\rho\tilde{c}(k)$ must approach unity from below but not pass into the region $\rho\tilde{c}(k) > 1$. The linear perturbation analysis of Lovett[97] does in fact push $\rho\tilde{c}(k)$ above unity—as any linear calculation would have to—so, although the identification of a bifurcation point is reasonable, the continuation to higher densities is unreasonable. Given these considerations, Lovett[97] concludes that, although the linear perturbative analysis cannot be strictly correct, bifurcation is certainly *plausible* when one considers an additional attractive pair interaction in the hard-sphere problem.

V. CONCLUDING REMARKS

From an examination of the classical theories of phase transitions (phenomenological and mean-field theories), it could be claimed that the presence of nonlinearity in the underlying structure of the problem was the necessary and sufficient mathematical ingredient for the successful description of a phase transition. Onsager's *tour de force*,[5] "Crystal Statistics, I, A Two-Dimensional Model with an Order-Disorder Transition," changed this picture, revealing subtleties in the mathematical structure of the problem previously unsuspected. We now know that the naive use of a classical theory in the description of phase transitions leads inevitably to a characterization of the singularity at the point of transition which is qualitatively incorrect. Indeed, a principal norm by which theories of phase transitions are assessed today is the qualitative and quantitative success with which the nature of the singularity is characterized. Yet the unmasking of singularity in the statistical-mechanical theory does not minimize the importance of nonlinearity; it does, however, refine the kind of question that is asked regarding its significance.

The objective of this review has been to examine the role of nonlinearity in lattice and distribution function theories of cooperative phenomena. We have focused on the underlying mathematical structure of several representative developments and have seen that uniqueness of the governing nonlinear equation is not always guaranteed; in some cases one can find bifurcating branches of solutions. When this occurs, one must address the statistical-mechanical consequences of this behavior and seek to understand its thermodynamic significance.

The classical distinction between second- and first-order phase transitions was preserved in this development, since theories based on a lattice model versus theories based on a distribution-function approach could be discussed conveniently within this framework. In considering the second-order ferromagnetic transition, the application of the methods of modern nonlinear functional analysis to the D-dimensional Ising model (in the continuum representation) revealed a very delicate relationship between the failure of uniqueness in the problem and the occurrence of a phase transition.[27] There the relationship between the point of bifurcation and the location of the thermodynamic singularity was clearly displayed. Bifurcation was found in all dimensions, but a phase transition only for $D \geq 2$. In the one-dimensional case, the "potential" singularity occurred beyond the bifurcation point on a bifurcating branch, but the singularity was not realized; that is, $(H, \lambda(H))$ did not lie on the path of steepest descent, $(H, \lambda_s(H))$, in the asymptotic evaluation of the partition function. An analysis of the two- and three-dimensional cases also revealed bifurcation, again with the possibility of a singularity occurring at a point somewhat displaced from the bifurcation point (namely, on one of the bifurcating branches), but here it was found that the "potential" singularity *did* lie on the steepest-descent path; a singularity in the specific heat was realized, and the qualitative characterization of the singularity in two and three dimensions was consistent with known results, although quantitative estimates of the critical temperature were not in good agreement with the accepted values. Since one can find bifurcation but no phase transition, we conclude that branching, that is, the failure of uniqueness in the underlying mathematical description of the problem, is a necessary but not sufficient condition for the occurrence of a second-order phase transition. This raises the following question: If branching does not guarantee a phase transition here, what significance, if any, does it have? A possible interpretation can be given if one applies to this problem the qualitative insights gained from a study of the theorem of Ruelle, and specifically the predictions of this theorem for the hard-rod system, a system that exhibits no phase transition. There it was found that the bound on the correlation functions determined by Ruelle corresponded to a *lower* bound on the limit of stability of the pure, low-density phase; evidentally, realization of this bound does not imply the existence of a phase transition. Using this insight as a guide in understanding the significance of bifurcation in the Ising model study, we may argue that a bifurcation point corresponds to a bound (namely, a lower bound in terms of $H=\beta J$) on the limit of stability of the high-temperature disordered phase, with the singularity actually achieved in two and three dimensions but at a temperature somewhat lower (or, a value of H somewhat higher) than that which marks the bifurcation point. This conclusion can be examined in light of Brout's analysis of short-range order and fluctuations in the Weiss mean-field theory of ferromagnetism.[76] Again, the essence of a mean-field description is the neglect of local order, or, alternatively, of local fluctuations. As stated by Brout, "in the Weiss approximation, once the magnetization disappears, the energy disappears with it", and since all the exchange energy is not exhausted in the *correct* description, "we may also anticipate that the true Curie point is lower than the Weiss-Curie point by $\mathcal{O}(1/z)$," where $z = 1/H$ in terms of our earlier notation, and the term $\mathcal{O}(1/z)$ reflects the importance of short-range correlations. In a discussion remark at the recent conference[21] on "Bifurcation Theory and Applications in Scientific Disciplines," Ruelle stated that a traditional use of bifurcation theory must lead inevitably to a classical description of phase transitions. If one regards the "soft" bifurcation uncovered in the Ising model as signaling a changeover in a mean-field description, then it is the local fluctuations (implicitly taken into account in the continuum representation of the Ising model) which "push" the singularity to a point somewhat displaced from the bifurcation point. In a theory in which fluctuations are suppressed from the very outset, for example a classical mean-field formulation, no such "gap" would occur, and a traditional bifurcation approach would predict the onset of a phase transition (if one occurs) *at* the bifurcation point. This is the behavior found in the study of Rice and Lemberg[42] on orientational transitions in crystalline methane. In the study of Tareyeva[51] cited in Section III, also based on a mean-field equation, the change in symmetry of the lattice occurs *at* the point of bifurcation.

Whereas the relationship between the onset of branching and the occurrence of a thermodynamic singularity is very clearly displayed in the Ising model study, the problem is less well exposed in the various distribution function theories discussed in Section IV. In fact, a central difficulty in studies based on equations obtained from the BBGKY hierarchy is to identify some statistical-mechanical, thermodynamical, or other criterion which reflects the (possible) presence of a singularity in the underlying mathematical structure of the problem. The insights of a number of authors were reviewed in this article, and the consequences of each criterion were explored within the context of a physically uninteresting, but mathematically tractable model—the one-dimensional line of hard rods. In the study based on the first equation in the Kirkwood-Salsburg hierarchy,[67] it was found that the possible bifurcation points $(\rho, z_0(\rho))$ did not lie on the steepest descent path $(\rho, z_s(\rho))$. In two studies based on the direct correlation function, the first involving the "indirect coupling kernel[78] and the second proceeding from a new integral equation relationship for $c(r)$,[97] no bifurcation was found in the hard-rod problem; in Lovett's interpretation, the low-density phase never becomes mechanically unstable, which (at least) assures the absence of a "soft" bifurcation point. In the study based on the first two equations in the YBG hierarchy, Raveché and Stuart[94] show that a proof of uniqueness for the hard-rod problem can be given by examining the class of admissible solutions to the underlying nonlinear problem in light of symmetry constraints. In fact, as we have seen, bifurcation in the hard-rod problem was predicted only when a certain, approximate closure was used in specifying the kernel or when the kernel itself was constrained via an averaging procedure á la Kirkwood. We believe the evidence favors the conclusion of no bifurcation in the one-dimensional hard-sphere problem.

In the three-dimensional hard-sphere system (a model for which no *exact* analytic results are available), both the IC analysis (with no kernel averaging) and the mechanical stability analysis of Lovett predict no bifurcation. However, Lovett's study also shows that an additional pair attraction may induce a mechanical instability, and for this reason his perturbative analysis (which suggests that bifurcation is *plausible*) is also relevant to the previously cited work on the square-well fluid.[71] For this model, the correlation functions and attendent thermodynamic properties were determined via numerical solution of the second equation in the YBG hierarchy, and a traditional bifurcation analysis led to the phase diagram displayed in Fig. 11. As is seen, both fluid-"periodic" and liquid-gas coexistence lines can be identified[120] (with the behavior in the neighborhood of the critical point characterized by several "nonclassical" exponents). Lovett[121] has pointed out that the identification at high densities of a "periodic" phase in these studies is the same as that which follows from a $c(r)$-based argument; that is, the square-well fluid becomes mechanically and thermodynamically unstable at the same density point. These arguments would suggest that the failure of uniqueness in a nonlinear equation obtained from the BBGKY hierarchy is in correspondence with the onset of a phase transition, that is, when branching occurs, it signals the onset of a phase transition. This conclusion is qualitatively different from the one that followed from our discussion of the second-order transition in ferromagnetic systems (Section III.B) and the (essentially) second-order, orientational phase transitions in solid methane (Section III.C); there it was found that one could find bifurcation but no phase transition. This difference in interpretation is important and requires further comment. Let us recall the nature of the theorems used to probe the possibility of bifurcation in the two cases. As stated several times in this review, the techniques and theorems of "soft" bifurcation are particularly well suited to the analysis of second-order phase transitions. However, from a rigorous mathematical point of view, these methods are less appropriate for studying the kinds of discontinuous change that characterize first-order transitions, and it may be that this distinction is crucial. In dealing with first-order transitions, rather than use bifurcation theorems which seek to establish the existence of new solutions of "small norm" relative to the basic solution of the problem, it may be necessary to use from the very outset a theory designed to handle discontinuous change. The obvious candidate here is the theory of singularities of stable differentiable mappings. The reader is reminded of the discussion in Section II, where this theory was used to elaborate some features of the Landau model, and, although Landau theory is certainly wide of the mark in its neglect of fluctuations, this illustration and the accompaning dialogue (which up until this moment may have seemed rather an excrescence) show clearly that the notions of bifurcation and singularity are very closely coupled in the theory of stable mappings and their singularities (recall Fig. 8). If this liaison were to be preserved in a formulation of the problem of phase transitions based on a nonlinear equation obtained from the BBGKY hierarchy, we believe that marked progress in understanding the relationship between nonlinearity and singularity in first-order transitions would follow. It is in this direction that the author looks for the most significant advances in understanding the role of nonlinearity in statistical-mechanical theories of phase transitions based on a distribution function approach.

Acknowledgments

The seminal contributions of Professor S. A. Rice and his colleagues to the development of the subject reviewed in this article are self-evident; the author gratefully acknowledges the opportunity of being associated with his group and of participating in the early stages of this development. The benefit of later discussions with J. D. Weeks, R. W. Zwanzig, H. Haken, R. Thom, R. Lovett, and H. Raveché is also gratefully acknowledged. Very special thanks are due to my colleagues and former colleagues at the University of Notre Dame, K. D. Luks, G. L. Jones, I.-Y. Cheng, and especially R. A. Goldstein.

References

1. D. Ruelle, *Statistical Mechanics*, Benjamin, New York, 1969.
2. H. G. Stanley, *Introduction to Phase Transitions and Critical Phenomena*, Oxford University Press, Oxford, 1971.
3. G. E. Uhlenbeck and G. W. Ford, *Lectures in Statistical Mechanics*, American Mathematical Society, Providence, 1963; for a discussion of the origins of Van der Waals theory, see S. G. Brush, *The Kind of Motion We Call Heat*, vol. I, North-Holland, Amsterdam, 1976.
4. Redrawn from A. Michels, B. Blaisse, and C. Michels, *Proc. R. Soc.*, **A160,** 367 (1937).
5. L. Onsager, *Phys. Rev.*, **65,** 117 (1944).
6. L. Van Hove, *Physica*, **15,** 951 (1949); **16,** 137 (1950).
7. C. N. Yang and T. D. Lee, *Phys. Rev.*, **87,** 404 (1952); T. D. Lee and C. N. Yang, *Phys. Rev.*, **87,** 410 (1952).
8. E. H. Lieb and D. C. Mattis, eds., *Mathematical Physics in One Dimension*, Academic, New York, 1966; see especially the papers by G. E. Uhlenbeck, M. Kac, and P. C. Hemmer.
9. K. G. Wilson, *Phys. Rev.*, **B4,** 3174 (1971); 3184 (1971).
10. C. Domb and M. S. Green, eds., *Phase Transitions and Critical Phenomena*, vol. 6, Academic, New York, 1976.

11. J. G. Kirkwood and E. Monroe, *J. Chem. Phys.*, **9,** 514 (1941).
12. A. A. Vlasov, *Many Particle Theory and Its Application to Plasma*, Gordon and Breach Scientific Publication, New York, 1961.
13. S. V. Tyablikov, *JETP*, **17,** 386 (1947).
14. H. Haken, *Synergetics*, Springer-Verlag, Berlin, 1977; H. Haken, *Rev. Mod. Phys.*, **47,** 67 (1975).
15. L. D. Landau and E. M. Lifshitz, *Statistical Physics*, 2nd ed., Pergamon Press, Oxford, 1970.
16. R. Thom, *Stabilité structurelle et morphogénèse*, Benjamin, New York, 1972; R. Thom, "Phase Transitions as Catastrophes," in *Statistical Mechanics*, S. A. Rice, K. F. Freed, and J. C. Light, eds., University of Chicago Press, Chicago, 1972
17. M. Golubitsky and V. Guillemin, *Stable Mappings and their Singularities*, Springer-Verlag, Berlin, 1973.
18. L. P. Kadanoff, W. Götze, D. Hamblen, R. Hecht, E. A. S. Lewis, V. V. Palciauskas, M. Rayl, J. Swift, D. Aspnes, and J. Kane, *Rev. Mod. Phys.*, **39,** 395 (1967).
19. R. Landauer, *Ber. Bunseges physik. chem.*, **80,** 1048 (1976); see also Ref. 21.
20. J. S. Langer, *Ann. Phys.*, **54,** 258 (1969); J. S. Langer, Metastable States, in *Van der Waals Centennial Conference on Statistical Mechanics*, North Holland, Amsterdam, 1973.
21. See the proceedings of the recent conference on *Bifurcation Theory and Applications in Scientific Disciplines, Ann. N.Y. Acad. Sci. 316* (1979).
22. I. Stakgold, *SIAM Rev.*, **13,** 289 (1971).
23. M. Vainberg and V. A. Trenogin, *Russian Math. Surveys*, **17,** 1 (1962).
24. J. B. Keller, Bifurcation Theory for Ordinary Differential Equations, in J. B. Keller and S. Antman, eds., *Bifurcation Theory and Nonlinear Eigenvalue Problems*, Benjamin, New York, 1969.
25. E. Ising, *Z. Phys.*, **31,** 253 (1925).
26. M. E. Fisher, *Rep. Prog. Phys.*, **30,** 615 (1967); also, M. E. Fisher, *Essays in Physics*, **4,** 43 (1972).
27. R. A. Goldstein and J. J. Kozak, *Physica*, **71,** 267 (1974); R. A. Goldstein and J. J. Kozak, Phase Transitions in D-Dimensional Ising Lattices, in *Global Analysis and its Applications*, vol. II, International Atomic Energy Agency, Vienna, 1974.
28. T. H. Berlin and M. Kac, *Phys. Rev.*, **86,** 821 (1952).
29. A. J. F. Siegert, Functional Integrals in Statistical Mechanics, in *Statistical Physics*, Benjamin, New York, 1963.
30. G. Gallavotti, private communication to R. A. Goldstein.
31. N. G. DeBruijn, *Asymptotic Methods in Analysis*, North Holland, Amsterdam, 1961.
32. F. Rellich, *Perturbation Theory*, Gordon and Breach, New York, 1968.
33. J. J. Stoker, *Nonlinear Vibrations*, Wiley, New York, 1950.
34. J. J. Stoker, *Nonlinear Elasticity*, Gordon and Breach, New York, 1968.
35. J. Keller and M. J. Millman, *J. Math. Phys.*, **10,** 342 (1969).
36. D. H. Sattinger, *Topics in Stability and Bifurcation Theory*, Springer-Verlag, Berlin, 1973.
37. P. Dennery and A. Krzywicki, *Mathematics for Physicists*, Harper and Row, New York, 1967; B. Friedman, *Principles and Techniques of Applied Mathematics*, Wiley, New York, 1956.
38. G. N. Watson, *Quart. J. Math.*, **10,** 266 (1939).
39. M. S. Green, A Van der Waals Fixed Point, in U. Landman, ed., *Statistical Mechanics and Statistical Methods in Theory and Application*, Plenum, New York, 1977.
40. A. R. Ubbelohde, *Melting and Crystal Structure*, Clarendon Press, Oxford, 1965.

41. H. M. James and T. A. Keenan, *J. Chem. Phys.*, **31,** 12 (1959).
42. H. L. Lemberg and S. A. Rice, *Physica,* **63,** 48 (1973).
43. A. Hammerstein, *Acta Math.*, **54,** 117 (1930).
44. F. G. Tricomi, *Integral Equations*, Interscience, New York, 1957.
45. C. L. Dolph, *Trans. Am. Math. Soc.*, **66,** 289 (1949); C. L. Dolph and G. J. Minty, On Nonlinear Integral Equations of the Hammerstein Type, in P. M. Anselone, ed., *Nonlinear Integral Equations*, University of Wisconsin Press, Madison, 1964.
46. M. A. Krasnosel'skii, *Am. Math. Soc. Transl.*, **10,** 345 (1958).
47. M. A. Krasnosel'skii, *Topological Methods in the Theory of Nonlinear Integral Equations*, Pergamon, London, 1964.
48. M. A. Krasnosel'skii and Y. B. Rutickii, *Convex Functions and Orlicz Spaces*, P. Noordhoff, Groningen, 1961; M. A. Krasnosel'skii and Y. B. Rutickii, *Am. Math. Soc. Transl.*, **60,** 51 (1967).
49. M. Vainberg, *Variational Methods for the Study of Nonlinear Operators*, Holden Day, San Francisco, 1964.
50. A. G. Khachaturyan, *Sov. Phys.-Solid State*, **5,** 16 (1963).
51. E. E. Tareyeva, *Phys. Lett.*, **49A,** 309 (1974).
52. E. Schmidt, *Math. Ann.*, **65,** 370 (1908).
53. R. Iglisch, *J. Reine Angew. Math.*, **164,** no. 3 (1931); *Monatsh. Math. Phys.*, **37,** 325 (1930); **39,** 173 (1932); **42,** 7 (1935).
54. E. E. Tareyeva and T. I. Trapezina, *Theoret. Math. Phys.* (*Moscow*), **26,** 269 (1976).
55. T. I. Trapezina, *Theoret. Math. Phys.* (*Moscow*), **29,** 136 (1976).
56. E. E. Tareyeva and T. I. Shchelkacheva, *Theoret. Math. Phys.* (*Moscow*), **31,** 359 (1977).
57. E. E. Tareyeva and T. I. Trapezina, *Phys. Lett.*, **60A,** 217 (1977).
58. P. A. Egelstaff, *An Introduction to the Liquid State*, Academic, New York, 1967.
59. T. L. Hill, *Statistical Mechanics*, McGraw-Hill, New York, 1956.
60. J. G. Kirkwood and Z. Salsburg, *Discussions Faraday Soc.*, **15,** 28 (1953); these equations are related to those derived earlier by J. E. Mayer and E. Montroll, *J. Chem. Phys.*, **9,** 2 (1941).
61. J. G. Kirkwood, *J. Chem. Phys.*, **3,** 300 (1935).
62. J. Yvon, *Actualités scientifiques et industrielles*, Hermann and Cie, 1935; M. Born and H. S. Green, *A General Kinetic Theory of Liquids*, Cambridge, London, 1949.
63. K. Huang, *Statistical Mechanics*, Wiley, New York, 1967.
64. J. E. Mayer, *J. Chem. Phys.*, **10,** 629 (1942).
65. D. Ruelle, *Helv. Phys. Acta*, **36,** 183 (1963).
66. I.-Y. S. Cheng and J. J. Kozak, *J. Math. Phys.*, **14,** 632 (1973); I.-Y. S. Cheng, Ph.D. Dissertation, University of Notre Dame, 1973.
67. J. Kozak, Phase Transitions as a Problem in Bifurcation Theory, in *Bifurcation Theory and Applications in Scientific Disciplines, Ann. N.Y. Acad. Sci. 316*, 417 (1979).
68. For background on this equations, see Refs. 46, 47, and 49.
69. S. A. Rice and P. Gray, *The Statistical Mechanics of Simple Liquids*, Interscience, New York, 1965.
70. O. Penrose, *J. Math. Phys.*, **4,** 1312 (1963).
71. K. D. Luks and J. J. Kozak, *Adv. Chem. Phys.*, **37,** 139 (1978).
72. L. Nirenberg, *Topics in Nonlinear Functional Analysis*, Courant Institute of Mathematical Sciences, New York, 1974.
73. R. Brout, *Physica*, **29,** 1041 (1963); **30,** 459 (1964).
74. B. Jancovici, *Physica*, **31,** 1017 (1965); **32,** 1663 (1966).

75. J. G. Kirkwood and E. M. Boggs, *J. Chem. Phys.*, **10,** 394 (1942).
76. R. H. Brout, *Phase Transitions*, Benjamin, New York, 1965.
77. V. N. Ryzhov and E. E. Tareyeva, *Sov. Phys. Dokl.*, **22,** 131 (1977).
78. J. D. Weeks, S. A. Rice, and J. J. Kozak, *J. Chem. Phys.*, **52,** 2416 (1970).
79. T. Morita and K. Hiroike, *Prog. Theoret. Phys. (Kyoto)* **25,** 537 (1961).
80. I.-Y. Cheng and J. J. Kozak, *J. Math. Phys.*, **13,** 51 (1972).
81. J. D. Weeks, S. A. Rice, and I. Katz, *J. Chem. Phys.*, **51,** 4414 (1969).
82. M. Girardeau, *J. Math. Phys.*, **3,** 131 (1962).
83. S. Gartenhaus and G. Stranahan, *Phys. Rev.*, **138A,** 1346 (1965); C. M. Anderson, S. Gartenhaus, and G. Stranahan, *Phys. Rev.*, **146,** 101 (1966).
84. T. L. Saaty, *Modern Nonlinear Equations*, McGraw-Hill, New York, 1967.
85. K. Hiroike, private communication to S. A. Rice (see following reference).
86. J. J. Kozak, S. A. Rice, and J. D. Weeks, *Physica*, **54,** 573 (1971).
87. H. S. Green, *Molecular Theory of Fluids*, North-Holland, Amsterdam, 1952, p. 112.
88. W. W. Lincoln, J. J. Kozak, K. D. Luks, *J. Chem. Phys.*, **62,** 2171 (1965).
89. H. J. Raveché and C. A. Stuart, *J. Chem. Phys.*, **63,** 1099 (1975).
90. H. J. Raveché and C. A. Stuart, *J. Chem. Phys.*, **65,** 2305 (1976).
91. H. J. Raveché and C. A. Stuart, *J. Math. Phys.*, **17,** 1949 (1976).
92. H. J. Raveché and R. F. Kayser, Jr., *J. Chem. Phys.* **68,** 3632 (1978).
93. H. J. Raveché and M. S. Green, *J. Chem. Phys.*, **50,** 5334 (1969).
94. H. J. Raveché and C. A. Stuart, *J. Stat. Phys.*, **17,** 311 (1977).
95. For a general review of the subject see J. K. Percus, The Pair Distribution Function in Classical Statistical Mechanics, in H. L. Frisch and J. L. Lebowitz, eds., *The Equilibrium Theory of Classical Fluids*, Benjamin, 1964.
96. E. T. Dean and P. L. Chambré, *J. Math. Phys.*, **11,** 1567 (1970).
97. (i) R. Lovett, *J. Chem. Phys.*, **66,** 1225 (1977). (ii) R. Lovett, private communication.
98. R. Lovett, C. Y. Mou, and F. P. Buff, *J. Chem. Phys.*, **65,** 570 (1976).
99. L. Tonks, *Phys. Rev.*, **50,** 955 (1936).
100. E. H. Hauge and P. C. Hemmer, *Physica*, **29,** 1338 (1963).
101. L. A. Pastur, preprint of the Academy of Sciences of the Ukranian SSR, Kiev, 1973.
102. H. Moraal, *Physica*, **81A,** 469 (1975).
103. Z. W. Salsburg, R. W. Zwanzig, and J. G. Kirkwood, *J. Chem. Phys.*, **21,** 1098 (1953).
104. F. Gürsey, *Proc. Cambridge Phil. Soc.*, **46,** 182 (1950).
105. J. G. Kirkwood and E. M. Boggs, *J. Chem. Phys.*, **10,** 394 (1942).
106. J. G. Kirkwood, in R. Smoluchowski, ed., *Phase Transformations in Solids*, Wiley, New York, 1951.
107. S. A. Rice and J. Lekner, *J. Chem. Phys.*, **42,** 3559 (1965).
108. W. Kunkin and H. L. Frisch, *J. Chem. Phys.*, **50,** 1817 (1969).
109. E. Meeron and A. J. F. Siegert, *J. Chem. Phys.*, **48,** 3139 (1968).
110. E. Thiele, *J. Chem. Phys.*, **39,** 474 (1963).
111. M. S. Wertheim, *Phys. Rev. Lett.*, **8,** 321 (1963).
112. W. Hoover and J. C. Poirier, *J. Chem. Phys.*, **37,** 1041 (1962).
113. F. H. Ree, R. N. Keeler, and S. L. McCarthy, *J. Chem. Phys.*, **44,** 3407 (1966).
114. W. W. Wood, Monte Carlo Studies of Simple Liquid Models, in H. N. V. Temperley, J. S. Rowlinson, and G. S. Rushbrooke, eds., *Physics of Simple Liquids*, Wiley Interscience, New York, 1968.
115. B. J. Alder and T. E. Wainwright, *Phys. Rev.*, **127,** 359 (1962).
116. P. M. Morse and H. Fesbach, *Methods of Theoretical Physics*, McGraw-Hill, New York, 1953.
117. D. A. Young and S. A. Rice, *J. Chem. Phys.*, **47,** 4228 (1967).

118. Ref. 95, p. II-33.
119. T. Schneider, R. Brout, H. Thomas, and J. Feder, *Phys. Rev. Lett.*, **25,** 1423 (1970).
120. K. U. Co, K. D. Luks and J. J. Kozak, "Solutions of the Yvon-Born-Green equation for a system of square-well molecules at a temperature below the triple point," *Mol. Phys.* **36,** 1883 (1978); for a recent determination of the critical exponent γ for the square-well fluid see, K. A. Green, K. D. Luks and J. J. Kozak, *Phys. Rev. Letts.* **42,** 985 (1979).
121. R. Lovett, private communication.

COHERENT PROCESSES IN MOLECULAR CRYSTALS

DONALD M. BURLAND
IBM Research Laboratory
San Jose, California 95193

AND

AHMED H. ZEWAIL*
A. A. Noyes Laboratory of Chemical Physics†
California Institute of Technology
Pasadena, California 91125

CONTENTS

* Alfred P. Sloan Fellow
† Contribution Number 5780; This work was supported in part by a grant from the National Science Foundation.

I. MOTIVATION AND GOALS

Since the early work of Frenkel[1] and Peierls[2] in the 1930s and Davydov in the 1960s[3] on the application of band theory to molecular crystals, there have been many books and hundreds of articles dealing with the excited states of this class of solids. Experimental efforts have focused on the determination of the band structure in a variety of systems, and only recently have questions relating to the dynamics of excitation transfer (coherent vs. incoherent motion) been explored in great detail. Molecular crystals, on one hand, provide a prototype system for investigating tightly bound excitons with well-defined intermolecular interactions. On the other hand, probing the excitation transfer in these systems is nontrivial, and in many cases important questions still remain unanswered.

One such incompletely understood area involves the nature of the energy transport in molecular crystals. One would like to know if the transport is coherent (wave-like), partially coherent, or incoherent (diffusive). The distinction between coherent and incoherent energy transfer in dimer and excition systems is the major focus of this chapter. To this end we do not attempt to survey all of the work that has been done in the area of excitons in molecular crystals. The reader is referred to the many review articles and books on this subject.[3–18] A reader who wants an even broader picture of recent work on molecular crystals should consult the useful bibliographies in the journal *Molecular and Liquid Crystals.*[19]

In addition we do not specifically discuss cases in which the coupling between exciton and photon is sufficiently strong that the polariton formalism[12] must be invoked. This essentially limits the scope of most of the discussions in this chapter to the treatment of triplet excitons and weakly absorbing singlet excitons.

We begin with a theoretical treatment of coherence in an ensemble of two-level systems and show how a system of dimers may be considered as such an ensemble. From this theoretical analysis we are able to extract a precise definition of coherence for the dimer system. In discussing the theory, we introduce the reader who is not already familiar with work on coherence to the density matrix approach, the concept of homogeneous and inhomogeneous line broadening, and to nonlinear effects that can result from the interaction between the radiation field and the solid.

After this theoretical discussion we focus on the application of the theoretical findings to both the optical and magnetic resonance work that has been done. We show, by a detailed analysis of a few key experiments, how the optical and magnetic resonance relaxation times (T_1, longitudinal relaxation time; T_2, transverse relaxation time) are related to specific dimer dephasing processes. Next a comparison of dimer and exciton dephasing is made. A discussion of the extent to which dimer processes can provide information about analogous exciton processes is also given.

In the second half of this chapter we examine the various experimental techniques that have been used to investigate coherence in excitons. We see that exciton coherence has had different meanings in different experimental contexts, and it is a major goal of this part of the chapter to expose these differences and thus remove possible sources of confusion.

We hope that this chapter will help to answer some of the following questions:

1. What does coherence mean in a two-level (dimer) system?
2. In what ways has the term coherence been used to describe processes in excitons?
3. What are the experimental criteria for distinguishing coherent from incoherent processes in dimers and excitons?
4. Do the available optical and magnetic resonance experiments on dimers and excitons support the presence or absence of coherence?
5. Can we explain optical and spin dephasing processes in dimers and excitons by a single general theory?
6. What are the different effects of exciton-phonon and exciton-impurity interactions on the overall exciton coherence?
7. Can we talk about coherence in disordered systems?

II. INTRODUCTORY REMARKS

A crystal is a collection of identical unit cells. The unit cells may consist of atoms, ions, or, as in the crystals of interest here, molecules. Peierls[2] and Frenkel[1] pointed out in the 1930s that the stationary excited states of such crystals do not consist of localized excitations. Because of the interactions between the molecules, the excitation can be transferred from molecule to molecule, and the stationary states, known as excitons, consist of excitation delocalized throughout the crystal.

Soon after the work of Peierls and Frenkel it was realized that these mobile electronic excitations in crystals provided a mechanism by which energy could be transported from the point of absorption to the place where useful photochemistry, for example, might occur. Franck and Teller[20] suggested along these lines that excitons might be important in photosynthesis and in the photographic process.

Within the exciton model, the process of absorption of light by a molecular crystal is envisioned to occur in the following way. The photon incident on the crystal has a well-defined momentum $\hbar\mathbf{Q}$ where $\mathbf{Q}$ is the photon wave vector. In the process of absorption an exciton wave packet with the same energy is created. This exciton has a momentum $\hbar\mathbf{k}$ where $\mathbf{k}$ is the photon wave vector, and for momentum to be conserved $\mathbf{k}$ must equal $\mathbf{Q}$. In the crystal we then have an exciton wave packet propagating with a well-defined direction and velocity.

In a perfect crystal the propagation is, in one sense of the word, *coherent*, that is, the exciton retains its initial momentum. This is clearly a very efficient means of energy transport, since the exciton travels undeviated through the crystal. Of course, all real crystals have imperfections, and completely coherent transport will probably never occur. Anything that disturbs the translational symmetry of the perfect lattice can scatter the exciton and destroy this coherence. Thus impurities, lattice defects, and phonons can all act as scatterers, changing the exciton's propagation from coherent and directional to incoherent and random. Random walk transport is a much less efficient means of energy transport.

It is obviously of importance in describing energy transport in molecular crystals to know whether the process is coherent or incoherent. For this reason the problem of exciton coherence has been the subject of a great many experimental and theoretical investigations through the years.

Recently, EPR, optically detected magnetic resonance (ODMR) experiments, careful measurements of exciton absorption lineshapes, and more detailed theoretical elaboration of exciton scattering processes have brought the question of exciton coherence into sharper focus. As is often the case, an increased understanding has also exposed previously unappreciated complexities. For example, it has been assumed in much of the past work that correlation times for exciton-phonon scattering measured by a variety of experimental techniques could be directly related to exciton coherence. As the result of recent work, however, it is now clear that one must consider how seriously the scattering process changes $\mathbf{k}$. Scattering that results in small changes of $\mathbf{k}$ that may give rise to a measurable correlation time should not be considered as having an important effect on energy transport. One must carefully define what is meant by exciton coherence and relate this definition in a consistent way to the experimental results.

In addition to these experimental and theoretical results directly on excitons, an increasing amount of recent work has been done on dimer coherence in isotopically mixed crystals. These dimers consist of isolated pairs of isotopically substituted guest molecules imbedded in a host crystalline lattice. Because the isotopic guest molecules fit substitutionally into the host lattice, the relative orientation of the molecules in the dimer pair is identical to the orientation of molecules in the pure crystal. In this sense an excited dimer may be considered to be a "mini-exciton" in which the energy is exchanged between the two constituent molecules.

Some of the interest in these dimer systems has been stimulated by the possibility that they may serve as simplified models for excitons. An exciton band consists of a near continuum of states with different wave vector values **k**. Scattering processes in this multilevel system can thus be quite complicated. An excited dimer system, on the other hand, may be viewed as consisting of only two excited states. It has been hoped that processes that destroy excited-state dimer coherence would be simpler to describe and yet still be relevant to the problem of exciton coherence.

III. COHERENCE IN DIMERS

A. The General Two-Level System

If two identical molecules (1 and 2) are interacting with each other by a potential V, the total wavefunctions describing the pair excited states may be written

$$\psi(\pm) = C_1\Phi_1 \pm C_2\Phi_2 \tag{1}$$

where C_1 and C_2 are coefficients and Φ_1 and Φ_2 are the wavefunctions describing the case in which excitation is localized on molecules 1 and 2, respectively, that is,

$$\Phi_1 = \phi_1^*\phi_2 \tag{2}$$

$$\Phi_2 = \phi_1\phi_2^* \tag{3}$$

ϕ_1 and ϕ_2 are isolated molecule wavefunctions, and ϕ_1^*, ϕ_2^* excited-state wavefunctions.

When the energies of the two molecules are the same, that is, $C_1 = C_2 = 1/\sqrt{2}$, one has the resonance coupling limit. On the other hand, if the energy difference between the two molecules is Δ, then in general

$$\psi(+) = \cos\theta\Phi_1 + \sin\theta\Phi_2 \tag{4}$$

$$\psi(-) = \sin\theta\Phi_1 - \cos\theta\Phi_2 \tag{5}$$

where

$$\tan 2\theta = 2\langle V\rangle\Delta^{-1} \tag{6}$$

where $\langle V\rangle$ is the matrix element for the potential energy operator V.

For the case in which $\psi(\pm)$ are equal mixture of Φ_1 and Φ_2, that is, dimers composed of identical molecules in identical environments, one frequently describes the excitation as oscillating back and forth between the two molecules. One must be careful, however, in the case of energy transfer after optical excitation, because the method of *preparation* of the dimer excited state is very important. In other words, when the light excites the two molecules one should ask, are we preparing $\psi(\pm)$ or the $\Phi_{1,2}$ states, and how does preparation of one or the other of these states affect the experimental observable? In the absence of damping, both $\psi(\pm)$ are stationary states and thus evolve in time with the following simple quantum mechanical phase factor (see Fig. 1).

$$\psi(\pm;\mathbf{r},t)=\psi(\pm;\mathbf{r})e^{-iE_{\pm}t/h}=\psi(\pm;\mathbf{r})e^{-i\omega_{\pm}t} \tag{7}$$

One sees that $\psi(\pm)\,\psi(\pm)^*$ is independent of time. The phase coherence of these $\psi(\pm)$ states is therefore infinite in time, and the two states have δ-function resonances in the frequency domain.

One can imagine an experiment in which optical excitation of a dimer system prepares one of the stationary states, say $\psi(+;\mathbf{r},t)$. As long as damping sources such as radiation and phonon scattering are absent, the system remains in the state $\psi(+;\mathbf{r},t)$. We might define such a dimer system as being *completely coherent.* Of course any real physical system is damped, and as a result the state $\psi(+,\mathbf{r},t)$ will decay into other states and/or have its phase interrupted. It is this coherence and its loss that we wish to describe in the dimer case and relate if possible to similarly defined coherence in the exciton case.

We next describe in some detail the optical excitation of dimers by both narrow- and broad-band sources and the effect of the manner of excitation on the time evolution of the excited states of the dimer pair. This discussion serves two purposes. First, because of the simplicity of the dimer system involving only three electronic states, the ground state and

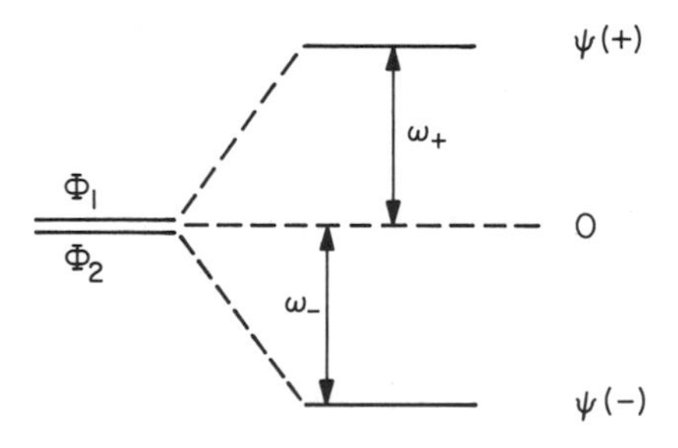

Fig. 1. The plus ($\psi(+)$) and minus ($\psi(-)$) states of a dimer obtained from the molecular wave functions Φ_1 and Φ_2. ω_+ and ω_- represents the shift in frequency of dimer plus and minus states, respectively, from the degenerate monomer level taken as the zero of energy.

two excited states, it is possible to give a precise definition of coherence. Second, the discussion suggests a group of experiments that could be performed on dimer systems to measure this kind of coherence.

B. Photon Absorption and Emission

As mentioned before, in the absence of damping, the ground and excited dimer states have δ-function resonances, that is, their lifetimes are infinite. In such a three-level system the absorption and emission of the photon depend on the linewidth of the excitation source. For example, for light with a band width ($\Delta\omega_L$) much smaller than ω_{+-}, $\psi(+)$ or $\psi(-)$ can be reached independently, as shown schematically in Fig. 2. Two cases can now be distinguished: narrow- and broad-band excitation.

1. *Narrow-band Excitation*

We first treat the narrow band case. In this case we are considering the interaction between the light source and the two levels of the pair, ground state $|g\rangle$ and excited state $\psi(+)(\equiv|+\rangle)$. In the following discussion, we restrict ourselves to a two-level system, that is, we assume that only the state $|+\rangle$ is accessible by photon absorption from the ground state. The validity of this assumption is discussed later. The light source is assumed to be coherent, and it is important to remember that this coherence is not directly related to the dimer coherence. The total Hamiltonian of the

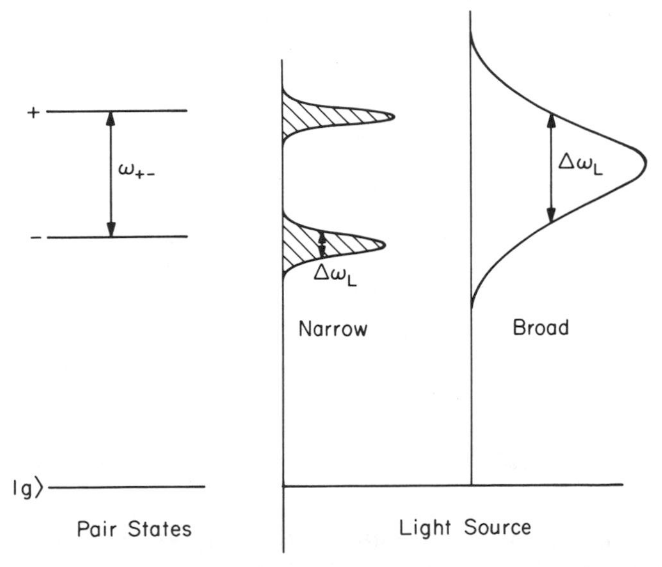

Fig. 2. Narrow- and broad-band excitation of $\psi(+)$ and $\psi(-)$. $\Delta\omega_L$ is the light bandwidth, and ω_{+-} is the frequency separation between + and − dimer states.

system (light+molecule) becomes

$$\mathcal{H} = \mathcal{H}_M + \mathcal{H}_L + \mathcal{H}_{\text{int}} \tag{8}$$

where the interaction Hamiltonian for light of frequency ω_L is given by

$$\mathcal{H}_{\text{int}} = -(\boldsymbol{\mu} \cdot \boldsymbol{\varepsilon}) \cos \omega_L t \tag{9}$$

and depends on the transition dipole moment operator $\boldsymbol{\mu}$ and the electric field amplitude $\boldsymbol{\varepsilon}$ of the light. Thus for the two-levels $|g\rangle$ and $|+\rangle$, we have

$$\begin{aligned}\mathcal{H} &= \hat{1}\mathcal{H}\hat{1} \\ &= |g\rangle\, \hbar\omega_g \,\langle g| + |+\rangle\, \hbar\omega_+ \langle +| + \{|g\rangle\, (\mu_{g+}\varepsilon) \cos \omega_L t \langle +| + c.c.\}\end{aligned} \tag{10}$$

where

$$\begin{aligned}\hat{1} &= |g\rangle\langle g| + |+\rangle\langle +| \\ \mu_{g+} &= \langle g|\, \boldsymbol{\mu}\, |+\rangle\end{aligned} \tag{11}$$

To find the time evolution of the system, we assume that the system is in the $|g\rangle$ state before the light is turned on, that is,

$$\Psi(t=0) \equiv |g\rangle \tag{12}$$

From the Schrödinger equation

$$i\hbar\dot{\Psi} = \mathcal{H}\Psi \tag{13}$$

the general solution for Ψ at any time t is given in terms of the zero-order basis set ($|n\rangle \equiv \{|g\rangle, |+\rangle\}$):

$$\Psi(t) = \sum_n C_n(t) e^{-i\omega_n t}\, |n\rangle \tag{14}$$

with the coefficients satisfying the following two differential equations.

$$\dot{C}_+(t) = -i(\mu_{+g}\varepsilon/\hbar) e^{-i(\omega_g - \omega_+)t} \cos \omega_L t C_g(t) \tag{15}$$

$$\dot{C}_g(t) = -i(\mu_{g+}\varepsilon/\hbar) e^{-i(\omega_+ - \omega_g)t} \cos \omega_L t C_+(t) \tag{16}$$

To solve these two equations one can utilize the rotating-wave approximation (RWA).[21] To see what this involves let us rewrite (15) in the following form:

$$\dot{C}_+(t) = -i\left(\frac{\mu_{+g}\varepsilon/\hbar}{2}\right)[e^{-i(\omega_g - \omega_+ - \omega_L)t} + e^{-i(\omega_g - \omega_+ + \omega_L)t}]C_g(t) \tag{17}$$

Since the near-resonance condition (i.e., $\omega_g + \omega_L \approx 0$) is satisfied in the second term, the oscillation of the first term is very rapid because the frequency term is far away from resonance. This rapidly oscillating term can be neglected. Thus we have

$$\dot{C}_+(t) = \frac{-i}{2}\,\omega_R e^{-i(\omega_g-\omega_+ +\omega_L)t} C_g(t) \tag{18}$$

$$\dot{C}_g(t) = \frac{-i}{2}\,\omega_R e^{-i(\omega_+-\omega_g-\omega_L)t} C_+(t) \tag{19}$$

and

$$\omega_R = (\mu_{+g}\varepsilon/\hbar) \tag{20}$$

Here ω_R is the Rabi frequency. We now define the following time dependent coefficients:

$$b_+(t) = C_+(t)e^{-i\omega_+ t} \tag{21}$$

$$b_g(t) = C_g(t)e^{-i(\omega_g+\omega_L)t} \tag{22}$$

Then (18) and (19) become, in the RWA approximation

$$i\begin{pmatrix}\dot{b}_g\\ \dot{b}_+\end{pmatrix} = \begin{pmatrix}\omega_g+\omega_L & \omega_R/2\\ \omega_R/2 & \omega_+\end{pmatrix}\begin{pmatrix}b_g\\ b_+\end{pmatrix} \tag{23}$$

or

$$i\dot{\mathbf{b}} = \mathbf{H}\mathbf{b} \tag{24}$$

The **H** matrix represents the energies of *dressed* (molecule-photon) states, as they are called in quantum optics. Furthermore, this matrix equation can be solved in a straightforward manner if ω_R is time independent, that is, if the amplitude of the light is not a function of time. In the latter case the new Rabi frequency becomes

$$\omega_R = \boldsymbol{\mu}_{+g}\cdot\int \boldsymbol{\varepsilon}(t')\,dt' \tag{25}$$

We now proceed to solve the **H** matrix *on* resonance ($\omega_L = \omega_+ - \omega_g$) to find the probability of light absorption and emission. Under the approximation that $\boldsymbol{\varepsilon}$ is time independent, the solution of (24) gives[22]

$$b_+(t) = A\sin\left(\frac{\omega_R t}{2}\right) + B\cos\left(\frac{\omega_R t}{2}\right) \tag{26}$$

$$b_g(t) = C\sin\left(\frac{\omega_R t}{2}\right) + D\cos\left(\frac{\omega_R t}{2}\right) \tag{27}$$

A, *B*, *C*, and *D* are coefficients to be determined by the initial conditions. With the initial condition that at $t=0$, $b_+(t)=0$ and $b_g(t)=1$, and satisfying the normalization condition

$$\sum_n |C_n(t)|^2 = \sum_n |b_n(t)|^2 = 1 \tag{28}$$

we obtain

$$|b_+|^2 = |C_+|^2 = \sin^2\left(\frac{\omega_R}{2}t\right) \tag{29}$$

$$|b_g|^2 = |C_g|^2 = \cos^2\left(\frac{\omega_R}{2}t\right) \tag{30}$$

It is interesting to note that in the absence of damping the system oscillates between ground and excited state with time. In other words, *there is no real absorption of the photon by the pair state.* Under this ideal condition of no damping in the system, 50% of the population is in the upper $|+\rangle$ state if $\omega_R t = \pi/2$. If the excitation is terminated when this equality holds, the system has experienced a so-called $\pi/2$ pulse. Pulsed excitation of this kind has been used extensively in coherent optical experiments.

At any time t, the wavefunction of the system [see (13)] is given by

$$\Psi(t) = \left(\sin\frac{\omega_R}{2}t\right)e^{-i\omega_+ t}\,|+) + \left(\cos\frac{\omega_R}{2}t\right)e^{-i\omega_g t}\,|g) \tag{31}$$

where the ket symbol $|\ )$ denotes the state $|\ \rangle$ *including* the appropriate phase factor.

A close examination of the above wavefunction indicates that the expectation value of the dipole moment operator is time dependent, that is,

$$\langle\Psi(t)|\,\hat{\mu}\,|\Psi(t)\rangle = \frac{\mu_{g+}}{2}\,e^{-i(\omega_+-\omega_g)t}\sin\omega_R t + \text{c.c.} \tag{32}$$

This means that when a $\pi/2$ pulse ($\omega_R t = \pi/2$) is performed, and, if the dipole matrix elements are real, the macroscopic moment $\langle\hat{\mu}\rangle$ has a maximum value of $\mu\cos(\omega_+ - \omega_g)t$. On the other hand, if $\omega_R t$ is 0 or π, then the moment vanishes. With this in mind, one can picture a geometrical representation for the phase coherence of the ensemble. When $\omega_R t = 0$, the system is in $|g\rangle$; when $\omega_R t = \pi/2$, the system is in a coherent (equal) superposition of $|g\rangle$ and $|+\rangle$; when $\omega_R t = \pi$, the system is in the $|+\rangle$ state (see Fig. 3).

In the language of magnetic resonance[23] one can define a population vector that is along the plus or minus z-axis when the system is in the excited or ground state, respectively. The system is in the xy-plane after a $\pi/2$ pulse (Fig. 3). This vector model of a two-level system has been

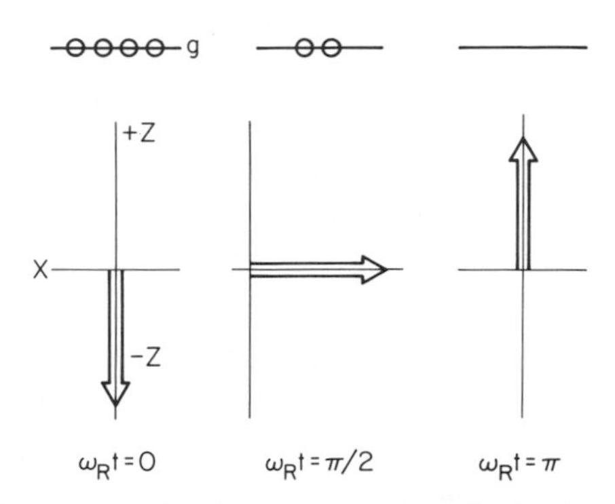

Fig. 3. A geometrical representation for the population distribution between the ground g and excited + dimer states. Left, population at $t = 0$; middle, coherent superposition; right, total inversion of population.

developed in detail by Feynman, Vernon, and Hellwarth.[24] These authors have shown that a two-level system obeys a dynamical equation that is mathematically equivalent to the equation of motion of a magnetic moment in an oscillating magnetic field. More is said about this vector model in Section III.D, where we consider damping processes.

Note that so far we are considering no decay channels that can lead to homogeneous line broadening, nor have we considered inhomogeneous broadening of the whole system. In the absence of any decay channels, the dimer system retains indefinitely the memory of its preparation by a coherent light source.

2. *Broad-band Excitation*

The preceding subsection considered the excitation of the dimer system by a narrow-band coherent light source. In this section we consider in some detail the excitation of the dimer by a broad-band source. We show that in this case one can gain information on a restricted kind of coherence that we call *transition coherence* to distinguish it from other kinds of coherence. To illustrate our point we use an incoherent source for the excitation.

Before discussing the broad-band excitation case, it is useful to relate the transition moment of the $|\pm\rangle$ states to the *single* isolated molecule moment. Naturally, the relationship depends on the relative geometry of the two molecules forming the dimer. For the general case discussed above [see (4) and (5)], the absorption intensity is

$$|\mu_{g+}|^2 = |\mu_{g1}\cos\theta + \mu_{g2}\sin\theta|^2 \equiv I_+ \tag{33}$$

When $\mu_{g1} = \mu_{g2} = \mu$ we have

$$I_+ = |\mu|^2(1+\sin 2\theta) \tag{34}$$

Again if $\theta = \dfrac{\pi}{4}$ (resonance equivalent pair), then

$$I_+ = 2\mu^2 \tag{35}$$

Similarly

$$I_- = 0 \tag{36}$$

and only *one* state $|+\rangle$ located at $\langle V\rangle$ from the monomer energy has an optically allowed transition to the ground state. We assumed this situation to hold in the discussion of the preceding subsection. Of course if the two molecular dipoles are at an angle other than zero, then both the $|+\rangle$ and $|-\rangle$ states are radiatively active.

As pointed out before, if there are no relaxation mechanisms (radiative and/or nonradiative) in the $\pm$ states, then there will be no real photon absorption or emission, and the energy is exchanged between molecule and radiation field in an alternating way. Irreversible relaxation leads to line broadening of the $\pm$ states. One example of these irreversible relaxation channels is spontaneous emission to the ground state. Imagine that one of the pair molecules absorbs light, whereas the other one does not, as depicted in Fig. 4. Consider the light source to be incoherent, that is, the output is purely thermal with no phase relationship among the emitters. If we assume that the light band width, $\Delta\omega_L$, is large enough to excite both the $+$ and $-$ states, then the description of the time evolution in the pair depends on the *relative* coherence of the source and the dimer.

For pulsed light sources with a pulse duration τ_p, the Fourier transform of the pulse imposes restrictions on the states prepared by the light. In Fig. 4 we describe a situation in which the oscillator strength of the

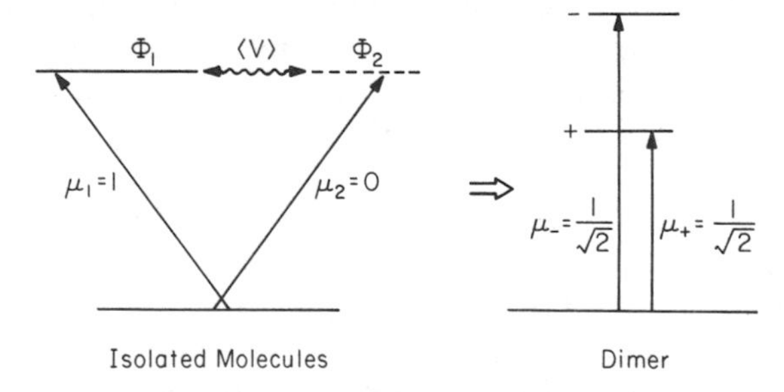

Fig. 4. Distribution of the transition moments of two interacting molecules, Φ_1 and Φ_2, among the $+$ and $-$ states of a dimer. $\langle V\rangle$ is the interaction energy between the molecular states. μ_i is the transition dipole moment of the ith state where $i = +, -, 1, 2$.

allowed molecular transition is divided equally between the + and − states. If the frequency width of the pulse is larger than the energy separation between the "stationary" ± states,[25] then the excitation prepares a coherent superposition of these states yielding a nonstationary state; this is the basis of the familiar beat phenomena.[26] It is known that in these quantum beat experiments the signal is not only sensitive to coherence in the light source, that is, optical coherence,[27] but also to transition coherence, that is it is necessary that both dimer states have transitions with the same polarization allowing the possibility of interference. The above picture, however, is oversimplified because, in addition to the pulse width consideration, we must consider, for a full treatment of the problem, the pulse correlation time and the pumping time, which is inversely proportional to the spectral density at the peak resonance frequency of the transition and to the transition oscillator strength. Nevertheless it is sufficient for our discussion to consider a simple situation in which a pulse of light excites both the + and − states that in turn decay to a final state $|f\rangle$ that need not be the ground state. The problem can now be handled using a quantum electrodynamical approach.[26,28] The spontaneous emission observed at right angles to the exciting beam is detected with a photomultiplier located at point r and only sees light of a specific polarization through a filter.

The wavefunction of the system at $t=0$ can be written as

$$|\Psi(0)\rangle = C_+ \, |+, 0\rangle + C_- \, |-, 0\rangle = \sum_i C_i \, |i, 0\rangle \tag{37}$$

where $|\pm, 0\rangle$ represent the dimer in the $\pm$ states with no *photons* present. The coefficients $C_\pm$ are the probability amplitudes for finding the system in the $\pm$ states. These probabilities ($|C_\pm|^2$) depend on the nature of the light pulse. At any time t, the wavefunction is

$$|\Psi(t)\rangle = \sum_i C_i \, |i, 0\rangle \, e^{-\omega_i t_e - \Gamma_s t/2} + \sum_{f, \mathbf{Q}\boldsymbol{\alpha}} \left\{ \sum_i C^{(i)}_{f, \mathbf{Q}\boldsymbol{\alpha}}(t) \right\} |f, \mathbf{Q}\boldsymbol{\alpha}\rangle \tag{38}$$

Because in our special case the excited states have the same oscillator strength, both the + and − states are damped by the same rate constant for spontaneous emission, Γ_s. Emission of a photon takes the system into the final states f, with the probability amplitude $C^{(i)}_{f, \mathbf{Q}\boldsymbol{\alpha}}$ ($\mathbf{Q}$ is the photon wave vector and $\boldsymbol{\alpha}$ is its unit polarization vector). These coefficients can be obtained by invoking the Wigner-Weisskopf approximation[29]:

$$\begin{aligned} C^{(i)}_{f\mathbf{Q}\boldsymbol{\alpha}} &= C_i \langle f | \, \boldsymbol{\mu}_i \cdot \boldsymbol{\alpha} \, | i \rangle \boldsymbol{\varepsilon}_{\mathbf{Q}} e^{-i\mathbf{Q}\cdot\mathbf{R}} \\ &\quad \times \frac{e^{-i(\omega_f + c\mathrm{Q})t} - e^{-i(\omega_i - i/2\Gamma_s)t}}{\hbar c \mathrm{Q} - (\omega_i - \omega_f)\hbar + i\hbar\Gamma_s/2} \end{aligned} \tag{39}$$

where c is the speed of light.

The light signal seen by the detector is proportional to the electric field strength. To obtain an expression for this field strength, it is standard practice in quantum electronics to quantize the radiation field[30] to yield the following equation which describes the electric field $\boldsymbol{\varepsilon}_{\mathbf{Q}}$ for the modes of the radiation field with wave vector $\mathbf{Q}$:

$$\boldsymbol{\varepsilon}_{\mathbf{Q}} = iA_{\mathbf{Q}}\boldsymbol{\alpha}_{\mathbf{Q}}\{a_{\mathbf{Q}}e^{-i\omega_{\mathbf{Q}}t+i\mathbf{Q}\cdot\mathbf{r}} - a_{\mathbf{Q}}^{+}e^{i\omega_{\mathbf{Q}}t-i\mathbf{Q}\cdot\mathbf{r}}\} \tag{40}$$

where $A_{\mathbf{Q}}$ is a constant that depends on $\hbar\omega_{\mathbf{Q}}$, and $a_{\mathbf{Q}}$ and $a_{\mathbf{Q}}^{+}$ are the annihilation and creation operators, respectively. The total field is therefore given by

$$\boldsymbol{\varepsilon}_T = \sum_{\mathbf{Q}} \boldsymbol{\varepsilon}_{\mathbf{Q}} \tag{41}$$

which can be written as the sum of positive and negative frequency parts of the electric field at point r_1 and time t_1, that is,

$$\boldsymbol{\varepsilon}_T(\mathbf{r}_1t_1) = \boldsymbol{\varepsilon}^{+}(\mathbf{r}_1t_1) + \boldsymbol{\varepsilon}^{-}(\mathbf{r}_1t_1) \tag{42}$$

where

$$\boldsymbol{\varepsilon}^{+}(\mathbf{r}_1t_1) = i\sum_{\mathbf{Q}} A_{\mathbf{Q}}\boldsymbol{\alpha}_{\mathbf{Q}}a_{\mathbf{Q}}e^{-i\omega_{\mathbf{Q}}t_1+i\mathbf{Q}\cdot\mathbf{r}_1} \tag{43}$$

and

$$\boldsymbol{\varepsilon}^{-}(\mathbf{r}_1t_1) = -i\sum_{\mathbf{Q}} A_{\mathbf{Q}}\boldsymbol{\alpha}_{\mathbf{Q}}a_{\mathbf{Q}}^{+}e^{i\omega_{\mathbf{Q}}t_1-i\mathbf{Q}\cdot\mathbf{r}_1} \tag{44}$$

We now suppose that the photons are in some unspecified state $|P\rangle$, and that after emission the system is in the final state $|P_f\rangle$. The matrix element that describes the emission process is

$$\begin{aligned}\langle\boldsymbol{\mu}\cdot\boldsymbol{\varepsilon}_T(\mathbf{r}_1t_1)\rangle &= \langle iP|\,\boldsymbol{\mu}\cdot[\boldsymbol{\varepsilon}^{+}(\mathbf{r}_1t_1)+\boldsymbol{\varepsilon}^{-}(\mathbf{r}_1t_1)]\,|f, P_f\rangle \\ &= \langle i|\,\boldsymbol{\mu}\,|f\rangle\cdot\langle P|\,\boldsymbol{\varepsilon}^{+}+\boldsymbol{\varepsilon}^{-}\,|P_f\rangle \end{aligned} \tag{45}$$

Since we are interested only in the rate of photon emission (i.e., creation of photons in the radiation field) the important term in (45) gives

$$\begin{aligned} I &= |\boldsymbol{\mu}_{if}|^2\,|\langle P|\,\boldsymbol{\varepsilon}^{+}\,|P_f\rangle|^2 \\ &= \sum_{P_f}\langle P|\boldsymbol{\varepsilon}^{+}|P_f\rangle\langle P_f|[\boldsymbol{\varepsilon}^{+}]^{*}|P\rangle \\ &\equiv \langle P|\,\boldsymbol{\varepsilon}^{+}(\mathbf{r}_1t_1)\boldsymbol{\varepsilon}^{-}(\mathbf{r}_1t_1)\,|P\rangle \end{aligned} \tag{46}$$

In other words, the rate at which the radiation field is populated is related to the matrix element of the operator $\boldsymbol{\varepsilon}^{+}\boldsymbol{\varepsilon}^{-}$. We now define a normalized signal intensity

$$S(\mathbf{r}, t) = \langle\Psi(t)|\,\boldsymbol{\varepsilon}^{-}(\mathbf{r})\boldsymbol{\varepsilon}^{+}(\mathbf{r})\,|\Psi(t)\rangle \tag{47}$$

Expanding $\boldsymbol{\varepsilon}^{\pm}$ in terms of the normal modes of the radiation field, substituting (38), and carrying out the energy and angular summations, one obtains

$$S(\mathbf{r}, t) = \sum_{ij} \sum_{f} A_{ij}\theta(t')e^{-i\omega_{ij}t'}e^{-\Gamma_s t'} \tag{48}$$

where

$$A_{ij} = a(\boldsymbol{\mu}_{fi} \cdot \boldsymbol{\alpha})(\boldsymbol{\mu}_{jf} \cdot \boldsymbol{\alpha})C_i C_j^* \tag{49}$$

and

$$t' = t - r_0 c^{-1} \tag{50}$$

In the above expressions, t' is the retarded time due to the distance, r_0, between the dimer and the detector. The matrix elements of the transition dipole operators are $\boldsymbol{\mu}_{fi}$ and $\boldsymbol{\mu}_{jf}$, and θ is the Heaviside function (1 if $t > r_0 c^{-1}$ and 0 if $t < r_0 c^{-1}$). The interesting feature of (48) when we consider a detector close to the sample (i.e., $t = t'$) is the presence of a modulation term that depends on the energy splitting between the + and − states and damps by Γ_s.

The process we have described in this subsection is the dimer analog of the quantum beat experiment. It is a consequence of the fact that both excited dimer states have the same polarization in absorption and emission and thus can share a "single photon." With broad-band excitation the two dimer states cannot be distinguished, and the emission must be considered to arise from a superposition of the two excited states. This transition coherence, as we have called it, is not to be confused with the coherence that is measured using coherent excitation sources.

Quantum beat experiments are of relatively limited use in investigating the excited states of dimers for several reasons. First, one must have the special circumstance that both excited dimer states must have transitions to a final state with the same polarization. Second, dimer splittings are generally on the order of a few wave numbers. A wave number splitting corresponds to a beat frequency of 30 GHz and might be very difficult to detect experimentally. Finally, the two pieces of information that one might obtain from a quantum beat experiment, namely, the separation between dimer excited states and the excited state lifetime, in general can be obtained more easily by other means.

C. Homogeneous and Inhomogeneous Broadening

In the discussion to this point, we have considered only radiative damping of the excited state. It is this damping process and others that we wish to study to investigate coherence in excited-state dimers. Damping processes result in a broadening of the absorption line. Measurements of this broadening can thus yield information about the dynamics of a system, provided that we can be sure that the line broadening is predominantly due to damping.

If the linewidth in absorption is due to damping, we say that the line is homogeneously broadened (HB). The other source of line broadening, inhomogeneous broadening (IB), arises because of inhomogeneity in the environment of the absorbers that make up the ensemble being investigated. This distinction is illustrated in Fig. 5. HB and IB may sometimes be distinguished by analyzing the absorption lineshape. HB usually (but not always, as discussed in Section V.C) results in a Lorentzian resonance. To see this, let us consider the radiated field from, say, the $|+\rangle$ dimer state. Phenomenologically, the electric field $E(t)$ can be written

$$E(t) = \varepsilon(\cos \omega_{g+}t)e^{-t/\tau_+}$$
$$= \frac{\varepsilon}{2}[e^{i(\omega_{g+}+i\Gamma_+/2)t} + \text{c.c.}] \tag{51}$$

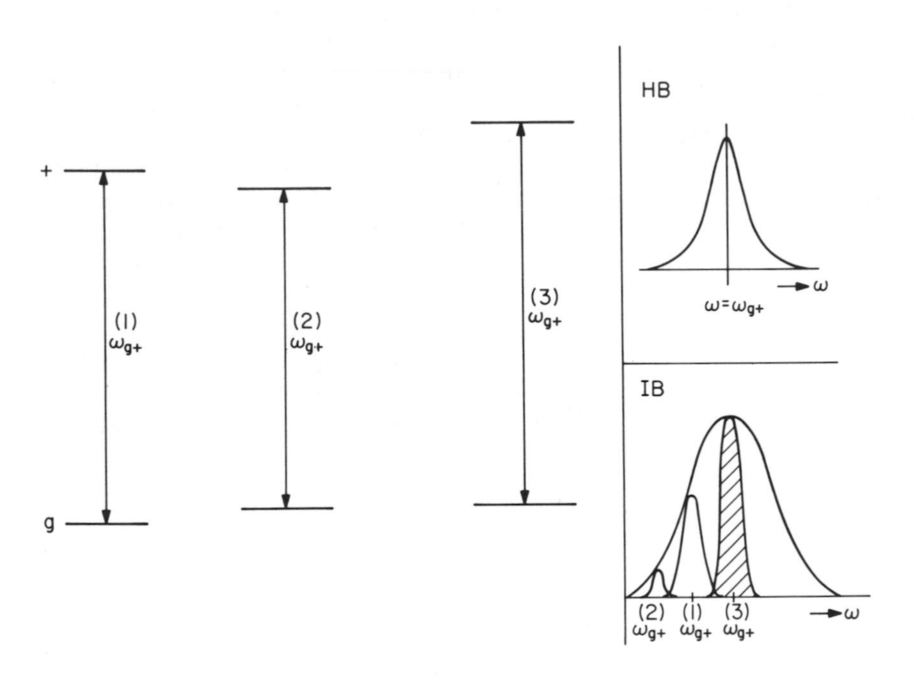

Fig. 5. Homogeneous (HB) and inhomogeneous (IB) broadening. On the left is shown the dimer energy level scheme for three different crystal environments. Note that the transition frequency from g to + states, given by $\omega_{g+}^{(i)}$ where $i = 1$, 2 or 3, varies with i because of differences in both ground- and excited-state energies. On the right we see examples of HB, when the differences between the $\omega_{g+}^{(i)}$ are small compared to the natural linewidth and IB when the linewidth is determined by the differences between the $\omega_{g+}^{(i)}$.

where ω_{g+} is the transition angular frequency and $\Gamma_+/2 = \tau_+^{-1}$. τ_+ is the assumed exponential damping of the dimer state. The Fourier transform of (51) gives[31]

$$E(\omega) = \int_0^\infty E(t)e^{-i\omega t}\,dt$$

$$= \frac{i\varepsilon}{2}\left[\frac{1}{(\omega_{g+} - \omega) + i\Gamma_+/2} - \frac{1}{(\omega_{g+} + \omega) + i\Gamma_+/2}\right] \tag{52}$$

If we ignore the term at about twice the frequency of the transition, then the spectral density, D, of the emission is

$$D \propto |E(\omega)|^2 \propto \frac{1}{(\omega - \omega_0)^2 + (\Gamma_+/2)^2} \tag{53}$$

which is Lorentzian around $\omega = \omega_0$. The linewidth of the transition is therefore

$$\Delta\omega = 2\pi\Delta\nu = \Gamma_+ = \frac{2}{\tau_+} \tag{54}$$

or

$$\Delta\nu = \frac{1}{\pi\tau_+} \tag{55}$$

HB transitions do exist in gases,[32] liquids,[33] and solids.[34] Broadening in gases is due to the *intrinsic* lifetime of the emitter and to collisions between molecules or atoms. In solids the situation is analogous—spontaneous lifetime broadening, dipolar broadening with neighboring molecules, and the broadening due to "collisions" with phonons provide the source of HB. On the other hand, shifts in the crystal field energy of the individual molecules in the lattice result in inhomogeneously broadened (IB) transitions. If the distribution of crystal fields is statistical, this kind of broadening results in a Gaussian lineshape.

D. Density Matrix Description of Dephasing

1. *General Remarks*

In addition to the above-mentioned relaxation processes, namely, radiative decay and nonradiative quenching, there exist processes by which excited and ground states lose their phase coherence with no change in the total population distribution. These processes are pure dephasing processes and are sometimes referred to as *optical dephasing* for optical transitions and *spin dephasing* for spin transitions. In this chapter we discuss both kinds of dephasing as they apply to dimer and exciton systems. The density matrix description of both optical and spin dephasing is outlined in what follows, and the reader is referred to several books[35] and articles[36] for a more complete formulation of density matrices. We begin by describing spin dephasing because of the illustrative conclusions about coherence that come from the well-known commutation relationships among spin angular momentum operators. We then make the transition to the optical case, pointing out connections to the problem of spin coherence.

2. *Spin Dephasing*

Consider a quantum mechanical system whose state cannot be described by a single wavefunction because our knowledge of the system is insufficient. In this case we must consider the system to consist of an ensemble of wavefunctions. The system is said to be in a *mixed* state, and it is described by providing the probability P_i of finding the system in the state $|\psi_i\rangle$. The system is in a pure state when one of the P_i's is unity and all the others are zero. Under these conditions, the density operator is simply

$$\rho = |\psi\rangle\langle\psi| \tag{56}$$

whereas for impure state

$$\rho = \sum_i P_i|\psi_i\rangle\langle\psi_i| \tag{57}$$

The density matrix operator has the following properties:

1. $Tr\rho = 1$
2. ρ is Hermitian
3. ρ is positive definite
4. for a pure state $Tr(\rho)^2 = Tr\rho = 1$
5. The expectation of value of operator $\hat{O}$ is $Tr(\rho\hat{O})$.
6. the equation of motion is

$$i\hbar\dot{\rho} = [\mathcal{H}, \rho] = \mathcal{H}\rho - \rho\mathcal{H} \tag{58}$$

For a spin $\frac{1}{2}$ problem, there are two spinors $|\alpha\rangle = \binom{1}{0}$, $|\beta\rangle = \binom{0}{1}$, and three operators L_x, L_y, L_z that describe the system. The matrices of these operators (in the representation that makes L_z diagonal) are

$$L_z = \frac{\hbar}{2}\begin{pmatrix} 1 & 0 \\ 0 & -1 \end{pmatrix} \tag{59}$$

$$L_y = \frac{\hbar}{2}\begin{pmatrix} 0 & -i \\ i & 0 \end{pmatrix} \tag{60}$$

$$L_x = \frac{\hbar}{2}\begin{pmatrix} 0 & 1 \\ 1 & 0 \end{pmatrix} \tag{61}$$

It is clear therefore that in this case ρ is a 2×2 matrix. For a pure ensemble the state function can be written as

$$\begin{aligned} |\psi\rangle &= C_\alpha\,|\alpha\rangle + C_\beta\,|\beta\rangle \\ &\equiv C_1\,|1\rangle + C_2\,|2\rangle \end{aligned} \tag{62}$$

Using (56) and (62) and taking the ensemble average indicated by a bar above the quantities to be averaged, we have

$$\rho_{11} = \langle 1|\,\{|\psi\rangle\langle\psi|\}\,|1\rangle = \overline{C_1C_1^*} \tag{63}$$

$$\rho_{12} = \langle 1|\,\{|\psi\rangle\langle\psi|\}\,|2\rangle = \overline{C_1C_2^*} \tag{64}$$

or in matrix form

$$\rho = \begin{pmatrix} \overline{|C_1|^2} & \overline{C_1C_2^*} \\ \overline{C_2C_1^*} & \overline{|C_2|^2} \end{pmatrix} \tag{65}$$

with

$$Tr\rho = \overline{|C_1|^2} + \overline{|C_2|^2} = 1 \tag{66}$$

The expectation values of L_x, L_y, and L_z can now be related to the matrix elements of ρ as follows:

$$\langle L_z\rangle = Tr(\rho L_z) = \frac{\hbar}{2}(\rho_{11} - \rho_{22}) \tag{67}$$

$$\langle L_y\rangle = Tr(\rho L_y) = \frac{\hbar}{2}\,i(\rho_{12} - \rho_{21}) \tag{68}$$

$$\langle L_x\rangle = Tr(\rho L_x) = \frac{\hbar}{2}(\rho_{12} + \rho_{21}) \tag{69}$$

Since $\rho_{11} + \rho_{22} = 1$, ρ takes the following form (letting $\hbar = 1$):

$$\rho = \begin{pmatrix} \frac{1}{2} + \langle L_z\rangle & \langle L_x\rangle - i\langle L_y\rangle \\ \langle L_x\rangle + i\langle L_y\rangle & \frac{1}{2} - \langle L_z\rangle \end{pmatrix} \tag{70}$$

$$= \begin{pmatrix} \langle L_z\rangle & \langle L_-\rangle \\ \langle L_+\rangle & \langle L_z\rangle \end{pmatrix} + \tfrac{1}{2}\mathbf{1} \tag{71}$$

If the spins of the particles making up the ensemble are aligned in the same direction as in a pure ensemble, then the system is completely polarized. For this reason one may refer to the above matrix as the *polarization matrix.* For a partially polarized ensemble, the coherence is not completely destroyed and the $Tr(\rho)^2 < 1$. For a completely random ensemble ρ is diagonal and $Tr(\rho)^2 = \frac{1}{2}$. As an example, consider

$$\rho = \begin{pmatrix} 1 & 0 \\ 0 & 0 \end{pmatrix} \tag{72}$$

For this case $Tr\rho = Tr(\rho)^2 = 1$, that is, the ensemble is in a pure state where all the particles (or molecules) have their spin up (α state) but with *no* net polarization. On the other hand, for

$$\rho = \begin{pmatrix} \frac{1}{2} & 0 \\ 0 & \frac{1}{2} \end{pmatrix} \tag{73}$$

$Tr\rho = 1$, $Tr(\rho)^2 = \frac{1}{2}$, and the ensemble (impure) is in an equal mixture of the $|\alpha\rangle$ and $|\beta\rangle$ states, but with *no* polarization. In this case ρ describes an incoherent state. If the off-diagonal elements of ρ are nonzero, for example,

$$\rho = \begin{pmatrix} \frac{1}{2} & \frac{1}{2} \\ \frac{1}{2} & \frac{1}{2} \end{pmatrix} \tag{74}$$

with $Tr\rho = Tr\rho^2 = 1$, then a pure state of the ensemble is formed (coherent state). Note that the difference between the two situations is that, although in both cases $\langle L_z \rangle = 0$ (i.e., the ensemble has an equal population distribution between levels α and β), $\langle L_x \rangle \neq 0 = \frac{1}{2}$ for the latter case. This in fact means that *a coherent state has a nonzero* x-y *polarization component, and hence the off-diagonal elements of* ρ *are nonzero.*

To find the degree of coherence one introduces two relaxation rate constants for the ensemble averaged decay of the diagonal and off-diagonal elements of ρ. These are the longitudinal (T_1) and transverse (T_2) relaxation rates, respectively. The latter is sometimes called the dephasing time, whereas the former gives the energy relaxation time.

3. *Optical Dephasing*

Consider the explicit example of a two-level system that describes an optical transition. As an example, let us consider the optical transition between the ground and the electronic excited + state of a dimer. By analogy to the density matrix of the spin $\frac{1}{2}$ problem, the 2×2 density matrix for the two level optical case is

$$\rho = \begin{pmatrix} \overline{|C_g|^2} & \overline{C_g C_+^*} \\ \overline{C_+ C_g^*} & \overline{|C_+|^2} \end{pmatrix} \tag{75}$$

where the subscript g refers to the ground state and the + to the excited state. For a completely random system, that is, a completely incoherent system, there is no correlation between the coefficients C_+ and C_g. As we have seen in this case, the off-diagonal terms are zero and $Tr(\rho)^2 = \frac{1}{2}$, for example,

$$\rho = \begin{pmatrix} \frac{1}{2} & 0 \\ 0 & \frac{1}{2} \end{pmatrix} \tag{76}$$

The states depicted in Fig. 3 are pure (coherent) states in the sense described above. To see this, consider the following three orientations of the vector described in the figure:

$$-z\text{-axis} \qquad \begin{pmatrix} 1 & 0 \\ 0 & 0 \end{pmatrix} \tag{77}$$

$$+z\text{-axis} \qquad \begin{pmatrix} 0 & 0 \\ 0 & 1 \end{pmatrix} \tag{78}$$

$$xy\text{-plane} \qquad \begin{pmatrix} \frac{1}{2} & \frac{1}{2} \\ \frac{1}{2} & \frac{1}{2} \end{pmatrix} \tag{79}$$

In all cases $Tr(\rho) = Tr(\rho)^2 = 1$

Similarly to (67) to (69) of the spin case, a three-component vector can be related to the components of the density matrix[24]:

$$r_x = \rho_{g+} + \rho_{+g} \tag{80}$$

$$r_y = i(\rho_{+g} - \rho_{g+}) \tag{81}$$

$$r_z = \rho_{++} - \rho_{gg} \tag{82}$$

This vector obeys the equation of motion of a vector of constant length precessing about an effective field. It is shown schematically in Fig. 6.

The formal introduction of damping into the density matrix formalism can be done in a variety of ways. One method[23,24] that we do not discuss here in detail is to note the formal similarity between the equation of motion for the vector **r** and for a magnetic moment precessing about an external magnetic field. One may then use the Bloch equations[35,37] with the two relaxation times T_1 (longitudinal or spin-lattice relaxation time) and T_2 (transverse or spin-spin relaxation time), redefining these quantities for the optical case.

We introduce damping directly into the density matrix in a manner consistent with the Bloch formalism. Assume that the amplitude of $\bar{C}_i$ decays with the time $2T_{1i}$ so that we may write

$$\rho_{ii} = |C_i|^2 e^{-t/T_{1i}} \tag{83}$$

where T_{1i} represents the decay of the population in the ith state. Assume that in addition to this decay, the relative phases of C_i and C_j decay with a time T_2'. We can then write a density matrix including damping for the two-level system of dimer ground state and + excited state as

$$\rho = \begin{pmatrix} |C_g^2| e^{-t/T_{1g}} & |C_g C_+^*| e^{-1/2[1/T_{1g} + 1/T_{1+}]} e^{-t/T_2'} \\ |C_+ C_g^*| e^{-1/2[1/T_{1g} + 1/T_{1+}]} e^{-t/T_2'} & |C_+|^2 e^{-t/T_{1+}} \end{pmatrix} \tag{84}$$

The ground state lifetime T_{1g} is nearly infinite in the case of interest, and we can thus define the total decay of the off-diagonal density matrix

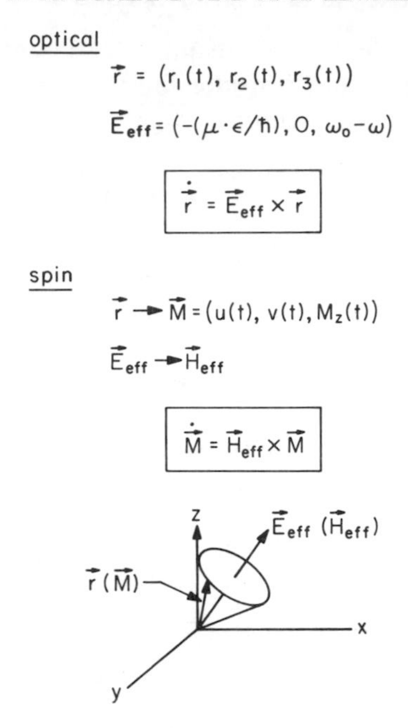

Fig. 6. The precession of the **r**-vector about an effective field for optical and magnetic resonance experiments. The components of the **r**-vector are given by (80) to (82). The field about which precession occurs in the optical experiment is a fictitious field $\mathbf{E}_{\text{eff}}$. In magnetic resonance it is an effective magnetic field $\mathbf{H}_{\text{eff}}$. In magnetic resonance the components of the **r**-vector are equal to the components of the magnetization **M**. The subscripts 1, 2, and 3 for the **r**-vector components correspond to x, y, z in the text.

elements as T_2 given by

$$\frac{1}{T_2}=\frac{1}{2T_{1+}}+\frac{1}{T_2'} \tag{85}$$

T_2 thus represents the characteristic time for the loss of phase coherence in the two-level dimer system. We should note at this point that (84) is not a rigorously correct method of introducing T_1 and T_2 into the density matrix. One should include diagonal terms that describe the build-up of population in one state as a result of decay in the other. Equation 84 is sufficient for our descriptive purposes here, however.

A direct measurement of the quantity T_2 would be the most unambiguous method of investigating the mechanisms by which phase coherence is lost in dimer systems. Fortunately, these experiments on solids are just becoming possible in the optical domain[38–41] and, as we shall see, have already been done on dimer systems in the microwave region (see Section III.F). As one example of a whole class of coherent optical experiments, we consider the photon echo experiment shown schematically in Fig. 7. In this figure the behavior of the vector **r** is depicted. From its definition [(80) to (82)] and the decay behavior of the density matrix [(84)], we see that r_z has a decay time of T_1 and r_x and r_y a decay time of T_2. If the system is initially along the $-z$ direction, that is, all molecules are in the ground state (see Fig. 3), a $\pi/2$ pulse places the system in a coherent state along the y-axis. The vector **r** will then begin to precess about the z-axis. Because of the presence of inhomogeneous broadening discussed in Section III.C, not all dimers will precess at the same frequency. If, after a time τ, a π-pulse is applied, the precessing vectors will be inverted through the origin. After another period τ they will have reformed along the $-y$-axis. At this point a coherent burst of light, the echo, is observed. By measuring the echo intensity as a function of the pulse separation τ, one obtains a direct measure of T_2. When the vector system is in the xy-plane, the irreversible loss of phase is due only to T_2 processes. Since the system dephases for a total time 2τ, the echo intensity $I(\tau)$ is given by

$$I(\tau) = I(0)e^{-2\tau/T_2} \tag{86}$$

from which T_2 is directly obtained.

Photon echo experiments are just one of a variety of types of experiments (e.g., nutation, free induction decay) in which the loss of coherence of a system is studied by exciting the system with a laser.[38–41]

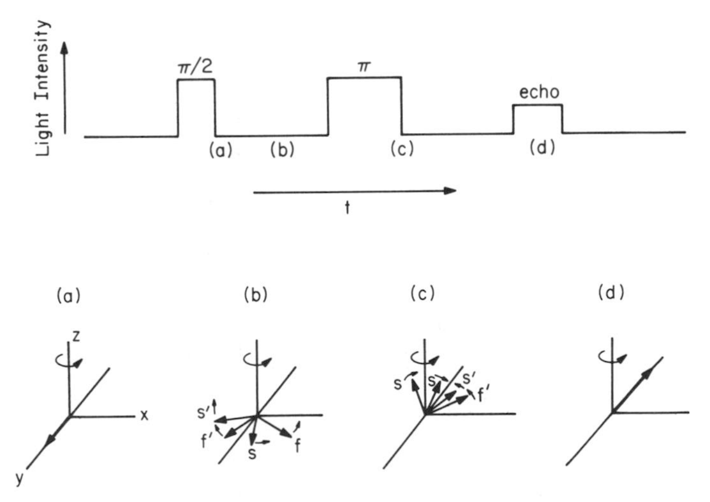

Fig. 7. A two-pulse photon echo. At the top is shown the sequence of optical pulses giving rise to the echo plus the echo itself. Below is shown the position of the **r**-vector (see Fig. 6) at various stages in the pulse sequence. At (b) the **r**-vectors for different components of the inhomogeneous line fan out due to their slightly different precession frequencies. After application of a π-pulse at (c) the **r**-vectors rephase yielding an echo at (d).

E. Optical Spectra of Dimers

During the 1960s several groups investigated the band structure of triplet and singlet excitons using optical absorption and emission techniques. The approaches were different and included direct measurement of the Davydov splitting,[42] the density of states function[43] and band-to-band (exciton band to ground state vibron band) transitions,[44] and the measurements of optical splittings in aggregates (that is, dimers and higher N-mers) in isotopically mixed crystals.[45] Nieman[46] in Robinson's group made an early attempt to observe the emission from isolated triplet dimers of perprotobenzene in a perdeuterohost crystal. The experiments were unsuccessful, perhaps because of the complicating presence of three sets of translationally inequivalent pairs of molecules in the crystallographic unit cell, and because of the fact that the dimer states (whose separation from the monomer are now known to be small)[47] are hidden under the inhomogeneous width of the $T \rightarrow S$ monomer resonance. Hanson[48] was later successful in identifying the naphthalene dimer splitting for the lowest triplet state of perprotonaphthalene in a perdeuterohost. Hanson's experiment gave a pairwise splitting ($\equiv 2\beta$) of $2.5 \pm 0.5\ \text{cm}^{-1}$ which, when multiplied by the required factor of 4 (two molecules in the unit cell), gave a triplet exciton band splitting ($10\ \text{cm}^{-1}$)

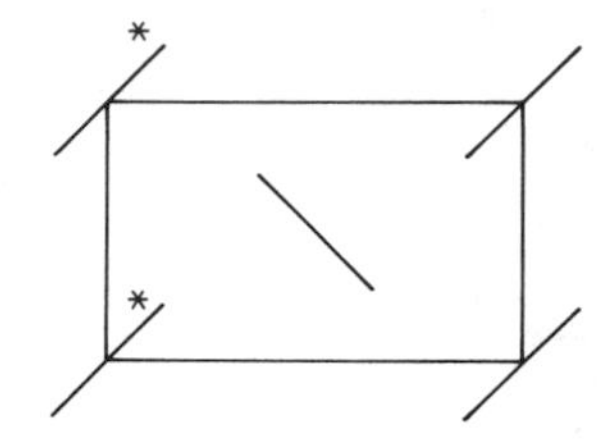

Translationally Equivalent Dimer

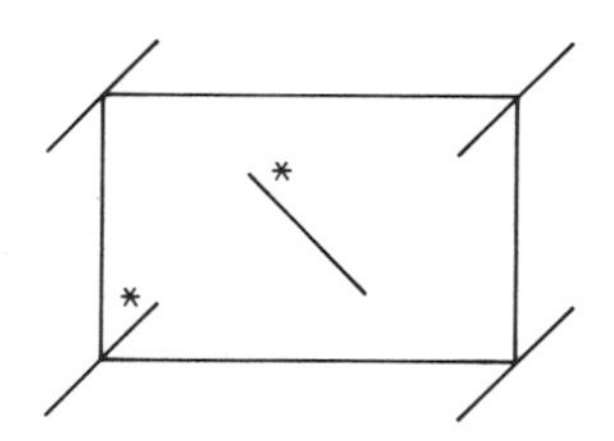

Translationally Inequivalent Dimer

Fig. 8. Translationally equivalent and translationally inequivalent molecules (denoted by a star) for crystals like naphthalene and phenazine.

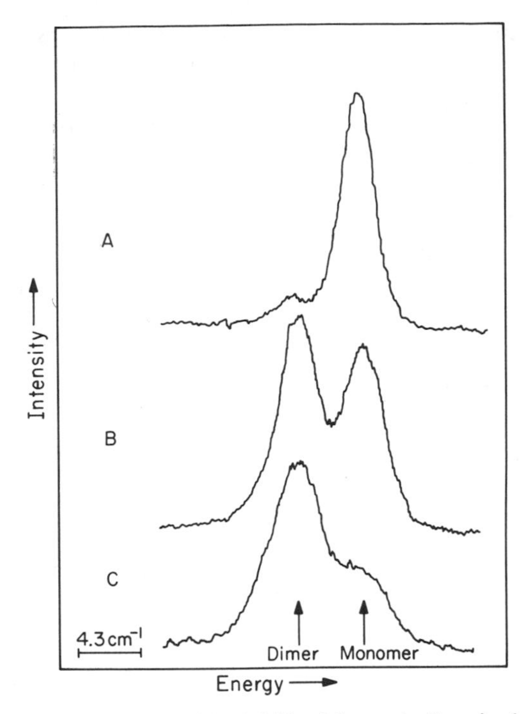

Fig. 9. The triplet emission spectra ($T = 1.5$ K) of isotropically mixed phenazine crystals (6463 Å) at different guest concentrations: $A = 0.5\%$, $B = 3.1\%$ and $C = 5.7\%$ (by weight). After Ref. 53.

that is consistent with the splitting measured for the 0, 0 band of pure naphthalene crystals.[49] Furthermore, the concentration and the light polarization dependence experiments of Braun and Wolf[50] have ruled out the possibility that the observed lines are due to site splittings and/or isotopic impurity transitions. The above experiments, because they involved translationally inequivalent pairs of molecules, did not give the contribution of the translationally equivalent interactions to the band splitting. The distinction between translationally equivalent and inequivalent dimers[51] is shown in Fig. 8.

Phenazine crystals, which have two molecules in the unit cell[52] and a similar crystal structure to naphthalene, exhibit a large triplet-state translationally equivalent interaction[14,53] in addition to the Davydov splitting arising from inequivalent interactions.[54] From the triplet dimer emission experiments of Zewail[53,55] on phenazine-h_8 doped into phenazine-d_8, one infers that the translationally equivalent excition band splitting is $17 \pm 1\ \text{cm}^{-1}$ (see Fig. 9) which is consistent with the large spectral broadening observed in the 0, 0 absorption spectrum of phenazine-h_8 when doped with relatively large amounts of phenazine-d_8.[14] The other excitation exchange interactions in phenazine must be much smaller than the inhomogeneous linewidth identified recently by Smith et al.[56]

Recently, high-resolution optically detected magnetic resonance spectroscopy[57] and laser excitation spectroscopy have helped in identifying aggregates hidden under the IB transitions. In Figs. 10 and 11 we show the naphthalene guest spectra obtained by Dupuy et al.[58] using direct narrow- and wide-band laser excitation into the triplet origin. The structure shown in Fig. 11 with narrow-band excitation clearly indicates that the naphthalene pair spectra of Hanson contained much more information than originally anticipated. This additional information was masked by the IB. The line narrowing experiments of Dupuy et al. circumvent the IB by selectively exciting only a subset of molecules. Although many of the lines in Fig. 11 are not assigned, they have identified the translationally equivalent pair state.

Energy transfer in the "two-dimensional" naphthalene and phenazine

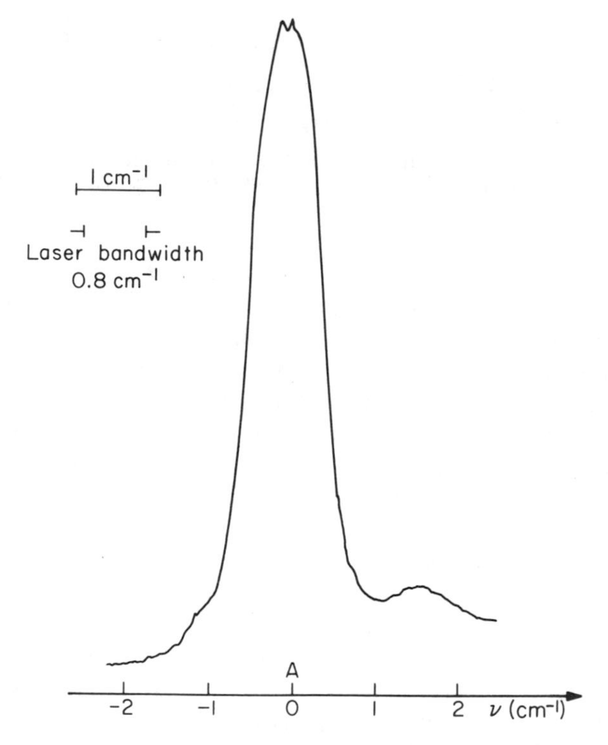

Fig. 10. Triplet excitation spectra of naphthalene isotopically mixed crystals ($T = 1.7$ K) with an excitation half bandwidth of 0.8 cm^{-1}. After Ref. 58.

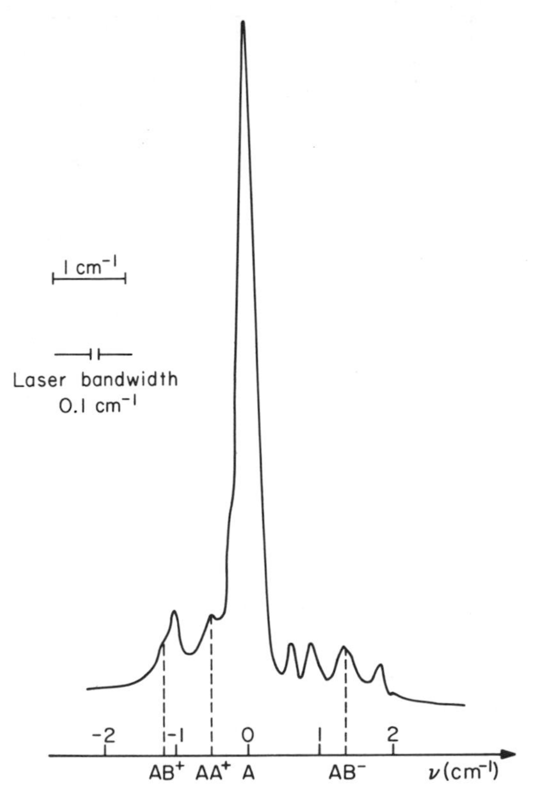

Fig. 11. Triplet excitation spectra of naphthalene isotopically mixed crystals ($T = 1.7$ K) with an excitation half bandwidth of 0.1 cm^{-1}. After Ref. 58.

systems, although providing information about the influence of the anisotropy of intermolecular interactions, are complicated to unravel. The analysis becomes somewhat simpler in a "one-dimensional" system like that discussed by Hochstrasser et al.[59,60] in 1,4-dibromonaphthalene (DBN) crystals. Aggregates in the 0,0 phosphorescence spectra of DBN at high resolution were seen up to tetramers. Figure 12 shows the high-resolution phosphorescence spectra as a function of guest concentration. The analysis of the resonance interaction splittings is consistent with a linear chain of molecules with a nearest-neighbor excitation exchange interaction energy of $6.2 \pm 0.3\ cm^{-1}$. Also, because DBN is essentially a linear chain, the intensity distribution of the emission among the various aggregate excited states can be predicted knowing the symmetry of the aggregate. Table I summarizes the intensities and energies of linear chain systems. The splittings mentioned above do not give accurate resonance interactions unless one corrects for solvent shifts and quasiresonance

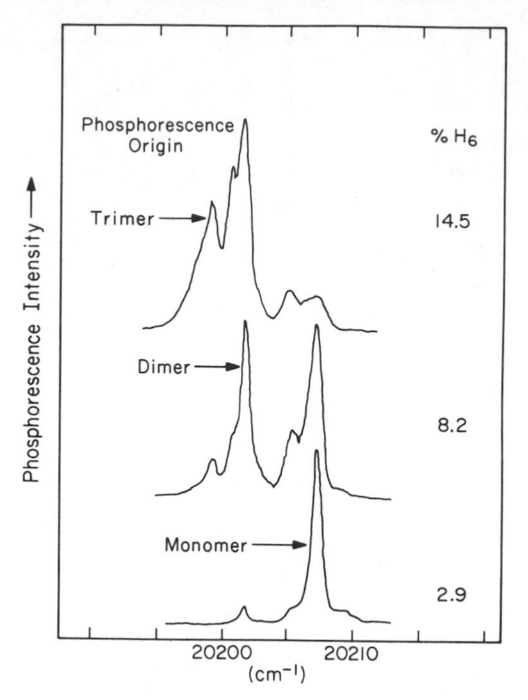

Fig. 12. Resonance multiplet emission of isotopically mixed DBN crystals at 1.46 K and at different guest concentrations. After Ref. 60.

interactions (i.e., the guest-host indirect interactions which depend on the trap depth for the guest). These corrections have been treated extensively for the naphthalene exciton system by Kopelman and his co-workers.[51,61] The question in the current context is, can the measurements of translationally equivalent and translationally inequivalent resonance pair interactions provide information about the coherence of the pair?

The answer to the above question is no, because in addition to these *stationary* interactions, one must know the degree to which nonstationary (e.g., exciton-phonon coupling) interactions influence the homogeneous width of both the + and − states. It is these nonstationary and random interactions that result in the destruction of dimer coherence.

A qualitative estimate of the dimer coherence can be obtained by comparing the energy splitting in the pair (singlet or triplet) to the homogeneous linewidth of the + and − states, provided one can separate homogeneous and inhomogeneous contributions to the linewidth (see Fig. 13). One can define a memory parameter that is given by

$$g_{\pm} = \tfrac{1}{2}\omega_{+-}\tau_{\pm} \tag{87}$$

TABLE I
Intensities, Wavefunction Coefficients and Energies of Linear Chain Aggregates

$$\varepsilon_K = 2\beta \cos (K\pi/(N+1))$$

$$C_{Kj} = \sqrt{\frac{2}{N+1}} \sin (Kj\pi/(N+1))$$

$$I_K(N) = \left[\frac{2}{N+1} \operatorname{ctg}^2(K\pi/(2N+2))\right]\delta_K$$

$$K = 1, \ldots, N$$

$$j = 1, \ldots, N$$

$$\delta_K = 1 \text{ for odd } K$$

$$= 0 \text{ for even } K$$

Monomer: $N=1$

$$\varepsilon_K = (0) \qquad C_{11} = (1) \qquad I_K = (1)$$

Dimer: $N=2$

$$\varepsilon_K = \begin{pmatrix} \beta \\ -\beta \end{pmatrix} \quad C_{Kj} = \begin{pmatrix} 0.707 & 0.707 \\ 0.707 & -0.707 \end{pmatrix} \quad I_K = \begin{pmatrix} 2 \\ 0 \end{pmatrix}$$

Trimer: $N=3$

$$\varepsilon_K = \begin{pmatrix} 1.414\,\beta \\ 0 \\ -1.414\,\beta \end{pmatrix} \quad C_{Kj} = \begin{pmatrix} 0.5 & 0.707 & 0.5 \\ 0.707 & 0 & -0.707 \\ 0.5 & -0.707 & 0.5 \end{pmatrix} \quad I_K = \begin{pmatrix} 2.91 \\ 0 \\ 0.09 \end{pmatrix}$$

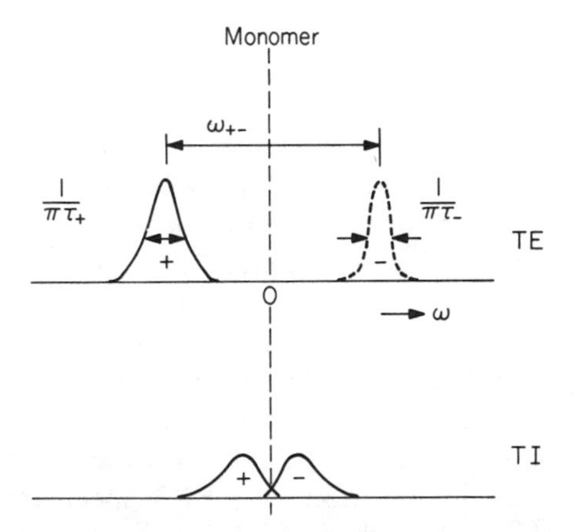

Fig. 13. Expected position of translationally equivalent and inequivalent dimer states in phenazine-type systems. Note that $\omega_{+-} > 1/\pi\tau$, implying the partial coherence of both types of dimer systems.

If $g_{\pm} \gg 1$, then the random coupling between the excitation and the lattice is not strong enough to completely destroy the coherence of the $\pm$ states. On the other hand, if $g_{\pm} \lesssim 1$, then the identity of + and − states is lost and hence the coherence is destroyed. We note that $\tau_{\pm}$ in the above expression is the total dephasing time obtained from the homogeneous linewidth, and it reflects the ensemble averaged cross-terms (between the ground state and the $\pm$ states) in the density matrix described previously by (84). For naphthalene and phenazine the dimer lines are inhomogeneously broadened, and thus no estimate of g can be obtained. It may be possible to determine the homogeneous linewidth in the dimer states by using narrow-band excitation techniques. So far, however, the optical experiments have only been able to provide a lower limit to the dimer dephasing time $\tau_{\pm}$.

F. Magnetic Resonance Spectra of Dimers

1. *Schwoerer-Wolf Experiments*[62]

In 1967 Schwoerer and Wolf observed new EPR transitions in isotopically mixed naphthalene crystals that were not due to monomer transitions. These new lines (the so-called *M* lines) were assigned as transitions within the excited triplet state of nearest neighbor, translationally inequivalent pairs of naphthalene-h_8 molecules. The spin Hamiltonian of this dimer system was found to be

$$\mathcal{H}_s^* = g_e^* \beta_e H_0 S + D^* S_{z^*}^2 + E^*(S_{x^*}^2 - S_{y^*}^2) \tag{88}$$

with

$$D^*/hc = -0.0059 \pm 0.0006\ \text{cm}^{-1} \tag{89}$$

$$E^*/hc = +0.0485 \pm 0.0006\ \text{cm}^{-1} \tag{90}$$

$$g^* = 2.0030 \pm 0.0010 \tag{91}$$

where g_e^* is the electron g factor, β_e the Bohr magneton, H_0 the applied magnetic field, and S the electron spin angular momentum. The components of S along the various directions x^*, y^*, and z^* are given by S_x^*, S_y^*, S_z^*. D^* and E^* are the zero-field splitting parameters. These values are different from those of the naphthalene monomer. Furthermore, the x^*, y^*, z^* principal magnetic axes are not the same as the molecular axes, but are related to the crystal axes: The z^* axis is identical to the **b** axis, and the x^* axis makes an angle of 22.4° with the **a**. Although hyperfine structure was resolved in the monomer EPR spectrum, the *M*-lines (at that time) showed no hyperfine structure in any orientation of the magnetic field. The above results of Schwoerer and Wolf agree well with the spin Hamiltonian deduced for triplet excitons by Sternlicht and

McConnell.[63] Furthermore, the integrated intensity of EPR monomer and dimer pairs is consistent with a complete statistical distribution of guest napthalene-h_8 in host naphthalene-d_8. Therefore the assignment of the M-line EPR spectrum as being due to dimer pairs was confirmed.

Knowing that the M-lines corresponded to the + and − optical transitions of the pair, Schwoerer and Wolf used Anderson's theory[64] of exchange narrowing and the linewidth of the M-lines to obtain information about the dynamics of the pair. This theory, which is discussed in detail later, can be summarized as follows. Suppose we have two transitions (say for two different molecules, 1 and 2) with Larmor frequencies ω_1 and ω_2. The EPR spectrum will therefore consist of two lines separated by $\omega_2 - \omega_1$. We assume that the linewidth for the two transitions is vanishingly small. If spin exchange between 1 and 2 is now turned on at a frequency $\omega_{ex} = 2\beta/\hbar$ (where β is the exchange integral), and with the condition

$$\omega_{ex} = 2\beta/\hbar \gtrsim (\omega_2 - \omega_1) \tag{92}$$

then the spectrum will contain only a single line at the frequency $(\omega_2 + \omega_1)/2$. Furthermore, if $\omega_{ex} \gg \omega_1, \omega_2$, the width of the line becomes

$$\Delta\omega_{ex} = \frac{10\hbar}{3\beta}\left[\tfrac{1}{4}(\omega_1 - \omega_2)^2\right] \tag{93}$$

and the lineshape function is approximately Lorentzian. It should be pointed out that (93) was derived for isotropic systems, and its application to anisotropic molecular crystals must be checked.

The problem of applying these equations to the naphthalene pair spectra is that in the zero exchange limit the two monomer lines have finite *inhomogeneous* widths. This point was realized by Schwoerer and Wolf. However, they argued that in the limit of infinitely slow exchange, the resulting linewidth of the M-lines will be related to the inhomogeneous width simply by the relationship

$$\Delta\omega_M(\infty) = \Delta\omega_{1,2}/\sqrt{2} \tag{94}$$

Therefore at finite exchange rate

$$\Delta\omega_M = \Delta\omega_M(\infty) + \Delta\omega_{ex} \tag{95}$$

Based on the above equations, they deduced the value $\beta = 5 \pm 0.7\ \text{cm}^{-1}$, which agreed with the value obtained theoretically by Jortner et al.[65] (4 to $4.6\ \text{cm}^{-1}$, depending on the wavefunctions chosen) for the exchange integral of the triplet exciton band in naphthalene. This number is different from the directly measured value of $1.25\ \text{cm}^{-1}$.[49] Assuming that the coherence time in the pair is determined by β (i.e., $\tau_c = h/4\beta = 1.7$ psec) and that the mean distance between the two molecules comprising the dimer is $\sqrt{(a/2)^2+(b/2)^2}$, they concluded that an exciton being scattered in a similar way would have a diffusion coefficient of 5×10^{-4} cm^2/sec. This value agrees with the results of Avakian and Merrifield[66] obtained for naphthalene crystals ($D=2\times10^{-4}$ cm^2/sec) at room temperature where one expects the "motion" of the exciton to be incoherent. It is not surprising that the room temperature result for naphthalene is characteristic of random-walk type migration, since essentially all low-frequency phonons are populated. What is surprising, however, is that the motion is, according to these results, still incoherent at very low temperatures. However, the problem of inhomogeneous broadening in the pair spectrum is a serious one, and, as we discuss later, the EPR results on other pairs have demonstrated the presence of coherence at low temperatures. Furthermore, as we see later, the relationship between dimer coherence and exciton transport is not as straightforward as this analysis makes it sound. We return to this naphthalene problem later in this section. Finally, we should add that these high field EPR experiments cannot distinguish the monomer spectrum from the spectrum of translationally equivalent pairs, and hence no information about coherence in these dimers can be obtained.

2. *Hutchison-King Experiments*[67]

As mentioned before, the EPR experiments on naphthalene pairs isolated in the perdeuterohost did not show any hyperfine splittings due to interactions between the triplet electrons and the surrounding nuclear spins of the protons. In 1973 Hutchison and King[67] reported proton hyperfine structure of the EPR lines and the ENDOR (electron-nuclear double resonance) spectra of the protons at 1.8 to 2.0 K at both X-band and K-band microwave frequencies. The spectra were seen at guest concentrations ranging from 0.02 to 0.10 mole fractions of naphthalene-h_8.

To analyze these results, the spin Hamiltonian of the pair now must contain the hyperfine term in addition to the electron spin terms described in the preceding section, that is,

$$\begin{aligned}\mathcal{H}_s = {}& +|\beta_e|\,\mathbf{H}_0\cdot\mathbf{g}_e^*\cdot\mathbf{S}+D^*S_{z^*}^2+E(S_{x^*}^2-S_{y^*}^2)\\ & -|\beta_n|\,g_n\mathbf{H}_0\cdot\sum_{k=1}^{8}\mathbf{I}_k+\sum_{k=1}^{8}\mathbf{S}\cdot\mathbf{A}_k\cdot\mathbf{I}_k\end{aligned} \tag{96}$$

The last two terms describe the nuclear Zeeman term and hyperfine term, respectively. In the above spin Hamiltonian, $S=1$ and $I_k=\frac{1}{2}$ for all k protons. g_n is the nuclear g-factor, β_n the nuclear magneton, and $\mathbf{A}_k$ is the hyperfine coupling tensor.

The degree of excitation delocalization can be inferred from these magnetic resonance experiments. With the exception of one measurement ($H_0//X$ of one of the molecules in the pair) Hutchison and King could fit the hyperfine structures, within the uncertainty of their measurements, by a spin Hamiltonian of the form of the last two terms of (96) *but with the* k-*sums running over the* 16 *indices of all the protons in the pair and with hyperfine tensor* ($\mathbf{A}_k$) *elements equal to half those of the isolated molecule.* The result is consistent with the notion that the excitation exchange matrix element is *much* larger than the D and E values of the isolated molecule, in agreement with the findings of the fine structure experiments of Schwoerer and Wolf. Again, as in the previous experiments, the degree of coherence in the pair cannot be obtained unless we know the coherence time. Further, because of the inhomogeneous broadening at high magnetic fields, the chances for separating the monomer resonance from the translationally equivalent resonances seems remote.

3. *Hochstrasser-Zewail Experiments*[60]

Around 1970 in Hochstrasser's laboratory Scott and Zewail[68] attempted to observe the zero-field EPR spectra of naphthalene pairs using isotopic mixed crystals and conventional optical detection of magnetic resonance (ODMR). The experiments were unsuccessful, perhaps due to the absence of spin polarization in the pair. The observation of the ODMR pair spectra was then made in 1,4-dibromonaphthalene (DBN).[60,69] DBN was chosen because (*a*) it was known by that time that the triplet exciton in DBN is quasi-one-dimensional[59]; (*b*) since the magnetic axes in the pair are expected to be close to those of the monomer, spin-lattice relaxation will not be enhanced in the dimer, and hence some spin polarization is expected; (*c*) the monomer optical transition is reasonably well separated from the dimer optical transition; (*d*) the monomer zero-field ODMR showed a large degree of spin polarization; (*e*) the lifetime of the triplet state is relatively short (~ 5 msec), and hence the detection of magnetic resonance transitions via the phosphorescence signal should be straightforward; and (*f*) one expects large differences in the zero-field splittings of the $\pm$ states (discussed in the next section) as a result of the large spin-orbit coupling. It was hoped that these differences due to spin-orbit anisotropy and other magnetic interactions (such as intermolecular electron spin-electron spin interactions) could be resolved in zero field because the linewidth of the ODMR spectra are typically 1 to 2 MHz. However, this turned out not to be the case in DBN.

DBN belongs to the space group P_1^2/a (C_{2h}^5) with eight molecules in the unit cell.[70] The projection of its crystallographic unit cell in the **ab** plane is shown in Fig. 14. The cell consists of four asymmetric units of two

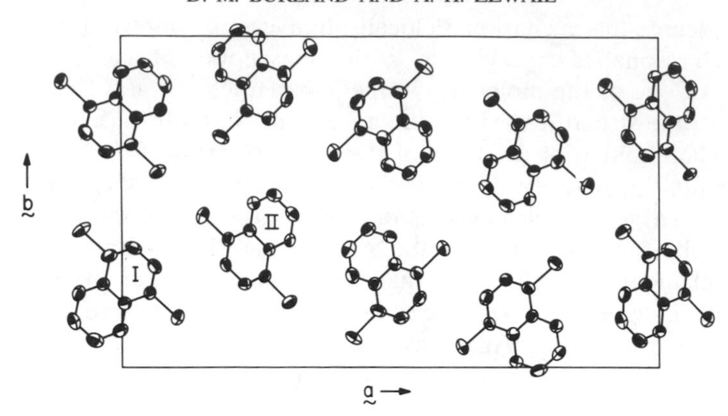

Fig. 14. The projection of the unit cell contents of 1,4-dibromonaphthalene on the **ab** plane. The crystal structure is monoclinic, $P2_1/a$, $a = 27.248$, $b = 16.477$, $c = 4.059$ Å, $\beta = 91.91°$, $Z = 8$. The two molecules labeled I and II are in inequivalent unit cell positions. After Ref. 70b.

molecules each, since the space group (C_{2h}^5) admits only four crystallographically equivalent molecules per unit cell. The dimer we discuss is the translationally equivalent pair along the **c** crystallographic axis. The molecular symmetry is C_{2v}. Because there is no center of inversion in the molecular point group, the two molecules in *the pair are not necessarily identical in the isotopically mixed crystals.* In other words θ in (6) is not necessarily 45°.

To find the degree of excitation delocalization in DBN, Hochstrasser and Zewail measured the emission intensity from the monomer and from the + state of the dimer to the zero point level and compared it to the intensity of emission to a vibrational level of the ground state. It was found that at 1.46 K, the ratio of dimer to monomer emission intensity for the 0,0 band is 1.9 ± 0.3 times larger than the same ratio for emission to the 310 cm^{-1} ground state vibrational level and 2.2 ± 0.4 times larger than emission to the 527 cm^{-1} level. Theoretically, if one uses weak coupling functions of the form $1/\sqrt{2}[\Phi_1 V_1 + \Phi_2 V_2]$, where V_1 is an excited vibrational wavefunction localized on molecule 1 and assumes that the ground state functions are simply $1/\sqrt{2}\Phi_0(V_1 \pm V_2)$, this ratio should be two.

Even if the excitation is shared equally[60,71] between the two DBN molecules that form the pair, the zero field splittings of the dimer and the monomer should be similar in a zero-order approximation. The spin Hamiltonian for the pair is simply

$$\mathcal{H}_s \approx \tfrac{1}{2}[\mathcal{H}_s^{(1)} + \mathcal{H}_s^{(2)}] \tag{97}$$

where $\mathcal{H}_s^{(1)}$ is the spin Hamiltonian of molecule 1. In DBN in addition to the protons there exist two Br atoms (labeled by k') which give rise to quadrupole satellites in the ODMR spectra. These satellites are split from the main peak of the "pure" electron-spin zero field transitions by ~270 MHz. The total Hamiltonian must therefore contain the quadrupole terms[72] which do not exist in the naphthalene case,[73] that is,

$$\mathcal{H}_s(\text{DBN}) = DS_z^2 + E(S_x^2 - S_y^2) + \sum_{k'} (\mathcal{H}_Q(k') + \mathbf{S}\cdot\mathbf{A}_{k'}\cdot\mathbf{I}_{k'}) + \sum_k \mathbf{S}\cdot\mathbf{A}_k\cdot\mathbf{I}_k \tag{98}$$

where $\mathcal{H}_Q$ is the quadrupole Hamiltonian given by

$$\mathcal{H}_Q = \frac{e^2qQ}{12}[(3I_y^2 - \tfrac{15}{4}) + \eta(I_z^2 - I_x^2)] \tag{99}$$

The quadrupole coupling constant ($\frac{1}{2}e^2qQ$) and the asymmetry parameter η are different for Br^{79} and Br^{81} isotopes even though both isotopes have $I = \frac{3}{2}$. The zero-field ODMR spectra[74] of monomer and dimer DBN are shown in Figs. 15 and 16. From the spectra it is clear that the linewidths are broad relative to typical ODMR spectra of molecules in other chemically mixed or isotopically mixed crystals. Second, within the experimental error the D and E terms in the monomer and the dimer are

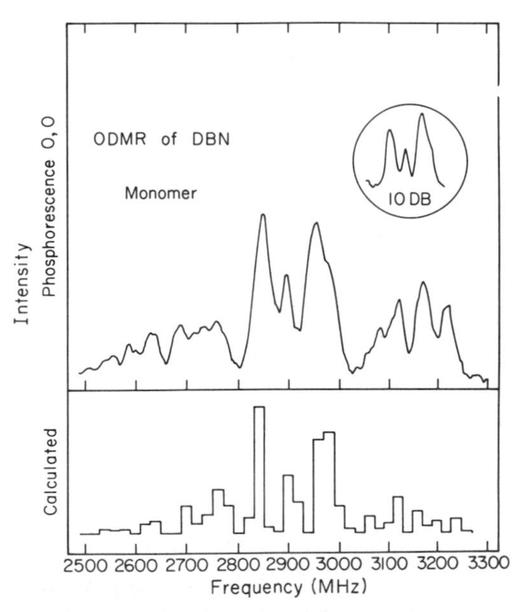

Fig. 15. Zero-field ODMR spectrum of DBN-h_6 monomer in DBN-d_6 at 1.5 K (above) and the calculated spectrum (below). The monomer emission was isolated using a spectrometer. After Ref. 74.

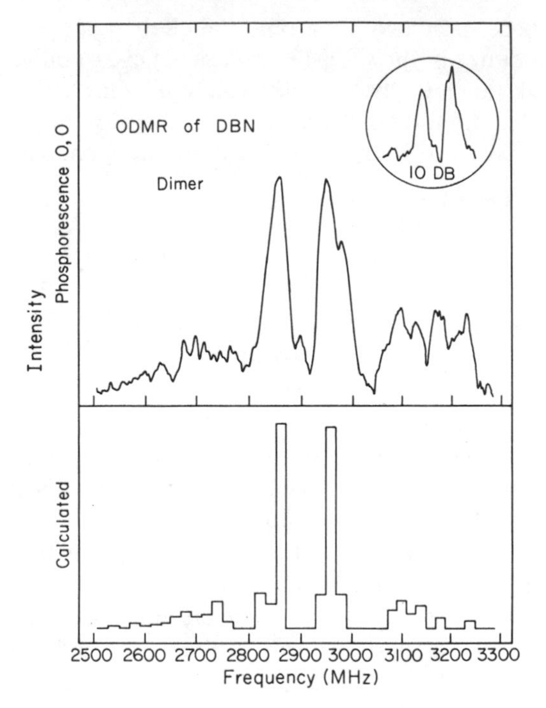

Fig. 16. Zero-field ODMR spectrum of DBN-h_6 dimer in DBN-d_6 at 1.5 K (above) and the calculated spectrum (below). The dimer emission was isolated using a spectrometer. After Ref. 74.

the same ($D+|E|=2952$ MHz, $D-|E|=2860$ MHz). This means that the principal axes are similar in the dimer and the monomer and that (97) adequately describes the spin Hamiltonian parameters given in (98).

To find the degree of excitation delocalization in the DBN pair using ODMR, Hochstrasser and Zewail made the following approximations: (*a*) the molecule is planar and has C_{2v} symmetry with z the C_2 axis and x out of the molecular plane; (*b*) the principal axes of the fine structure tensor and the quadrupole tensor coincide consistent with previous work on dihalobenzenes, (*c*) the y and z components of the Br hyperfine Hamiltonian and all components of the proton hyperfine interaction may be neglected. The neglect of the proton hyperfine interaction is justified on the grounds that these interactions are small[72,75] compared to the Br hyperfine and quadrupole coupling. The neglect of all but A_{xx} for Br is based on previous studies[72,75] of $^3\pi\pi^*$ states for which it was found that $A_{xx} \gg A_{yy}, A_{zz}$; (*d*) there is a cylindrically symmetric field gradient tensor ($\eta=0$). It is known that η is comparatively small for aryl-bromine compounds. All these assumptions have been successfully used previously in the interpretation of the ODMR spectra of haloaromatics and azines.[76,77] On the basis of these assumptions the spectra of the dimer were fit using the average values for the quadrupole coupling constant in the ground and excited states and half the hyperfine value in the monomer. The results are represented by histograms in Figs. 15 and 16 and are consistent with the procedure of Hutchison and King described before. Table II summarizes the computer-fitted parameters for DBN monomers and dimers and compares the results to those of the exciton. Based on these observations and interpretations, it was concluded that the hyperfine attenuation in the dimer is due to the redistribution of the spin density that approximates an average over a ground- and excited-state molecule. Furthermore, the ODMR spectra are consistent with the pair structure deduced from optical experiments, as is the fact that the spin sublevels are partially isolated at low temperature (this would not be expected if the fine structure axes of the molecule of the pair were other than parallel).[78] What the ODMR spectra did not show was the splitting between the + and − dimer states of the EPR resonances. Thus one could not deduce the coherence time in the DBN pair. However, if one assumes, as seems reasonable, that $\Delta\omega_{+-}$ is less than the inhomogeneous width of the transition, then the coherence time (~20 nsec) is orders of magnitude longer than the transfer time between the two molecules.

TABLE II
Spin Hamiltonian Parameters for the Lowest Triplet State of DBN Monomers, Dimers, and Excitons

MHz	Monomer[a]	Dimer[a]	Exciton[b]
$D+\|E\|$	2952	2952	2934±30
$D-\|E\|$	2860	2860	2832[d]
$2\|E\|$	92[c]	92	102±30
$\frac{1}{2}e^2qQ\ {}^{79}Br$	270	275	—
$\frac{1}{2}B^2qQ\ {}^{81}Br$	220	225	—
$A_{XX}\ {}^{79}Br$	65	32.5	—
$A_{XX}\ {}^{81}Br$	60	30	—

[a] Taken from Ref. 60. The monomers and dimers were isolated in isotopically mixed DBN Crystals.
[b] From the pure crystal data (Ref. 158) for the lowest energy sublattice at 2 K.
[c] Observed directly by electron–electron double resonance technique (Ref. 60).
[d] The D parameter is positive.

4. *Zewail-Harris Experiments*[79]

In 1974 Zewail and Harris observed the zero-field ODMR spectra of 1,2,4,5-tetrachlorobenzene (H_2) dimers in perdeuterated tetrachlorobenzene (D_2) host. In contrast to phenazine and naphthalene, the emission spectra of isotopically mixed tetrachlorobenzene (TCB) crystals showed no dimer peaks around the monomer spectrum (the inhomogeneous linewidth of the transition, approximately $2\,cm^{-1}$, is larger than the expected spectroscopic dimer splittings). However, in these isotopically mixed crystals, emission from H_2, HD, and D_2 can be seen, and all ODMR transitions can be observed by "sitting" on an emission line from the particular isotopic species of interest.

The zero-field transitions of the H_2 monomer at low concentrations in D_2 can be observed together with their quadrupole satellites in isotopically mixed crystals at temperatures between 1.4 and 2.1 K. A typical ODMR spectrum is shown in Fig. 17. The analysis of the spectrum reveals all of the features to be expected of an isolated molecule spectrum with the characteristic quadrupole coupling constants for ^{35}Cl and ^{37}Cl isotopes. One additional feature appears in the spectrum of H_2 traps, namely, the shallow trap (HD) resonance transition at 3560.3 MHz. In this experiment one is monitoring the emission from H_2 while sweeping

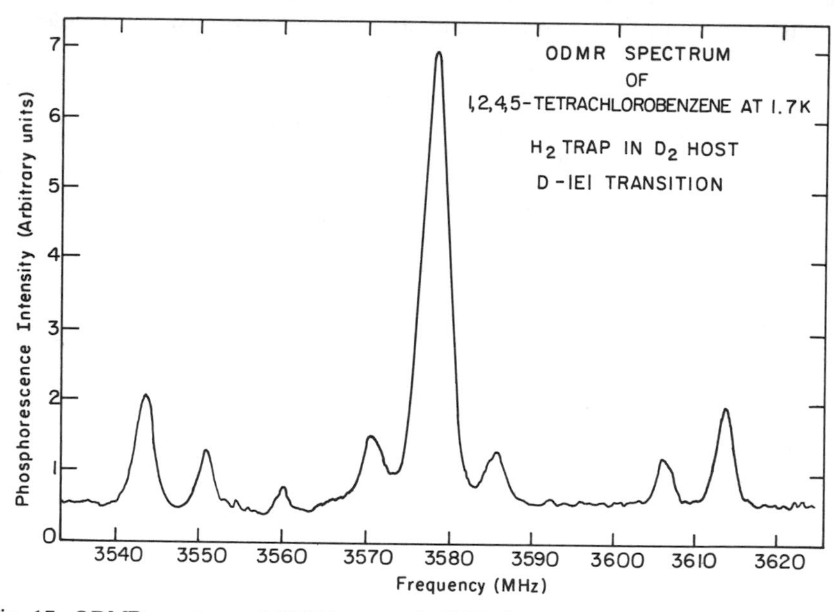

Fig. 17. ODMR spectrum of TCB-h_2 traps in TCB-d_2-host at 1.7 K ($D-|E|$ transition). Note the quadrupole satellites flanking the pure spin transition (most intense peak). After Ref. 79.

the microwave frequency through the region of a zero-field transition. The presence of the HD resonance implies that there is energy transfer between the isotopic traps below the exciton band, since if HD and H_2 traps were completely isolated from each other, a microwave field perturbing the spin of the HD trap should not influence the emission from H_2.[80]

At relatively low concentration (0.06% H_2), both the $D+|E|$ and $D-|E|$ transitions have reasonably sharp lines with a half band width of ~3 MHz, at very low power (0.1 mw). (For definitions of the terms D and E see Fig. 18.) Furthermore, at low powers no quadrupole satellites were observed. With higher power levels (62 mw) the lines become somewhat broader, and the forbidden satellites due to nuclear ^{35}Cl and ^{37}Cl quadrupole transitions appear. At low power levels (0.6 mw), for crystals of low concentration, no ODMR dimer signals are observed around the monomer line. Increasing the guest concentration, however, results in the appearance of new satellites on the monomer electron spin transition that exhibit different behavior as the power increases.[79] Figure 19 shows the H_2 $D+|E|$ and $D-|E|$ transitions at high guest concentration. The results indicated that the saturation behavior of the satellites is different from that of the central resonance line. From the power and concentration dependence studies the following was concluded:

1. Satellites are absent from the zero-field ODMR spectrum of the crystals at low concentration even though the power may be relatively high.
2. The separation between the two dimer states on the $D-|E|$ transition is 5.7 ± 1 MHz and on the $D+|E|$ transition 7.6 ± 1 MHz.
3. The linewidth of the $D-|E|$ transition in the dimer state is approximately 2 MHz.

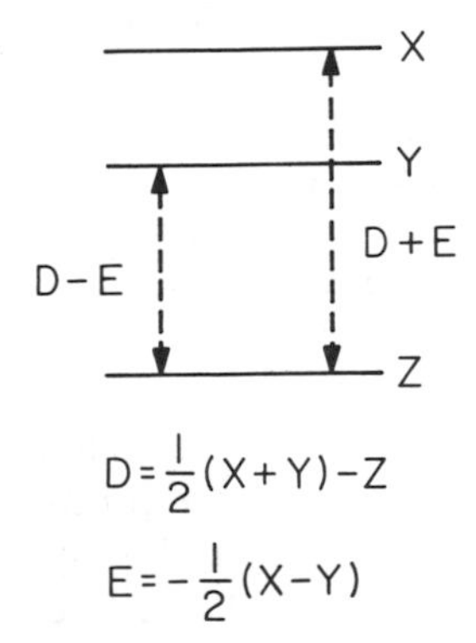

Fig. 18. Zero-field splitting of a triplet state, together with the conventional definitions of the parameters D and E.

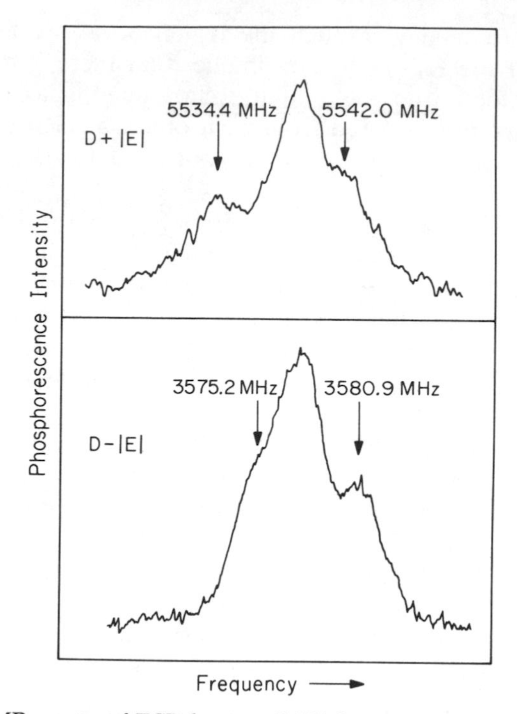

Fig. 19. The ODMR spectra of TCB-h_2 traps (11%) in TCB-d_2 host, for both the $D+|E|$ and $D-|E|$ transitions, at 1.7 K. After Ref. 79.

Two important questions arise from these high-resolution ODMR studies. First, how can one be sure that the satellites on the monomer line are really due to dimer states, as we have assumed, and second, what can be learned about coherence in dimers from these ODMR transitions? These questions are discussed in the remainder of this section.

First as background let us consider the crystal structure of TCB in a little more detail. There are two molecules in the unit cell of TCB which belongs to the $P2_1/c$ space group.[81] At 188 K a phase transition occurs and the lattice becomes triclinic,[82] with a structure closely related in molecular orientations and unit cell dimensions to the room temperature monoclinic structure. Translationally equivalent molecules stack very nearly along the **a**-crystallographic axis in both phases (monoclinic and triclinic). The out-of-plane molecular axis is almost parallel to axis **a**. This fact, together with the small length of **a** compared to the **b** and **c** lattice constants, led to the conclusion that this system could essentially be considered a linear chain exciton with the exciton motion dominated by the translationally equivalent interaction along the **a** axis. The manifestations of this one dimensionality for excitons are discussed in greater detail in later sections; here we consider only those aspects of the problem necessary for understanding dimers.

Next let us consider what we might expect for the zero-field magnetic splittings in dimer triplet states. In zero magnetic field the triplet spin of an isolated molecule is quantized in planes parallel to the molecular axes.[83] The molecular axes themselves will be coincident with the symmetry axes provided the molecular point group has C_{2v} symmetry or higher. Thus the total triplet wavefunction in zero-field can be written as $\phi_{\text{orbital}} \times T_i$ where ϕ_{orbital} is the orbital wavefunction and T_i is the spin wavefunctions ($i = x, y,$ or z). The corresponding wavefunctions for a dimer can be chosen to be a linear combination of isolated molecule wavefunctions in the following way:

$$\psi_i(\pm) = \frac{1}{\sqrt{2}}[|\phi_1^* T_i \phi_2 S^\circ\rangle \pm |\phi_1 S^\circ \phi_2^* T_i\rangle] \tag{100}$$

where S° is the singlet spin function. There are six distinct basis functions that can be obtained for the dimer triplet state. In general, the matrix elements of the total Hamiltonian can be written as

$$\langle\psi_i(\pm)|\,\mathcal{H}_1 + \mathcal{H}_2 + V + \mathcal{H}_s^{(1)} + \mathcal{H}_s^{(2)}\,|\psi_j(\pm)\rangle$$
$$= \delta_{\pm\mp}[-\delta_{ij}D_{ij} \pm \langle V\rangle\langle T_i^{(1)}\,|\,T_j^{(2)}\rangle] \tag{101}$$

In this expression $\mathcal{H}_s^{(1)}$ and $\mathcal{H}_s^{(2)}$ are spin Hamiltonian for molecules 1 and 2, respectively, and D_{ij} are the fine structure tensor elements, that is,

$$\begin{aligned} D_{xx} &= -X \\ D_{yy} &= -Y \\ D_{zz} &= -Z \end{aligned} \tag{102}$$

where X, Y, and Z are the zero-field energies of the triplet spin sublevels and are defined in Fig. 18. The magnetic quantization in the dimer is determined by the spin projection factor $\langle T_i^{(1)}|T_j^{(2)}\rangle$. This quantity measures the projection of a spin axis of molecule 1 onto a spin axis of molecule 2 and is strictly a geometrical factor that can be obtained by knowing the relative orientation in the dimer pair.[84] For translationally equivalent dimers these projections are either one or zero, since all axes are either parallel or perpendicular, and the magnetic axes of the dimer are expected to be coincident with those of the monomer, as discussed in the section on DBN. Under the above approximations, this will be strictly true for any value of $\langle V\rangle \equiv \beta$ provided the molecules are centrosymmetric. On the other hand, the zero-field splittings for translationally *inequivalent* dimers are very sensitive to β, even in the case in which the molecules have a center of symmetry. This is simply because translationally inequivalent molecules in the unit cell have different orientations. In TCB the energies of the six spin states of the translationally inequivalent dimer can be obtained using (101). For the geometry where $X_1//Y_2$; $Y_1//Y_2$; $Z_1//Z_2$ (1 and 2 refer to the two different molecules), one has the new zero-field energies X^*, Y^* and Z^* given by

$$Z^* = Z \pm \beta \tag{103}$$

and

$$Y^*, X^* = \frac{X+Y}{2} \pm \tfrac{1}{2}[(X-Y)^2 + 4\beta^2]^{1/2} \tag{104}$$

In the limit where β is very small compared to the zero-field splittings of the molecule, then

$$X^*, Y^*, Z^* \simeq X, Y, Z \tag{105}$$

and the dimer and molecular splittings are nearly identical. However, in the limit of large β, there will be only one transition in both the + and − dimer states and for both states it will be located at $Z-(X+Y/2)$. Note that when $\beta \sim 0$, the situation is analogous to the case of translationally equivalent dimers. These cases are summarized in Fig. 20.

In general for translationally inequivalent dimers, the dimer zero-field transition occurs at a different microwave frequency from the monomer

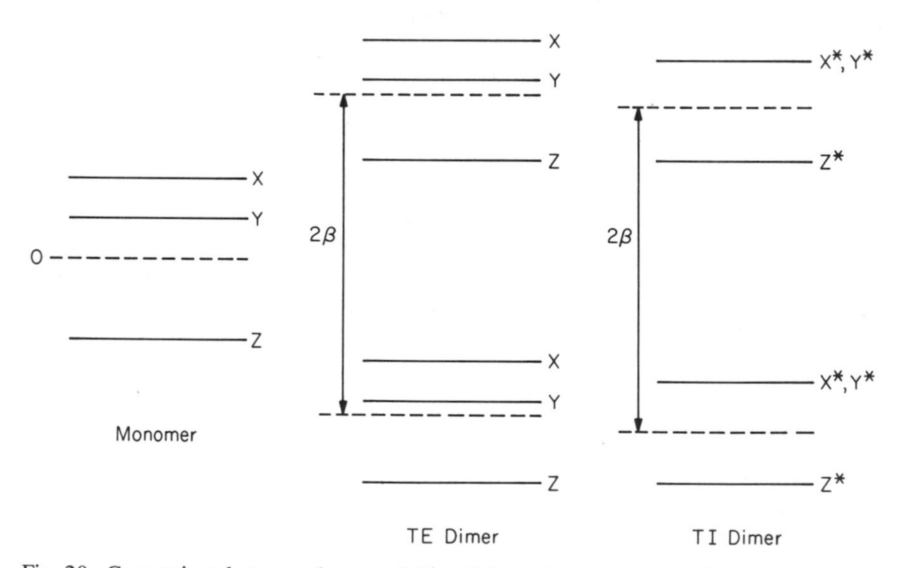

Fig. 20. Comparison between the zero-field splittings of a monomer and those translationally equivalent and inequivalent dimers in systems like TCB.

transition. As Fig. 20 illustrates, it is not possible, in the absence of intermolecular magnetic interactions and intramolecular spin-orbit coupling, to observe the translationally equivalent dimer transitions, since they fall at the same energy as the monomer transitions.

The translationally equivalent dimer transitions become observable if one introduces these neglected interactions into the problem, however. To see how this arises we next discuss in some detail the theory of spin-orbit coupling in dimers and excitons. The analysis is useful for the assignment of the ODMR spectra of dimers and excitons and follows closely the results of Ref. 79 which was based on an earlier analysis of TCB data by Francis and Harris.[85]

We begin by first considering the effect of spin-orbit coupling on the zero-field transitions of excitons. This theoretical treatment is useful when we discuss ODMR studies of exciton coherence. By comparing the theoretical results for excitons and dimers at this point, we also facilitate the discussion in Section IV, where we investigate the extent to which one can learn about exciton dephasing by studying dimer dephasing.

The stationary properties of triplet exciton bands can be understood by a simple dispersion relationship provided the coupling between the linear chains is neglected, that is, if one considers a one-dimensional exciton. The one-site exciton functions can be written explicitly as

$$\psi^{\circ} = A \prod_{u\alpha} (\psi^{\circ}_{u\alpha} S^{\circ}) \tag{106}$$

for the ground singlet state, and as

$$\psi^{f_i}_{m\beta} = A(\psi^{f}_{m\beta} T_i) \prod_{u\alpha \neq m\beta} (\psi^{\circ}_{u\alpha} S^{\circ}) \tag{107}$$

for the ith spin state of the fth excited state (u and m label the unit cell, α and β the particular molecule in the unit cell). $\psi^{f}_{m\beta}$ and $\psi^{\circ}_{u\alpha}$ are the antisymmetrized molecular wavefunctions, and A is the electron permutation operator[86] effecting an interchange of electrons between *molecules*. The crystal eigenfunctions may be written

$$\phi^{f_i}(\mathbf{k}) = \frac{1}{\sqrt{N}} \sum_{m\beta}^{N} \exp\left[\mathbf{k} \cdot \mathbf{r}_{m\beta}\right] \psi^{f_i}_{m\beta} \tag{108}$$

where $\mathbf{k}$ (the one-dimensional wave vector) determines both the symmetry and the energy of the states of the exciton band. Here N is the number of molecules per unit volume.

In zeroth order the state $\psi^f T_i$ has no electric dipole strength to the ground state because of the spin orthogonality between S and T_i. The transition becomes allowed, however, when one introduces spin-orbit coupling. Since the magnitude of the spin-orbit coupling $|\langle H_{SO}\rangle|$ is much less than the singlet-triplet splittings,[87] the molecular wavefunction can be adequately approximated in first order by

$$|\psi^f T_i\rangle^{(1)} = a\,|\psi^f T_i\rangle^{(0)} + \sum_r \frac{\langle \psi^f T_i|\, H_{SO}\, |\psi^r \sigma\rangle}{(E_f^0 - E_r^0)}\, |\psi^r \sigma\rangle \tag{109}$$

where $|\psi^r \sigma\rangle$ (space $\otimes$ spin) is either a singlet or a triplet state, b is a normalization constant, and the superscripts (0) and (1) label the order of the wavefunction or energy.

The magnetic anisotropy introduced into the three spin exciton bands by the molecular spin-orbit coupling depends on the nature of the interaction β between molecules in the lattice. Two extreme cases could arise.

(*a*) *β is relatively large and there is negligible spin-orbit anisotropy:* k-*independent microwave dispersion.*

When $|\langle H_{SO}\rangle|$ is exactly zero, there is no spin-orbit anisotropy, and a one-dimensional band is composed of three parallel spin bands with separations equal to the molecular zero-field splittings. In this case the microwave band-to-band transition is a single line whose frequency is independent of the energy of the k-state in the band.

(*b*) *β is relatively large and there is a finite spin-orbit anisotropy:* k-*dependent microwave dispersion.*

To see what happens in this case we combine (107) and (109) to obtain

$$\psi_{m\beta}^{f_i(1)} = A \prod_{u\alpha \neq m\beta} \psi_{u\alpha}^0 S^0 \Big[a\psi_{m\beta}^f T_i + \sum_r \langle H_{SO}\rangle_{fr} \Delta E_{fr}^{-1} |\psi^r \sigma\rangle \Big] \tag{110}$$

$$= a\,\psi_{m\beta}^{f_i(0)} + \sum_r \langle H_{SO}\rangle_{fr} \Delta E_{fr}^{-1}\, |\psi^r \sigma\rangle \Big(A \prod_{u\alpha \neq m\beta} \psi_{u\alpha}^0 S^0\Big) \tag{111}$$

The corresponding Bloch functions that describe the exciton band are now given by

$$\phi^{f_i(1)}(k) = a\phi^{f_i(0)}(k) + \sum_r \langle H_{SO}\rangle_{fr} \Delta E_{fr}^{-1} \phi^{f_{r\sigma}}(k) \tag{112}$$

where a is

$$\Big(1 - \sum_r |\langle H_{SO}\rangle_{fr}|^2\, \Delta E_{fr}^{-2}\Big)^{1/2} \tag{113}$$

The energy spectrum of the crystal can be determined from the above equations by using the crystal Hamiltonian. For a one-dimensional singlet exciton, the total energy of the ith spin state taking account of spin-orbit coupling is thus given by

$$\begin{aligned} E_i^t(k) &= (E_g^t + D^t) + 2\beta_t \cos ka \\ &\quad - |\langle H_{SO}\rangle_{ts}|^2 \, \Delta E_{ts}^{-1} - |\langle H_{SO}\rangle_{ts}|^2 (D^t - D^s) \Delta E_{ts}^{-2} \\ &\quad - |\langle H_{SO}\rangle_{st}|^2 \, \Delta E_{ts}^{-2} (2\beta_t - 2\beta_s) \cos ka + D_{ii} \end{aligned} \tag{114}$$

where for triplet states $f = t$ and for singlets $f = s$. $(E_g^t + D^t)$ is the energy of the molecular triplet state in the crystal field; E_g^t is the gas phase excitation energy, and D^t is the gas-crystal shift. D_{ii} is again the molecular fine structure constant of (102), and a is the lattice constant.

We return to a more detailed discussion of the gas-crystal shift D terms and the βs in Section IV. At this point it seems appropriate to mention some notational problems that may arise. First the letter D is used here in three very different ways; it may refer to an element of the fine structure tensor, the gas-crystal energy shift, and in Section V the diffusion coefficient. The particular meaning should, however, be clear from the context. We apologize to the reader for this possible source of confusion, but it seems better to stay with traditional notations than to introduce new ones. Similarly the intermolecular interaction in this section is referred to as β, as is traditional for dimers. In later sections we refer to the same interaction by the letter M, as is traditional in discussions of exciton-phonon coupling.

Returning to (114), we see that the spin-orbit interaction introduces an additional k-dependence into the energy of the exciton spin state. Because of the selectivity of the spin-orbit coupling interaction, different spin subbands have slightly different k-dependences. This means that the microwave transition between the spin-sublevels is slightly different depending on the value of k. Using (114) the following equation can be obtained for the k-dependent microwave transition between, say the x and the z spin sublevels:

$$\begin{aligned} \Delta E_{xz}(k) &= |E_x(k) - E_z(k)| \\ &= [X - Z] - |\langle H_{SO}^{(z)}\rangle|^2 \, \Delta E^{-1}\left(1 + \frac{\Delta D}{\Delta E} + \frac{(2\beta_t - 2\beta_s)}{\Delta E} \cos ka\right) \end{aligned} \tag{115}$$

where $\Delta E \equiv \Delta E_{ts}$ and

$$\Delta D = D^t - D^s \tag{116}$$

To display the nature of the exciton zero-field transition more clearly, let us write the k-independent contribution to (115) as

$$\Delta E_{xz} = (X - Z) - |\langle H_{SO}^{(z)}\rangle|^2 \, \Delta E^{-1}[1 + (\Delta D/\Delta E)] \tag{117}$$

The k-dependent term is then

$$(\Delta E_{xz}(k) - \Delta E_{xz}) = |\langle H_{SO}^{(z)}\rangle|^2 \, \Delta E^{-2}(2\beta_s - 2\beta_t) \cos ka \tag{118}$$

Now by defining an attenuation factor in the following way

$$f = \frac{|\langle H_{SO}^{(z)} \rangle|^2 (2\beta_s - 2\beta_t) \Delta E^{-2}}{2\beta_t} \tag{119}$$

(118) may be rewritten

$$(\Delta E_{xz}(k) - \Delta E_{xz}) = f 2\beta_t \cos ka \tag{120}$$

Equation 120 clearly shows that the microwave transition energy is linearly related to cos ka. The right-hand side of (120) is the dispersion relation for a one-dimensional exciton attenuated by the factor f. Thus the microwave band-to-band transition, derived from the above equations, has a shape that is related to the exciton density of states function observed in optical band-to-band transitions but attenuated by the factor f which depends on the spin-orbit coupling. This attenuation of the exciton dispersion in the microwave transition is illustrated in Fig. 21.

The treatment of the dimer case is essentially identical to that developed above for excitons, except that the complications introduced by the many k-states of the exciton band are now absent, since there are only two states in the dimer. The relationship in this context between dimers and excitons is simple if the band dispersion is one dimensional. For this reason, as we see later, the observation of dimer microwave absorption can be very useful in determining the exciton band dimensionality.

To form a dimer, we might imagine somehow shortening the infinite

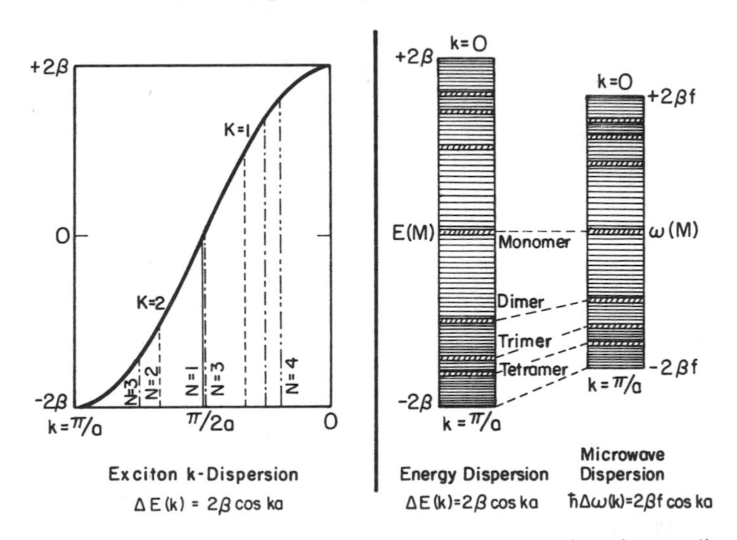

Fig. 21. The relationships between exciton and N-mer dispersions in one-dimensional crystals. The $k=0$ exciton level is assumed to be at the top of the band. After Ref. 79.

chains of a one-dimensional crystal, for example, by introducing barriers at certain sites in the crystal. However, in all experiments performed to date one does not have direct control over the statistical distribution of chain lengths.[88,89] Thus the treatment of N-mers is necessary. The energies[59,90] of an N-mer (i.e., dimer, trimer, tetramer, etc.) in a given chain is given by (see Table I)

$$\varepsilon_K = \varepsilon_0 + 2\beta \cos [K\pi/(N+1)] \tag{121}$$

where $K = 1, \cdots, N$ and should not be confused with exciton quasimomentum k since for dimer there is no translational symmetry.

The energies of these N-mers and their relationship to the energies of k states in the band is shown in Fig. 21 for a one-dimensional system. Ideally, the monomer, $K = 1$, is at the center of the band ($k = \pm\pi/2a$), and the infinite chain level is at the band edge. It follows from (120) that the two dimer states, which are located at $\pm\beta$ relative to the monomer energy, should have a microwave transition frequency reduced by the attenuation factor f, that is,

$$(\omega_{xz}(\pm) - \omega_{xz}) = \pm \beta_t f/\hbar \tag{122}$$

Therefore the frequency spread across the exciton band (that is, the frequency difference between the transitions at $k = 0$ and $k = \pi/a; \Delta_k$) is related to the dimer frequency spread (the frequency difference between transitions at $K = 1$ and $K = 2$; Δ_K) by

$$\Delta_k = 2\Delta_K \tag{123}$$

Similar equations can be derived for higher members of the chain (trimer, tetramer, etc.). Figure 22 shows the microwave frequencies calculated for the N-mers of translationally equivalent molecules compared to the calculated exciton resonances.

The above treatment was used for the assignment of the ODMR spectra of TCB dimers[79] in the following way. The fact that the experimental splitting between the + and − state ODMR resonances is half that of the exciton splitting between the $k = 0$ and $k = \pi/a$ levels (see Fig. 23) in both the $D + E$ and $D - E$ transitions have lead Zewail and Harris to conclude that (a) the two resonances observed in the highly concentrated crystals are due to the + and − dimer states, and (b) the system, TCB, is one dimensional, since the dimer results are in very good agreement with the exciton results of Francis and Harris.[85] These results answer the question raised earlier regarding the assignment of observed ODMR resonances to dimer transitions.

Recently Sheng and Hanson[91] have investigated the effect of intermolecular couplings on the band-to-band transitions in TCB. They concluded that the effect of a small but significant between-chain interaction

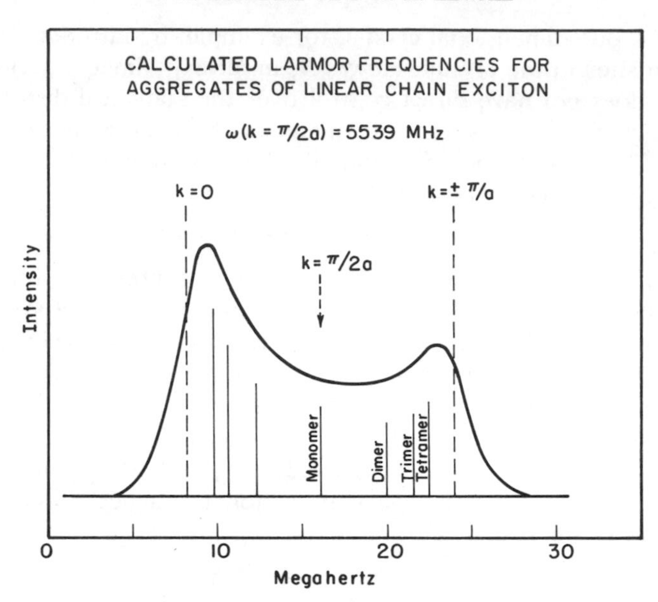

Fig. 22. Calculated Larmor frequencies for aggregates of one-dimensional systems. The continuous double-humped curve is the exciton band-to-band transition calculated[85] for a Boltzmann distribution among the k-states and $\omega(k = \pi/2a) = 5539$ MHz. The solid vertical lines represent the frequency positions of the different N-mer resonances, and the dashed lines the locations of the indicated k-states. After Ref. 79.

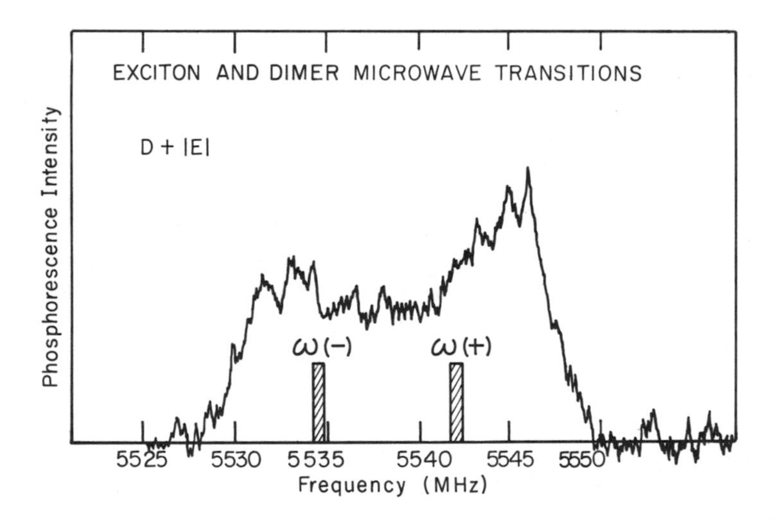

Fig. 23. Dimer and exciton $D + |E|$ transition of TCB crystals. The vertical bars represent the experimentally determined frequencies of the $\psi(\pm)$ dimer states. The width of these bars is the estimated error in these frequencies. The exciton spectrum is the band-to-band microwave transition of Ref. 79.

is to make the density of states function smooth by removing the band edge singularities of a purely one-dimensional triplet exciton band and to spread the ODMR transition over a considerable range with the introduction of new structure. With the help of numerical simulation, they produced in a multidimensional system a density of exciton states that looked like the density of states expected in a one-dimensional system. The message from the work of Sheng and Hanson is that the determination of the dimensionality of an exciton system should not be based solely on the shape of the density of states.

In Ref. 79 the Davydov splitting, which is a measure of the interchain coupling strength, was estimated to be $\simeq 1$ MHz for TCB, that is, orders of magnitude less than the intrachain excitation exchange matrix element. This estimate, along with the relationship between the dimer and exciton ODMR frequencies, the consistency of the results with the theory of spin-orbital coupling in one-dimensional systems, and the effect of temperature on the band-to-band ODMR transition, all make it likely that the exciton in TCB is very nearly one dimensional.

We have seen how an investigation of the ODMR dimer transitions can help in an assignment of the exciton dimensionality. We now turn to a discussion of the relationship between these ODMR transitions and coherence in dimers. The first reports on coherence in dimers were made in papers by Zewail and Harris on TCB[79] and by Zewail[55] on phenazine. They used the following criterion to establish a limit on the coherence time. Excitation "moves" back and forth between the two molecules comprising the dimer in a time $t_{\text{jump}} = \hbar/2\beta$. If scattering occurs on a time scale fast with respect to this motion of the excitation, then we might describe the motion as completely incoherent; the phase of the wavefunction would be randomized at each hop of the excitation. The randomization of the wavefunction is of course related to the homogeneous linewidth of the ODMR transition, as discussed in Sections III.C and III.D and summarized by (55). In this respect one may use the ODMR experiments to investigate coherence in exactly the same way as one uses the optical spectra of dimers (Section III.E). If the + and − dimer state ODMR linewidth is much less than the splitting between them, then the states are considered to be coherent. The fact that Zewail and Harris[79] could distinguish the ± dimer resonances in ODMR provides a lower limit on the coherence time. Because of the lower frequency used, the microwave ODMR experiments measure dimer coherence on a longer time scale than the optical experiments. One must be careful in comparing optical and magnetic resonance coherence measurements, however. In the optical experiments the dephasing time of an electric transition dipole moment is measured; in the microwave experiments magnetic dipole dephasing is measured. We return to this important point again when we discuss optical and magnetic resonance lineshape measurements for excitons in Section V.C.

To obtain a more quantitative estimate of dimer coherence Zewail and Harris[79] turned to the exchange theory of Kubo,[92] Anderson,[93] and McConnell.[94] Mathematically one can distinguish three limits for the exchange of energy between the molecules of the dimer: fast, intermediate, and slow exchange. One begins with the well-known Bloch equations in the rotating frame. In the presence of a weak oscillating *rf* field of the form

$$H(t) = -\gamma H_1 \cos \omega t \tag{124}$$

where γ is the gyromagnetic ratio and H_1 the magnetic field amplitude, the Bloch equations may be written[37]

$$(du/dt) + (u/T_2) - \Delta\omega v = 0 \tag{125}$$

$$(dv/dt) + (v/T_2) + \Delta\omega u - \gamma H_1 M_z = 0 \tag{126}$$

$$(dM_z/dt) + (M_z - M_0)T_1^{-1} + \gamma H_1 v = 0 \tag{127}$$

where $\Delta\omega = \omega_0 - \omega$ with ω_0 the resonance frequency. The quantities u, v, and M_z are defined as components r_1, r_2, and r_3, respectively, of the vector $\mathbf{r}$ in Fig. 6. M_0 is the equilibrium magnetization.

Equations 125 and 126 can be combined into a single equation by defining a complex moment G by

$$G = u + iv \tag{128}$$

Since the macroscopic pseudomagnetization M_z of the triplet ensemble is not strongly disturbed by the external field,[95] we approximate it by its equilibrium value M_0. Equations 125 and 126 may then be written

$$(dG/dt) + i(\Delta\hat{\omega})G - i\gamma H_1 M_0 = 0 \tag{129}$$

where $\Delta\hat{\omega} = \hat{\omega}_0 - \omega$, and the complex frequency $\hat{\omega}_0$ includes the transverse relaxation time $\hat{\omega} \equiv \omega_0 - iT_2^{-1}$.

For both dimer states one may write an equation of the form of (129). In the absence of any scattering between dimer states we thus have

$$(dG_+/dt) + i(\Delta\hat{\omega}_+)G_+ - i\gamma H_1 M_0 = 0 \tag{130}$$

$$(dG_-/dt) + i(\Delta\hat{\omega}_-)G_- - i\gamma H_1 M_0 = 0 \tag{131}$$

For dimers isolated in molecular crystals, the resonance frequencies, ω_0^+ and ω_0^- in the two plus and minus dimer "stationary" states could be different due, for example, to the spin-orbit coupling discussed above.

At low temperatures (≤ 1.5 K) the longitudinal relaxation time T_1[96] is expected to be equal to the lifetime of the excited dimer state. However, at somewhat higher temperatures coupling between the dimer states and lattice phonons (see Section IV) may cause exchange of energy between the two dimer states (a T_1 process). In addition, this coupling may cause dephasing of the wavefunction (a T_2 process) and thus contribute to the magnetic resonance linewidth. The magnitude of the scattering can, in principle, be obtained from the magnetic resonance spectra of dimers by two means: linewidth measurements and measurements of Larmor frequency differences. Using (130) and (131) following the formalism of Kubo,[92] Anderson,[93] and McConnell[94] for chemical exchange, and defining τ_{+-} and τ_{-+} as the scattering times between the two states, the modified Bloch equations are given by[79]

$$(dG_+/dt) = i[N_+\omega_1 M_0 - \Delta\hat{\omega}_+ G_+] + (G_-/\tau_{-+}) - (G_+/\tau_{+-}) \tag{132}$$

and

$$(dG_-/dt) = i[N_-\omega_1 M_0 - \Delta\hat{\omega}_- G_-] + (G_+/\tau_{+-}) - (G_-/\tau_{-+}) \tag{133}$$

The power factor, γH_1, is abbreviated by ω_1, and $N_\pm$ is the fraction of spins in the *plus* or *minus* state.

In steady state, where $dG_\pm/dt = 0$, the solution is particularly simple, since we are dealing with only two states (*plus* and *minus*). The lineshape function $g_D(\omega)$ of the microwave transition in the dimer is simply given by

$$g_D(\omega) = \text{Im } G_D = \text{Im } (G_+ + G_-) \tag{134}$$

whereas in the case of one-dimensional excitons it is given by

$$g_E(\omega) = \text{Im } G_E = \text{Im} \sum_k G_k \tag{135}$$

One can see from (134) and (135) that the solution of the Bloch equations for the two-level dimer problem is relatively straightforward, whereas for exciton states, where scattering among all the k-states must be considered, the equations become much more difficult to solve. A sum over all k-states must in this case be included:

$$(dG_k/dt) + i\Delta\hat{\omega}_k G_k + i\omega_1 M_0^k - \sum_{k'} [(G_{k'}/\tau_{k'k}) - (G_k/\tau_{kk'})] = 0 \tag{136}$$

The $(2k+1)$ equations, obtained from (136) under the steady-state approximation, have been discussed[97] for the slow, intermediate, and fast exchange limits using the experimentally[85] determined values for ω_0^k and for the population distribution among the k-states of the one-dimensional triplet exciton band of TCB. Because both the energy dispersion of the band and the population distribution between the different k-states determine the magnitude of M_0^k, the analysis of the lineshape of the exciton resonance could, in principle, give the band dispersion, density of states, and coherence time.

In the dimer case the relative absorption intensities and energies in the *plus* and *minus* states can directly give information regarding the physics of scattering and the influence of the scattering amplitudes on the population of the two states. If there are no large host influences on the dimer states, the scattering probabilities $1/\tau_{+-}$ and $1/\tau_{-+}$ should be determined by the form of the exciton-phonon interaction. On the other hand, the two states of the dimer may scatter differently to states in the host crystal exciton band. In this case the resonance lineshape will depend on $1/\tau'_{\pm\mp}$ where

$$1/\tau'_{\pm\mp} = 1/\tau_{\pm\mp} + \sum_{k_h} 1/\tau_{k_h} \tag{137}$$

and the sum over host exciton levels k_h will be different for the *plus* and *minus* states.

Two specific scattering cases are considered here, a Boltzmann distribution between the + and − states and and non-Boltzmann distribution with $N_+ = N_- = \frac{1}{2}$. The contribution to the linewidth due to processes other than exchange is neglected.

Non-Boltzmann Distribution; $N_+ = N_- = \frac{1}{2}$ *and* $\tau_{+-} = \tau_{-+} = \tau$. In this case the lineshape function is given by

$$\text{Im } G_D = (\omega_1/2)M_0 \frac{R(\omega_+ - \omega_-)^3}{(\omega-\omega_+)^2(\omega-\omega_-)^2 + 4R^2(\omega_+ - \omega_-)^2(\omega - \bar{\omega})^2} \tag{138}$$

where $\bar{\omega}$ is the average frequency and R is the ratio of the scattering rate to the difference in Larmor frequencies between *plus* and *minus* dimer states, that is,

$$(\omega_+ - \omega_-) = 1/\tau \times 1/R \tag{139}$$

From (139) one sees that a small value of R means that both transitions of the dimer are sharp and well separated. This in turn means that if the exciton-phonon scattering rate is much slower than the rate corresponding to the difference in Larmor frequencies, the spin in each state can absorb microwave energy as if the two states were not connected. On the other hand, if the two dimer states are strongly coupled via the relaxation Hamiltonian, the spin can no longer distinguish between the two subsystems, and averaging takes place. Increasing the value of R results in overlap between the two microwave transitions, since the spin can no longer effectively be assigned to one state or the other. Figure 24 shows the dimer resonance spectra for different values of R, covering the fast, intermediate, and slow exchange limits.[79] Notice that for small R when $\omega \approx \omega_+$, a single Lorentzian line is obtained with a width given by

$$1/T_{2e} = 1/\tau \tag{140}$$

The linewidth thus gives directly the time that the spin "spends" in the *plus* state during the exchange. This of course implies that the linewidth of the *plus* and *minus* state would be zero in the absence of exchange. If there is a residual inhomogeneous linewidth (T_{2+}^{*-1} and T_{2-}^{*-1}) due, for example, to crystal and/or hyperfine fields, the total width must be modified to take into account these broadening mechanisms.

Boltzmann Distribution. In this case the exchange is between states of unequal population, and the whole thermalization mechanism depends on the magnitude of the resonance interaction, β, and the temperature of the bath. Utilizing the (statistical) partition function, Z, of the system, the

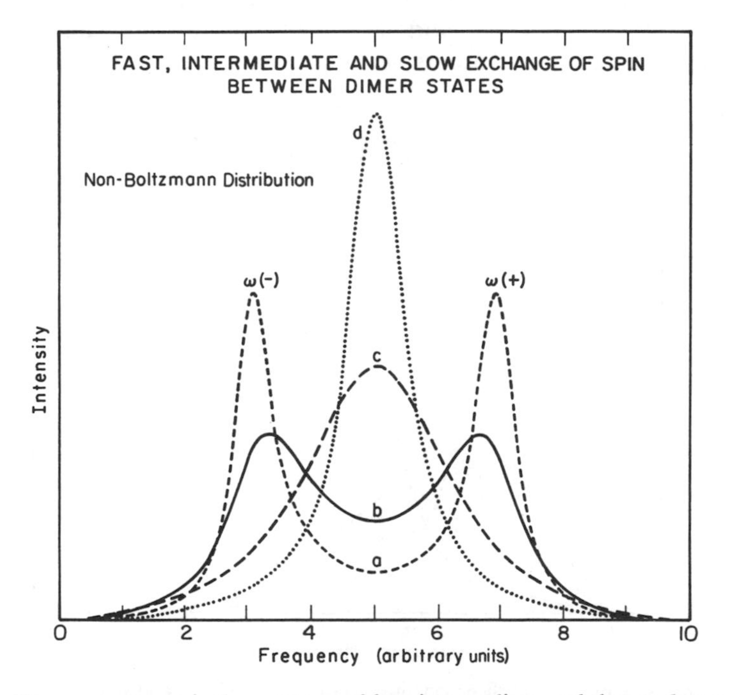

Fig. 24. The consequences in the spectrum of fast, intermediate, and slow exchange of spin between the dimer states. The two states are assumed to be equally populated, and the transfer time for the two channels is also taken to be equal. (a) $R = 0.1$; (b) $R = 0.2$; (c) $R = 0.5$; (d) $R = 1$. After Ref. 79.

imaginary part of the magnetization is given by

$$\operatorname{Im} G_D = \omega_1 M_0(\varepsilon/Z) \frac{R(\omega_+ - \omega_-)^3}{(\omega-\omega_+)^2(\omega-\omega_-)^2 + R^2(\omega_+ - \omega_-)^2[\varepsilon(\omega-\omega_-)+(\omega-\omega_+)]^2} \tag{141}$$

where $\varepsilon = \exp(-2\beta/k_BT)$, and T is the temperature. Again if R is small, the characteristic resonances in the *plus* and *minus* states can be resolved. However, unlike the case of non-Boltzmann distribution, here the relative intensities are different. Figure 25 shows the dependence of a typical magnetic spectrum on the magnitude of R for a fixed temperature and constant value of β.

Figures 24 and 25 illustrate how measurements of the width and separation of dimer resonance lines can lead to a quantitative measure of the exchange rate between the dimer states. For slow exchange the $\omega(+)$ and $\omega(-)$ resonances are distinguishable, but as exchange becomes more rapid the distinction between dimer states disappears and a single line emerges in between $\omega(+)$ and $\omega(-)$. In this fast exchange case the scattering between states occurs so rapidly that it exceeds the time scale

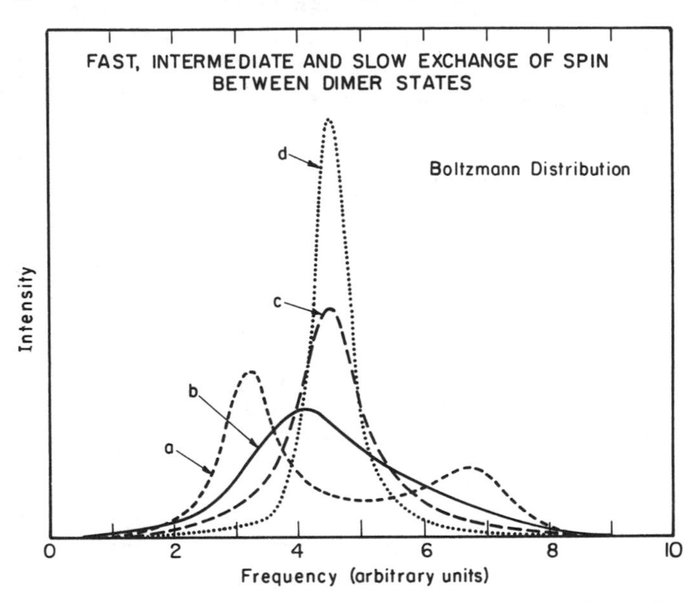

Fig. 25. The consequences in the spectrum of fast, intermediate, and slow exchange of spin between the dimer states. The two states are assumed to be in thermal equilibrium with $|\beta| = 0.25\ \text{cm}^{-1}$. (a) $R = 0.2$; (b) $R = 0.5$; (c) $R = 1.0$; (d) $R = 2.0$. R is now defined for one-way transfer, since the system is in Boltzmann equilibrium. Notice the shift of the peak in the fast exchange limit away from $[(\omega(+)+\omega(-)]/2$. After Ref. 79.

of the experiment. The resonance line begins to narrow, and its width is no longer a direct measure of the scattering rate. Another interesting observation resulting from exchange is the shift of the resonance frequency as the magnitude of exchange increases (Fig. 25). Thus by, say, increasing the temperature, one expects to see the frequency of the transition change. This is discussed in later sections for both dimers and excitons.

We have assumed that the intensity of the microwave resonances in the above expressions is given by the imaginary part of G. This is strictly true only for a conventional EPR experiment.[23] In an ODMR experiment the signal is proportional, not to G (i.e., to r_1 or r_2 in Fig. 6) but to the population difference r_3.[98] It can be shown, however, that the frequency spectra in both the EPR and ODMR experiments are related. The results of this section thus apply to ODMR as well as to EPR experiments.

Using (141), Zewail and Harris[79] obtained a coherence time on the order of 10^{-7} sec from the ODMR of TCB dimers by extracting an estimate of R from the spectrum. An exact value for the coherence time could not be obtained because of the lack of exact values for the intensity ratios and linewidths of the $\pm$ resonances. The lines in these cases are at least partially inhomogeneously broadened due to hyperfine interactions and perhaps also to crystal field effects. In TCB the estimated ratio of the ODMR intensity in the + and − states is 1.3, which gives in a Boltzmann regime a β of 0.15 cm^{-1}, hence an exciton bandwidth of 0.6 cm^{-1}. This is almost a factor of 2 less than that obtained from the exciton band-to-band ODMR absorption results.[85] Hochstrasser et al.[69] have carried out careful measurements of the optical lineshape of neat, isotopically mixed and heavily doped crystals. Their conclusions were that if the system is a linear chain and the site shifts of isotopes are negligible, then the bandwidth must be less than 0.8 cm^{-1}.

Because β in phenazine[53] is much larger than in TCB, the ODMR dimer lines can be resolved by monitoring only the dimer emission. The studies of Zewail[55] have shown a shift between the zero-field ODMR resonance of the monomer and the resonance of the dimer. Only one ODMR transition for the dimer was seen, and the shift of the $D+E$ and $2E$ transition is predicted using the above treatment of spin-orbit coupling theory. The appearance of one band in the ODMR is consistent with the above theory of exchange for a Boltzmann distribution between + and − states, since β is large compared to k_BT. Based on the linewidth measurement (which in this case can be measured accurately, in contrast to the TCB case) and the frequency shift (2.8 MHz for the $2E$ transition and 1.3 MHz for the $|D|+|E|$ transition) in the Larmor frequencies going from the monomer to the dimer a value of 10^{-6} sec for the coherence

TABLE III
Experimentally Determined Coherence Parameters for Dimer Systems

Dimer System	Temperature (K)	τ (nsec)	$\|\beta\|$ (cm^{-1})	t_{jump} (psec)[a]
TCB[b]	1.7	~100	~0.15	18
Phenazine[c]	1.6	~1000	~4.3	0.6
Naphthalene				
inequivalent[d]	1.18	10	1.25	2
	4.2	1	1.25	2
Equivalent	4.2	300	0.5	5

[a] $1/t_{jump} = (2\pi c)(2\beta)$.
[b] A. H. Zewail and C. B. Harris, *Phys. Rev.*, **B11,** 952 (1975).
[c] A. H. Zewail, *Chem. Phys. Lett.*, **33,** 46 (1975).
[d] B. J. Botter, C. J. Nonhof, J. Schmidt, and J. H. Van der Waals, *Chem. Phys. Lett.*, **43,** 210 (1970); B. J. Botter, A. J. van Strien, and J. Schmidt, *Chem. Phys. Lett.*, **49,** 39 (1977).

time was extracted. In both the phenazine and the TCB studies the important message, more important than the precise values of the coherence time, is that the measured *coherence time is much longer than the single-jump time,* thus establishing that coherence exists in these systems on the time scale of a magnetic resonance experiment. These results are summarized in Table III.

5. *Botter-van Strien-Schmidt-Van der Waals Experiments*[99,100]

Botter et al. have recently measured the dimer dephasing time T_2 of translationally inequivalent[99] and equivalent[100] pairs of naphthalene-h_8 molecules in naphthalene-d_8 crystals; the same system studied by Schwoerer and Wolf[62] and discussed in Section III.F.1. Botter et al. measured the dephasing time directly using spin echo techniques (see Section III.D) and avoided the problems of inhomogeneous broadening that plague attempts to measure T_2 from microwave absorption linewidths.

The spin echo measurements made in an external magnetic field were combined with zero-field ODMR measurements of the shift in resonance frequency to obtain a quantitative measure of the degree of coherence. To extract the degree of coherence from the experimental results, they used a modification of the exchange theory described in the preceding section. This model follows closely that of van't Hof and Schmidt[101] used to describe the loss of phase coherence in triplet monomer states. A similar model is used in Section V.C to describe the dephasing of excitons.

The model is summarized in Fig. 26. Scattering occurs between the

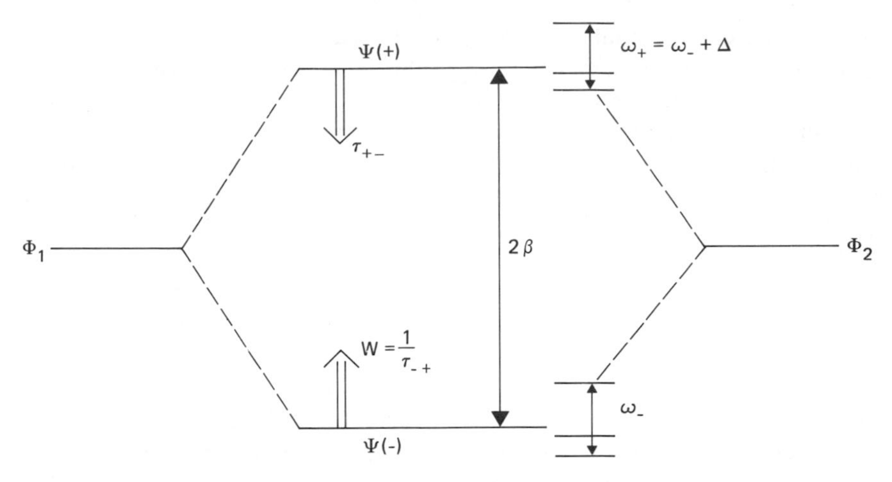

Fig. 26. The energy level diagram for a dimer system $\Psi(\pm)$ formed from the monomer states Φ_1 and Φ_2. After Ref. 99.

dimer states. The upward rate is W and the downward rate $1/\tau$, and thus these rates are assumed to be related. The criterion for coherence is the same as that used by Zewail and Harris,[79] namely,

$$W \gg 2\beta \qquad \text{(incoherent hopping limit)} \tag{142}$$

$$W \ll 2\beta \qquad \text{(coherent limit)} \tag{143}$$

One assumes that the transitions between plus and minus dimer states originate from a phonon bath and that the entire system is in thermal equilibrium so that

$$W\tau = \exp\left[-2\beta/k_B T\right] \tag{144}$$

From (144) and the Bloch equations (132) and (133), one can derive expressions for the dephasing time (assuming *only* the above process) and the resonance frequency shift as follows (the + state energy is assumed to be higher):[99,100]

$$\frac{1}{T_2} = \frac{1}{\tau}\frac{\Delta^2\tau^2}{1+\Delta^2\tau^2}\exp\{-2\,|\beta|/k_B T\} \tag{145}$$

$$\omega_-(T) - \omega_-(0) = \frac{\Delta}{1+\Delta^2\tau^2}\exp\{-2\,|\beta|/k_B T\} \tag{146}$$

where Δ is the difference in Larmor frequency between plus and minus states.

Note that the temperature dependence of $1/T_2$ and the energy shift should enable one to obtain an estimate of β. This is an important feature of the theoretical model. In all our previous discussions of exchange theory we *assumed* that the predominant scattering process is between dimer plus and minus states. Although this is a very logical assumption, it is by no means the only scattering process that could occur. Equations 145 and 146 give an important means of testing this assumption. If plus-minus scattering predominates, then the values of β obtained from measurements of the temperature dependence of $1/T_2$ and the shift should agree with other independent measurements of this quantity.

Turning to the experimental results, we see in Fig. 27 the measured values of $1/T_2$ from the spin echo experiments as a function of temperature for the translationally inequivalent dimer pair. The low temperature results are in agreement with (145) with a value of 2β equal to $2.5 \pm 0.1\ \text{cm}^{-1}$, in good agreement with the results of direct optical measurements of the dimer splitting.[48,50] Similarly the ODMR results for the frequency shift, also shown in Fig. 27, yields a value of 2β from (146), in agreement with these other experiments.

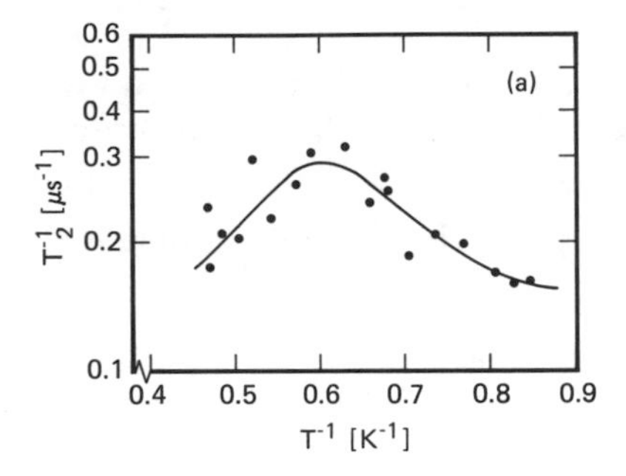

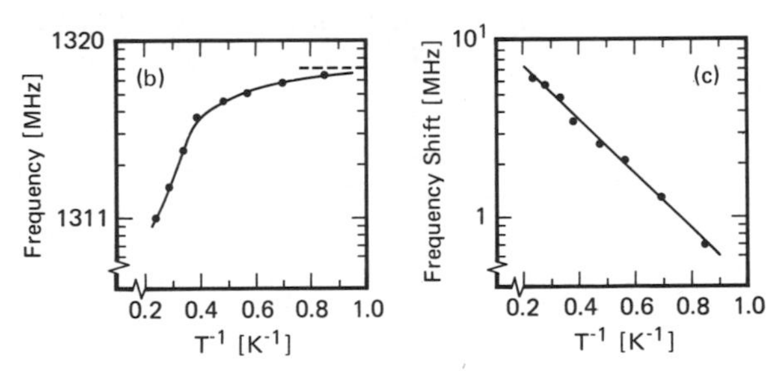

Fig. 27. Magnetic resonance of the translationally inequivalent triplet dimer in the system naphthalene-h_8 in naphthalene-d_8. (a) The decay rate T_2^{-1} of the high-field ESR transition as a function of inverse temperature. (b) The zero-field resonance transition frequency versus inverse temperature. The dotted line represents the extrapolated $T=0$ frequency. (c) The frequency shift of the zero-field resonance from its $T=0$ value versus inverse temperature. After Ref. 99.

One can see that (145) and (146) are considerably simplified when $\Delta^2\tau^2 \ll 1$. In fact, in this limit the frequency shift measurement yields Δ directly. Using this value of Δ and knowing the value of β in the exponential from the temperature dependence experiments, the lifetime of the + state can be obtained. Botter et al.[99] obtained a τ of $(4.9 \pm 1.6) \times 10^{-10}$ sec. Moreover, their measured values of Δ (which ranged from 4 to 23 MHz depending on the particular spin transition investigated) are in "agreement" with the frequency shift expected as a result of the calculation of King.[102] This calculation involves a complete diagonalization of a 6×6 matrix similar to the procedure discussed in connection with (101).

The rate of the phonon-induced relaxation W can now be obtained knowing β and τ and using (144). Experimental values of τ at 1.18 and 4.2 K are listed in Table III. Comparing the relaxation time with the excitation jump time in the table, we see that (143) is satisfied and the system is coherent, as was the case for TCB and phenazine at low temperatures. This naphthalene result is not consistent with the earlier conclusion of Schwoerer and Wolf[62] and does not support the incoherent hopping picture of Schwoerer[103] that was based on the analysis of the ESR lineshapes using the theory of Reineker and Haken.[104] As mentioned in Section III.F.1, these lineshape measurements are plagued by complications from inhomogeneous broadening. It is interesting to note that the measured values of W apparently explain the maximum of T_2^{-1} that occurs at 1.7 K in Fig. 27. At this temperature $\Delta/W \approx 1$. For higher temperatures W becomes so large, that is, Δ/W becomes so small, that an effect analogous to motional narrowing occurs. This means that the spin jumps between the two states so fast that the phase losses average out and cannot be observed on the time scale of the experiment. The effect is well known for NMR experiments in solution.

In similar experiments Botter et al.[100] have measured T_2^{-1} for the translationally equivalent dimer pair of naphthalene along the **b** crystal axis. Utilizing the above mentioned theoretical formalism, they again obtained the energy splitting between the $\pm$ states as well as the lifetime of the upper state (considered to be the $-$state). From the temperature dependence of $1/T_2$ and the frequency shift, Botter et al. obtained a value of 1 cm^{-1} for 2β, in good agreement with previously measured values for the translationally equivalent dimer splitting.[58] For τ, a value of 7×10^{-8} sec was obtained, *two orders of magnitude longer than the comparable value for the translationally inequivalent pair.* The difference between the τ's may be explained in part by differences in the acoustic phonon density of states available for scattering in the two system. Assuming a Debye model where the density of states depends quadratically on the energy separation, one expects the lifetime of the translationally equivalent dimer state to be longer than the lifetime of the translationally inequivalent one by a factor of about 6. This of course assumes that the lifetime broadening is due to transitions between the dimer states induced by the phonons and that the matrix elements for connecting these states are independent of the energy separation between $\pm$ states. It may also be that the scattering matrix element itself depends on the relative orientation of molecules in the dimer pair. One also expects multiphonon interactions, guest-host interactions, and other processes to contribute to the overall lifetime of the dimer state. More experiments are clearly needed to clarify this point.

The frequency shift Δ between plus and minus dimer states was found to be small (less that 1 MHz) but nonzero for the translationally equivalent dimer pair. For translationally inequivalent dimers the finite value of Δ results from the different relative orientations of the two constituent molecules. Since the molecules of the translationally equivalent dimer have the same orientation, this explanation cannot explain the finite value of Δ in this case. One possible source of a nonzero Δ is the effect of spin orbit interactions described by Zewail and Harris[79] and discussed in Section III.F.4.

The value of τ determined from these experiments for the translationally equivalent dimer is given in Table III. From the table we see in this case, as in all other cases investigated, that the dimer state is coherent at low temperatures.

IV. RELATIONSHIP BETWEEN DIMER AND EXCITON COHERENCE

A. Exciton-Phonon Interaction

Much of the recent work on dimer coherence has been done, as we have indicated, with the hope that one can learn something from this work about exciton coherence. In this regard it seems useful to compare the scattering or dephasing processes that both excitons and dimers can undergo from a unified point of view to define those areas in which the study of coherence in one can illuminate coherent phenomena in the other. This section also serves the purpose of allowing us to introduce some of the mathematical formalism used in later sections.

One difference between exciton and dimer is obvious from the beginning. Whereas a dimer involves excitation localized on two adjacent lattice sites, an exciton represents energy that is free to move through the entire crystal. Consequently, excitons may be scattered by phonons, impurities, and lattice imperfections. Dimers, on the other hand, are primarily dephased by phonons. In comparing dimer and exciton coherence in this section we only consider dephasing by phonons.

We use a formulation of the exciton-phonon interaction problem originally outlined for excitons in molecular crystals by Davydov[3,105] and later modified to describe mixed molecular crystals by Hochstrasser and Prasad.[106] This formulation has the advantage, for our purposes, that both exciton-phonon and dimer-phonon interactions can be displayed in a parallel fashion, facilitating the comparisons we wish to make.

First consider the exciton-phonon interaction. Within the occupation number representation, the Hamiltonian corresponding to the electronic states in a molecular crystal may be written

$$H_{\text{ex}}(R) = \sum_n [\varepsilon + D_n(R)]B_n^+ B_n + \sum_{n,m}{}' M_{nm}(R) B_m^+ B_n \tag{147}$$

In this expression B_n^+ and B_n are the operators that create and annihilate, respectively, an electronic excitation at the lattice site n. $M_{nm}(R)$ is the intermolecular interaction responsible for the exchange of excitation between lattice sites n and m. The prime on the sum indicate that $m \neq n$. This quantity depends on the distance between the two sites and the relative orientation of the molecules at the sites as indicated by its explicit dependence on the set of lattice coordinates $\{R\}$. Note that M_{nm} is equivalent to β, as discussed in Section III. $D_n(R)$ is defined by $D_n(R) = \sum_{n \neq m} D_{nm}(R)$ where $D_{nm}(R)$ is the change in the Van der Waals interaction between molecules m and n when the nth molecule is excited. $D_n(R)$ is the term primarily responsible for the difference in excitation energy between the molecule in the gas phase and the molecule in the crystal—what we might call the "solvent" shift.

The coupling between the electronic excitations of the crystal and the lattice coordinates or phonons is contained in the R-dependence of M_{nm} and D_n. The set of coordinates represented by $\{R\}$ may include both intra- and intermolecular motions. For most organic molecules, intramolecular motion generally results in vibrations with frequencies greater than $100\ \text{cm}^{-1}$, whereas for a given molecular crystal there are generally several intermolecular normal modes with frequencies below $50\ \text{cm}^{-1}$.[107] Because of the greater thermal occupation of these intermolecular modes, it has generally been considered that they are more effective in dephasing dimer and exciton excited states than the higher-frequency intramolecular modes, especially at low temperatures. In our discussion of phonon dephasing we primarily think of these intermolecular motions when we use the word phonon.

Some molecules do have low-frequency intramolecular modes.[108] Molecules like toluene, with methyl groups that can undergo torsional motion, are examples. It is an interesting but so far untreated problem how effectively these intramolecular methyl group vibrations can scatter excitons.

In the treatment of exciton-phonon interactions outlined by Davydov[3,105] one begins with a zeroth order Hamiltonian for a rigid lattice where all of the molecules are at their equilibrium positions, that is, $R=0$ for all of the normal coordinates. In this case the Hamiltonian is

$$H_{ex}(0)=\sum_{n}[\varepsilon+D_n(0)]B_n^+B_n+\sum_{n,m}{}' M_{nm}(0)B_m^+B_n \tag{148}$$

By transforming from the local operators B_n, B_n^+ to operators representing phased excitations throughout the crystal, that is excitons, (148) may be diagonalized. The new operators are defined by

$$B_n=\frac{1}{\sqrt{N}}\sum_{\mathbf{k}} B(\mathbf{k})\exp(i\mathbf{k}\cdot\mathbf{n}) \tag{149}$$

and the Hamiltonian becomes

$$H_{ex}(0)=\sum_{\mathbf{k}} E(\mathbf{k})B^+(\mathbf{k})B(\mathbf{k}) \tag{150}$$

and the energy for the case of one molecule per unit cell is given by

$$E(\mathbf{k})=\varepsilon+D(0)+\sum_{m}{}' M_{nm}(0)\exp\{i\mathbf{k}\cdot(\mathbf{n}-\mathbf{m})\} \tag{151}$$

In Fig. 28 we represent this energy expression schematically. The D term can be seen here and in (114) to be responsible, as we have

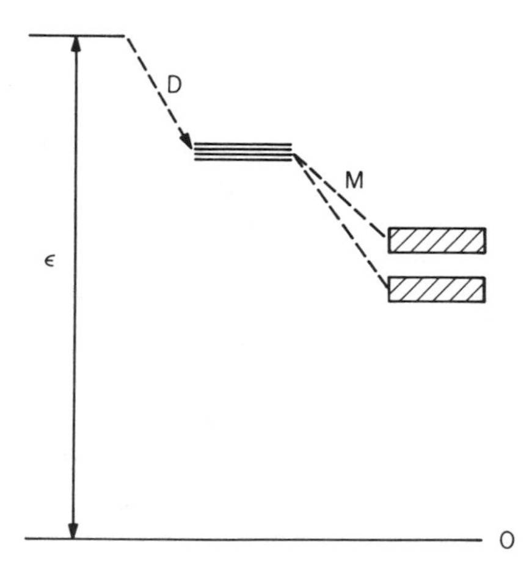

Fig. 28. The development of exciton bands from an isolated molecule transition at an energy ε. The D-term causes a gas to crystal "solvent" shift, and the M-term causes a further small shift and a break-up of the transition into exciton bands.

mentioned, for a "solvent" shift in the energy. The M term causes an additional energy shift, but, more important, also causes the single electronic transition to split into discrete exciton bands. There will be as many bands as there are molecules in the crystallographic unit cell.[3,4]

To include exciton-phonon interactions, indeed, to account for phonons at all, one must go beyond the rigid lattice approximation and incorporate the $\{R\}$ dependence of the Hamiltonian given by (147). This can be done by expanding $H_{ex}(R)$ in powers of the displacement from equilibrium and, assuming that this displacement is small, truncating the series after the quadratic terms. In most treatments all the linear terms are retained, but only those quadratic terms are kept which are necessary to describe the harmonic phonons. The Hamiltonian in this approximation is

$$H_{ex}(R) = H_{ex}(0) + W(R) + H_M^{ex}(R) + H_D^{ex}(R) \tag{152}$$

The operator $W(R)$ contains some of the quadratic terms in the displacements $\{R\}$ and provides the potential energy that is responsible for the phonons in the crystals.[3] $H_M^{ex}(R)$ and $H_D^{ex}(R)$ are linear exciton-phonon coupling terms given by

$$H_M^{ex}(R) = \sum_{n,m}' B_n^+ B_m \sum_j \left\{ \left(R_n^j \frac{\partial}{\partial R_n^j} + R_m^j \frac{\partial}{\partial R_m^j} \right) M_{nm}(R) \right\}_0 \tag{153}$$

$$H_D^{ex}(R) = \sum_{n,m}' B_n^+ B_n \sum_j \left\{ \left(R_0^j \frac{\partial}{\partial R_0^j} + R_m^j \frac{\partial}{\partial R_m^j} \right) D_{0m}(R) \right\}_0 \tag{154}$$

where R_n^j is the jth coordinate for the molecule at lattice site n. The superscript j describes both translational and librational molecular motion.

If we transform (153) and (154) from the localized to the exciton representation we obtain[3,105]

$$H_M^{ex} = \frac{1}{\sqrt{N}} \sum_{\mathbf{k},\mathbf{q}s} F_s(\mathbf{k}, \mathbf{q}) B^+(\mathbf{k}+\mathbf{q}) B(\mathbf{k}) (b_{s,\mathbf{q}} + b_{s,-\mathbf{q}}^+) \tag{155}$$

$$H_D^{ex} = \frac{1}{\sqrt{N}} \sum_{\mathbf{k},\mathbf{q}s} \chi_s(\mathbf{q}) B^+(\mathbf{k}) B(\mathbf{k}) (b_{s,\mathbf{q}} + b_{s,-\mathbf{q}}^+) \tag{156}$$

where the subscript s refers to a particular phonon branch, and $b_{s,\mathbf{q}}^+$ and $b_{s,\mathbf{q}}$ are the creation and annihilation operators for an sth branch phonon of wave vector $\mathbf{q}$. The strength of the coupling is given by

$$F_s(k, q) = \sum_{j,m(\neq 0)} e_s^j(\mathbf{q}) \beta_s^j(\mathbf{q}) \left\{ \left(\frac{\partial}{\partial R_0^j} + e^{i\mathbf{q}\cdot\mathbf{m}} \frac{\partial}{\partial R_m^j} \right) M_{0m} \right\}_0 e^{i\mathbf{k}\cdot\mathbf{m}} \tag{157}$$

$$\chi_s(\mathbf{q}) = \sum_{j,m(\neq 0)} e_s^j(\mathbf{q}) \beta_s^j(\mathbf{q}) \left\{ \left(\frac{\partial}{\partial R_0^j} + e^{i\mathbf{q}\cdot\mathbf{m}} \frac{\partial}{\partial R_m^j} \right) D_{0m} \right\}_0 \tag{158}$$

In these expressions $e_s^j(\mathbf{q})$ is the jth component of a unit polarization vector for the sth phonon branch with wave vection $\mathbf{q}$. $\beta_s^j(\mathbf{q})$ is the *rms* amplitude of the corresponding motion.[3,109]

Equations 155 and 156 refer specifically to a crystal with only one molecule per unit cell and consequently only a single exciton branch. Furthermore, we consider only one isolated electronic excited state. Within these approximations we see that the term H_D^{ex} is diagonal in the exciton representation. Recall from Fig. 28 that the D-term is responsible for the gas to crystal shift in excitation energy. Fluctuations in this term due to lattice vibrations cause fluctuations in the center of gravity of the exciton band. These fluctuations do not, however, result in scrambling of k-states; rather they result in a renormalization[110] of the exciton's energy. The resulting change in excited state energy due to H_D^{ex} results in a shift in the equilibrium configuration of the intermolecular coordinates, that is, a local lattice relaxation about the excited molecule. If this relaxation is great enough, it can even result in localization of the exciton, as we see later. The H_M^{ex} term, in contrast, does give rise to scattering of the exciton from one k-value to another. It is thus the term responsible for exciton damping.

In most cases in molecular crystals $D \gg M$ (see Table IV), and it has thus been assumed that H_D^{ex} is more important than H_M^{ex}. If the effect of H_D^{ex} on the exciton energy is not too great, one can remove H_D^{ex} by a renormalization procedure[110] and then proceed to calculate the effect of H_M^{ex} on this renormalized entity called a *polaron.*[111] A polaron (or, more

TABLE IV
Exciton Parameters for Typical Molecular Crystals

Crystal	State	D_{crystal} (cm^{-1})	$M(\beta)$ (cm^{-1})
Benzene	$^3B_{1u}$	$+131 \pm 11$	1[a]
	$^1B_{2u}$	-248 ± 1	4[b]
Napthalene	$^3B_{2u}$		1[c]
	$^1B_{3u}$	-464 ± 10	20[d]
Anthracene	$^3B_{2u}$		3[e]
	$^1B_{2u}$	-2400	25[f]

[a] D. M. Burland, G. Castro, and G. W. Robinson, *J. Chem. Phys.*, **52,** 4100 (1970).
[b] E. R. Bernstein, S. D. Colson, R. Kopelman, and G. W. Robinson, *J. Chem. Phys.*, **48,** 5596 (1968).
[c] D. M. Hanson and G. W. Robinson, *J. Chem. Phys.*, **43,** 4174 (1965).
[d] D. M. Hanson, R. Kopelman, and G. W. Robinson, *J. Chem. Phys.*, **51,** 212 (1969).
[e] R. H. Clarke and R. M. Hochstrasser, *J. Chem. Phys.*, **46,** 4532 (1967).
[f] L. B. Clark and M. R. Philpott, *J. Chem. Phys.*, **53,** 3790 (1970).

precisely, an exciton polaron to distinguish it from an electron polaron) is a lattice excitation plus its accompanying lattice deformation. If the lattice deformation becomes large enough, the polaron becomes localized and no longer moves in a band.[112,113] In this limit one may have to include the effect of the lattice distortion on the symmetry of the crystal. The polaron problem is discussed in more detail in Section V.A.

When a crystal has more than one molecule per unit cell along with the intraband exciton scattering described by (155), interband scattering is also possible.[109,114] This scattering between neighboring exciton bands can involve coupling terms such as (157) that not only scatter the exciton to a new band but also change the exciton's quasimomentum. In addition, coupling terms similar to (158) can cause scattering with no change in **k**. In Fig. 29 the various exciton-phonon scattering processes are shown in a schematic fashion.

We have not considered the possibility that another excited electronic or vibronic state of the molecule may be close to the one in which we are interested. If this is the case, the exciton can of course be scattered to this state as well. In general, neighboring electronic states are separated by several thousand wave numbers and thus are not of great importance in exciton damping processes at low temperatures.

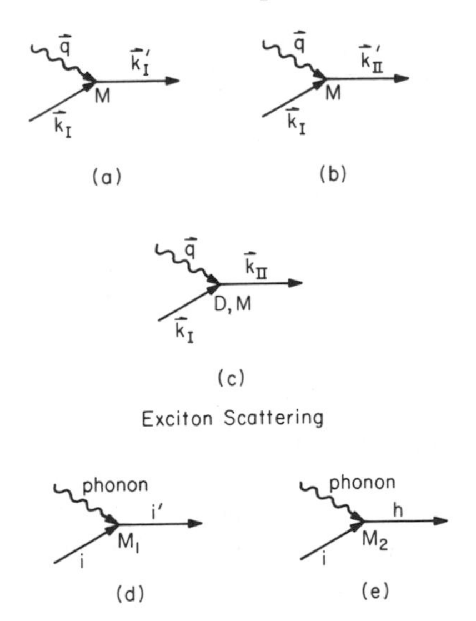

Fig. 29. A schematic representation of some of the possible phonon scattering processes for dimers and excitons. (a) Intraband exciton-phonon scattering involving a change in exciton k-value. (b) Interband exciton-photon scattering involving a change in k-value. (c) Interband scattering with no change in k-value. I and II in these examples represent the different exciton branches. (d) Dimer-phonon scattering from one dimer component i to another i'. (e) Dimer-photon scattering from a dimer state i to a state of the host crystal h.

B. Dimer-Phonon Interaction

The Hamiltonian for dimer-phonon interactions can be written by analogy with (147):

$$H_{\text{dimer}}(R) = \sum_{n=1,2} (\varepsilon_n + D_n(R)) a_n^+ a_n + \sum_{m,n}{}' M_{mn}(R) a_n^+ a_m + \sum_n M_{nh}(R)(a_n^+ A_h + A_h^+ a_n) \tag{159}$$

In this expression a_n^+ and a_n are creation and annihilation operators for electronic excitation of the nth molecule of the dimer pair. A_h^+ and A_h are similar operators for excitation of a molecule of the host crystal lattice. Recall that these isolated dimer pairs are imbedded in a host lattice and that generally the host excited states are above those of the dimer under consideration. Since host and guest dimer electronic states are energetically close to each other, one must include the possibility of phonon-assisted transfer of energy from a dimer state to the host exciton band. The third term in (159) is responsible for this process. M_{nh} is the excitation exchange interaction between the nth molecule of the dimer and a host molecule h.

The first term in (159) is the diagonal energy of the dimer with D_n as in the exciton case, being responsible for the shift in energy between an isolated dimer in the gas phase and one imbedded in a crystal. The second term allows for exchange of energy between the two molecules constituting the dimer.

As in the exciton case, we begin by ignoring the coordinate dependence of the Hamiltonian and write for the rigid lattice

$$H_{\text{dimer}}(0) = \sum_n (\varepsilon_n + D_n(0)) a_n^+ a_n + \sum_{m,n}{}' M_{mn}(0) a_n^+ a_m + \sum_n M_{nh}(0)(a_n^+ A_h + A_h^+ a_n) \tag{160}$$

Ignoring the last term in this expression, we can diagonalize the Hamiltonian by transforming from a representation with the energy localized on a single molecule to a delocalized representation with energy spread between the two dimer states. The energy in this dimer representation then becomes

$$E_i = \varepsilon + D(0) + (-1)^i M_{+-} \tag{161}$$

Here i takes the value 1 or 2, depending on the particular dimer state under consideration. Using the Hamiltonian (160) as a zero-order Hamiltonian, we adopt a similar procedure to the one used to arrive at (152).

For dimers this leads to

$$H_{\text{dimer}}(R) \approx H_{\text{dimer}}(0) + H_D^{\text{dimer}}(R) + H_{M1}^{\text{dimer}}(R) + H_{M2}^{\text{dimer}}(R) \quad (162)$$

In this equation we have three linear dimer-phonon interaction terms:

$$H_D^{\text{dimer}}(R) = \sum_{K,n,j} a_K^+ a_K (\partial D_n(R)/\partial R_j)_0 R_j \quad (163)$$

$$H_{M1}^{\text{dimer}}(R) = \sum_{K,l}{}' a_K^+ a_l \sum_{n,m}{}' \sum_j (\partial M_{nm}(R)/\partial R_j)_0 R_j \quad (164)$$

$$H_{M_2}^{\text{dimer}}(R) = \sum_{K,n,j} (a_K^+ A_h + A_h^+ a_K)(\partial M_{nh}(R)/\partial R_j)_0 R_j \quad (165)$$

The operators a_K^+, a_K create or annihilate a localized dimer state. $H_D^{\text{dimer}}(R)$ is similar to $H_D^{\text{ex}}(R)$ given by (154). It is responsible for a relaxation of the dimer in the excited state. A term that we have not included but that is related to the third term in (155) causes the host molecules to relax about the excited dimer. Neither of these terms alone causes dephasing of the dimer excited state.

Destruction of excited state dimer coherence is brought about by the interaction terms given by (164) and (165). $H_{M1}^{\text{dimer}}(R)$ causes scattering between the two plus and minus dimer excited states, and $H_{M2}^{\text{dimer}}(R)$ causes scattering from an excited dimer level into the host exciton band. These scattering processes are summarized in Fig. 29.

The set of coordinates $\{R\}$ is a bit more complex in the dimer case because of the localized nature of the dimer. $\{R\}$ can of course refer to delocalized phonons as it does for excitons, but one must also consider localized lattice vibrations in the vicinity of the dimer.[115] Although in our theoretical treatment we make no distinction between these two kinds of modes, it should be remembered when analyzing experimental data that the active mode in destroying dimer coherence need not be a mode of the pure crystal lattice.

C. A Comparison

If one is going to study dimer-phonon scattering processes with the hope of inferring something about exciton-phonon scattering, it seems useful to compare the two processes. We begin with the differences that must be considered. If one ignores for the moment dimer scattering into host excited states, the dimer scattering problem is essentially a two-level problem, whereas exciton scattering involves a nearly infinite set of levels. Scattering of excitons can involve very small changes of energy and wave vector. For the dimer the only process that can occur involves scattering between the two dimer states. Even if we include the host excited states, dimer scattering involves only a few well-separated excited states, whereas exciton scattering can involve scattering within a quasicontinuum of states.

The consequences of this difference may be important. Suppose that an exciton with a wave vector **k** is scattered to another state in the band **k**′. Since there are a quasicontinuum of states for **k**′ to be scattered into, the probability that it returns to the initial state **k** is small unless the matrix element coupling **k** and **k**′ are much larger than other matrix elements. In the dimer case one must always consider the scattering back and forth between the two states. This is an important point that is discussed in more detail in Section V.C.

Another difference is involved with the localized nature of the dimer excited state. The dimer states that we are considering are actually energy traps in the lattice. An excited dimer state can be created by annihilation of a host state, or it can be destroyed and a host excited state created. This process is not possible in the exciton case unless another excited electronic state is nearby, and this is not the case in the systems in which exciton scattering has been considered.

The similarities between exciton-phonon and dimer-phonon scattering processes can be seen by looking in more detail at the terms $H_M^{\text{ex}}(R)$ given by (155) and $H_{M1}^{\text{dimer}}(R)$ given by (164). In both cases the scattering strength is determined by the derivative of $M_{mn}(R)$. $M_{mn}(R)$ may be defined in the following way:

$$\beta = M_{nm} = \int \psi_{n\alpha}^{0*} \psi_{m\alpha}^{f*} V_{nm} \psi_{m\alpha}^{0} \psi_{n\alpha}^{f} \, d\tau \tag{166}$$

where, as in (107), $\psi_{n\alpha}^0$, $\psi_{n\alpha}^f$ are ground- and excited-state wavefunctions for the αth molecule of the nth unit cell. V_{nm} is the interaction energy between molecules $m\alpha$ and $n\alpha$. For allowed singlet excited states this term is well approximated by a dipole interaction.[116] For forbidden singlet excited states, higher multipole terms have been considered.[117] For triplet excitons the interaction is due to electron exchange.[49,118]

In comparing exciton-phonon and dimer-phonon scattering it would be useful to be able to relate the dimer states with the exciton k-states. Unfortunately, this is not as simple as it might seem at first. The concept of a k-vector arises from the assumption of periodic boundary conditions or an infinite crystal for the exciton.[119] The N-mers, on the other hand, are finite in extent, and different boundary conditions must be used. Equation (121) gives the expression for the energy of an N-mer, and Fig. 21 illustrates how the N-mer states merge into the exciton band states as N increases.

We can compare the two systems in a qualitative way by examining the nodal character of the dimer and exciton wavefunctions. The nodal character of one of the dimer states corresponds closely to that of the $k=0$ exciton level and the other one to the $k=\pi/a$ level. Scattering depends on the change of M as the distance between molecules is varied slightly. To the extent that the nodal character of the wavefunction determines the amplitude of the scattering, we might expect scattering between the two dimer states to be related to exciton scattering from the zone center ($k=0$) to the zone boundary ($k=\pi/a$).

We explicitly assumed in the discussion above that the interaction M involves nearest-neighbor molecules only. This is certainly a reasonable assumption for triplet excitons because of the very short-range nature of the exchange interaction. It may also be useful for dipole-forbidden singlet excited states where the higher multipole interactions have a relatively short range. For dipole-allowed transitions, however, the interaction is long range, and in general at least next nearest-neighbor interactions must be considered.[12] For the dimer of course the next nearest-neighbor interactions involve molecules of the host and may be completely different from guest-guest interactions. Thus for dipole-allowed transitions, comparison between dimer and exciton scattering processes should be done with even greater caution.

From the above discussion it seems clear that one may compare dimer-phonon and exciton-phonon scattering processes in systems in which the intermolecular interactions are short range, that is, for triplet excitons and dipole-forbidden singlet excitons. Also the scattering from one dimer state to another may be related to scattering of the exciton from the zone center to the zone boundary.

V. EXCITON COHERENCE

A. Diffusion Coefficient Measurements

Much work has been carried out in an attempt to relate the experimentally measured diffusion coefficient to coherent or incoherent exciton transport processes. In many ways this seems like a logical place to begin. The exciton diffusion coefficient is a direct consequence of energy transfer by excitons and thus should have a close relationship to the coherent or incoherent nature of the transport. Unfortunately, as we shall see, the measurement of exciton diffusion coefficients is not a simple experimental matter, nor is it easy to relate the observed coefficients to exciton transport theories because of the complicated nature of the theories themselves.

Chandrasekhar[120] has provided the theoretical link between exciton transport and diffusion. Suppose that an exciton in an isotropic crystal hops in a random fashion from site to site in the lattice. If the lattice spacing is a and the time between hops is τ, then the probability that the exciton finds itself in the volume element defined by $\mathbf{R}$ and $\mathbf{R}+d\mathbf{R}$ at time t is

$$W(\mathbf{R})\,d\mathbf{R}=\frac{1}{(2\pi a^2/3\tau)^{3/2}}\exp\left[-3\,|\mathbf{R}|^2\,\tau/2a^2\right]d\mathbf{R}$$

$$=\frac{1}{(4\pi Dt)^{3/2}}\exp\left[-|\mathbf{R}|^2/4Dt\right]d\mathbf{R} \tag{167}$$

provided that the exciton was at the origin $\mathbf{R}=0$ at $t=0$. D here is $a^2/6\tau$, and Chandrasekhar[120] has shown that in the limit of a large number of steps it behaves like a diffusion coefficient.

At this point it might seem that a system that could be described by a diffusion coefficient must by definition be an example of incoherent energy transport. This need not be the case. Energy transport can be coherent, in some sense of the word, and still exhibit a random walk character. This brings up again the important point mentioned in the introduction regarding the definition of coherence. If, as is frequently done, we consider exciton motion coherent when the mean free path, that is, the distance that the exciton travels between scattering events, is greater than the lattice constant, then even when the energy transport is coherent, a diffusion coefficient may be defined. Note that this definition of exciton coherence is not nearly as precise as the definition we have given for dimer coherence. In the coherent exciton case

$$D=\langle r^2\rangle_{\mathrm{AV.}}/6\tau \tag{168}$$

where $\langle r^2\rangle_{\mathrm{AV.}}$ is the mean square distance that the exciton travels between scattering events. The motion of the exciton for long times is then given by (167).

In this sense, whether coherent exciton motion results in a diffusion coefficient depends on the experiment being performed. The diffusion coefficient in the coherent case can be defined by[121]

$$D=\langle v^2\tau\rangle \tag{169}$$

where v is the exciton velocity and τ is the scattering time. In the absence of any scattering, of course, the exciton motion is completely coherent, τ is infinite and so is the diffusion coefficient. In this limit it is inappropriate to speak of exciton diffusion. The exciton travels in an undeviated "line" and it is not possible to describe its motion as being a random walk.

To see how the kind of experiment being done can in the sense described above determine whether we are observing coherent or incoherent transport, we must examine in a general way the manner in which one measures the exciton diffusion coefficient. All of the experimental techniques have in common the feature that the exciton is created at a particular site and then diffuses to another site where an event occurs such as trapping,[15,122,123] charge carrier production,[124] or exciton-exciton annihilation.[121,125,126] This event is monitored by following, for example, the resulting emission. Then one describes the exciton dynamics for one-dimensional motion with a diffusion equation such as

$$\frac{dn}{dt} = \alpha I_0 - \beta n - \gamma n^2 + D\frac{d^2 n}{dx^2} - kn \tag{170}$$

The first term on the right-hand side of this equation represents the excitation process, β is the emission rate, γ the bimolecular annihilation rate, and k the trapping rate which, as we shall see later, need not be time independent. Solving this equation for $n = n(x, t)$ using various assumptions and boundary conditions, one can then obtain from the experimentally measured quantity the diffusion coefficient D.

Obviously if the exciton hops randomly from site to site the description given by (170) is appropriate. If, however, the mean free path is long compared to the distance traveled from the point of exciton creation to the point at which it is destroyed, then the exciton motion is not correctly described as diffusive.[127] One must thus be careful in extracting diffusion coefficients from experimental data that the diffusion picture really does apply on the time scale defined by the experiment.

It is clear from the above discussion that the mere observation of a diffusion coefficient is not sufficient for us to distinguish between coherent and incoherent motion. It may even be possible to obtain a meaningless numerical value for D when the diffusion process itself is inapplicable.

One way to determine the character of the exciton motion is to measure the temperature dependence of the diffusion coefficient. From such measurements it should be possible to make a distinction between coherent and incoherent motion. In practice this has not been so simple, however. There are both theoretical and experimental obstacles in the way. First consider the theoretical problem. Here there are two kinds of barriers to a straightforward interpretation of the experiments. First, there are a number of possible scattering processes that can contribute to D as given by (169). Both optical and acoustic phonons can scatter the exciton, and in general one expects a different temperature dependence for the two kinds of phonons. Defects and impurities can scatter excitons as well. The defects can be "intrinsic" or thermally created.

The second theoretical complication is that in no case is the theory so well developed that one can obtain the temperature dependence of the diffusion coefficient exactly. One must make many approximations to obtain a useful result. The applicability of these approximations to a given set of experimental data is not obvious in many cases.

Agranovich and Konobeev[128] have considered the temperature dependence of the diffusion coefficient that results from exciton-phonon scattering. They begin with the expression for the relaxation time τ_D[129]

$$\frac{1}{\tau_D} = -\sum_{\mathbf{q}} \frac{\Delta k_z}{k_z}(W_a^{\mathbf{k},\mathbf{k}'} + W_e^{\mathbf{k},\mathbf{k}'}) \tag{171}$$

where $W_a^{\mathbf{k},\mathbf{k}'}$ is the rate at which exciton $\mathbf{k}$ is scattered to $\mathbf{k}'$ by absorption of a phonon with wave vector $\mathbf{q} = \mathbf{k}' - \mathbf{k}$. $W_e^{\mathbf{k},\mathbf{k}'}$ is the corresponding rate for phonon emission. $\Delta k_z = k_z' - k_z$ is the change in the zth component of the quasimomentum due to scattering. The z-direction is assumed to be the crystal direction in which the diffusion coefficient is measured. The factor $\Delta k_z/k_z$ reduces the contribution of scattering processes that do not result in significant changes in k_z. The diffusion coefficient is obtained by using τ_D in (169).

The rates can be obtained in a straightforward way by using the Hamiltonian (155) in the golden rule expression. We then obtain

$$\frac{1}{\tau_D} = -\frac{2\pi}{Nh}\sum_{\mathbf{q}s} \frac{\Delta k_z(\mathbf{q})}{k_z} |F_s(\mathbf{k},\mathbf{q})|^2 \{\bar{N}_{\mathbf{q}s}\,\delta(E(\mathbf{k}) - E(\mathbf{k}+\mathbf{q}) + \hbar\omega_s(\mathbf{q})) + (\bar{N}_{\mathbf{q}s}+1)\,\delta(E(\mathbf{k}) - E(\mathbf{k}+\mathbf{q}) - \hbar\omega_s(\mathbf{q}))\} \tag{172}$$

where $\bar{N}_{\mathbf{q}s}$ is the Plank distribution function, and $\hbar\omega_s(\mathbf{q})$ is the phonon energy. The expression for $F_s(\mathbf{k},\mathbf{q})$ is given in (157).

To evaluate the temperature dependence of τ_D one needs an explicit expression for $F_s(\mathbf{k},\mathbf{q})$. Davydov[105] has evaluated this expression for the specific case of a linear exciton system with dipolar intermolecular coupling. His expression thus strictly applies only to linear singlet excitons. Davydov considers two kinds of phonons; optical phonons involving librations and vibrations of molecules within a unit cell and acoustic modes that involve phased vibrations of entire unit cells with respect to each other. In Fig. 30 we see the dispersion of these two kinds of phonons for a typical molecular crystal.[130] Note that the acoustic phonon frequency goes to zero as the wave vector goes to zero.[119]

The expressions that Davydov obtains are modestly complicated, but if we consider the limit in which k, $q \ll 1/a$ where a is the lattice constant, the interaction for acoustic phonons then becomes

$$|F_s^{\mathrm{ac}}(\mathbf{k},\mathbf{q})|^2 = qa\,|F_0^{\mathrm{ac}}|^2 \tag{173}$$

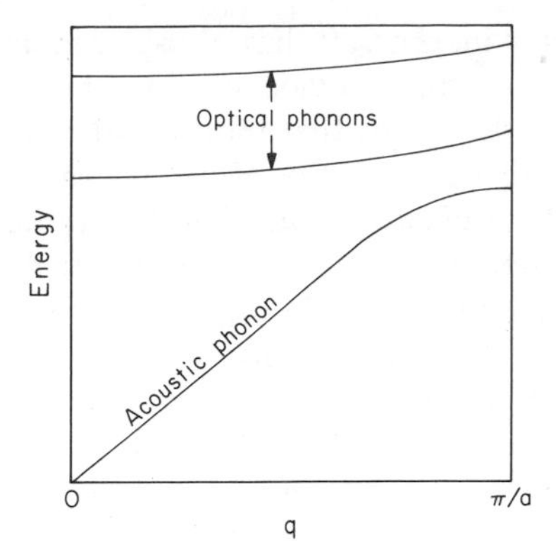

Fig. 30. The energy dispersions for optical and acoustic photons.

and for optical phonon it is

$$|F_s^{\text{op}}(\mathbf{k}, \mathbf{q})|^2 = |F_0^{\text{op}}|^2 \tag{174}$$

where the F_0's are constants independent of **k** and **q**.

Agranovich and Konobeev[128] use the expressions (173) and (174) for the exciton-phonon coupling strength in (172) and derive the temperature dependence of D for both acoustic and optical phonons. For acoustic modes when $\mathbf{k}=0$ is at the bottom of the band and when

$$\frac{B}{k_B} \gg T \gg 0.1\,\text{K}\left(\frac{m^*}{m}\right) \tag{175}$$

the diffusion coefficient has the temperature dependence

$$D(T) \propto 1/\sqrt{T} \tag{176}$$

Here B is the exciton bandwidth, k_B the Boltzmann constant, m the electron mass, and m^* the exciton effective mass given by

$$m^* = 2\hbar^2(a^2B)^{-1} \tag{177}$$

K refers to the temperature in degrees Kelvin.

For optical phonons a similar inverse square root T-dependence is obtained in the limit

$$\frac{B}{k_B} \gg T \gg \frac{\hbar\omega_{\text{op}}}{k_B} \tag{178}$$

where $\hbar\omega_{\text{op}}$ is the energy of the optical phonon responsible for the coupling.

Along with all of the other explicit assumptions, Agranovich and Konobeev[128] have implicitly assumed by their use of (169) that the exciton transport is coherent. It has thus been assumed that a $1/\sqrt{T}$ dependence for D is strong evidence for the presence of coherent excitons, coherent in the sense that motion takes place in a band.

As mentioned in Section IV, the $H_D^{\mathrm{ex}}(R)$ term in the Hamiltonian of (152) does not result in exciton scattering, but rather is responsible for a distortion of the lattice around the excited molecule. The excitation plus its accompanying distortion, a polaron, can move about the lattice in a random incoherent fashion if the distortion is great enough. Polarons arising from excess electrons or holes have been treated by Holstein.[111] His results can also be applied to excitons and are summarized here.

One can relate the polaron theory of Holstein to the treatment given here by considering in greater detail the expression for $W(R)$ given in (152). In treating lattice vibrations one generally expands $W(R)$ in a Taylor series in R and arrives at the harmonic expression

$$W(R) = W(0) + \tfrac{1}{2} \sum U_{nm}^{ij} R_n^i R_m^j \tag{179}$$

where the sum is over all lattice sites n and m and all molecular degrees of freedom i and j. The coefficient U_{nm}^{ij} is the second derivative of W with respect to the coordinates evaluated at the equilibrium position. When the term in the Hamiltonian involving $D_n(R)$ is large, the expansion given by (179) is no longer adequate and one must write $W(R)$ as

$$W(R) = W(0) + \tfrac{1}{2} \sum U_{nm}^{ij} R_n^i R_m^j + \sum_n D_n(R) B_n^+ B_n \tag{180}$$

If $D_n(R)$ is expanded in a power series, the linear term results in a shift in the equilibrium position of the molecule, and the quadratic term results in a change in the phonon frequency. Holstein[111] considers only the linear term, so that (180) is approximated by

$$W(R) = \tfrac{1}{2} M\omega_0^2 R^2 - CR \tag{181}$$

where M is a mass coefficient, ω_0 the vibrational frequency and C is proportional to $(\partial D_n(R)/\partial R)_0$. Here for simplicity we consider only one lattice normal coordinate.

The shift of the equilibrium position as a result of this linear coupling is given by

$$\delta = \frac{C}{M\omega_0^2} \tag{182}$$

and the binding energy of the polaron is found to be

$$E_B = \tfrac{1}{2} M\omega_0^2 \delta^2 \tag{183}$$

These quantities are schematically shown in Fig. 31.

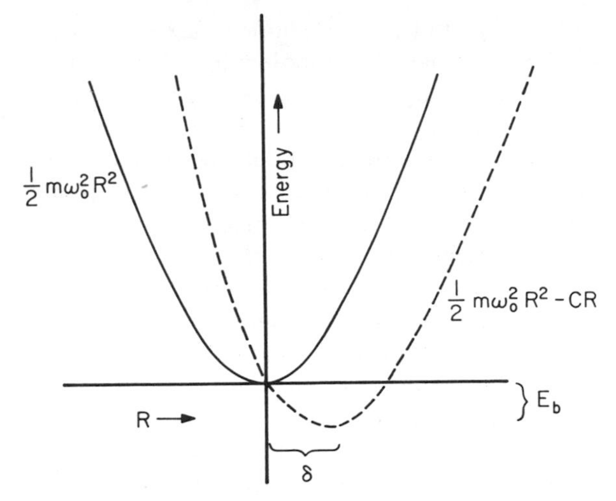

Fig. 31. Potential energy curve for a molecule in a crystal as a function of the lattice coordinate R. The solid line represents the curve for a harmonic lattice and the dotted line the curve in the presence of a linear phonon coupling.

Holstein[111] has shown that in the limit where k_BT is much larger than the vibrational frequencies involved, the temperature dependence of the diffusion coefficient exhibits an activated type of behavior

$$D \propto \left(\frac{1}{T}\right)^{1/2} e^{-E_B/2k_BT} \tag{184}$$

For the polaron to diffuse through the crystal, it must acquire enough thermal energy to overcome the polaron binding energy.

Munn and Siebrand[131] have pointed out that in some cases the quadratic term in the expansion of $D_n(R)$ may be as large as, if not larger than, the linear term. They have thus treated the case of quadratic exciton-phonon coupling. As mentioned quadratic coupling results in a change $\delta\omega$ in the lattice vibrational frequency. The polaron binding energy is then approximately

$$E_b = (n + \tfrac{1}{2})\hbar\omega_0\left(\frac{\delta\omega^2}{\omega_0^2}\right) \tag{185}$$

where n is the level of excitation of the phonon under consideration.

This expression for the polaron binding energy differs in a significant way from the binding energy for linear coupling. Because it depends on the level of phonon excitation n, the binding energy itself is temperature dependent, increasing as the temperature increases. This results in a relatively complicated temperature dependence for the diffusion coefficient. Munn and Siebrand have obtained explicit expressions for D in two limits. In their language in the slow phonon-fast exciton limit the transfer of the excitation from molecule to molecule is limited by the transfer of the lattice distortion accompanying the polaron, and in the fast phonon-slow exciton limit the motion is limited by the transfer of electronic excitation. Within both of these limits the polaron motion may be coherent or incoherent. Incoherent motion in this context refers to a random hopping of the localized polaron. In the coherent case the polaron is not completely localized but moves about the lattice within a band that has been narrowed by the exciton-phonon interaction.

In Table V we list the temperature dependences of D found by Munn and Siebrand[131] for the various limits discussed above. In addition the table also includes the temperature dependences for the other scattering mechanisms discussed in this section.

Not only may excitons be scattered by phonons, they may also be scattered by imperfections in the lattice such as defects or chemical impurities. Scattering by lattice imperfections may be described by the following relationship between the scattering cross-section σ and the scattering time τ

$$1/\tau = \langle N\sigma v \rangle \tag{186}$$

TABLE V
Temperature Dependences of the Diffusion Coefficient

Scattering Regime	Temperature Dependence
Phonon	
Acoustic	$1/\sqrt{T}$
Optical	$1/\sqrt{T}$
Polaron	
Linear coupling	$e^{-\mathrm{E}_a/\mathrm{kT}/\sqrt{\mathrm{T}}}$
Quadratic coupling	
Slow exciton limit	
Incoherent	$e^{-\hbar\omega_0/\mathrm{kT}}/(1+e^{-\hbar\omega_0/\mathrm{kT}})$
Coherent	$(1-e^{-\hbar\omega_0/\mathrm{kT}})^2(1+e^{-\hbar\omega_0/\mathrm{kT}})e^{-\hbar\omega_0/\mathrm{kT}}$
Slow phonon limit	
Incoherent	$(1+e^{-\hbar\omega_0/\mathrm{kT}\hbar})^{-2}$
Coherent	$(1-e^{-\hbar\omega_0/\mathrm{kT}})^2(1+e^{-\hbar\omega_0/\mathrm{kT}})/e^{-\hbar\omega_0/\mathrm{kT}}$
Defects, impurities	
Intrinsic	$\sqrt{T}$
Thermal	$\sqrt{T}e^{\mathrm{E}_D/\mathrm{kT}}$

and then using (169) to write

$$D = \langle v/N\sigma \rangle \tag{187}$$

Consider first chemical impurities and lattice defects such that the number of defects or impurities N is independent of temperature. Assume that σ is independent of the exciton velocity and thus of temperature. (In fact σ will depend slightly on exciton velocity[132] but not enough to affect the overall dependence of D.) We also consider, from the elementary kinetic theory of moving particles, $v \propto \sqrt{T}$. In this case (187) reveals that

$$D \propto \sqrt{T} \tag{188}$$

It is also possible that defects may be created thermally with a certain activation energy E_D. In this case N depends on temperature, and (187) yields

$$D \propto \sqrt{T} e^{E_D/k_B T} \tag{189}$$

One can easily see the problem awaiting the experimentalist who wants to compare his data with theory by considering the various expected temperature dependences for D in Table V. It is quite possible that several of the functional forms may describe the experimental results. Furthermore since all these expressions involve approximations, one expects an even greater variety of temperature dependences when one considers the limits where the approximations we have used do not apply.

A variety of techniques have been used to measure exciton diffusion coefficients. Powell and Soos[15] have recently reviewed these techniques for singlet excitons, supplementing and bringing up to date the earlier excellent review by Wolf.[6] Triplet exciton diffusion has been reviewed by Wolf[6] and for anthracene by Avakian and Merrifield.[8]

There have not, however, been a large number of measurements of the temperature dependence of the exciton diffusion coefficient. In fact much of the work has been directed at obtaining a consistent set of diffusion coefficients for a fixed temperature. The problems involved in obtaining values for the diffusion coefficient have been discussed by Powell and Soos.[15] These authors emphasize that the techniques used to measure diffusion coefficients involve two steps. The first step is the diffusion of the exciton to a particular point in the crystal where it is detected by some means. The second step is the detection process itself. As mentioned before, the detection process may involve trapping of the exciton at a chemical impurity and subsequent emission,[15,122,123] it may involve the production of a charge carrier,[124] or it may involve bimolecular annihilation of the exciton.[121,125,126] One must unravel the two processes to obtain a diffusion coefficient. Even more important for our discussion here, if one is going to measure the temperature dependence of the diffusion coefficient, one must be sure that the temperature dependence of the second process, the detection step, does not interfere.

Despite these difficulties, interesting experimental results have been obtained. Hammer and Wolf[133] have measured the diffusion coefficient for excitons in very pure naphthalene crystals doped with a small amount of anthracene that acts as a trap. By measuring the ratio of the guest fluorescence to the host fluorescence, they obtain the diffusion coefficient from the expression

$$I_G/I_H = 4\pi DRN_G\tau_H \tag{190}$$

where N_G is the number of guest molecules per cubic centimeter, τ_H is the host exciton decay time and R is the capture radius of the trap. One assumes that τ_H and R are independent of temperature and obtains the temperature dependence of I_G/I_H which is thus the temperature dependence of D.

In Fig. 32 we have plotted the values of D obtained by Hammer and Wolf assuming that $R = 10^{-7}$ cm and $\tau_H = 10^{-7}$ sec. The slope of the line

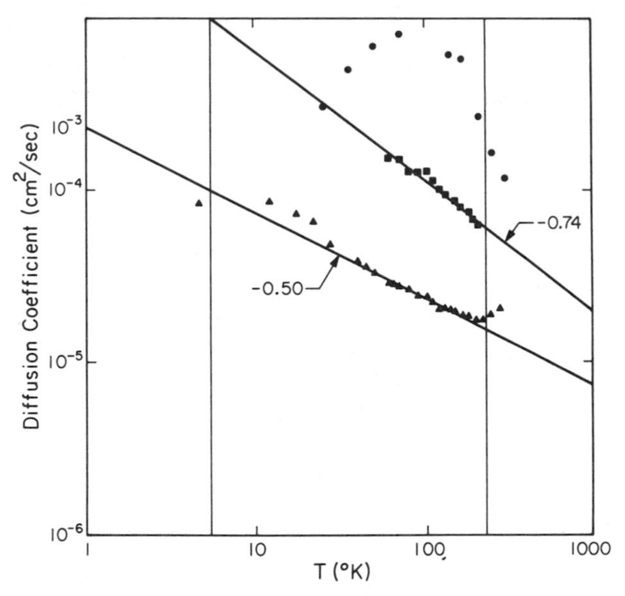

Fig. 32. Experimental values of the diffusion coefficient for singlet excitons in naphthalene as a function of temperature. ▲ from Ref. 133 assuming $R = 10^{-7}$ cm and $\tau_H = 10^{-7}$ sec.; ■ from Ref. 136; ● from Ref. 135. The value of the slopes of the two straight lines are indicated. The two vertical lines indicate the temperature range over which the theory of Agranovitch and Konobeev[128] is valid.

is -0.5 in good agreement with Agranovich and Konobeev's prediction for scattering of coherent excitons by either acoustic or optical phonons. The vertical lines in the figure delineate the limits of the theory for optical phonons. It would seem from these results that singlet exciton transport within this temperature range is coherent with phonons being the dominant scattering mechanism. Similar $1/\sqrt{T}$ dependences have been found for singlet excitons in anthracene crystals.[134] It should be noted that excitons in naphthalene and anthracene crystals are two dimensional, whereas the theory leading to the $1/\sqrt{T}$ dependence strictly applies to linear excitons. One should also note that other functional dependences can explain the results. For example, referring to Table V, thermal defect scattering also agrees with the experimentally observed temperature dependence if the energy necessary to create a defect is assumed to be $32\,\text{cm}^{-1}$.

Also shown in Fig. 32 are measurements of the naphthalene singlet exciton diffusion coefficient made using other techniques. The lack of agreement in the magnitude of the diffusion coefficient is not as serious as it may seem at first if one recalls that D is obtained from experiments by estimating the trap radius R, among other parameters. It is possible by suitably adjusting these parameters to obtain agreement at a given temperature. What is a bit disconcerting is the difference in the temperature dependences observed. The experimental results of Powell and Soos[135] bear no relationship to a $1/\sqrt{T}$ dependence. In fact, the process appears to go like $e^{-\Delta E/k_B T}$ at low temperatures and $e^{\Delta E/k_B T}$ at higher temperatures.

An explanation for the differences between these two experiments does not seem to lie in the differences in the experimental technique used. (Powell and Soos measured the build-up and decay of guest and host fluorescence.) Rather, the difference most likely involves the relative purity of the crystals. Powell and Soos did not remove chemical impurities from the naphthalene they used, whereas Hammer and Wolf used extensively purified naphthalene.

The results of Uchida and Tomura[136] obtained from fluorescence decay time measurements can also be reconciled with the results of Hammer and Wolf. In Fig. 32 we see that they obtain a temperature dependent of $T^{-3/4}$, which does not correspond to any of the functional dependences in Table V. Uchida and Tomura assume that energy transfer takes place by dipole-dipole interaction between the diffusing excitons and the trap, rather than assuming, as Hammer and Wolf do, that the diffusing excitons transfer their energy on direct collision with the trap. If one analyzes the Uchida–Tomura data using the same assumption that Hammer and Wolf use, one obtains a $1/\sqrt{T}$ dependence as well.

A consistent interpretation of all of the experimental results shown in

Fig. 32 can be obtained assuming that the exciton transport in pure naphthlene crystals is coherent and limited by phonon scattering. One must also assume that the diffusing excitons transfer their energy to the traps only by "collisions."

There are, as we have indicated, other scattering mechanisms, such as scattering by thermally activated traps, that can explain the temperature dependence. It would be useful in clarifying this situation to have more measurements of exciton diffusion coefficients in naphthalene using a variety of experimental techniques. In this way, if consistent results could be obtained, one could remove any uncertainties concerning interference from the second step in the measurement of D, the step involving transfer to traps, bimolecular quenching, charge carrier production, and so on.

There are even fewer measurements of the temperature dependence of triplet exciton diffusion coefficients. Durocher and Williams[137] have measured the triplet exciton diffusion length $\ell = (2D\tau)^{1/2}$ for various temperatures in anthracene crystals by following the delayed fluorescence intensity as a function of the singlet excitation wavelength.[126] In Fig. 33 we have plotted the values of D obtained from these measurements versus temperature. One interpretation of these results is that the triplet exciton behaves like a polaron with $E_a = 110\ \text{cm}^{-1}$. Also included in the figure are the diffusion coefficient values obtained by Durocher and Williams[137] using a spectroscopic technique[121,138] that we discuss later. There seems to be no correlation between the temperature dependence of the diffusion coefficient using the two methods.

An approach that has frequently been used to relate exciton diffusion to the coherent or incoherent transport of triplet energy has been called the spectroscopic approach.[121,138] Avakian and co-workers,[121] in initially outlining this approach, began by considering (169). If one can measure the exciton velocity v and the scattering time τ, one should, according to (169), be able to obtain an expression for the diffusion coefficient.

Of course if one knows the exciton velocity and τ, one can easily obtain the mean free path $\ell = v\tau$, which is the distance traveled between scattering events. If ℓ is larger than a lattice constant, one may consider the transport as coherent. This approach does not necessarily involve the measurement of a diffusion coefficient, and we discuss it in more detail in the next sections, where we consider the measurement of τ from magnetic resonance and optical experiments.

Avakian et al.[121] were primarily concerned with developing a simple method of measuring the diffusion coefficient. In their work they obtained the exciton velocity v from a measurement of the anthracene triplet Davydov splitting. The Davydov splitting can be simply related to the excitation transfer matrix element which in turn can be related to the time

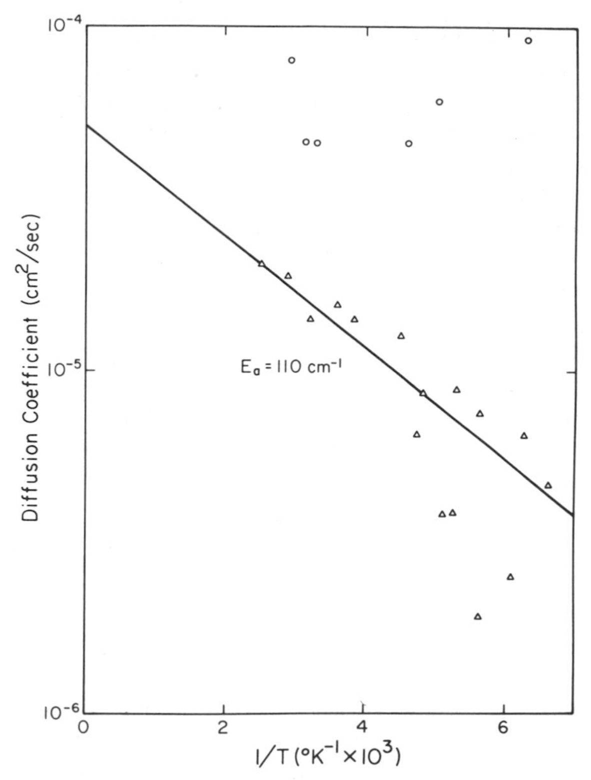

Fig. 33. Temperature dependence of the diffusion coefficient for triplet excitons in anthracene. △ from delayed fluorescence intensity measurements; ○ from the spectroscopic technique. Both sets of data are from Ref. 137. The straight line fits the form $e^{-E_a/kT}$ with E_a as indicated.

that the excitation spends on each molecule. Knowing this time and the lattice constant in the direction of interest, one obtains the exciton velocity. Implicit in this analysis is that all exciton k-states be equally populated so it is valid only for k_BT large compared to the exciton bandwidth. If the k-states are not equally populated and, for example, are in Boltzmann equilibrium, one calculates the velocity by taking the gradient of the band dispersion with respect to k and then thermally averages over all k-states[80,97].

The scattering time τ is obtained from a measurement of the triplet exciton absorption linewidth. The absorption line is assumed to be homogeneous and its width related to the scattering of the exciton. All excitons are assumed to have equal scattering times τ, implying that all exciton scattering mechanisms are localized. With these assumptions, a value for D_{aa}, that is the component of the diffusion tensor in the

a-direction, of 0.2 to $0.9 \times 10^{-4}\ \text{cm}^2\ \text{sec}^{-1}$ was obtained at room temperature, in reasonable agreement with directly measured values of D.

A more detailed analysis of exciton scattering reveals that the situation is a little more complicated than outlined by Avakian et al. In particular, the restriction to local scattering alone, comparable to considering scattering only by H_D^{ex} in (156), seems overly restrictive. A more detailed theory that considers both local (H_D^{ex}) and nonlocal (H_M^{ex}) scattering terms has been developed by Haken and co-workers[104,139,140] and specifically applied to the measurement of the triplet exciton diffusion coefficient by Ern et al.[138]

This approach, which is quite different in philosopy from the approach outlined in Section IV.A, has recently been reviewed by Silbey[18] and has been discussed in some detail by Kenkre and Knox.[141] In the approach described in Section IV, some of the terms in the Hamiltonian responsible for exciton-phonon scattering are displayed explicity and an attempt is made to solve the problem to some order of perturbation theory. Because the phonons are included explicitly, this approach can be used at low or high temperatures, but has the disadvantage, as we have seen, that many approximations must be made to obtain tractable expressions.

In the approach used by Haken and co-workers[104,139,140] one divides the Hamiltonian into two parts: a part H_1 describing the coherent motion of the exciton, corresponding to (148), and a part H_2 that describes the temporal fluctuations in the exciton Hamiltonian and results in incoherent motion. Instead of writing equations like (153) and (154) that rely on a linear exciton-phonon interaction, Haken et al. write in the localized representation

$$H_2 = \sum_{n,n'} h_{nn'}(t) B_n^+ B_{n'} \tag{191}$$

If $n = n'$, (191) describes local scattering and can be related to terms like H_D^{ex}. When $n \neq n'$ the equation describes nonlocal scattering and is related to H_M^{ex}. We have seen that, in the linear exciton-phonon interaction approximation, H_D^{ex} does not give rise to exciton scattering. Since (191) includes local scattering, it implicitly treats exciton-phonon interactions to higher order.

The phonon bath in this approach is treated classically, and the term $h_{nn'}(t)$ is assumed to describe a Gaussian-Markov process. In the nearest-neighbor approximation, that is, $h_{nn'}(t) = 0$ when $n' > n+1$, the following correlation functions are assumed:

$$\langle h_{nn'}(t) \rangle = 0 \tag{192}$$

$$\langle h_{01}(t) h_{01}(t') \rangle = 2\hbar\gamma_1\, \delta(t-t') \tag{193}$$

$$\langle h_{00}(t) h_{00}(t') \rangle = 2\hbar\gamma_0\, \delta(t-t') \tag{194}$$

The exciton absorption linewidth is given in this approximation by

$$\Gamma = (\gamma_0 + \gamma_1) \tag{195}$$

and the diffusion coefficient by

$$D = \frac{a^2}{\hbar}\left(2\gamma_1 + \frac{M_{01}^2}{\gamma_1 + \Gamma}\right) \tag{196}$$

where a is the lattice constant. In these equations γ_0 arises from local fluctuations and has its origins in the D-term discussed in Section III.A. γ_1 arises from nonlocal scattering and is related to the M-term.

Using this approach Haken and Strobl[139] and Grover and Silbey[18,112b] were able to derive a single equation that gave absolutely coherent wavelike motion in one limit (no exciton-phonon coupling) and incoherent motion in another limit (exciton-phonon coupling $\gg$ exciton bandwidth). Ern et al.[138] noted that by measuring the diffusion coefficient and the exciton absorption linewidth, and knowing M_{01} from the Davydov splitting, one could, using (195) and (196), obtain values of γ_0 and γ_1. In this way one could hope to decide the relative importance of local versus nonlocal scattering. Furthermore, by comparing measured values of Γ with the Davydov splitting, one might be able to make a distinction between coherent and incoherent motion.

The experimental results of Ern et al.[138] are summarized in Fig. 34.

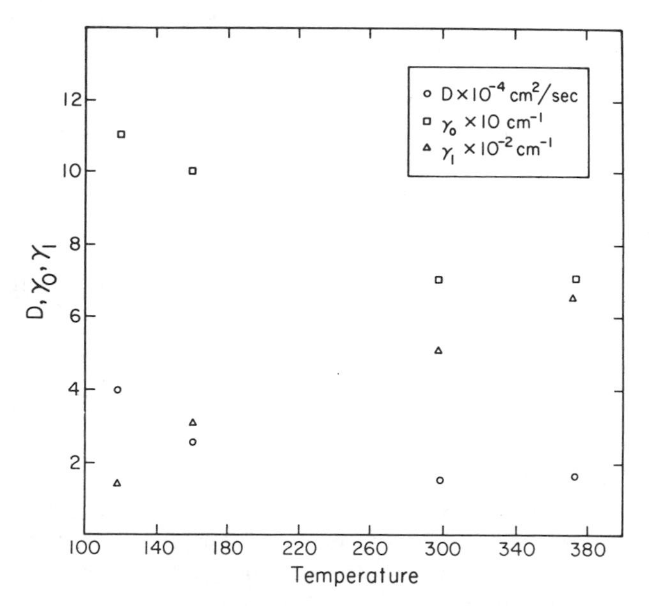

Fig. 34. Measurements of the diffusion coefficient (D) and local (γ_0) and nonlocal (γ_1) scattering rates for triplet excitons in anthracene.

Measurements at four different temperatures indicate that $\gamma_0 \gg \gamma_1$ and that local scattering is dominant. This is contrary to what has been assumed in most other discussions of exciton-phonon scattering. By comparing Γ to the exciton splitting, Ern et al. showed that the triplet exciton transport in the temperature region studied was incoherent.

One big difficulty in the spectroscopic method of studying exciton transport is the assumption that the absorption linewidth is actually a measure of Γ. As we see in Section V.C, high-temperature lineshapes may be Gaussian as well as Lorentzian. If the lines are Gaussian, then Γ and the measured absorption linewidth are not simply related to each other.

B. Band-trap Equilibrium and Disorder

In this section we discuss two experimental techniques that have been used to distinguish coherent and incoherent exciton transport in molecular crystals and that bear some relationship to the diffusion coefficient measurement technique. Both techniques involve the artificial incorporation of an impurity into the crystal to act as an exciton trap or scatterer.

1. *Time-Resolved Experiments*

In the first of these methods one considers the exciton transport in one-dimensional crystals.[142] When an exciton is propagating coherently, that is, when it has a well-defined group velocity, the rate at which it is localized by a trap is given by[142]

$$K_t = \alpha \langle v_g \rangle / d \tag{197}$$

where $\langle v_g \rangle$ is the average exciton group velocity, d is the average distance between traps, and α is the trapping probability per exciton collision with the trap.

On the other hand, when the exciton motion is completely incoherent, that is, when its motion may be assumed to be a random walk from lattice site to lattice site, the trapping rate constant is on the order of

$$K_t = a^2 / t\, d^2 \mathscr{L} \tag{198}$$

where a is the lattice constant and t is the average time per step. $\mathscr{L}$ is a function that depends on the localization energy resulting from self-trapping of the exciton by phonon emission and absorption.[18]

From (197) and (198) we can immediately see a way of distinguishing between coherent and incoherent exciton transport. For coherent transport K_t depends on temperature via $\langle v_g \rangle$ and perhaps even α. In contrast, in the incoherent case K_t should be nearly independent of temperature assuming that $\mathcal{L}$ is insensitive to temperature. This is probably a reasonable assumption for low temperatures. Thus by measuring K_t versus temperature one should be able to distinguish these two limiting cases.

It might seem that the restriction of (197) and (198) to one-dimensional excitons would severely limit this method, but there are a few well-studied exciton systems that are very nearly one dimensional. The triplet excitons of 1,4 dibromonaphthalene (DBN)[59,60] and 1,2,4,5 tetrachlorobenzene (TCB)[79,85] are perhaps the best known of these systems.

In practice this technique is limited to simple cases like those illustrated in Fig. 35 (one trap-one exciton band). In addition to the trapping rate constant K_t, there are rate constants for detrapping K_d and for decay to the ground state from both exciton and trap levels. Therefore if the system has more than one trap, there are even more rate constants to be considered, and the analysis becomes difficult. In principle, many of these rate constants can be extracted from conventional optical[143] and magnetic resonance techniques[144] and from optically detected spin coherence experiments.[145,146]

The results on DBN indicate the difficulties of the problem. Shelby et al.[142] found five different low-temperature traps for the DBN triplet exciton. It is impossible under these circumstances to sort out all of the

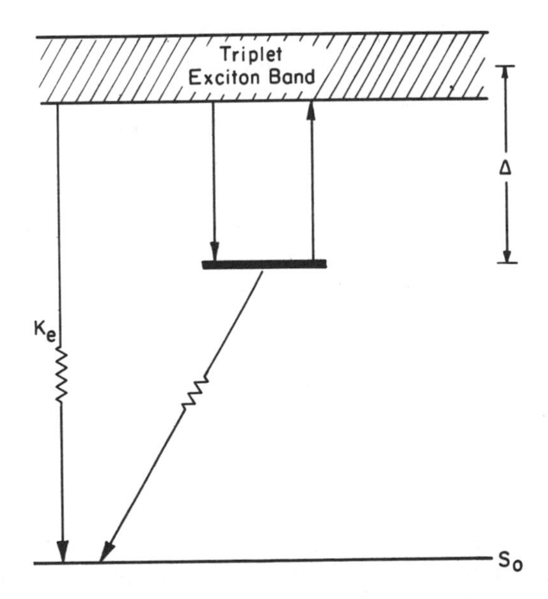

Fig. 35. Kinetics of band-trap interactions for TCB crystals. The arrow from the triplet exciton band to the trap represents trapping and the one from the trap to the band represent detrapping. Δ is the trap depth, and the jagged arrows indicate band and trap emission.

rate constants for trapping and detrapping of the excitation, and one gains little information on the coherence of exciton motion in DBN.

In TCB, on the other hand, one has a single narrow exciton band and an x-trap 17 cm^{-1} [147] below the band. x-traps are molecules that have slightly different orientation or environments from the rest of the molecules in the crystal.[148] Figure 35 corresponds well to the situation in TCB if we treat the three spin sublevels of the triplet as a single radiating state. From measurements of the build-up and decay of x-trap phosphorescence following an 8 nsec laser pulse as a function of temperature, Shelby et al. were able to obtain an estimate for the trapping rate constant K_t of 100 to 300 sec^{-1}. Using (197) and values of $\langle v_g \rangle$ and d known from other experiments, one obtains a value of $\sim 10^{-3}$ for α, assuming completely coherent exciton transport. This means that 1000 "collisions" between exciton and trap are necessary before the excitation is trapped. This seems an unrealistically large number, and on this basis Shelby et al. have concluded that the transport is not completely coherent.

In the incoherent limit, (198) is applicable and since one knows t, d, and a, one can estimate K_t to be 0.46 sec^{-1}, inconsistent with the measured value. The transport thus seems to be intermediate between complete coherence and incoherence. The minimum coherence length consistent with the measured K_t is 700 Å. This means that the exciton travels at least 700 Å (186 molecules) before being scattered by an impurity or phonon.

Guttler et al.[149] have questioned the assumption made by Shelby et al.[142] that the trapping rate constant K_t is limited only by the diffusion of the exciton to the trap. They have concluded that K_t is actually limited by the trapping step itself and therefore that nothing definite can be said about exciton coherence from these measurements.

Recently Fayer and co-workers[150] analyzed in detail the build-up and decay after impulse optical excitation of x-trap emission in TCB. In contrast to previous workers, they included the scattering of the one-dimensional exciton by the naturally occurring isotopic impurity hd-TCB. Their analysis assumes that the exciton transport is coherent. With no adjustable parameters, they are able to fit the experimental trap intensity versus time results with their model. The model uses a value for the in-chain resonance interaction and an x-trap concentration obtained from trap intensity versus temperature measurements. It is assumed in obtaining the x-trap concentration that the trap is in thermal equilibrium with the band and that the exciton k-states are in equilibrium among themselves. They conclude that the experiments are consistent with coherent exciton motion at 1.25 K, although they cannot rule out the possibility that the results could also be explained within an incoherent random walk model. Two important points emerge from their work; the necessity of considering naturally occurring isotopic impurities as scatterers, at least at low temperatures and the fact that the trapping rate constant K_t may not in fact be constant with time.

The value of β obtained by Fayer and co-workers[150] is consistent with values obtained from optical spectra[69] and dimer ODMR experiments,[79] but different from an early result[80] obtained from x-trap emission intensity versus temperature measurements. This discrepancy has been attributed[147] to differences in the lifetimes used for the guest-host system and to the use of an earlier value for the trap depth. The value used here for the x-trap depth is $17.3\ \text{cm}^{-1}$. Based on these new results together with the exciton ODMR linewidth,[85] Dlott and Fayer[147] have concluded that the coherence length of an exciton in TCB at low temperatures is 60,000 lattice sites, that is, that in these crystals the exciton propagates in the chain for about 2×10^5 Å without losing its coherence.

2. *Disorder: Percolation and Anderson Localization*

A new experimental technique that has been used for deciding between coherent and incoherent exciton transport was recently developed by Kopelman and co-workers.[17] This technique, known as exciton percolation, has its origins in the study of electrons in disordered solids,[151] gases,[152] and solutions.[153] The basic principle is illustrated in Fig. 36. A broad-band light source primarily excites the host exciton. The host exciton is then quenched by both the trap (usually an isotopic modification of the host) or the supertrap (a chemical impurity at very low concentration). The concentration of the trap is varied.

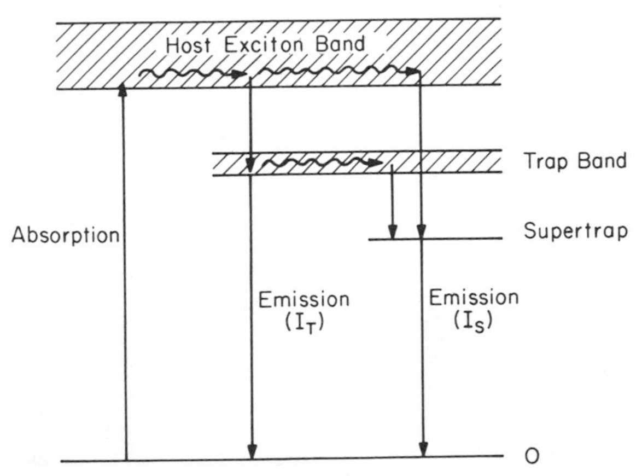

Fig. 36. Kinetics of energy transfer between a host band, a trap band, and a supertrap state.

When the trap concentration is low, there is very little transfer of excitation from trap to supertrap, and the ratio of the supertrap emission intensity I_s to the total trap plus supertrap intensity I_s+I_T reflects the concentration differences between trap and supertrap. When the trap concentration becomes high enough, there are islands of trap molecules in the host crystal that include supertrap molecules. It is thus possible for the trap molecules to feed the supertrap molecules. At this point the ratio I_s/I_s+I_T begins to increase. In Fig. 37 we see a plot of the observed fluorescence intensity ratio versus the trap concentration for the system naphthalene(trap)-perdeuteronaphthalene(host)-β-methylnaphthalene (supertrap). The problem is to relate the mechanism of exciton transport to the shape of this experimental curve. Several factors must be considered, including the dimensionality of the transport and whether the transport is coherent or incoherent. The critical behavior, that is, the sudden change in intensity ratio with concentration seen in Fig. 37, was explained by Kopelman et al. using percolation theory. They analyzed the data in terms of "partially" coherent singlet exciton transport and inferred an exciton coherence length from an application of the theory. Similar results have also been found in benzene by Colson, et al.[156c]

An important point regarding the applicability of percolation theory to triplet exciton systems has been discussed by Klafter and Jortner.[154] These authors raise the possibility that the rapid change in I_s/I_s+I_T shown in Fig. 37 is not due to the opening up of percolation channels in the mixed

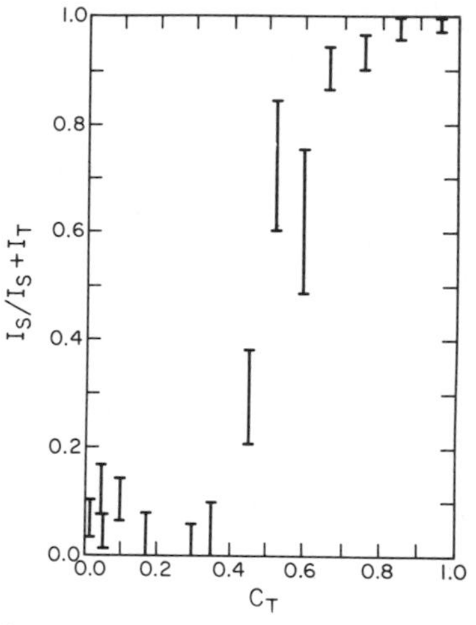

Fig. 37. Experimental exciton percolation for the naphthalene singlet exciton. I_S is the β-methylnaphthalene supertrap fluorescence intensity and I_T the naphthalene trap intensity. C_T is the mole fraction of trap present. The supertrap concentration is about 10^{-3}. The temperature of the sample was 2 K. After Ref. 17.

crystal, but rather to a transition from a localized to a delocalized excitation in the sense described by Anderson.[155] In this model at high guest concentration a narrow trap exciton band is formed. As the trap concentration is reduced, the band narrows because of the increasing distance between resonant trap molecules. At some point the trap band becomes much narrower than the inhomogeneous spread of trap excited-state energies. At this point the excitation is localized, and the trap states can no longer feed the supertrap states. Thus the transition from a localized to an extended state depends critically on the ratio of the trap bandwidth to the inhomogeneous linewidth. The inhomogeneous broadening in molecular crystals is a measure of the disorder in these systems. In a macroscopically disordered system classical percolation theory is not adequate, as pointed out by Mott,[155b] because tunneling among sites with slightly different energies prevails. Moreover, the concept of a coherent exciton propagating in a disordered lattice seems ill defined.

Smith et al.[156] recently examined the applicability of the Anderson localization model to isotopically mixed crystals of DBN and phenazine. The experimentally observed emission intensity ratio as a function of isotopic composition can be explained quite well using the Anderson theory as modified by Klafter and Jortner.[154] The results of Smith et al.[156] for the phenazine triplet exciton system are shown in Fig. 38. In the isotopically mixed phenazine crystals used, the trap was the perprotonated phenazine monomer and the supertrap the corresponding dimer. As the isotopic composition of the crystal is changed, both trap and supertrap concentrations vary. Since only isotopic species are used, changes in trap-supertrap concentrations do not perturb the lattice as much as they would if chemically different traps or supertraps were used.

In Fig. 38, at 1.38 K one sees the same kind of critical behavior in phenazine that Kopelman, et al.[17] observed in naphthalene. At very low temperatures the transition from localized to delocalized states is sharp, whereas at 1.92 K and above there is no evidence of critical behavior. These experimental results can be explained by the simple kinetic scheme[156] outlined in Fig. 39. The monomer and dimer states are populated via the host exciton band at rates given by Γ_{HM} and Γ_{HD}, respectively. It is also possible for excitation to leave the monomer and dimer trap states at the rates Γ_{MH} and Γ_{DH}. This detrapping process is a thermally activated process and should be dependent on the trap depth. Furthermore, when the monomer concentration is high enough extended monomer states exist, and monomer trap and dimer supertrap may communicate via the trap band states (Γ_{MD} and Γ_{DM}).

Raising the temperature has the effect of increasing the detrapping

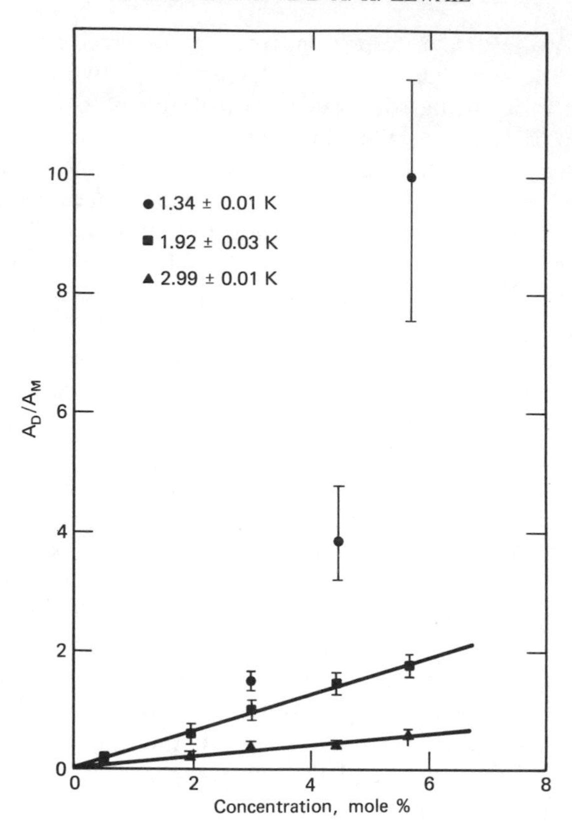

Fig. 38. Ratio of the integrated emission from the two impurities traps (dimers and monomers) of isotopically mixed phenazine crystals as a function of concentration and at three different temperatures. The concentrations are determined from mass spectrometry. After Ref. 156.

rates Γ_{DH} and Γ_{MH}, thus permitting indirect contact between trap and supertrap via the host exciton band. This indirect communication occurs with an activation energy determined by the difference in trap depths between monomer and dimer and has been observed[156b] (see Fig. 40, curves A and B). As the concentration of traps and supertraps increases, direct communication channels open up (Γ_{DM} and Γ_{MD}), so that even at low temperatures there is communication among traps and supertraps (Fig. 40, curves E and D). The temperature dependence in the extended state regime is still governed by the energy difference between monomer and dimer.

Smith et al.[156] concluded from their work on phenazine and on DBN that the results are consistent with the Anderson localization model[154].

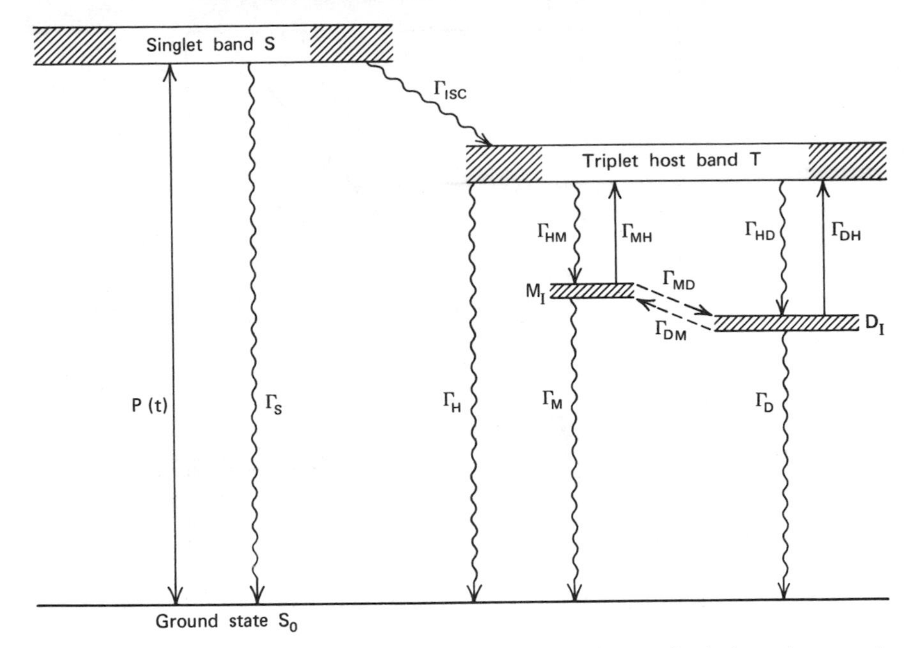

Fig. 39. A diagram of the energy transfer routes in isotopically mixed phenazine crystals. $P(t)$ is the light-pumping rate and the Γs denote the different rate processes (see text). After Ref. 156.

Furthermore, by comparing the phenazine and DBN results they observed evidence of the crucial role that superexchange among monomers plays in determining the concentration at which localization occurs. The question of whether, for a particular exciton system, percolation or Anderson localization theory is applicable is still unsettled.

C. Lineshape Measurements

1. *Correlation Time and Dephasing: Preliminary Remarks*

Recently two experimental techniques have been used in an attempt to measure the exciton scattering time directly. Both techniques involve the measurement of absorption lineshapes. In one case the magnetic resonance lineshape is measured and related to the correlation time for triplet exciton scattering. Wolf and co-workers[157,158] and Sharnoff and co-workers[159] have developed this method in EPR and Harris and co-workers[85,159] in zero-field ODMR (optically detected magnetic resonance). Similarly, optical frequency exciton absorption lineshape measurements have been made on singlet[160–162] and triplet excitons.[163,164] Here as well the attempt has been made to relate the observed optical linewidths with exciton relaxation processes. These optical measurements

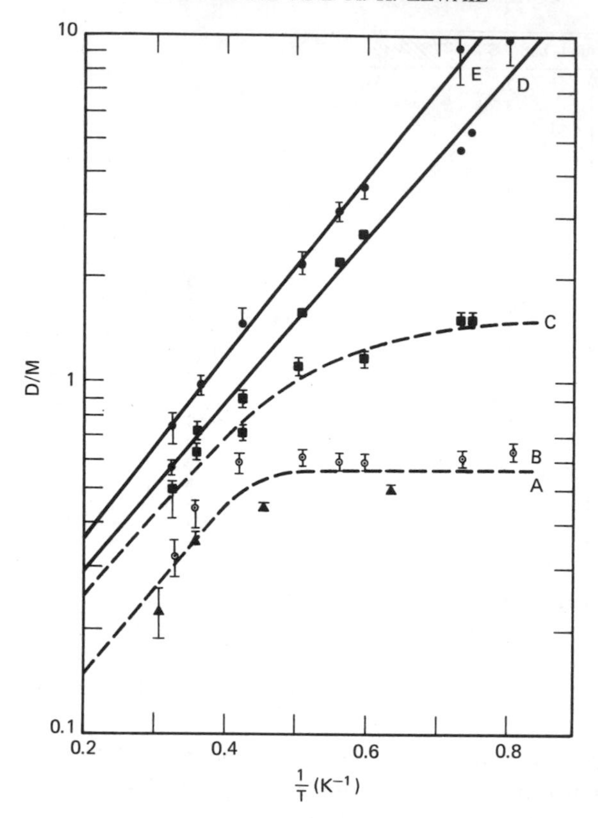

Fig. 40. Effect of temperature on the ratio of dimer-to-monomer emission intensities in isotopically mixed phenazine crystals. A, B = 2% crystal (two independent runs); C = 3 ± 0.2%; D = 4.4%, E = 5.65%. Taken from Ref. 156.

of course have their antecedent in the work of Avakian and coworkers[121,138] discussed in Section V.A.

Before discussing these techniques in detail, it would perhaps be useful at this point to discuss in general terms the relationship between a measured linewidth in absorption and scattering processes. As seen in Section III.F, a measurement of an absorption linewidth, even when the line is definitely homogeneous, does not yield direct information about scattering processes. The value of T_2 determined experimentally depends not only on the particular process responsible for the dephasing but also on the time scale of the experiment.

To make this point a bit clearer, we turn to a simple formulation of the problem outlined by Kubo.[165] These remarks apply both to optical and magnetic resonance lineshape measurements. Consider an oscillator whose motion can be described by the equation

$$\dot{x} = i\omega(t)x \tag{199}$$

x may represent a magnetic moment in the magnetic resonance case or a transition dipole moment in the optical absorption case. $\omega(t)$ is the fluctuating frequency, the fluctuations being due to the scattering processes under investigation. These fluctuations are assumed stationary in time and random so that

$$\omega(t) \equiv \omega_0 + \omega_1(t) \tag{200}$$

and

$$\bar{\omega}_1 = 0 \tag{201}$$

According to the fluctuation dissipation theorem, the absorption intensity at the frequency ω is given by

$$I(\omega - \omega_0) = \frac{1}{2\pi} \int_{-\infty}^{\infty} e^{-i(\omega-\omega_0)t} \phi(t)\, dt \tag{202}$$

where $\phi(t)$ is the relaxation function defined by

$$\phi(t) = \left\langle \exp i \int_0^t \omega_1(t')\, dt' \right\rangle \tag{203}$$

The absorption line (202) is broadened about ω_0 by the random modulation $\omega_1(t)$.

In the simplest case, where the probability of finding a particular value of ω_1 has only one peak at $\omega_1 = 0$, the stochastic process represented by $\omega_1(t)$ can be described by only two parameters, the amplitude of modulation

$$\Delta^2 = \langle \omega_1^2 \rangle \tag{204}$$

and the correlation time

$$\tau_c = \frac{1}{\Delta^2} \int_0^{\infty} \langle \omega_1(t)\omega_1(t+\tau) \rangle\, dt \tag{205}$$

τ_c is a measure of the speed of the modulation.

For slow modulation $\Delta \cdot \tau_c \gg 1$, the absorption lineshape is Gaussian:

$$I(\omega - \omega_0) = \frac{1}{\sqrt{2\pi}\Delta} \exp\left\{-\frac{(\omega - \omega_0)^2}{2\Delta^2}\right\} \tag{206}$$

In this case the modulation is so slow that the intensity distribution accurately reflects the statistical nature of the fluctuations. For fast modulation when $\Delta\tau_c \ll 1$, the absorption line is Lorentzian:

$$I(\omega-\omega_0)=\frac{1}{\pi}\frac{\gamma}{(\omega-\omega_0)^2+\gamma^2} \tag{207}$$

The half-width of this line γ is given by[165]

$$\gamma=\Delta^2\tau_c \tag{208}$$

and taking into account the condition for fast modulation we have that

$$\gamma \ll \Delta \tag{209}$$

and the line will be narrowed by the fast modulation. Such narrowing is known in magnetic resonance as motional narrowing[166] and exchange narrowing[167] and is discussed in Section III.F.

The major point to be made here is that the linewidth γ does not measure the effect of the scattering process directly. The amplitude of the modulation Δ depends in general on the frequency ω_0 about which the modulation occurs; thus the linewidth measured depends on the frequency at which the experiment is performed (see Fig. 41). The problem facing the experimentalist who wants to relate exciton absorption lines to exciton scattering processes is then to develop a model of the scattering in sufficient detail for one to extract Δ or τ_c or both from the measured γ.

From the above discussion one clearly expects that the measurement of magnetic resonance linewidths and optical linewidths will illuminate different aspects of the exciton scattering problem, first, because of the

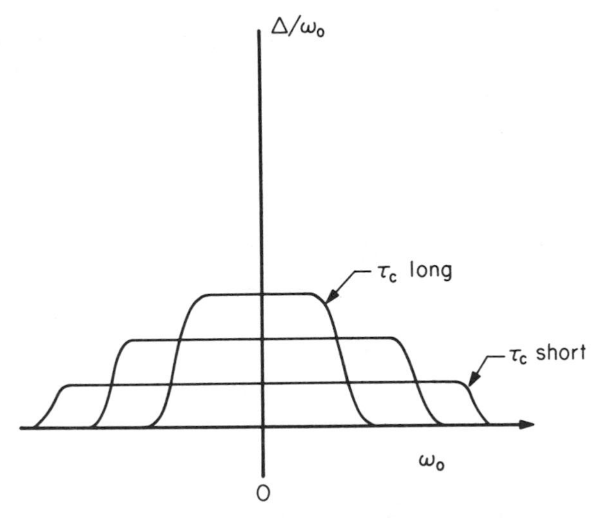

Fig. 41. The frequency modulation amplitude Δ as a function of the average frequency ω_0 for several different values of the correlation time τ_c. After Ref. 35a.

different frequencies used, and second, because of the difference in the parameter measured. In magnetic resonance, the scattering is affecting the magnetic dipole moment, whereas in optical absorption the electric dipole moment is affected. The problem arises in relating measured values of γ or equivalently T_2 with theoretical models of exciton scattering.

One other point must be mentioned before moving on to a discussion of specific experimental lineshape measurements. In discussing diffusion coefficient, trapping, and percolation experiments we have considered exciton coherence to mean transport in a specific direction in the crystal. *Scattering processes that do not disrupt this unidirectional exciton transport cannot lead to a destruction of coherence in this sense.* This point is emphasized by the factor $\Delta k_z/k_z$ in the expression (172) for $1/\tau_D$. This factor reduces the contribution of scattering processes that result in only a small change Δk_z in the direction of exciton transport.

In contrast, both magnetic resonance and optical lineshape measurements yield coherence properties of excitons that are much closer to the definition of dimer coherence given in Section III.D. Lineshape measurements can be related to the dephasing of the exciton wavefunction and are in this sense direct measurements of T_2. *They are sensitive to all dephasing processes no matter how small the change* Δk_z.

2. *Magnetic Resonance Experiments*

The first observation of an EPR signal from a triplet exciton in a molecular crystal was reported by Haarer and Wolf.[157] These authors observed exciton magnetic resonance in pure crystals of anthracene and naphthalene. The resonance lineshapes were quite accurately Lorentzian, suggesting that one could obtain information about exciton dephasing processes from such experiments.

The problem, as we have noted, is to relate the observed linewidth to a scattering time. Haarer and Wolf assumed that the observed EPR linewidth was due to the incoherent hopping of the exciton from one molecule in the unit cell to another (see Fig. 42). In general, the magnetic field orientation with respect to the molecular axes will differ for the two molecules in the anthracene or naphthalene unit cell. A triplet exciton on one molecule will respond at a slightly different Larmor frequency when it hops to a neighboring molecule. The random change of frequency due to exciton hopping through the crystal is assumed responsible for the measured EPR linewidth. These assumptions lead one to expect a strong dependence of the observed linewidth, that is, of T_2, on the orientation of the crystal in the magnetic field. Such a dependence is in fact observed.

Using the above model, Haarer and Wolf have been able to extract a

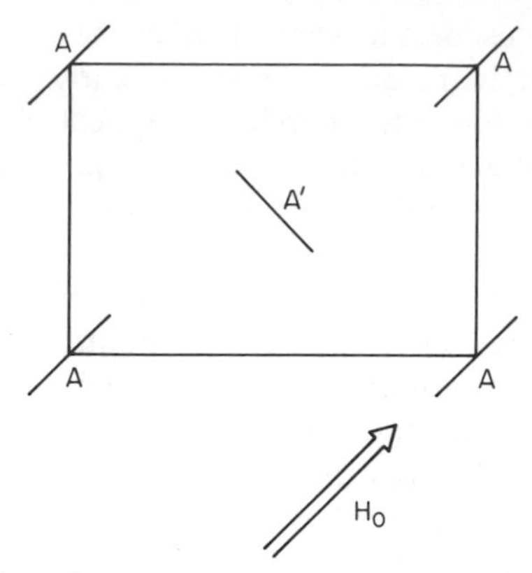

Fig. 42. The different orientations of translationally inequivalent molecules in the unit cell (A and A') with respect to an external magnetic field H_0.

correlation time τ_c for the scattering process from measured values of T_1 and T_2. In their model this correlation time is twice the exciton hopping time. The values of τ_c obtained are listed in Table VI. The values of τ_c displayed in the table are dependent on the particular model described above. In particular, the exciton transport must be incoherent and the exciton spin must be completely dephased with each hop. Haken and Reineker[104,168] have attempted to enhance the applicability of the model of Haarer and Wolf by allowing coherent motion of the exciton. They find, however, in agreement with Haarer and Wolf, that the exciton transport in anthracene and naphthalene at 300 K is adequately described

TABLE VI
Experimentally Determined Coherence Parameters for Exciton Systems

Crystal	Temperature (K)	τ_c (psec)	$8\beta(8M)$ (cm^{-1})	t_{jump} (psec)[a]
Anthracene	300	3.6[b]	22[c]	0.2
Naphthalene	300	9.0[b]	10[d]	0.5
DBN	1.4	5000[e]	<1	>1

[a] $1/t_{jump} = (2\pi c)(4\beta)$.
[b] D. Haarer and H. C. Wolf, *Mol. Cryst. Liq. Cryst.*, **10**, 359 (1970)
[c] R. H. Clarke and R. M. Hochstrasser, *J. Chem. Phys.*, **46**, 4532 (1967).
[d] D. M. Hanson and G. W. Robinson, *J. Chem. Phys.*, **43**, 4174 (1965).
[e] R. Schmidberger and H. C. Wolf, *Chem. Phys. Lett.* **16**, 402 (1972).

as incoherent. From the theory of Haken and Reineker, one obtains an expression relating the measured EPR linewidth to the quantity γ_1 defined in (193). The value obtained for anthracene from EPR measurements (0.05 cm^{-1}) is in good agreement with values obtained from optical linewidth and diffusion coefficient measurements.[121]

These EPR measurements on naphthalene and anthracene were done at above 100 K, and it is not surprising that at these temperatures the lattice vibrations are large enough in amplitude to render the exciton motion incoherent. It is at cryogenic temperatures (4.2 K and below) that we expect to find evidence for coherent exciton motion. Schmidberger and Wolf[158] have examined the low-temperature triplet exciton EPR of DBN which we have seen to be an example of a one-dimensional exciton. Sharnoff and Iturbe[159] have studied the ODMR of benzophenone crystals that are not one dimensional. The DBN molecule has a relatively complex crystal structure,[70] as shown in Fig. 14. There are eight molecules in the unit cell. The molecules separate into two crystallographically inequivalent sets of four molecules each. Exciton transport is predominantly along the c-axis, although an exciton occasionally may hop between translationally inequivalent chains.

Schmidberger and Wolf[158] assumed that the EPR lineshape and spin-lattice relaxation were determined by the interchain hops, since jumps between translationally equivalent molecules would not change the Larmor frequency of the spin. At 1.4 K a correlation time of $5 \pm 3 \times 10^{-9}$ sec was found for these interchain hops (see Table VI). There are orientations of the DBN crystal in the external magnetic field where all chains become equivalent. In these directions, neglecting nonsecular sources of broadening, the exciton line should become extremely narrow, since for these orientations interchain hopping does not dephase the spin. In fact, the line should be much narrower than the observed 3 to 10G residual width. A similar problem arises for the translationally equivalent dimer case as well.[100]

Schmidberger and Wolf[158] have also reported careful measurements of the EPR linewidth versus temperature. Their results are shown in Fig. 43 where we see that the linewidth increases with temperature up to 16 K, at which point the width decreases again. If one subtracts out the constant low- and high-temperature linewidths, one finds that at low temperatures the EPR line increases like T^3. At 16 K the temperature abruptly changes to a $1/T^3$ behaviour. These authors attribute this puzzling change in the temperature dependence to a change in the exciton migration from coherent to incoherent motion, but the theoretical connection between the nature of the exciton motion and this kind of temperature dependence has not yet been made in detail.

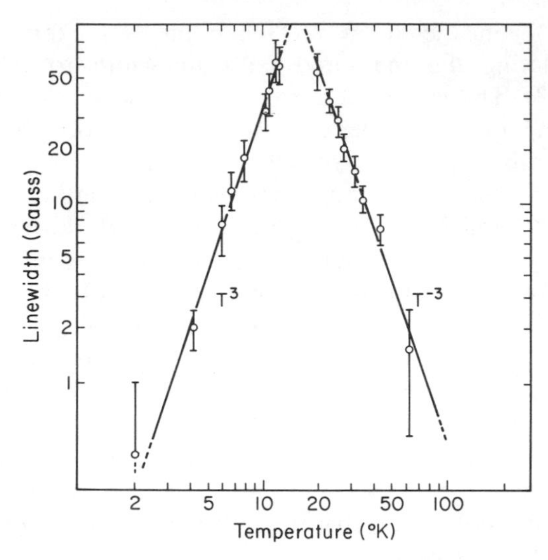

Fig. 43. EPR linewidth versus temperature for DBN showing the change from T^3 to T^{-3} behavior. After Ref. 158.

This observed temperature dependence is reminiscent of the temperature dependence one expects for $1/T_1$ in a two-level system.[35a] We write $1/T_2$ which is proportional to the linewidth as

$$\frac{1}{T_2} = \frac{1}{T_2'} + \frac{A\tau_c}{1+\omega^2\tau_c^2} \tag{210}$$

and assume that $1/T_2'$ is negligible for all temperatures. We also assume a fixed microwave frequency ω. Increasing the temperature is equivalent to decreasing the correlation time. One can see that for low temperatures when $\omega^2\tau_c^2 \gg 1$, $1/T_2 \propto 1/\tau_c$ and for high temperatures when $\omega^2\tau_c^2 \ll 1$, $1/T_2 \propto \tau_c$. This nicely explains the experimental results, but there is a problem. Triplet excitons are not two- but three-level systems. For a three-level system no such simple relationship as (210) exists. Nevertheless, it is worth considering that an explanation of the temperature dependence may involve, not the nature of the exciton motion, but the relationship between the scattering frequency and the characteristic frequency of the experiment.

Another one-dimensional exciton system that has been studied in detail is TCB. All the magnetic resonance measurements to date on this crystal have been made at zero magnetic field. Francis and Harris[85] were first to stimulate interest in this system from the point of view of excitons when they found that one could observe the exciton density of states directly from the optically detected magnetic resonance (ODMR) experiments.

Experimentally they observed the ODMR signal shown in Fig. 23. These curves have the shape expected for a slightly broadened one-dimensional exciton density of states. The explanation given by Francis and Harris for the ODMR lineshape shown in Fig. 23 is discussed in detail in Section III.F.

Harris and Fayer[97] have carried these theoretical ideas out in detail and obtained an expression relating the ODMR lineshape to, among other parameters, the Lorentzian width $1/T_2(k)$ of each of the individual k-states. Dlott and Fayer[147] have obtained a value for T_2 assuming that it is independent of k. Their value of T_2 is obtained by fitting the band-to-band curve (Fig. 23) and using a value for the exciton bandwidth obtained from careful measurements of x-trap intensity versus temperature. At 3.2 K they obtain a value of 9×10^{-7} sec. We emphasize that this is a measurement of T_2 and not the correlation time for exciton scattering. To relate these two times we need a model of the scattering process. In particular, we need to know how a change of exciton k-vector can dephase the magnetic dipole moment.

Recently Botter et al.[169] examined the temperature dependence of T_2 for the TCB triplet exciton system. They used a combination of pulsed magnetic resonance techniques (two and three pulse echoes) at zero magnetic field as well as ODMR measurements. These authors prefer to call the quantity that they measure T_M, the phase memory decay rate, to emphasize the fact that more than one dephasing process may be operative and the echo decays need not be exponential.

From the dephasing of the three-pulse echoes as a function of the pulse separation, Botter et al. infer that the change in k-vector amplitude due to exciton scattering at 1.13 K is small, amounting to only a few percent of the total bandwidth. To extract this information from their data they must assume that the measured T_M is the average lifetime of a k-state. This is of course an assumption that needs experimental verification.

The temperature dependence of T_M determined by Botter et al. is shown in Fig. 44. The upper curve refers to measurements of the ODMR linewidth and the lower curve to the pulse echo experiments. At very low temperatures the ODMR linewidth reflects the exciton bandwidth and density of states, as Francis and Harris[85] have pointed out. The echo decay, however, yields the k-state dephasing time averaged over all populated k-states at low temperatures. Below 1.6 K T_M has a constant value of 1 μsec. Botter et al.[70] consider this as evidence that the exciton is scattered by impurities and defects. From the three-pulse echo work mentioned above it is clear that this involves only small changes in k. Above 1.6 K T_M begins to decrease presumably as phonon scattering becomes increasingly important. Also included in Fig. 40 is the value of

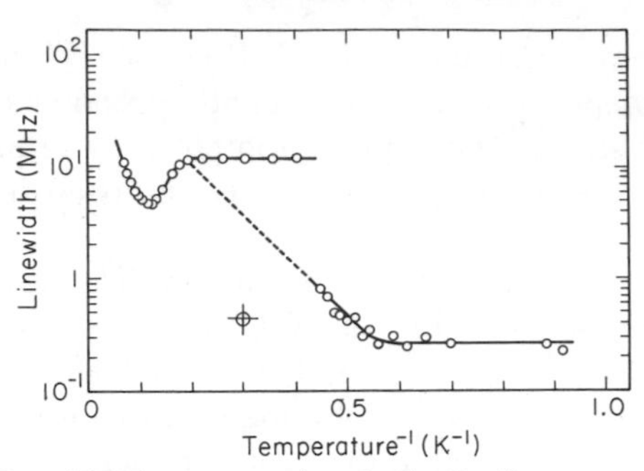

Fig. 44. The ODMR linewidth (upper curve) and $.1/\pi T_M$ (lower curve) measured from the two pulse- echo experiments for triplet excitons in TCB. The dotted line is an extrapolation of the lower curve. The crossed circle indicates the single value measured by Dlott and Fayer.[147] After Ref. 170.

T_2 measured by Dlott and Fayer.[147] There is an order of magnitude difference between this value of T_2 and the extrapolated value of T_M measured by Botter et al. Both experiments are presumed to measure an average k-state dephasing time. It is not clear whether the difference is due to the theoretical model used to relate T_2 or T_M to the exciton scattering process or whether the two processes actually measure a different aspect of the scattering process. It should be emphasized, however, that the T_2 obtained from the echo experiments is measured directly and does not rely, as does the method of Dlott and Fayer,[147] on assumptions about the exciton band k-state structure.

At 5 K the ODMR line begins to narrow. Botter et al. explain this narrowing as being due to motional narrowing due to rapid scattering of the exciton across the exciton band. This is close to the explanation we have suggested for a similar phenomenon in DBN. As one moves still higher in temperature, the ODMR line begins to broaden again above 8 K. At this temperature, according to Botter et al., the scattering rate approaches the magnitude of the intermolecular interaction ($\sim 10^{10}$ sec^{-1}), and the exciton transport becomes incoherent.

3. *Optical Linewidth Measurements*

In many ways, measurements of exciton absorption lineshapes should be the most direct method of obtaining information on the exciton dephasing process. Optical absorption, as we saw in Section II, corresponds to the creation of the $\mathbf{k}=\mathbf{Q}$ exciton. The homogeneous linewidth should thus be directly related to the dephasing time of this specific optically created exciton. Unlike the magnetic resonance experiments that measure a dephasing time averaged over a thermal distribution of excitons, optical linewidth measurements measure the dephasing of the $\mathbf{k}=\mathbf{Q}$ state only.

The optical linewidth is a measure of the dephasing time for the transition electric dipole moment.[170,171] When the absorption line is Lorentzian, one can obtain the dephasing time $[1/2\pi c\Gamma(\text{cm}^{-1})]$ directly from the linewidth. To relate T_2 to a scattering process that can destroy exciton coherence, we need a model of the scattering process. We emphasize again that the relationship between T_2 and exciton scattering processes depends on the specific model chosen.

The model most widely used [10,105,106,109,172–174] assumes that every scattering event scatters the optically created exciton, $\mathbf{k}_Q$, to another k-value, say $\mathbf{k}'$ (see Fig. 45). Since there are a semi-infinite number of k-states, the probability that the $\mathbf{k}'$ exciton will return to $\mathbf{k}_Q$ within the time scale of the experiment is assumed to be negligible. Thus each scattering event is looked on as removing population from the $\mathbf{k}_Q$ state. Referring to the density matrix formulation of (84), population changes are related to T_1. In this model phase changing scattering processes are not included, so T_2' is infinite. We are thus in a regime where (85) becomes

$$\frac{1}{T_2} = \frac{1}{2T_1} \tag{211}$$

If we could measure the decay of the $\mathbf{k}_Q$ exciton in emission, for example, we would expect to observe a decay time of T_1.

For an exciton of wave vector $\mathbf{k}$ and energy $\hbar\omega$ and assuming exciton scattering to be due solely to a linear exciton-phonon coupling, the above model yields the following relationships for the exciton absorption linewidth (full width at half maximum) $\Gamma(\mathbf{k}, \omega)$ and the shift in the energy of the absorption maximun $\Delta(\mathbf{k}, \omega)$[10]:

$$\begin{aligned}\Gamma(\mathbf{k}, \omega) = \sum_{\mathbf{q}s} |F_s(\mathbf{k}, \mathbf{q})|^2 \Big\{ &\bar{N}_{\mathbf{q}s}\, \delta(E(\mathbf{k}) - E(\mathbf{k}+\mathbf{q}) + \hbar\omega_s(\mathbf{q})) \\ &+ (\bar{N}_{\mathbf{q}s} + 1)\, \delta(E(\mathbf{k}) - E(\mathbf{k}+\mathbf{q}) - \hbar\omega_s(\mathbf{q})) \Big\}\end{aligned} \tag{212}$$

$$\begin{aligned}\Delta(\mathbf{k}, \omega) = \sum_{\mathbf{q}s} |F_s(\mathbf{k}, \mathbf{q})|^2 \Bigg\{ &\frac{(\bar{N}_{\mathbf{q}} + 1)}{\delta(E(\mathbf{k}) - E(\mathbf{k}+\mathbf{q}) - \hbar\omega_s(\mathbf{q}))} \\ &+ \frac{\bar{N}_{\mathbf{q}}}{\delta(E(\mathbf{k}) - E(\mathbf{k}+\mathbf{q}) + \hbar\omega_s(\mathbf{k}))} \Bigg\}\end{aligned} \tag{213}$$

Note the similarity between the above expression for $\Gamma(\mathbf{k}, \omega)$ and the expression for $1/\tau_D$ given by (172). Except for the factor $\Delta k_z(\mathbf{q})/k_z$, they are identical. This is a manifestation of the fact that the optical linewidth is sensitive, as we have mentioned, to any change of $\mathbf{k}$ no matter how small.

Recently Harris proposed a completely different model for the exciton-phonon scattering process.[175] This model is related to the model that

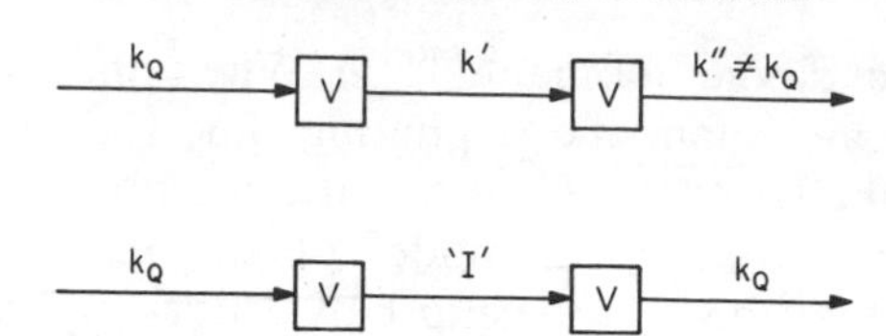

Fig. 45. Two different models for exciton-photon scattering. The upper model describes the case where the exciton is scattered from k_0 to k' by the potential V. k' is then scattered to a completely different k-value. The lower model illustrates the case where, after scattering to an intermediate state I, the exciton returns to its initial k-value.

Botter, van Strien, and Schmidt[99,100] used to describe the dephasing of excited dimer states (see Section III.F.5). A simplified picture of this process is given in Fig. 45. The exciton is scattered from its initial $\mathbf{k}_Q$ state to some intermediate state. While in the intermediate state, the exciton wavefunction is dephased, the extent of the dephasing being related to the length of time τ the system spends in this state. After a time τ the system returns to the $\mathbf{k}_Q$ state.

To understand this model in more detail consider Fig. 46. The figure displays the excited states of a system of excitons and phonons. In addition to the pure phonon and exciton states, there are one- and two-particle exciton-phonon compound states.[13] In the two-particle states the exciton and phonon motions are uncorrelated, whereas in the one-particle states the exciton and phonon are bound so that energy transport of the compound particle requires simultaneous transport of electronic and lattice excitations. Because the energy of a phonon in a one particle compound state is different from its isolated (ground state) energy, there will be a difference in energy $\delta\omega$ between transitions (a) and (b). One can now see the formal similarity between this exciton scattering mechanism and the Botter et al.[99,100] mechanism for dimer scattering. In both cases a rapid exchange of energy between the two transitions results in a broadening of the absorption line and a shift in energy.

The difference in energy between transitions (a) and (b) is due to the change in phonon energy in the compound exciton-phonon state. The energy change results from quadratic terms in the exciton-phonon coupling. This is to be contrasted with the linear exciton-phonon coupling mechanism that gives rise to (212) and (213).

Harris has shown that the linewidth and energy shift resulting from the above model are given by

$$\Gamma = \frac{(\delta\omega\tau)^2 W^+}{1+(\delta\omega\tau)^2} \tag{214}$$

$$\Delta = \frac{\delta\omega\tau W^+}{1+(\delta\omega\tau)^2} \tag{215}$$

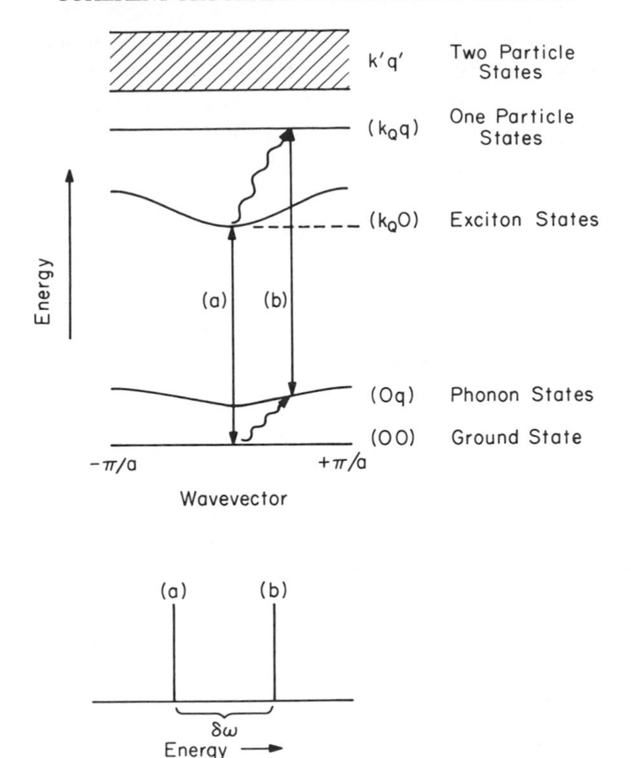

Fig. 46. A schematic diagram of the exchange model for exciton-dephasing. The various energy levels involved in the process are illustrated above. In particular, the two transitions between which exchange occurs are labeled (a) and (b). Below the spectrum of these two transitions is indicated. After Ref. 175.

where τ is the lifetime of the $(0q)$ state and W^+ is the transition probability for the transitions $(00) \rightarrow (0q)$ and $(k_Q 0) + (0q) \rightarrow (k_Q q)$.

Note that if the exciton band is broad enough so that the one- and two-particle states are within the exciton band itself, it seems likely that the $(k_Q q)$ state will decay into other states of the band rather than return to $(k_Q 0)$. This means that the exchange narrowing theory is only applicable for exciton bands that are narrower in energy than the energy of the phonon primarily responsible for scattering. The applicability of this model to singlet excitons is thus uncertain.

Recently Jones et al.[176,177] used a density matrix approach to describe dephasing in solids. Several points of relevance to the discussion here arise from their work. First they show that without invoking the exchange theory mechanism it is still possible to obtain an exponential temperature dependence for the linewidth Γ. In fact, a variety of temperature dependences may be obtained for various limits of the theory and depending on whether one considers the scattering to be due to acoustic or optical phonons. This underscores the point made in connection with diffusion coefficients in Table V and with linewidths in Ref. 163, that it is difficult to prove a particular scattering mechanism solely on the basis of the functional form of the temperature dependence.

The second point of interest here, arising from the work of Jones et al.,[176,177] involves the relationship between T_2's measured in optical and magnetic resonance experiments. As mentioned before, there is no reason to believe that a process that dephases spins will affect an optical transition moment in the same way. Jones et al. have explicitly shown that even when the agent for dephasing (e.g., optical or acoustic phonons) is the same, the matrix element determining the value of T_2 could be different in the two cases of spin and optical dephasing.

Measurements of absorption lineshapes as a means of obtaining information about singlet exciton scattering processes have been used for many years.[178] Recently Morris and Sceats[160] investigated the exciton absorption lineshape in the 4000 Å $^1B_{2u} \leftarrow {}^1A_{1g}$ transition of anthracene. These authors also considered in detail the theoretical relationship between the various spectral moments of the absorption line and the parameters involved in the scattering process. They consider scattering to be due to the single phonon process summarized in (212) and (213).

Because singlet states in general and this transition in anthracene in particular have large oscillator strengths, it is difficult to obtain reliable lineshapes in direct absorption. For this reason Morris and Sceats have obtained the absorption lineshape from measurements of reflection spectra. In an attempt to understand these experimental results, Dissado[114] extended the theoretical treatment of exciton-phonon scattering to the case of two molecules per unit cell. This extension permits one to invoke inter- as well as intraband scattering processed in explaining the anthracene results.

Morris and Sceats[160] were able to conclude from their analysis of the spectral moments of the absorption line that acoustic phonons are a factor of 10^3 less efficient than optical phonons at scattering anthracene singlet excitons. Furthermore, Dissado[114] has shown theoretically that in anthracene the librational phonons are about 100 times more efficient than translational phonons at scattering the excitions.

In Fig. 47 we exhibit the linewidth results of Morris and Sceats[160] along with the analysis of Dissado. The upper exciton components (**ac**-polarized) temperature-dependent linewidth can be explained adequately by assuming a downward interband scattering by 136 cm^{-1} phonons and an upward intraband scattering by 52 cm^{-1} phonons. (Note that these frequencies correspond to measured librational phonon frequencies.) The

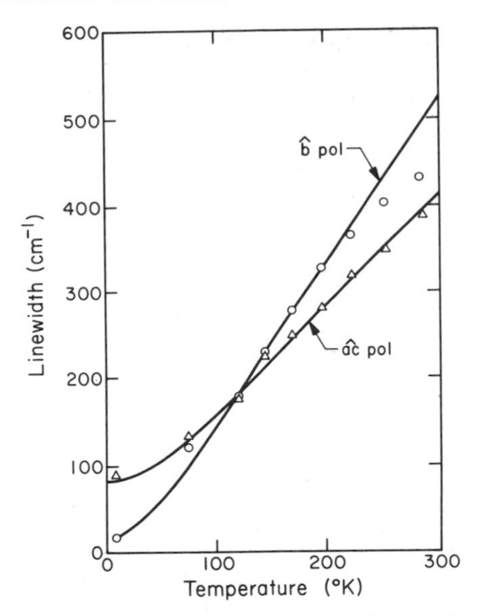

Fig. 47. The absorption linewidth versus temperature for singlet excitons in anthracene. After Ref. 160.

lower energy **b**-polarized component is not, however, well described by librational phonon scattering processes. The theoretical curve shown in the figure is for an *upward* intraband scattering by a 52 cm^{-1} phonon and a *downward* scattering by the same phonon. This downward scattering, if it is real, must be to a state 52 cm^{-1} below the bottom of the exciton band.

Similar analyses of singlet absorption lineshapes have been made for napthalene, phenanthrene[162] and thiophene.[161b] One should emphasize that in all these works the validity of (212) and (213) and thus linear exciton-phonon coupling is assumed over the entire temperature range studied. Furthermore, the line broadening is assumed to be homogeneous and thus directly related to scattering rates. We cite below an example of exciton absorption lines that become inhomogeneously broadened at higher temperatures.

Despite these assumptions, the validity of which must be checked in detail, it seems reasonable to conclude that the scattering of singlet excitons is due to optical phonons and that because the higher energy exciton component has an additional decay channel available at low temperatures, namely decay by phonon emission to the lower component, it will in general have a broader low temperature absorption linewidth.

Scattering of the singlet exciton according to the above analysis is very rapid. For anthracene Morris and Sceats[160] found scattering times at 10 K of 0.4 psec for the lower exciton component and 0.1 psec for the upper one. Note that in this analysis it is not possible to determine how much the exciton wave vector is changed during the scattering process so that it is not possible to determine how the exciton transport is affected.

Similar lineshape measurements have been made for triplet excitons. In some ways the study of triplet exciton absorption lineshapes is experimentally more straightforward than the singlet exciton studies discussed above. Because of the strength of singlet exciton absorption, very thin crystals must be used. Strain effects that can inhomogeneously broaden the absorption lines thus become a problem. Furthermore, one must be very careful to ensure that the weak transmitted light detected at the absorption maximum is really due to light that has passed through the bulk of the crystal.

Triplet exciton absorption measurements present the opposite kinds of problems. Many triplet absorptions are too weak to see with reasonable intensity in direct absorption. This is true of benzene,[47,179] naphthalene,[49] and anthracene crystals.[180] We will discuss lineshape measurements on excitons in DBN and TCB. The triplet absorption in these crystals is considerably enhanced by the heavy atom effect,[181] and absorption can be observed in crystals of reasonable thickness (~ 1 mm). Furthermore, the excitons in both DBN and TCB have been shown to be nearly one-dimensional considerably simplifying the theoretical analysis.

The triplet exciton structure of DBN is complicated by the fact that there are eight molecules in the crystallographic unit cell divided into two equal sets of inequivalent molecules.[70] There are thus two separate exciton systems, one for set I and the other for set II. Figure 48 shows the resulting triplet exciton structure.[59]

The absorption lineshape of DBN obtained by Burland et al. at 2 and 25 K is shown in Fig. 49. At 25 K the line is very accurately a Lorentzian and the width is thus a measure of T_2 for the exciton system. However, at low temperatures the line is asymmetric[163] but Lorentzian on the high energy side. Burland et al.[163] suggest that the 2 K line is homogeneously broadened by scattering of the optically created k_Q exciton by naturally occurring ^{13}C isotopic impurities. Such broadening of exciton transitions was earlier observed by Hanson[182] in the benzene singlet exciton system. The asymmetry is, according to this interpretation, due to the fact that k_Q is near the bottom of the exciton band. The exciton can only be scattered to higher energy band states and thus the Lorentzian broadening is restricted to the high energy side. Klafter and Jortner[183] have recently analyzed this asymmetric broadening in terms of strains setup in the crystal. Such strain broadening is certainly present in DBN crystals.[184]

To investigate the exciton-phonon scattering processes in DBN, the

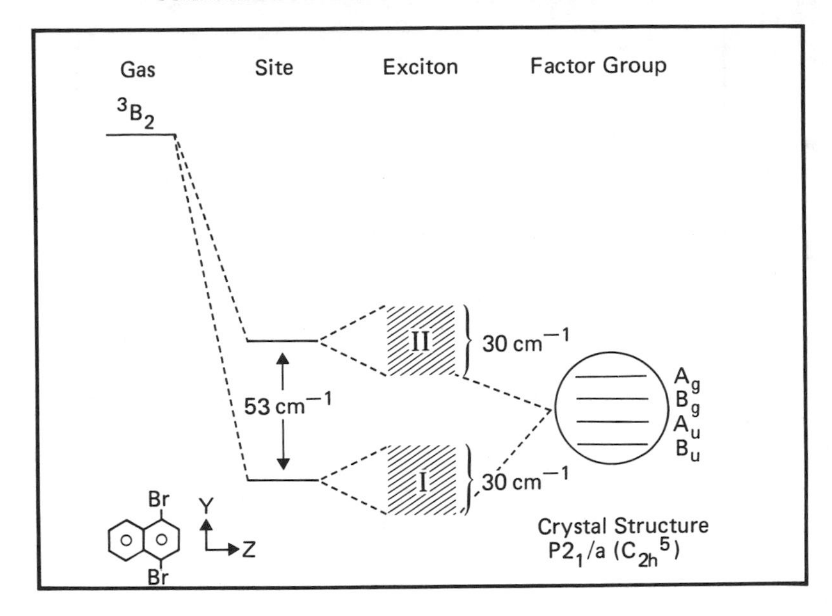

Fig. 48. The triplet exciton band structure of DBN. After Ref. 163.

linewidth and line broadening were investigated as a function of temperature. The temperature-dependent total linewidth Γ_T may be written as

$$\Gamma_T = \Gamma_0 + \Gamma(T) \tag{216}$$

Here Γ_0 is the temperature-independent residual low-temperature linewidth due to impurity or defect scattering as discussed above. $\Gamma(T)$ is the contribution of exciton-phonon scattering to the total linewidth.

Burland et al.[163] attempted to analyze the temperature dependent linewidth in terms of (212) for a linear exciton-phonon scattering process. They found that the data are consistent with scattering by a 50 cm^{-1} optical phonon. The data, however, are equally well explained by a quadratic ($\bar{N}(\bar{N}+1)$) exciton-phonon interaction with an optical phonon at 17 cm^{-1}. Harris[175] fit the $\Gamma(T)$ data to a simple exponential function $e^{-\Delta E/k_B T}$ and found ΔE to be equal to 38 cm^{-1} for the exciton site I. Clearly the data do not extend over a wide enough temperature range for one to unequivocally determine the exciton-phonon scattering mechanism from $\Gamma(T)$ measurements alone. Harris[175] also interpreted the DBN linewidth and frequency shift measurements in terms of the exchange model summarized in (214) and (215).

Both the linear exciton-phonon coupling model and the exchange model can explain the DBN linewidth and frequency shift data satisfactorily. How then is one to determine which of these models, if either,

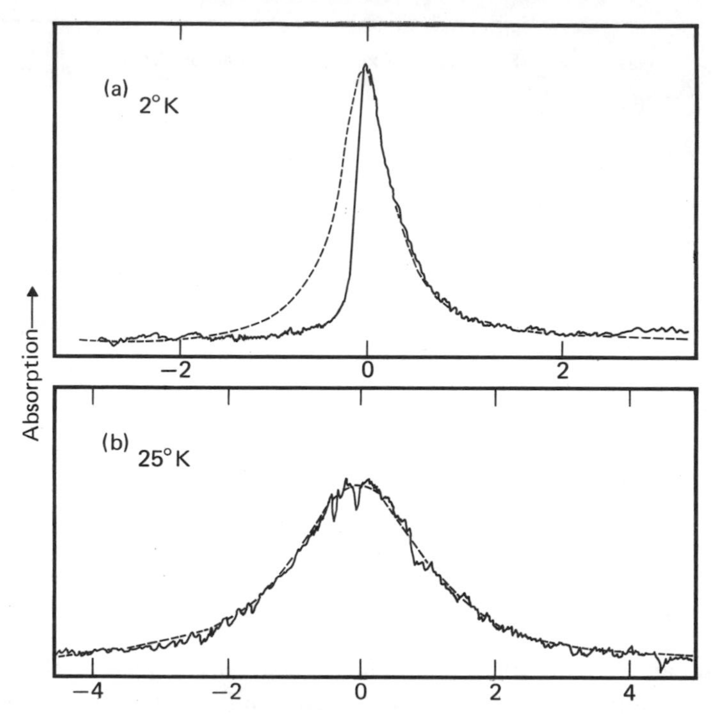

Fig. 49. Absorption lineshape for the 20192 cm^{-1} triplet exciton band in DBN at (a) 2 K and (b) 25 K. The dashed line in both cases is a Lorentzian. After Ref. 163.

correctly describes triplet exciton scattering? One major difference between the two models is that the linear coupling model is essentially a T_1 process, whereas the quadratic model is a T_2' process. If the linear model is valid, one should be able to measure T_1 by following the short time decay of the exciton emission, and since $T_2 = 2T_1$, one should be able to relate this temporal behaviour directly to the linewidth.

Is this experiment possible? At 25 K the absorption line is Lorentzian (see Fig. 49) and its width is dominated by phonon scattering. From the linewidth we should expect to observe a decay of 3 psec if the linear coupling model correctly describes the scattering. This decay time is just on the borderline of detectability using a picosecond dye laser. Of course the decay time will lengthen as the temperature decreases, but at low temperatures the temperature-independent contribution to the linewidth interferes with the analysis.

Similar lineshape and frequency shift measurements have been made for TCB.[149,164] The exciton band structure of TCB and DBN are quite different. Whereas the DBN exciton bandwidth is 26 cm^{-1} the TCB bandwidth is only about 1 cm^{-1}. Furthermore, the optically accessible $\mathbf{k}_Q$ level is at the top of the exciton band. These features are reflected in several aspects of the TCB triplet exciton lineshape.[164] First, as might be expected, the 2 K absorption line is asymmetric, but, unlike the DBN case, the asymmetric broadening is to the low energy side of the line. Second, at about 20 K the lineshape changes from Lorentzian to Gaussian. This result was predicted by Toyozawa[172] many years ago and is a consequence of the localization of the exciton.[112] As the temperature increases, the thermal fluctuations in the exciton energy become larger and larger until at some point they exceed the exciton band halfwidth. At this point the concept of exciton band motion breaks down and the excitation must be considered localized. The absorption of this localized exciton is given by the expression

$$I(E) = C \exp[-(E - E_a)^2/2D^2] \tag{217}$$

where E_a is the position of the absorption maxima, C a constant, and D the thermally averaged value of the change in the exciton energy due to random lattice fluctuations. Toyozawa has shown that (217) describes the exciton lineshape in the region where $D \gg B$, where B is the exciton band half-width. In Fig. 50, we have extracted the values of D from the Gaussian absorption lines. It is found that in the region where $B \approx D$, the line changes shape from Gaussian to Lorentzian. This result emphasizes the importance of actually measuring the absorption lineshape before one attempts to relate linewidths to scattering times. Clearly, in the Gaussian region of Fig. 50 the linewidth is not related to T_2. This localization can occur for singlet excitons as well as triplet excitons.

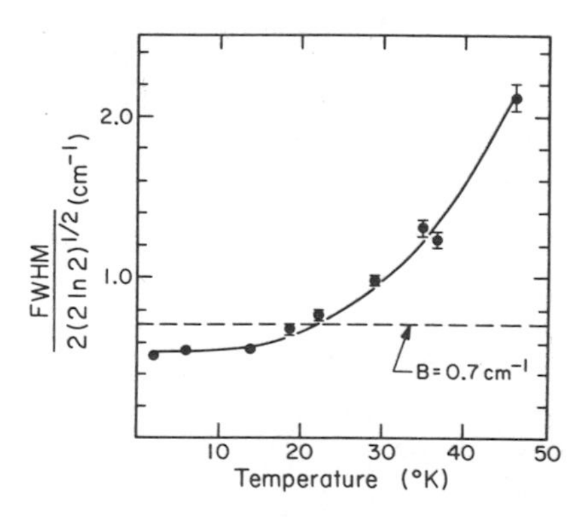

Fig. 50. The temperature dependence of the full width at half-maximum (FWHM) of the absorption origin for the triplet exciton in TCB. The dashed line locates the energy of the exciton band half width. Above the dashed line the absorption lineshape is Gaussian; below it is Lorentzian.

VI. CONCLUDING REMARKS

In this chapter we attempted to discuss the problem of dimer and exciton coherence within a framework established by available experimental and theoretical results. We have seen that for the dimer system coherence can be defined reasonably precisely. There are in fact experimental measurements of coherence times in a variety of dimer systems.

Coherence in excitons has not been defined as precisely. The problem for excitons lies in the multilevel nature of the k-states. One can ask two kinds of question. First, one might ask how rapidly a k-state prepared in a particular way dephases. In this case we are not interested in the magnitude of the change, if any, in the k-vector, but merely in the dephasing of the exciton wavefunction. This type of coherence is analogous to the kind of coherence measured, in the dimer experiments.

The second kind of question one might ask is, how rapidly is an exciton originally created with a particular k-vector scattered away from this k-state into other states with significantly different magnitudes and directions for k. In this case, scattering processes that result in only small changes in k are not important. We consider this question when we discuss the problem of coherent versus random walk energy transport in molecular crystals. Measurements of diffusion coefficients and trapping times yield information of this type of coherence.

The important point we wish to emphasize is that either definition of exciton coherence is appropriate provided one makes clear for a given case which definition one is using and provided one does not try to compare measurements made using different definitions.

We began this paper by asking several very ambitious questions. We certainly do not feel that we have provided answers to all of them. This is partially due to a lack of experimental data and in part to disagreements on how existing data is to be theoretically interpreted. We firmly believe that, because there remain many exciting questions to answer, the field of exciton dynamics in molecular crystals is still very much alive. If this article can serve as a stimulus to the development of the field, it will have served its purpose.

Acknowledgments

The authors are grateful to M. D. Fayer, D. Haarer, R. Kopelman, R. Silbey and J. D. Swalen for their constructive criticisms of the manuscript.

References

1. J. Frenkel, *Phys. Rev.*, **37,** 17 (1931); 1276 (1931).
2. R. Peierls, *Ann. Phys.*, **13,** 905 (1932).

3. A. S. Davydov, *Theory of Molecular Excitons*, translated by M. Kasha and M. Oppenheimer, McGraw-Hill, New York, 1962.
4. R. S. Knox, *Theory of Excitons*, Academic, New York, 1963.
5. S. A. Rice and J. Jortner, Comments of the Theory of the Exciton States of Molecular Crystals, in D. Fox, M. M. Labes, and A. Weissberger, eds., *Physics and Chemistry of the Organic Solid State*, vol III, Wiley-Interscience, New York, 1967.
6. H. C. Wolf, *Adv. At. Mol. Phys.*, **3,** 119 (1967).
7. D. P. Craig and S. H. Walmsley, *Excitons in Molecular Crystals*, Benjamin, New York, 1968.
8. P. Avakian and R. E. Merrifield, *Mol. Cryst.*, **5,** 37 (1968).
9. G. W. Robinson, *Ann. Rev. Phys. Chem.*, **21,** 429 (1970).
10. A. S. Davydov, *The Theory of Molecular Excitons*, Plenum, New York, 1971.
11. E. F. Sheka, *Usp. Fiz. Nauk*, **104,** 593 (1971).
12. M. R. Philpott, *Adv. Chem. Phys.*, **23,** 227 (1973).
13. Y. B. Levinson and E. I. Rashba, *Rep. Prog. Phys.*, **36,** 1499 (1973).
14. R. M. Hochstrasser, Triplet Exciton States of Molecular Crystals, D. A. Ramsay, ed., *International Review of Science, Physical Chemistry*, Series 2, Vol. 3, *Spectroscopy*, Butterworths, London, 1976, p. 1.
15. R. C. Powell and Z. G. Soos, *J. Luminesc.*, **11,** 1 (1975).
16. R. G. Kepler, *Treatise Sol. State Chem.*, **3,** 615 (1976).
17. R. Kopelman, Exciton Percolation in Molecular Alloys and Aggregates, in F. K. Fong, ed., *Radiationless Processes in Molecules and Condensed Phases*, Springer-Verlag, Berlin, 1976.
18. R. Silbey, *Ann. Rev. Phys. Chem.*, **27,** 203 (1976).
19. F. R. Lipsett, *Mol. Cryst.*, **3,** 1 (1967); **5,** 9 (1968); *Mol. Liq. Cryst.*, **6,** 175 (1969); A. I. Morozova and E. F. Sheka, *Mol. Cryst. Liq. Cryst.*, **14,** 329 (1971); V. S. Makarova and E. F. Sheka, *Mol. Cryst. Liq. Cryst.*, **17,** 55 (1972); E. F. Sheka and A. I. Morosova, **19,** 331 (1973); **21,** 87 (1974); E. F. Sheka, V. S. Makarova, and E. D. Simonovskaya, *Mol. Cryst. Liq. Cryst.*, **30,** 239 (1975); **33,** 261 (1976); **39,** 259 (1977).
20. J. Franck and E. Teller, *J. Chem. Phys.*, **6,** 861 (1938).
21. I. I. Rabi, N. F. Ramsey, and J. Schwinger, *Rev. Mod. Phys.*, **26,** 167 (1954).
22. On resonance $\omega_L = \omega_+ - \omega_g$, and therefore, both of the diagonal elements of H are equal to ω_+ which when subtracted from the matrix leaves us with the simple solution shown in (26) and (27).
23. A. Abragam, *The Principles of Nuclear Magnetism*, Oxford Univ. Press, London, 1961.
24. R. P. Feynman, F. L. Vernon, Jr., and R. W. Hellworth, *J. Appl. Phys.*, **28,** 49 (1957).
25. Because the $\pm$states are coupled to the radiation field, spontaneous emission is possible and the states are no longer truly stationary. Perhaps quasistationary would be a more accurate expression in this context.
26. For review see S. Haroche, *Tropics Appl. Phys.*, **13,** 253 (1976).
27. In the literature there is sometimes confusion between optical coherence and phase coherence. In quantum optics, optical coherence describes the correlation time of the radiation field, whereas phase coherence describes atomic or molecular correlation effects. Recently, however, optical coherence has also been used to describe coherence of optical transitions.
28. W. W. Chow, M. O. Scully, and J. O. Stoner, *Phys. Rev.*, **A11,** 1380 (1975).
29. W. Weisskopf and E. Wigner, *Z. Physik*, **63,** 54 (1930). The result can be found in (4) of Ref. 26.
30. M. Sargent, M. O. Scully, and W. Lamb, Jr., *Laser Physics*, Addison-Wesley, Reading, Mass., 1974.

31. The lower limit, $t = 0$, of the integration is not the usual $t = -\infty$ limit. It is assumed that $t = 0$ starts with the period of observation.
32. K. Shimoda, *Topics Appl. Phys.*, **13,** 1 (1976).
33. A. Laubereau and W. Kaiser, *Ann. Rev. Phys. Chem.*, **26,** 83 (1975).
34. See for example A. H. Zewail, ed., *Advances in Laser Chemistry*, Springer Series in Chemical Physics, Springer, Berlin, 1978).
35. a. C. Slichter, *Principles of Magnetic Resonance*, Harper and Row, New York, 1963. b. R. H. Pantell and H. E. Puthoff, *Fundamentals of Quantum Electronics*, Wiley, New York, 1969.
36. U. Fano, *Rev. Mod. Phys.*, **29,** 74 (1957); R. McWeeny, *Rev. Mod. Phys.*, **32,** 335 (1960); J. P. Barrat, *Proc. R. Soc.*, **A263,** 371 (1961).
37. F. Bloch, *Phys. Rev.*, **70,** 460 (1946).
38. N. A. Kurnit, I. D. Abella, and S. R. Hartmann, *Phys. Rev. Lett.*, **13,** 567 (1964); N. Takeuchi and A. Szabo, *Phys. Lett.*, **50A,** 361 (1974); J. Morsink, T. Aartsma, and D. A. Wiersma, *Chem. Phys. Lett.*, **49,** 34 (1977); A. H. Zewail, T. E. Orlowski, K. E. Jones, and D. E. Godar, *Chem. Phys. Lett.*, **48,** 256 (1977).
39. A. H. Zewail and T. E. Orlowski, *Chem. Phys. Lett.*, **45,** 399 (1977); A. H. Zewail, D. E. Godar, K. E. Jones, T. E. Orlowski, R. R. Shah, and A. Nichols, *Advances in Laser Spectroscopy* I, SPIE, **113,** 42 (1977).
40. A. Genack, R. M. Macfarlane, and R. Brewer, *Phys. Rev. Lett.*, **37,** 1078 (1976); D. A. Wiersma, Proceedings of the Eighth Molecular Crystal Symposium, Santa Barbara, Ca. (1977); T. E. Orlowski, K. E. Jones, and A. H. Zewail, *Chem. Phys. Lett.*, **54,** 197 (1978).
41. A. Szabo, *Phys. Rev.*, **B11,** 4512 (1975); A. P. Marchetti, W. C. McColgin, and J. H. Eberly, *Phys. Rev. Lett.*, **35,** 387 (1975).
42. See Refs. 9 and 14 and references therein.
43. See Refs. 9 and 17 and references therein.
44. S. D. Colson, D. M. Hanson, R. Kopelman, and G. W. Robinson, *J. Chem. Phys.*, **48,** 2215 (1968); E. I. Rashba, *Fiz. Tver. Tela*, **5,** 1040 (1963); *Sov. Phys.-Solid state*, **5,** 757 (1963).
45. In most cases, in isotopically mixed crystals the excited state of the fully protonated molecules lies at a lower energy than the fully deuterated excited state, primarily due to differences in zero-point energies. Furthermore, in these crystals one can obtain high guest concentrations and therefore form dimers, trimers, and so on. However, because the trap depth is typically 20 to 200 cm^{-1}, the guest states are perturbed by the so-called quasiresonance interactions (i.e., mixing of guest-host wavefunctions through the potential V). Hence, when we mention the word dimer, we do not imply a truly isoated dimer.
46. G. C. Nieman, *Triplet Exciton Phenomena in Benzene Crystals*, Ph.D thesis, California Institute of Technology, Pasadena, 1965.
47. D. M. Burland and G. Castro, *J. Chem. Phys.*, **50,** 4107 (1969).
48. D. M. Hanson, *J. Chem. Phys.*, **52,** 3409 (1970).
49. D. M. Hanson and G. W. Robinson, *J. Chem. Phys.*, **43,** 4174 (1965).
50. C. L. Braun and H. C. Wolf, *Chem. Phys. Lett.*, **9,** 260 (1971).
51. H. K. Hong and R. Kopelman, *J. Chem. Phys.*, **55,** 724 (1971).
52. F. H. Herbstein and G. Schmidt, *Acta Cryst.*, **8,** 399, 406 (1955).
53. A. H. Zewail, *Chem. Phys. Lett.*, **29,** 630 (1974).
54. R. H. Clarke and R. M. Hochstrasser, *J. Chem. Phys.*, **47,** 1915 (1967).
55. A. H. Zewail, *Chem. Phys. Lett.*, **33,** 46 (1975).
56. D. D. Smith, R. Mead, and A. H. Zewail, *Chem. Phys. Lett.*, **50,** 358 (1977).

57. For a review on the optical detection of magnetic resonance see M.A.El-Sayed, *Accts. Chem. Res.*, **4,** 23 (1971).
58. F. Dupuy, Ph.Pee, R. Lalanne, J. P. Lemaistre, C. Vaucamps, H. Port, and Ph. Kottis, *Mol. Phys.*, **35,** 595 (1978).
59. R. M. Hochstrasser and J. D. Whiteman, *J. Chem. Phys.*, **56,** 5945 (1972).
60. R. M. Hochstrasser and A. H. Zewail, *Chem. Phys.*, **4,** 142 (1974).
61. H. K. Hong and R. Kopelman, *J. Chem. Phys.*, **57,** 3888 (1972).
62. M. Schwoerer and H. C. Wolf, *Mol. Cryst.*, **3,** 177 (1967).
63. H. Sternlicht and H. McConnell, *J. Chem. Phys.*, **35,** 1793 (1961).
64. P. W. Anderson and P. R. Weiss, *Rev. Mod. Phys.*, **25,** 269 (1953).
65. J. Jortner, S. Rice, J. Katz and S. Choi, *J. Chem. Phys.*, **42,** 309 (1965).
66. P. Avakian and R. Merrifield, *Phys. Rev Lett.*, **13,** 541 (1964).
67. C. A. Hutchison, Jr. and J. S. King, Jr., *J. Chem. Phys.*, **58,** 392 (1973).
68. R. M. Hochstrasser, G. W. Scott, and A. H. Zewail, unpublished results.
69. R. M. Hochstrasser, T. Y. Li, H. N. Sung, J. Wessel, and A. H. Zewail, *Pure Appl. Chem.*, **37,** 85 (1974).
70. a. J. Trotter, *Can. J. Chem.*, **39,** 1574 (1961) b. J. J. Mayerle, IBM Research Report #RJ2121, 1977.
71. R. M. Hochstrasser, in R. D. Levine and J. Jortner eds. *Molecular Energy Transfer*, Wiley, New York, 1976.
72. M. J. Buckley and C. B. Harris, *J. Chem. Phys.*, **56,** 137 (1972); G. Kothandaraman and D. S. Tinti, *Chem. Phys. Lett.*, **19,** 225 (1973); J. Schmidt and J. H. van der Waals, *Chem. Phys. Lett.*, **3,** 546 (1969); D. S. Tinti, G. Kothandaraman, and C. B. Harris, *J. Chem. Phys.* **59,** 190 (1973).
73. C. A. Hutchison, Jr., J. V. Nicholas, and G. W. Scott, *J. Chem. Phys.*, **53,** 1906 (1970).
74. These spectra are taken from the unpublished work of A. H. Zewail and are obtained with better resolution than the spectra reported earlier (Ref. 60).
75. J. Vincent and A. H. Maki, *J. Chem. Phys.*, **39,** 3088 (1963).
76. M. J. Buckley, Ph.D thesis, University of California, Berkeley, 1971.
77. A. H. Francis and C. B. Harris, *J. Chem. Phys.*, **57,** 1050 (1972).
78. Z. G. Soos, *J. Chem. Phys.*, **51,** 2107 (1969).
79. A. H. Zewail and C. B. Harris, *Chem. Phys. Lett.*, **28,** 8 (1974); *Phys. Rev.*, **B11,** 935, 952 (1975).
80. M. D. Fayer and C. B. Harris, *Phys. Rev.*, **B9**, 748 (1974).
81. C. Dean, M. Pollak, B. M. Craven, and G. A. Jeffrey, *Acta Cryst.*, **11,** 710 (1958).
82. A. Monfils, *Compt. Rend. Acad. Sci. (Paris)*, **241,** 561 (1955); G. Gafner and F. H. Herbstein, *Acta Cryst.*, **13,** 702, 706 (1960); F. H. Herbstein, *Acta Cryst.* **18,** 997 (1965).
83. J. H. Van der Waals and M. S. de Groot, in A. B. Zahlan, ed., *The Triplet State*, Cambridge Univ. Press, London, 1967.
84. R. M. Hochstrasser and T. S. Lin, *J. Chem. Phys.*, **49,** 4929 (1968).
85. A. H. Francis and C. B. Harris, *Chem. Phys. Lett.* **9,** 181, 188 (1971).
86. See for example H. Hameka, *Introduction to Quantum Theory*, Harper and Row, New York, 1967.
87. D. S. McClure, *J. Chem. Phys.*, **20,** 682 (1952).
88. H. K. Hong and G. W. Robinson, *J. Chem. Phys.*, **52,** 825 (1970); **54,** 1369 (1971).
89. E. Economou and M. Cohen, *Phys. Rev.*, **B4,** 396 (1971).
90. A. Bierman, *J. Chem. Phys.*, **43,** 1675 (1961).
91. S. Sheng and D. M. Hanson, *Chem. Phys. Lett.*, **33,** 451 (1971).
92. R. Kubo and K. Tomita, *J. Phys. Soc. Jap.*, **9,** 888 (1954).

93. P. Anderson, *J. Phys. Soc. Jap.*, **9,** 316 (1954).
94. H. McConnell, *J. Chem. Phys.*, **28,** 430 (1958).
95. C. B. Harris, *J. Chem. Phys.*, **54,** 972 (1971).
96. See for example D. A. Antheunis, J. Schmidt, and J. H. van der Waals, *Mol. Phys.*, **27,** 1521 (1974).
97. C. B. Harris and M. D. Fayer, *Phys. Rev.*, **B10,** 1784 (1974).
98. W. Breiland, H. Brenner, and C. B. Harris, *J. Chem. Phys.*, **62,** 3458 (1975).
99. B. J. Botter, C. J. Nonhof, J. Schmidt, and J. H. Van der Waals, *Chem. Phys. Lett.*, **43,** 210 (1976).
100. B. J. Botter, A. J. van Strien, and J. Schmidt, *Chem. Phys. Lett.*, **49,** 39 (1977).
101. C. A. van't Hoff and J. Schmidt, *Chem. Phys. Lett.*, **36,** 460 (1975).
102. J. S. King, Jr., Ph.D thesis, University of Chicago, 1973.
103. M. Schwoerer, Habilitationsarbeit, University of Stutgart (1972).
104. P. Reineker and H. Haken, *Z. Physik*, **250,** 300 (1972).
105. A. S. Davydov, *Phys. Stat. Sol.*, **20,** 143 (1967).
106. R. M. Hochstrasser and P. N. Prasad, *J. Chem. Phys.*, **56,** 2814 (1972).
107. M. Ito, M. Suzuki and T. Yokoyama, in A. B. Zahlan, ed., *Excitons, Magnons and Phonons in Molecular Crystals*, Cambridge Univ. Press, 1968.
108. E. B. Wilson, J. C. Decius, and P. C. Cross, *Molecular Vibrations* (McGraw-Hill, New York, 1955.
109. D. P. Craig and L. A. Dissado, *Chem. Phys.*, **14,** 89 (1976).
110. A. S. Davydov and G. M. Pestryakov, *Phys. Stat. Sol.* (*b*), **49,** 505 (1972).
111. T. Holstein, *Ann. Phys.*, **8,** 325; 343 (1959).
112. a. D. Yarkony and R. Silbey, *J. Chem. Phys.*, **65,** 1042 (1976); **67,** 5818 (1977). b. M. K. Grover and R. Silbey, *J. Chem. Phys.*, **54,** 4843 (1971).
113. D. P. Craig, L. A. Dissado, and S. H. Walmsley, *Chem. Phys. Lett.*, **46,** 191 (1977).
114. L. A. Dissado, *Chem. Phys.*, **8,** 289 (1975).
115. K. K. Rebane, *Impurity Spectra of Solids*, Plenum, New York, 1970; A. A. Maradudin and G. F. Nardelli, eds., *Elementary Excitations in Solids*, Plenum, New York, 1968.
116. R. Silbey, J. Jortner, S. Rice, and M. Vala, *J. Chem. Phys.*, **42,** 2948 (1965).
117. D. Fox and O. Schnepp, *J. Chem. Phys.*, **23,** 767 (1955).
118. S. Choi, J. Jortner, S. Rice and R. Silbey, *J. Chem. Phys.*, **41,** 3294 (1964).
119. J. M. Ziman, *Principles of the Theory of Solids*, (Cambridge Univ. Press, 1965) pp. 19–26.
120. S. Chandrasekhar, *Rev. Mod. Phys.*, **15,** 1 (1943).
121. P. Avakian, V. Ern, R. E. Merrifield, and A. Suna, *Phys. Rev.*, **165,** 974 (1968).
122. O. Simpson, *Proc. R. Soc.*, **A238,** 402 (1956).
123. G. Gallus and H. C. Wolf, *Z. Naturf.*, **23,** 1333 (1968).
124. M. Silver, D. Olness, M. Swicord, and R. C. Jarnagin, *Phys. Rev. Lett.*, **10,** 12 (1963).
125. M. Levine, J. Jortner, and A. Szoke, *J. Chem. Phys.*, **45,** 1591 (1966).
126. D. F. Williams, J. Adolph, and W. G. Schneider, *J. Chem. Phys.*, **45,** 575 (1966); D. F. Williams and J. Adolph, *J. Chem. Phys.* **46,** 4252 (1967).
127. T. Markvart, *Phys. Stat, Sol.* (*b*), **73,** 689 (1976); **74,** 135 (1976).
128. V. M. Agranovich and Y. V. Konobeev, *Opt. Spect.*, **6,** 155 (1959); *Phys. Stat. Sol.*, **27,** 435 (1968).
129. H. Froehlich, *Proc. R. Soc.*, **A160,** 230 (1957).
130. G. S. Pawley, *Phys. Stat. Sol.*, **20,** 347 (1967).
131. R. W. Munn and W. Siebrand, *J. Chem. Phys.*, **52,** 47 (1970).
132. D. M. Burland and U. Konzelmann, *J. Chem. Phys.*, **67,** 319 (1977).
133. A. Hammer and H. C. Wolf, *Mol. Cryst.*, **4,** 191 (1968).

134. A. Inoue, S. Nagakura and K. Yoshihara, *Mol. Cryst. Liq. Cryst.*, **25,** 199 (1974).
135. R. C. Powell and Z. G. Soos, *Phys. Rev.*, **B5,** 1547 (1972).
136. K. Uchida and M. Tomura, *J. Phys. Soc. Jap.*, **36,** 1358 (1974).
137. G. Durocher and D. F. Williams, *J. Chem. Phys.*, **51,** 1675 (1969).
138. V. Ern., A. Suna, Y. Tomkiewicz, P. Avakian, and R. P. Groff, *Phys. Rev.*, **B5,** 3222 (1972).
139. H. Haken and G. Strobl, in A. B. Zahlan, ed., *The Triplet State*, Cambridge Univ. Press, 1967; *z. Phys.*, **262,** 135 (1973).
140. P. Reineker, *Phys. Stat. Sol.* (*b*), **52,** 439 (1972).
141. V. M. Kenkre and R. S. Knox, *Phys. Rev.*, **B9,** 5279 (1974); V. M. Kenkre, *Phys. Rev.*, **B11,** 1741 (1975); **12,** 2150 (1975).
142. R. M. Shelby, A. H. Zewail, and C. B. Harris, *J. Chem. Phys.*, **64,** 3192 (1976).
143. M. A. El-Sayed, *J. Chem. Phys.*, **54,** 680 (1971).
144. J. Schmidt, D. A. Antheunis, and J. H. Van der Waals, *Mol. Phys.* **22,** 1(1971).
145. C. B. Harris, R. L. Schlupp, and H. Schuch, *Phys. Rev. Lett.*, **30,** 1019 (1973).
146. M. D. Fayer and C. B. Harris, *Chem. Phys. Lett.*, **25,** 149 (1974); M. Lewellyn, A. H. Zewail, and C. B. Harris, *J. Chem. Phys.*, **63,** 3687 (1975).
147. D. D. Dlott and M. D. Fayer, *Chem. Phys. Lett.*, **41,** 305 (1976).
148. G. A. George and G. C. Morris, *Mol. Cryst. Liq. Cryst.* **11,** 61 (1970).
149. W. Guttler, J. U. von Schutz, and H. C. Wolf, *Chem. Phys.*, **24,** 159 (1977).
150. D. D. Dlott, M. D. Fayer, and R. D. Wieting, *J. Chem. Phys.*, **67,** 3808 (1977); **69,** 2752 (1978); R. D. Wieting, M. D. Fayer and D. D. Dlott, *J. Chem. Phys.*, **69,** 1996 (1978).
151. J. M. Ziman, *J. Phys.* **C1,** 1532 (1968).
152. T. P. Eggarter and M. H. Cohen, *Phys. Rev. Lett.*, **25,** 807 (1970); **27,** 129 (1971).
153. J. Jortner and M. H. Cohen, *J. Chem. Phys.*, **58,** 5170 (1973).
154. J. Klafter and J. Jortner, *Chem. Phys. Lett.*, **49,** 410 (1977).
155. a. P. W. Anderson, *Phys. Rev.*, **109,** 1492 (1958).
b. N. F. Mott, *Phil. Mag.*, **29,** 613 (1974); *Comm. Phys.*, **1,** 203 (1976).
156. a. D. D. Smith, R. D. Mead, and A. H. Zewail, *Chem. Phys. Lett.*, **50,** 358 (1977).
b. D. D. Smith, D. Millar, and A. H. Zewail, *J. Chem. Phys.* submitted c. S. Colson, S. George, T. Keyes and V. Vaida, *J. Chem. Phys.* **67,** 4941 (1977); **66,** 2187 (1977).
157. D. Haarer and H. C. Wolf, *Mol. Cryst. Liq. Cryst.*, **10,** 359 (1970).
158. a. R. Schmidberger and H. C. Wolf, *Chem. Phys. Lett.*, **16,** 402 (1972); **25,** 185 (1974); **32,** 18 (1975); **32,** 21 (1975).
b. R. Schmidberger, Ph.D. thesis, University of Stuttgart, 1974.
159. M. Sharnoff, *Symp. Far. Soc.* **3,** 137 (1969); M. Sharnoff and E. Iturbe, **1,** *Phys. Rev. Lett.*, **27,** 576 (1971).
160. G. C. Morris and M. G. Sceats, *Chem. Phys.*, **1,** 120 (1973); 259 (1973); 376 (1973); **3,** 332 (1974); 342 (1974).
161. a. L. A. Dissado, *Chem. Phys. Lett.*, **33,** 57 (1975).
b. A. Brillante and L. A. Dissado, *Chem. Phys.*, **12,** 297 (1976).
162. L. A. Dissado and A. Brillante, *J. Chem. Soc. Far. II*, **73,** 1262 (1977).
163. D. M. Burland, *J. Chem. Phys.*, **59,** 4283 (1973); D. M. Burland and R. M. Macfarlane, *J. Luminesc.*, **12–13,** 213 (1976); D. M. Burland, U. Konzelmann, and R. M. Macfarlane, *J. Chem. Phys.*, **67,** 1926 (1977).
164. D. M. Burland, D. E. Cooper, M. D. Fayer, and C. R. Gochanour, *Chem. Phys. Lett.*, **52,** 279 (1977).
165. R. Kubo, in D. ter Haar, ed., *Fluctuation, Relaxation and Resonance in Magnetic Systems*, Plenum, New York, 1962.

166. N. Bloembergen, E. M. Purcell, and R. V. Pound, *Phys. Rev.*, **69,** 37 (1946).
167. C. J. Gorter and J. H. van Vleck, *Phys. Rev.*, **72,** 1128 (1947).
168. H. Haken and P. Reineker, *Z. Physik*, **249,** 253 (1972).
169. B. J. Botter, A. I. M. Dicker, and J. Schmidt, *Mol. Phys.*, **36,** 129 (1978); B. J. Botter., Ph.D. thesis, University of Leiden, 1977.
170. R. G. Gordon, *Adv. Magnet. Res.* **3,** 1 (1968).
171. A. Ben-Reuven, *Adv. Chem. Phys.*, **33,** 235 (1975).
172. Y. Toyozawa, *Prog. Theoret. Phys.*, **20,** 53 (1958).
173. H. Sumi, *J. Chem. Phys.*, **67,** 2943 (1977).
174. S. F. Fischer and S. A. Rice, *J. Chem. Phys.*, **52,** 2089 (1970); M. K. Grover and R. Silbey, *J. Chem. Phys.*, **52,** 2099 (1970).
175. C. B. Harris, *Chem. Phy. Lett.*, **52,** 5 (1977); *J. Chem. Phys.*, **67,** 5607 (1977); C. B. Harris, P. Cornelius, and R. Shelby in Ref. 34.
176. K. E. Jones and A. H. Zewail in Ref. 34, p. 196; K. E. Jones, A. H. Zewail, and D. Diestler in Ref. 34, p. 258.
177. T. E. Orlowski, K. E. Jones, and A. H. Zewail, *Chem. Phys. Lett.*, **50,** 45 (1977) T. E. Orlowski and A. H. Zewail, *J. Chem. Phys.* **70,** 1390 (1979).
178. A. F. Prikhot'ko and M. S. Soskin, *Opt Spect.*, **13,** 291 (1962); H. Maria and A. Zahlan, *J. Chem. Phys.*, **38,** 941 (1963); H. J. Maria, *J. Chem. Phys.*, **40,** 551 (1964).
179. D. M. Burland, G. Castro, and G. W. Robinson, *J. Chem. Phys.*, **52,** 4100 (1970).
180. R. H. Clarke and R. M. Hochstrasser, *J. Chem. Phys.*, **46,** 4532 (1967).
181. D. S. McClure, *J. Chem. Phys.*, **17,** 905 (1949).
182. D. M. Hanson, *J. Chem. Phys.*, **51,** 653 (1969).
183. J. Klafter and J. Jortner, *J. Chem. Phys.*, **68,** 1513 (1978).
184. R. M. Macfarlane, U. Konzelmann, and D. M. Burland, *J. Chem. Phys.*, **65,** 1022 (1976).

AUTHOR INDEX

Numbers in parentheses are reference numbers and indicate that the author's work is referred to although his name is not mentioned in the text. Numbers in italics show the pages on which the complete references are listed.

SUBJECT INDEX